International Series in Operations Research & Management Science

Founding Editor
Frederick S. Hillier

Volume 276

The book series **International Series in Operations Research and Management Science** encompasses the various areas of operations research and management science. Both theoretical and applied books are included. It describes current advances anywhere in the world that are at the cutting edge of the field. The series is aimed especially at researchers, advanced graduate students, and sophisticated practitioners.

The series features three types of books:

• Advanced expository books that extend and unify our understanding of particular areas.

• Research monographs that make substantial contributions to knowledge.

• Handbooks that define the new state of the art in particular areas. Each handbook will be edited by a leading authority in the area who will organize a team of experts on various aspects of the topic to write individual chapters. A handbook may emphasize expository surveys or completely new advances (either research or applications) or a combination of both.

The series emphasizes the following four areas:

Mathematical Programming: Including linear programming, integer programming, nonlinear programming, interior point methods, game theory, network optimization models, combinatorics, equilibrium programming, complementarity theory, multiobjective optimization, dynamic programming, stochastic programming, complexity theory, etc.

Applied Probability: Including queuing theory, simulation, renewal theory, Brownian motion and diffusion processes, decision analysis, Markov decision processes, reliability theory, forecasting, other stochastic processes motivated by applications, etc.

Production and Operations Management: Including inventory theory, production scheduling, capacity planning, facility location, supply chain management, distribution systems, materials requirements planning, just-in-time systems, flexible manufacturing systems, design of production lines, logistical planning, strategic issues, etc.

Applications of Operations Research and Management Science: Including telecommunications, health care, capital budgeting and finance, economics, marketing, public policy, military operations research, humanitarian relief and disaster mitigation, service operations, transportation systems, etc.

This book series is indexed in Scopus.

Dmitry Ivanov • Alexandre Dolgui • Boris Sokolov
Editors

Handbook of Ripple Effects in the Supply Chain

Second Edition

Springer

Editors
Dmitry Ivanov
Department of Business and Economics
Berlin School of Economics and Law
Berlin, Germany

Alexandre Dolgui
Department of Automation, Production and
Computer Sciences
IMT Atlantique, LS2N-CNRS UMR 6004
Nantes, France

Boris Sokolov
St. Petersburg Federal Research Center of
the Russian Academy of Sciences
(SPC RAS)
St. Petersburg, Russia

ISSN 0884-8289 ISSN 2214-7934 (electronic)
International Series in Operations Research & Management Science
ISBN 978-3-031-85507-8 ISBN 978-3-031-85508-5 (eBook)
https://doi.org/10.1007/978-3-031-85508-5

This Springer imprint is published by the registered company Springer Nature Switzerland AG
The registered company address is: Gewerbestrasse 11, 6330 Cham, Switzerland

Preface

Purpose and Content of the Book

This handbook comprises recent developments in a new research field of the ripple effects in supply chains (SC). Building upon success of the 1st edition of this handbook published in 2019, we developed this 2nd edition. A lot has changed in our lives and SCs since 2019. The COVID-19 pandemic has changed the world. SCs have been exposed to hyper shocks, which have widely destabilized demand, supply, and logistics. Ripple effects have been rolling through SC networks at the unprecedented scale unlike any seen before. The semiconductor and container crises along with increasing geopolitical tensions (e.g., global and reciprocal tariffs in March- April 2025) came on top of the pandemic, making the ripple effect and SC stress testing important topics in SC and operations management. At the same time, rapid developments in digital technologies and artificial intelligence (AI) made SCs adapt not only to negative but also to positive, technological disruptions. In this setting, it is not surprisingly that the interest in the ripple effect has been exponentially growing. New theories such as SC viability, intertwined supply networks, and reconfigurable SC have been developed. The ripple effect analysis using digital twins became a topic of a growing interest. AI applications to managing SC resilience and ripple effect provide new interesting research contributions.

The 2nd edition of the handbook addresses both novel developments encountered in 2019–2024, adding several new chapters, and presents the fundamental knowledge about the ripple effect building on the chapters of the 1st edition. The chapters of this handbook are written by leading experts in SC risk management and resilience. The chapters present a multiple-faceted view of the ripple effect in SCs, while considering management, optimization, and technology perspectives.

The ripple effect occurs when an SC disruption cascades downstream, rather than remaining localized, and impacts the performance of the SC. The ripple effect considers structural network dynamics in the SC, initiated by a severe disruption (or a series of disruptions), and describes the propagation of this disruption downstream in the SC in terms of switching off some nodes and arcs in the network, e.g., due to material shortage. The impacts of the ripple effect might include lower revenues, delivery delays, loss of market share and reputation, or decreases in stock returns—the costs of these negative impacts can be devastating.

This book offers an introduction to the ripple effect in the supply chain for larger audience. The book delineates major features of the ripple effect and methodologies to mitigate the supply chain disruptions and recover in case of severe disruptions. The book reviews recent quantitative literature that tackled the ripple effect and gives a comprehensive vision of the state of the art and perspectives. The methodologies comprise mathematical optimization, simulation, control theory, game theory, network science, AI, and complexity and reliability research. The book observes the reasons and mitigation strategies for the ripple effect in the supply chain and presents the ripple effect control framework that is comprised of redundancy, flexibility, and resilience. Even though a variety of valuable insights has been developed in the said area in recent years, some crucial research avenues have been identified for the near future.

The book is expected to furnish fresh insights for supply chain management and engineering regarding the following questions:

- In what circumstance does one failure cause other failures?
- Which structures of the supply chain are especially prone to the ripple effect?
- What are the typical ripple effect scenarios and what is the most efficient way to respond to them?

Given these reflections, numerous ways to apply quantitative analysis to ripple effect modelling arise. Several research gaps might be addressed by the ability to dynamically change parameters during experiments and to observe how these changes impact performance in real time, e.g., considering:

- disruption propagation in the supply chain;
- dynamic recovery policies;
- gradual capacity degradation and recovery;
- multiple performance impact dimensions including financial, customer, and operational performance.

Distinctive Features

- It considers ripple effect in the supply chain from interdisciplinary perspective.
- It offers an introduction to the ripple effect mitigation and recovery policies in the framework of disruption risk management in the supply chains for larger audience.
- It integrates management and engineering perspectives on disruption risk management in the supply chain.
- It presents innovative optimization and simulation models for real-life management problems.

- It considers examples from both industrial and service supply chains.
- It reveals decision-making recommendations for tackling disruption risks in the supply chain in proactive and reactive domains.

Target Audience

Management and engineering graduate and Ph.D. students, supply chain and operations management professionals, industrial engineers, operations and supply chain risk researchers.

Berlin, Germany — Dmitry Ivanov
Nantes, France — Alexandre Dolgui
St. Petersburg, Russia — Boris Sokolov

Acknowledgments

The author gratefully acknowledges all those who have helped bring this book to publication. First and foremost, we have greatly benefited from the wealth of a vast array of published materials on the subject of supply chain risk management and associated topics. We would also like to thank the authors of each of the chapters. The content of this book has benefited immensely from their valuable insights, comments, and the suggestions of many reviewers. With regard to manuscript preparation, we thank Phu Nguyen, Hiran Prathapage, and Vivian Samstag for their help and technical assistance. Finally, we wish to thank the editors, Dr. Camille Price and Jialin Yan, and the entire Springer production team for their assistance and guidance in the successful completion of this book. Last, but not least—we thank our families for their enormous support during the writing and development of this book.

Introduction

Chapters in This Book

The Chapter "Ripple Effect in the Supply Chain: Definitions, Frameworks and Future Research Perspectives" by Dmitry Ivanov, Alexandre Dolgui, and Boris Sokolov begins the book. This chapter aims to delineate both major features of the ripple effect and methodologies for mitigating SC disruptions and recovering from severe disruptions. It presents an overview of the causes of the ripple effects and mitigation strategies. A framework for ripple effect control, comprised of redundancy, flexibility, and resilience, is developed. In addition, though a variety of valuable insights has been garnered in recent years, new research avenues and ripple effect taxonomies have been identified for the near future. Special focus is directed toward SC risk analytics and the ripple effect in SCs.

Next, in the Chapter "A Multi-portfolio Approach to Integrated Risk-Averse Planning in Supply Chains under Disruption Risks," Tadeusz Sawik suggests a methodical approach to time- and space-integrated decision-making. In the context of SC disruptions, the portfolio is defined as the allocation of demand for parts among suppliers or the allocation of demand for products among production facilities. A disruptive event is assumed to impact both primary suppliers of parts and the firm primary assembly plant. Considering the integration of mitigation and recovery decisions over time and space, the author shows that the primary portfolios to be implemented before a disruptive event are optimized simultaneously via recovery portfolios for the aftermath period as well as the portfolios of both parts suppliers and product manufacturers in different geographic regions. Risk-averse solutions are obtained through conditional cost-at-risk and conditional service-at-risk measures. The findings indicate that when the objective is to optimize service level with no regard to costs, both supply and demand portfolios are more diversified. The author concludes that the proposed multi-portfolio approach enables time- and space-integrated decision-making that may help to better mitigate the impact of disruption propagation on SC performance, i.e., the ripple effect.

Virginia L.M. Spiegler, Mohamed M. Naim, and Junyi Lin focus their Chapter on "The rippling effect of supply chains as complex adaptive non-linear systems." Using control theoretic tools, they show that nonlinearities can lead to unexpected dynamic behaviors in the SC that could then either trigger disruptions or make the response and recovery process more difficult. This chapter is particularly relevant for researchers wanting to learn more about the different types of nonlinearities that can be found in the SC, the existing analytical methods for dealing with each type of nonlinearity, and the potential direction of future research based on current knowledge in this field.

In their Chapter "Bullwhip Effect of Multiple Products with Interdependent Product Demands," Srinivasan Raghunathan, Christopher S. Tang, and Xiaohang Yue present a study that extends current theory to provide insights for a firm that manufactures multiple products in a single product category with interdependent demand streams. This study finds that interdependency between demand streams plays a critical role in determining the existence and magnitude of the bullwhip effect. More importantly, the authors show that interdependency impacts whether the firm should manage ordering and inventory decisions at the category level or at the product level, and whether the bullwhip effect measure computed at the category level is informative or not.

The Chapter "Demand Propagation in Supply Networks: Ripple Effects" by Nikolay Osadchiy explores the rippling of demand shocks in supply networks, focusing on the mechanisms that drive the shock propagation and the factors that can either amplify or mitigate their impact. Considering idiosyncratic shocks and the bullwhip effect, correlated shocks and systematic risks, and the role of community structure in shock propagation, cation of systematic shocks, the author examines how (i) small fluctuations in consumer demand lead to increasingly large variations in orders upstream in the supply chain; (ii) shocks impact multiple nodes simultaneously, leading to systematic risks; and (iii) community structures within supply networks influence the propagation of shocks.

The Chapter "Performance Impact Analysis of Disruption Propagations in the Supply Chain" by Dmitry Ivanov, Alexander Pavlov, and Boris Sokolov develops a method for quantifying the ripple effect in the SC with consideration of recovery policy. The performance impact index developed is then used to compare sales (revenue) in different SC designs to measure the estimated annual magnitude of the ripple effect. First, optimal SC recovery for two disruption scenarios is computed. Second, the performance impacts of disruptions for six proactive SC designs are assessed. Finally, the performance impact indexes of different SC designs are compared and conclusions are drawn about the ripple effect in these SC designs along with recommendations for the selection of a proactive strategy. The performance impact index developed can be used to assess how different markets are exposed to the ripple effect and how different SC designs can be compared according to their resilience to severe disruptions.

Yuhong Li, Kedong Chen, and Jiawei Zhang wrote a chapter "AI-Powered Supply Chain and Operations Management (SCOM): Capabilities and Challenges." This chapter explores the capabilities and challenges of AI in SCOM, focusing on data management, decision-making under uncertainties, and buyer–supplier relationship management in the context of resilience and ripple effect. Through a systematic review of the literature, we aim to provide a comprehensive understanding of AI's role in SCOM and present frameworks for managing the complexities and risks associated with AI-powered operations.

Ellie C. Falcone, Laharish Guntuka, and Steven Carnovale propose in their Chapter "Supply chain plasticity: reshaping the future of risk management" a novel approach to SC resilience. Supply chain plasticity refers to a firm's capability to rapidly and significantly alter its supply chain network structure to accommodate substantial shifts and disruptions in the business environment. Unlike conventional methods that aim to return to a pre-disruption state, plasticity involves making dramatic changes to how a firm's supply chain operates that often resulting in a new protocol that better aligns with the post-disruption environment. When implemented correctly, supply chain plasticity has the potential to significantly improve recovery, resilience, and responsiveness of supply chains.

In their Chapter "A Model of an Integrated Analytics Decision Support System for Situational Proactive Control of Recovery Processes in Service-Modularized Supply Chain," Dmitry Ivanov, Boris Sokolov, and Valerii Zakharov consider the challenge recovery process, a disruptive event, planning of the recovery control policy, and implementation of this policy in the SC. These events are distributed in time and subject to SC structural and parametrical dynamics. In other words, environment, SC structure, and the SC's operational parameters may change in the period between the planning of the recovery control policy and its implementation. As such, situational proactive control with combined use of simulation optimization and analytics is proposed to improve processes of transition between a disrupted and a restored SC state. Implementation of situational proactive control can reduce investments in robustness and increase resilience by obviating time traps in problems of transition process control. This chapter presents a decision support system model for situational proactive control of SC recovery processes based on a combination of optimization and analytics techniques. More specifically, three dynamic models are developed and integrated with each other, i.e., a model of SC material flow control, a model of SC recovery control, and a model of SC recovery control adjustment. The given models are developed within a cyber-physical SC framework based on an approach of service modularization.

Dmitry Ivanov's Chapter "Entropy-Based Analysis and Quantification of Supply Chain Recoverability" addresses the problem of designing resilient SCs at the semantic network level. The entropy method is used to show the interrelations between SC design and recoverability. Easy-to-compute quantitative measures are proposed to estimate SC recoverability. For the first time, an entropy-based SC analysis is brought into correspondence with consideration of SC structural recoverability and flexibility downstream in the SC. Exact and heuristic computation algorithms are suggested and illustrated. This approach and recoverability measure can be applied in selecting a resilient SC design according to potential recoverability.

Dmitry Ivanov and Maxim Rozhkov study capacity disruption and recovery policy impacts on SC performance in their Chapter "Disruption Tails and Revival Policies in the Supply Chain." A discrete-event simulation methodology is used for analysis with real company data and real disruptions. Two novel findings are presented. First, disruption-driven changes in SC behavior may result in backlogs and delayed orders, the accumulation of which in the post-disruption period is referred to as "disruption tails." The transition of these residues into the post-disruption period causes post-disruption SC instability, resulting in further delivery delays and non-recovery of SC performance. Second, a smooth transition from a contingency policy through a special "revival policy" to the normal operation mode enables partial mitigation of the negative effects of the disruption tails. These results suggest three managerial insights. First, contingency policies need to be applied during the disruption period to avoid disruption tails. Second, recovery policies need to be extended toward integrated consideration of both the disruption and the post-disruption periods. Third, revival policies need to be developed for the transition from the contingency to the disruption-free operation mode. A revival policy is intended to mitigate the negative impact of the disruption tails and stabilize SC control policies and performance. The experimental results suggest a revival policy should be included in an SC resilience framework if performance cannot be recovered fully after capacity recovery.

The chapter "Epidemiological Models to Predict Infection Epidemic: A Literature Review" by Fatemeh Mirsaeedi, Mohammad Sheikhalishahi, Mehrdad Mohammadi, and Dmitry Ivanov presents state-of-the-art in epidemiological prediction models. In particular, the authors focus on Susceptible-Infected-Recovered (SIR) models and their various facets. The objectives of this review are threefold: (1) the assumptions, equations, and methodologies for estimating critical parameters within these models are evaluated; (2) the challenges associated with implementing disease control interventions, including both medical and non-medical strategies, and their integration into the models are explored; and (3) relation of SIR models and optimization models are investigated. Additionally, a systematic review at the micro-level has identified significant research gaps within the existing literature, leading to recommendations for future research directions. One of the most important findings is that many assumptions need to be reviewed to be better aligned with real conditions, which should hopefully stimulate some useful inquiries into how to model epidemic diseases.

In their Chapter "Managing Disruptions and the Ripple Effect in Digital Supply Chains: Empirical Case Studies," Ajay Das, Simone Gottlieb, and Dmitry Ivanov analyze the impact of accelerating digitalization on SC risk management. Digital technologies, such as big data analytics, Industry 4.0 applications, additive manufacturing, blockchain, advanced tracking and tracing technologies, and enterprise resource planning software systems are considered. Empirical evidence on the interrelations between digital technologies and the risk of SC disruptions, as well as the influence of the one on the other, is analyzed based on the findings from the multiple case studies. These findings are comprised of the insights and managerial recommendations of experts from multiple industries. The empirical analysis is guided by hypotheses and a conceptual framework based on extant theory.

Chapter "Contingent sourcing ripples considering hoarding behaviour under supply disruptions in a three-tier supply chain" by Shanshan Li and Yong He considers a make-to-order system where the production is disrupted due to a supply failure. To avoid a potential stockout risk and a responding price increase during disruption, customers might decide to stockpile extra units for future consumption, which may trigger procurement ripples. The authors investigate optimal sourcing decisions for both customers and the manufacturer to cope with random disruptions. To this end, both the once-off stockpiling and the continuous stockpiling (two-stage) are considered. The optimal stockpiling behaviors in these two modes are analytically provided. Next, a model is formulated for the manufacturer to optimize the contingent sourcing quantities under different patterns of customers hoarding behavior. Finally, by conducting numerical analysis, further insights are developed considering the role of relative factors and providing specific managerial suggestions on how to adapt contingent sourcing to alleviate different random disruptions, under different patterns of customers' hoarding behaviors.

Rameshwar Dubey devotes his Chapter "Healthcare Supply Chain Agility and Resilience: A Relational View Perspective" to establishing a link between relational competencies, such as transparency, trust, and collaboration, and supply chain capabilities, including agility and resilience, through empirical research. The chapter specifically concentrates on the healthcare sector, with a focus on the impact of the COVID-19 crisis, which laid bare the vulnerabilities of the healthcare supply chain and resulted in significant human casualties. The theoretical model is rooted in the relational view, and a meticulously tested survey instrument was developed using a seven-point Likert scale. Data were gathered from retail pharmacies in France utilizing a database acquired from a reputable healthcare sector consulting firm, resulting in 129 usable responses. Hypotheses were scrutinized using WarpPLS 7.0 commercial software, and the analysis uncovered that transparency and trust play pivotal roles as drivers of collaboration. Furthermore, collaboration among partners emerged as a significant driver of agility and resilience. These findings suggest that relational competencies wield a positive and significant influence on agility and resilience during crises. This bolsters the relational view and provides insight into how these competencies underpin supply chain capabilities. Additionally, the chapter recognizes its limitations and puts forward recommendations for future research areas.

The chapter "Digital Supply Chain Twins: Managing the Ripple Effect Resilience, and Disruption Risks by Data-Driven Optimization, Simulation, and Visibility" is written by Dmitry Ivanov, Alexandre Dolgui, Ajay Das, and Boris Sokolov. The impact of digital technology, Industry 4.0, blockchain, and data analytics on the ripple effect and disruption risk management in SCs is studied in this chapter. This chapter does not pretend to be encyclopedic, but rather seeks to advance the knowledge we have to further research on the relationship between digitalization and SC disruptions risks based on recent literature and case studies. It then presents an SC risk analytics framework and explains the concept of digital SC twins. It analyzes perspectives and future transformations that can be expected in the transition toward cyber-physical SCs. It shows how digital technologies and smart operations can help to integrate resilience and lean thinking into a *resileanness* framework for a "Low-Certainty-Need" (LCN) SC.

Contents

About the Editors

Dmitry Ivanov Prof. Dr. Dr. habil. Dmitry Ivanov teaches Supply Chain and Operations Management at Berlin School of Economics and Law (HWR Berlin) where he leads the Digital Supply Chain Lab and is academic director of M.A. Global Supply Chain and Operations Management.

Prof. Ivanov is an internationally renowned expert in supply chain and operations management, industrial and control engineering, and artificial intelligence. He has made fundamental and influential contributions, particularly exploring structural dynamics and control in complex networks with applications to supply chain resilience, Industry 4.0, supply chain simulation, risk analytics and digital supply chain twins. He is co-author of structural dynamics control method. His research coined several seminal academic and practical directions such as the ripple effect in supply chains and supply chain viability. He has applied different methodologies such as optimization, simulation, control theory, and artificial intelligence to a variety of supply chain and operations management problems. Most of his research stems from real practical context and focuses on the interface of supply chain management, operations research, industrial engineering, and digital technology.

He has been teaching courses in operations management, supply chain management, logistics, management information systems, and strategic management at undergraduate, master's, PhD, and executive MBA levels at different universities worldwide for about 25 years. He has given guest lectures and webinars, presented scholarly papers and has been a visiting professor at numerous universities across the globe. Author of anyLogistix educational complex for supply chain optimization and simulation that is in use at more than 300 universities worldwide. He enjoys being part of students' learning experience and creating a safe and active learning environment. He sees his job as an educator to equip future industry leaders with working knowledge and skills in management and technology that would shape more resilient, adaptable, and sustainable supply chains and operations.

His multi-disciplinary academic background includes industrial engineering and management, operations research, and applied control theory. His educational background includes industrial engineering and management during five years of diploma studies along with supply chain management, artificial intelligence, information systems, simulation, and control theory during PhD and habilitation. He gained PhD (Dr.rer.pol.), Doctor of Science (ScD), and Habilitation (Dr. habil.) degrees. Prior to becoming an academic, he was mainly engaged in industry and consulting, especially for process optimization in manufacturing and logistics and ERP systems. His practical expertise includes numerous projects on the application of operations research and process optimization methods to operations design, logistics, scheduling, and supply chain management. Prior to joining the Berlin School of Economics and Law, he was professor and interim chair of Operations Management at the University of Hamburg.

Professor Ivanov's research makes impact. His papers belong to the most cited worldwide in the areas of supply chain resilience and digital supply chain. For example, he has been the most cited author of premier journals International Journal of Production Research and Transportation Research: Part E in 2020–2024. His research record includes over 460 publications, with over 175 papers in prestigious academic journals, multiple editions of the leading textbooks "Global Supply Chain and Operations Management," "Introduction to Supply Chain Resilience," and "Introduction to Supply Chain Analytics," and several research books such as "Digital Supply Chain," "Supply Networks Dynamics and Control," "Structural Dynamics and Resilience in Supply Chain Risk Management," "Scheduling in Industry 4.0 and Cloud Manufacturing," and "Handbook of Ripple Effects in the Supply Chain." Professor Ivanov's research has been published in various academic journals, including Annals of Operations Research, Annual Reviews in Control, Computers and Industrial Engineering, European Journal of Operational Research, Expert Systems with Applications, IEEE Transactions on Engineering Management, IISE Transactions, International Journal of Information Management, International Journal of Integrated Supply Management, International Journal of Inventory Research, International Journal of Physical Distribution & Logistics Management, International Journal of Production Research, International Journal of Production Economics, International Journal of Technology Management, International Journal of Systems Science, International Transactions in Operational Research, Journal of Intelligent Manufacturing, Journal of Purchasing and Supply Management, Journal of Operational Research Society, Journal of Scheduling, Omega, Production Planning and Control, Supply Chain Management: International Journal, Transportation Research: Part E, etc.

He is a recipient of IISE Transactions Best Paper Award 2021, Best Paper and Most Cited Paper Awards of International Journal of Production Research (2018–2024), OMEGA Best Paper Award 2022, Annual Reviewer Award of International Journal of Production Economics (2020), Clarivate Highly Cited Researcher Award (2021–2024). Ranked #1 in the "Top Scientists in Business and Management in Germany, Austria, and Switzerland" ranking in 2024 published in the Wirtschaftwoche. Ranked #1 in Operations Research in Standford-Elsevier 2% Top Scientists List (2022–2024). In 2024 and 2025, ScholarGPS ranked Dr. Ivanov a Highly Ranked Scholar #1 in the "Prior 5 Years" category in Supply Chain, Business and Management, and Management Information Systems areas.

His research projects have been supported by funding from European Commission (Horizon 2020), DFG (German Research Foundation), IFAF Berlin, and industry. Principal investigator in several research projects on resilience and digital twins including European projects ACCURATE and CERERE (2023–2027) and the DFG Collaborative Research Cluster on Resilience of Global Supply Chains at HWR Berlin (2024–2029). Dr. Ivanov is profiled in supply chain and operations management, operations research, and industrial engineering. He is passionate about bridging the knowledge of different disciplines and applying it to solution of practically relevant problems. He delivered numerous invited plenary, keynote and panel talks at the conferences and global webinars of INFORMS, POMS, IFAC, IFIP, IFPR, IEEE, CIE, DSI, and Academy of Management. He is leading working groups, tracks, and sessions on the Digital Supply Chain, Supply Chain Risk Management and Resilience in global research communities. He is Editor of International Journal of Integrated Supply Management, Editor of Annals of Operations Research, and an Associate Editor in International Journal of Production Research, International Transactions in Operational Research, International Journal of Systems Science, and OMEGA. He is an editorial board member, associate and guest editor in different journals, including Annals of Operations Research, Annual Reviews in Control, IISE Transactions, International Journal of Production Economics, International Journal of Production Research, International Transactions in Operational Research, International Journal of Information Management, International Journal of Inventory Research, International Journal of Physical Distribution and Logistics Management, International Journal of Systems Science, and OMEGA. He is Chairman of IFAC CC5 "Cyber-Physical Manufacturing Enterprise," former Chairman of IFAC TC 5.2 "Manufacturing Modelling for Management and Control" (2017–2023), and Co-Chairman of the IFAC TC 5.2 Working group "Supply Network Engineering." He regularly presents his research results and has been Chairman, IPC and Advisory Board member of over 70 international conferences in supply chain and operations management, industrial engineering, control and information sciences where he has organized numerous tracks and sessions (including IFAC MIM, IFAC INCOM, EURO, INFORMS, IISE, IFORS, DSI, POMS, OR, IFAC World Congress, IFIP PRO-VE, LDIC, to name a few). He has been General Conference Chair of IFAC MIM 2019—one of the worldwide largest conferences in manufacturing, industrial engineering, operations and supply chain management (750 participants).

Alexandre Dolgui is a Distinguished Professor (Full Professor of Exceptional Class in France) and the Head of Automation, Production, and Computer Sciences Department at the IMT Atlantique, France. His research focuses on manufacturing line design, production planning and supply chain optimization. His research is mainly based on exact mathematical programming methods and their intelligent coupling with heuristics and metaheuristics algorithms.

He is the co-author of 5 books, the co-editor of 26 books for conference proceedings, the author of 325 refereed journal papers, 32 editorials, and 30 book chapters, as well as the author of over 500 papers written for conference proceedings. He is the Editor-in-Chief of the *International Journal of Production Research*, an Area Editor of *Computers & Industrial Engineering*, and an Associate Editor of *Journal Européen des Systèmes Automatisés*. He is also past Associate *Editor of International Journal of Systems Science* (2005–2008), *IEEE Transactions on Industrial Informatics* (2006–2009), and *Omega-the International Journal of Management Science* (2009–2012), as well as having been the consulting Editor of the International Journal of Systems Science (2009–).

He is a member of editorial boards for 25 other journals, including the *International Journal of Production Economics*, *International Journal of Manufacturing Technology & Management*, *International Journal of Simulation & Process Modelling*, *International Journal of Engineering Management & Economics*, *Journal of Decision Systems*, *Journal of Mathematical Modelling & Algorithms*, *Journal of Operations and Logistics*, *Journal of Industrial Engineering and Management & Production Engineering Review*, *Decision Making in Manufacturing and Service*, *Risk and Decision Analysis*, etc. He is also a fellow of the European Academy for Industrial Management, a member of the board of the International Foundation for Production Research, Vice-Chair of IFAC TC 5.2 Manufacturing Modelling for Management and Control, a member of IFIP WG 5.7 Advances in Production Management Systems, IEEE System Council Analytics, and Risk Technical Committee, and Guest Editor of special issues of European Journal of Operational Research, Annals of Operations Research, International Journal of Production Research, International Journal of Production Economics, Omega-the International Journal of Management Science, Journal of Intelligent Manufacturing, Journal of Mathematical Modeling and Algorithms, Engineering Applications of Artificial Intelligence, Algorithms and Annual Reviews in Control. He was General Scientific Chair of the 12th IFAC symposium INCOM'06, Chairman of International Program Committee of SCM'02, MOSIM'04, INCOM'09, INCOM'12, INCOM'15, IESM'13, MIM'13, IESM'17, GSC'18, MIM'19, and Chairman of Steering committee of MIM'16 and General chair of MIM'22 and APMS'21. He was also Chairman of the Organizing Committee of the International Conference MOSIM'01 and ROADEF'2011. In the last 20 years, he was a member of the Program Committees of over 250 International Conferences, etc. He has been responsible for the French national CNRS working group on Design of Production Systems (with about 336 individual members) and the regional project on Design and Management of Reconfigurable Manufacturing Systems.

Boris Sokolov born in 1951, Head of the Information Technologies in Systems Analysis and Modeling Laboratory at the St. Petersburg Federal Research Center of the Russian Academy of Sciences (SPC RAS). From 2006 to 2017, he was Deputy Director for Research at SPIIRAS (currently SPC RAS). In 2008, he was awarded the title of Honored Scientist of Russia. Laureate of two Russian Government Prizes in Science and Technology (2013, 2022). Since 2023, Professor B.V. Sokolov has been included in the list of outstanding scientists of the international scientific portal Research.com in the field of Engineering and Technology.

He received his Master's, PhD, Doctor of Engineering Sciences, and Professor degrees in 1974, 1983, 1993, and 1994, respectively. He is the founder of a new scientific direction in the field of automation and intellectualization of complex technical objects (CTO) proactive management processes, associated with complex analysis and management of distributed processes in critical applications. The areas of scientific interests of Professor B. Sokolov are: fundamental and applied interdisciplinary research in the field of complex modeling, simulation modeling and mathematical methods in scientific research, optimal control theory and mathematical models and methods for supporting decision-making in complex organizational and technical systems under conditions of uncertainty and multicriteria, optimization of the RFID, Wi-Fi, Bluetooth, and other mobile IT technologies implementation in supply chain management processes.

Speaking about the applied results obtained by Doctor of Technical Sciences, Professor B.V. Sokolov, it is necessary to point out that, since 1978, he has been directly involved in the implementation of 120 research and development projects. From 1999 to the present, he has been implementing numerous projects financed by the Russian Academy of Sciences, the Russian Foundation for Basic Research, the Russian Science Foundation, the Section of Applied Problems of the Presidium of the Russian Academy of Sciences and international organizations (EOARD, CRDF, EC). To date, he and his laboratory staff have completed 26 grants from the Russian Foundation for Basic Research, 4 grants from the Russian Science Foundation, and 11 international projects. He was the scientific supervisor in 45 works. Professor Sokolov and representatives of his scientific school have developed analytical and simulation methods, techniques, algorithms and methods for complex automated proactive planning and management of their dynamics, solved with minimal resource costs. Over the past few years, Professor Sokolov and his students have been intensively developing an original applied theory of proactive monitoring and control structural dynamics objects and systems proactive monitoring and control which oriented to implementation in the Industrial Internet of Things. The reliability and validity of his conclusions and developments have been confirmed both internationally and in Russia by numerous publications, implementations and tests in the rocket and space, transport and logistics, and industrial fields of application. Recently, the new scientific and practical results obtained have been implemented in the agriculture logistics sphere. Professor Sokolov has been actively teaching since 1982. Since 1999, he has been a professor at the St. Petersburg State University of Aviation Instrument Making. He has developed several original lecture courses on complex modeling of proactive control processes for the structural dynamics of complex objects in various subject areas. He has supervised 15 candidates of technical sciences (Ph.D.) and 5 doctors of technical sciences (Dr. Habil.). Professor Sokolov is a member of the Academic Council of SPIIRAS and two dissertation councils, and has repeatedly served on the program committees of prestigious Russian and international conferences. He is a member of the International Journal of Instrumentation, International Journal of Information Technology, the journal Reliability, Issues of Radio Electronics, the Federation of Astronautics, and the Academy of Navigation and Motion Control.

He is the (co)author of 15 monographs and books on system analysis, decision support systems, supply chain management, and control systems and theory, 5 textbooks, 11 manuals, 23 patents, and more than 720 scientific papers published in various academic journals, including Annals of Operations Research, International Journal of Production Research, International Journal of Information Management, International Journal of Manufacturing Technology and Management, Journal of Computer and Systems Sciences International, Differential Equations, Automation and Remote Control, and Annual Reviews in Control.

Ripple Effect in the Supply Chain: Definitions, Frameworks and Future Research Perspectives

Dmitry Ivanov, Alexandre Dolgui, and Boris Sokolov

Abstract This chapter aims at delineating major features of the ripple effect and methodologies to mitigate the supply chain disruptions and recover in case of severe disruptions. It observes the reasons and mitigation strategies for the ripple effect in the supply chain and presents the ripple effect control framework that is comprised of redundancy, flexibility and resilience. Even though a variety of valuable insights have been developed in the given area in recent years, new research avenues and ripple effect taxonomies are identified for the near future.

1 Ripple Effect in the Supply Chain: Basic Definitions

1.1 Supply Chain Risks and Ripple Effect

Disruptions are considered high-impact-low-frequency events (e.g. fire or tsunami) in the supply chain (SC) that change the SC's structural design and significantly impact performance. The propagation of a disruption through an SC and its associated impact is called the *ripple effect*. A ripple effect is distinct from the well-known bullwhip effect. It manifests when the impact of an SC disruption cannot be

D. Ivanov (✉)
Department of Business and Economics, Berlin School of Economics and Law, Berlin, Germany
e-mail: divanov@hwr-berlin.de

A. Dolgui
Department of Automation, Production and Computer Sciences, IMT Atlantique, LS2N-CNRS UMR 6004, Nantes, France
e-mail: alexandre.dolgui@imt-atlantique.fr

B. Sokolov
St. Petersburg Federal Research Center of the Russian Academy of Sciences (SPC RAS), St. Petersburg, Russia
e-mail: sokol@iias.spb.su

D. Ivanov et al. (eds.), *Handbook of Ripple Effects in the Supply Chain*, International Series in Operations Research & Management Science 276, https://doi.org/10.1007/978-3-031-85508-5_1

localized or contained to one part of the SC and cascades downstream, resulting in a high-impact effect on SC performance (Dolgui et al., 2018; Dolgui & Ivanov, 2021).

The ripple effect considers structural network dynamics in the SC while bullwhip effect characterizes the oscillations in operational parameters. The ripple effect is initiated by a severe disruption and describes the propagation of this disruption downstream of the SC, e.g., in terms of propagation of the demand fulfilment downscaling as a result of a severe disruption (Brusset et al., 2023; Deng et al., 2019). In more severe cases, the ripple effect can be manifested in temporary switching off some nodes and arcs in the network, e.g., due to material shortage (Dolgui et al., 2020, 2024). The bullwhip effect, on the contrary, is launched by a small operational deviation and is expected to be amplified in the upstream direction.

While the reasons for the bullwhip effect have been extensively studied over the past two decades, the ripple effect is quite a new phenomenon and analysis of its impacts deserves more research attention. These impacts might include lower revenues, delivery delays, loss of market share and reputation and stock return decreases—the cost of all of which could be devastating.

Consider an example. On 17 October 2016 as a result of an incorrect maintenance operation on a pipeline at the BASF facility in Ludwigshafen (Germany), there was an explosion and subsequent fires at North Harbor, a terminal for the supply of raw materials such as naphtha, methanol and compressed liquefied gases. More than 2.6 million tons of goods are handled there each year and an average of seven ships a day moor at its docks. Two steam crackers, the starting point for producing basic chemicals, needed to be stopped because they could no longer be supplied, and 22 were only partially working. The two steam crackers could have been restarted 2 days later, but only in May 2017 was the concept for reconstruction released whereby the reconstruction should be completed by September 2017. Restricted production output, a daily revenue decrease of 10–15% as compared to the previous year during the disruption period, impact on the basic chemicals division (about 21% of sales), delivery delays, limited access to key raw materials, exhausted product inventories and a forecasted impact on 6% of BASFs annual earnings were some of the consequences of this incident (Dolgui et al., 2018, and references within). Logistics was temporarily shifted from ships and pipelines to trucks and trains. BASF was in close contact with its customers to keep them informed about the current availability of products to minimize the impact on customer deliveries. Because of BASF's integrated "Verbundsystem" (networking system), comprised of various plants and delivery systems for feedstocks, the incident had an impact along the global SC. This high and long-term impact is the so-called ripple effect (Ivanov et al., 2014a, b).

BASF built a resilient SC, which is why the economic consequences of the afore-mentioned incident were considerably smaller than expected. BASF took process safety and risk prevention measures that included globally valid guidelines and requirements for buildings, etc. and practical security training for employees and support staff. Along with process safety and risk prevention measures, BASF has global emergency response management. This management consists of the integration of worldwide group companies, joint ventures, partners, suppliers and customers. Emergency phones and an integrated network of control centres (e.g. internal/external fire departments and rescue service) also enable this global emergency response management to work even more closely together. BASF was prepared for the incident in October 2016, but there is still a long-term impact.

The BASF example shows the importance of SC risk management and the threats that severe disruptions may influence SC performance. *Risk management* in the SC has become one of the most important topics in research and practice over the last decade. A number of books (Handfield & McCormack, 2008; Kouvelis et al., 2012; Waters, 2011; Gurnani et al., 2012; Heckmann, 2016; Khojasteh, 2017; Ivanov, 2018b; Sawik, 2018) and literature review papers (Dolgui et al., 2018; Fahimnia et al., 2015; Ho et al., 2015; Klibi et al., 2010; Simangunsong et al., 2012; Snyder et al., 2016) provide insightful overviews and introductions to different aspects of this exciting field.

Recent literature introduced different classifications of SC risks (Chopra & Sodhi, 2004; Tang & Musa, 2011; Ho et al., 2015; Quang & Hara, 2018) (see Fig. 1).

Risks of demand and supply uncertainty are related to random uncertainty and business-as-usual situation. Such risks are also known as *recurrent* or *operational risks. SC* managers achieved significant improvements in managing global SCs and mitigating recurrent SC risks through improved planning and execution (Chopra & Sodhi, 2014).

From 2000 thru 2018, SC disruptions (e.g. because of both natural and man-made disasters, such as on 11 March 2011 in Japan, floods in Thailand in 2011, fire in the Phillips Semiconductor plant in New Mexico, etc.) occurred in greater frequency

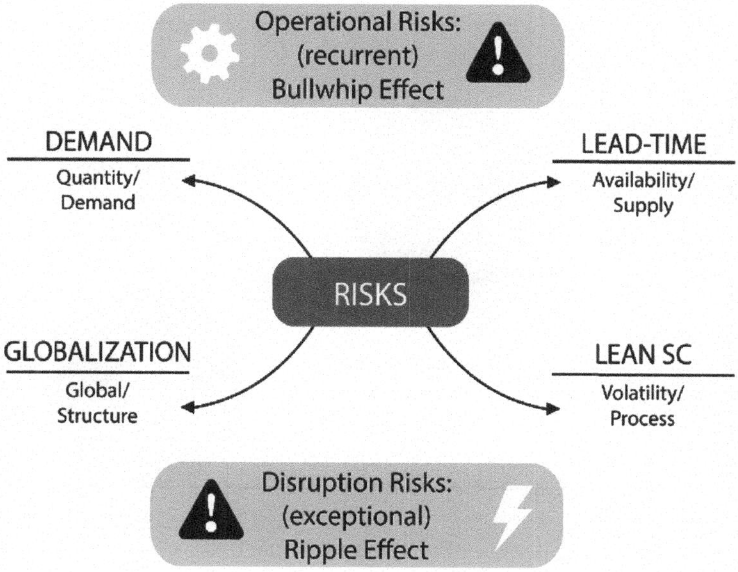

Fig. 1 Supply chain operational and disruption risks (Ivanov, 2018b)

and intensity, and thus with greater consequences (Chopra & Sodhi, 2014; Simchi-Levi et al., 2014). Hendricks and Singhal (2005) quantified the negative effects of SC disruption through empirical analysis and found 33–40% lower stock returns relative to their benchmarks over a 3-year time period that started one year before and ended two years after a disruption.

1.2 Disruption Risks and the Ripple Effect

Disruption risks represent a new challenge for SC managers who face the *ripple effect* (Liberatore et al., 2012; Ivanov et al., 2014a, b; Levner & Ptuskin, 2018; Ivanov, 2018a; Dolgui et al., 2018; Ivanov et al., 2018; Ivanov et al., 2019a; Hosseini et al., 2019, Ghadge et al., 2022) subject to *structural disruptions* in the SC, unlike the *parametrical deviations* in the bullwhip effect (Fig. 2).

In the last two decades, considerable advancements have been achieved in research regarding the mitigation of inventory and production shortages and response to demand fluctuations. In particular, the *bullwhip effect* in the SC has been extensively considered in this domain subject to *randomness uncertainty* with the help of stochastic and simulation models.

The differences between the bullwhip effect and ripple effect are presented in Table 1 (Dolgui et al., 2018).

The Bullwhip effect considers weekly/daily demand and lead-time fluctuations as primary drivers of the changes in the supply chain which occur at the parametric level and can be eliminated in a short-term perspective. In recent years, the research

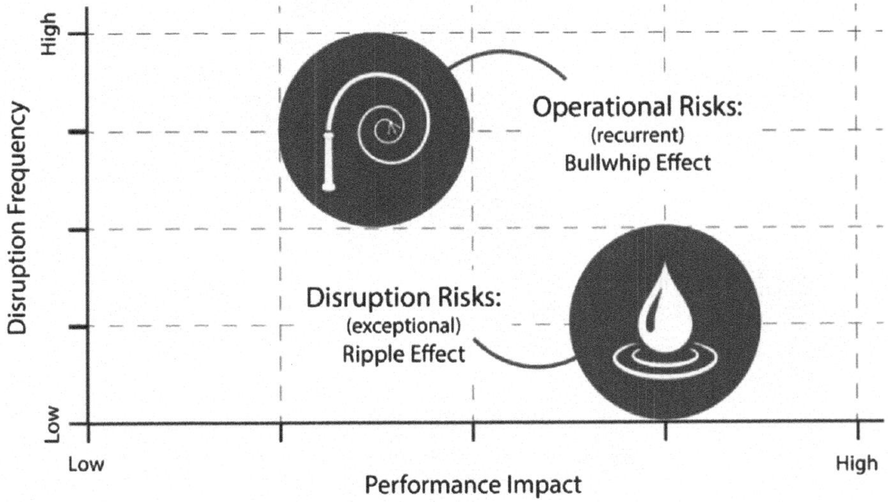

Fig. 2 Operational and disruption risks in supply chains (Ivanov et al., 2019a, b)

Table 1 Ripple effect and bullwhip effect (Dolgui et al., 2018)

Feature	Ripple effect	Bullwhip effect
What uncertainty?	Hazard, deep uncertainty	Random uncertainty
What risks?	Disruption, exceptional risks (e.g. a plant explosion)	Operational, recurrent risks (e.g. demand fluctuation)
What can be disturbed?	Structures and critical performance (such as supplier unavailability or revenue)	Operational parameters such as lead time and inventory
How are deviations prevented?	Proactive redundancy and flexibility	Information coordination
What happens after the disturbance?	Short-term stabilization and middle- and long-term recovery; high coordination efforts and investments	Short-term coordination to balance demand and supply
What is performance impact?	Output performance can decrease, such as in annual revenues or profits	Current performance can decrease such as in daily or weekly stock out/overage costs

community has started to investigate severe supply chain disruptions with long-term impacts that can be caused, for example, by natural disasters, political conflicts, terrorism, maritime piracy, economic crises, destroying of information systems, or transport infrastructure failures. We refer to these severe natural and man-made disasters as the ripple effect in the supply chain where changes in the supply chain occur at the structural level and recovery may take mid- and long-term periods of time with significant impacts on output performance such as annual revenues. In this setting, supply chain disruption management can be considered a critical capability which helps to create cost-efficient supply chain protection and implement appropriate actions to recover supply chain disruptions and performance.

Most studies on supply chain disruption consider how changes to some variables are rippling through the rest of the supply chain and impacting performance. Studies by Ivanov et al. (2014a, b) and Dolgui et al. (2018) suggest considering this situation as *the ripple effect in the supply chain*, as an analogy to computer science, where the ripple effect determines the disruption-based scope of changes in the system (Goldbeck et al., 2020).

The *ripple effect* in the supply chain occurs if a disruption cannot be localized and cascades downstream, impacting supply chain performance such as sales, stock return, service level and costs (Dolgui et al., 2018; Hosseini & Ivanov, 2022; Ivanov, 2018a; Ivanov et al., 2015; Ivanov & Dolgui, 2021; Ivanov et al., 2014a). The methodical elaborations on the evaluation and understanding of low-frequency-high-impact disruptions are therefore vital for understanding and further development of network-based supply concepts (Ivanov, 2024a, d; Liberatore et al., 2012; Sawik, 2016; Tomlin, 2006).

Details of empirical or quantitative methodologies differ across the works on supply chain disruption management, but most share a basic set of attributes (Li et al., 2020; Llaguno et al., 2022; Pavlov et al., 2022):

- A disruption (or a set of disruptions)
- Impact of the disruption on operational and strategic economic performance
- Stabilization and recovery policies

Within this set of attributes, most studies on supply chain disruption consider how changes to some variables are rippling through the rest of the supply chain and impacting performance. We suggest considering this situation, *the ripple effect in the supply chain*, as an analogy to computer science, where the ripple effect determines the disruption-based scope of changes in the system (Hosseini et al., 2020; Proselkov et al., 2024).

The ripple effect is a phenomenon of disruption propagations in the supply chain and their impact on output supply chain performance (e.g. sales, on-time delivery and total profit). It may have more serious consequences than just short-term performance decrease. It can result in market share losses (e.g. Toyota lost its market leader position after the tsunami in 2011 and needed to redesign its supply chain coordination mechanism). The ripple effect is also known as "domino effect" or "snowball effect." The reasons for ripple effect are not difficult to find. With increasing supply chain complexity and consequent pressure on speed and efficiency, an ever-increasing number of industries come to be distributed worldwide and concentrated in industrial districts. In addition, globalized supply chains depend heavily on permanent transportation infrastructure availability.

The ripple effect describes disruption propagation in the supply chain, the impact of a disruption on supply chain performance and disruption-based scope of changes in supply chain structures and parameters.

Following a disruption, its effect ripples through the supply chain. The missing capacities or inventory at the disrupted facility may cause missing materials and production decrease at the next stages in the supply chain. Should the supply chain remain in the disruption model longer than some critical period of time (i.e. *time-to-survive* (Simchi-Levi et al., 2015)), critical performance indicators such as sales or stock returns may be affected.

Ripple effects are not an infrequent occurrence. In many examples, supply chain disruptions go beyond the disrupted stage; i.e. the original disruption causes disruption propagation in the supply chain, at times still higher consequences are caused (Fig. 3).

The studies by Liberatore et al. (2012), Ivanov et al. (2014a, b, 2016, 2017a), Han and Shin (2016), Sokolov et al. (2016), Mizgier (2017), Schmitt et al. (2017), Ivanov (2017, 2018b), Levner and Ptuskin (2018), Dolgui et al. (2018), Pavlov et al. (2018), Scheibe and Blackhurst (2018) and Akkermans and van Wassenhove (2018) extensively analysed SC ripple effect, its reasons and efficient countermeasures. These findings are summarized in Table 2.

First, the literature provides evidence that disruption duration and propagation impact SC performance. Second, proactive strategies such as backup facilities and inventory have positive impacts concerning both performance and prevention

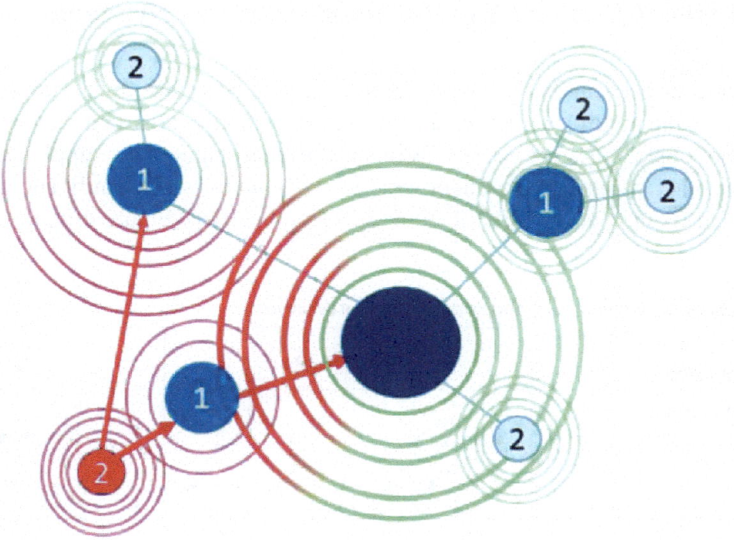

Fig. 3 Disruption propagation in the supply chain

Table 2 Ripple effect reasons and countermeasures (based on Dolgui et al., 2018)

Reason	SCM impact	Ripple effect impact	Countermeasures
Leanness	Single sourcing	In the non-disrupted scenario, it is irrational to avoid lean practices. At the same time, a capacity disruption may result in the ripple effect and a performance decrease.	Multiple/dual sourcing/backup suppliers
	Low inventory	Recommendation to use capacity buffers or a backup facility as additional capacity reserves	Risk mitigation inventory
	Inflexible capacity		Postponement
Complexity	Globalization	Without a coordinated contingency policy, disruption recovery and performance impact estimation can be very long-lasting and expensive. Coordinated control algorithms are needed to monitor SC behaviour, identify disruptions and adjust order allocation rules using a coordinated contingency policy	Geographical sourcing diversification
	Decentralization		Global SC contingency plans
	Multi-stage SCs		Supplier segmentation according to disruption risks

of disruption propagation. Third, the speed of recovery plays an important role in mitigating the performance impact of disruptions. Fourth, an increase in SC resilience implies significant cost increases in the SC.

1.3 Ripple Effect and Supply Chain Structural Dynamics

The ripple effect causes structural changes in the SC. The main supply chain features are the multiple structure design and changeability of structural parameters because of objective and subjective factors at different stages of the supply chain life cycle. In other words, supply chain *structural dynamics* is constantly encountered in practice (Ivanov et al., 2010; Ivanov & Sokolov, 2010). Figure 4 depicts major

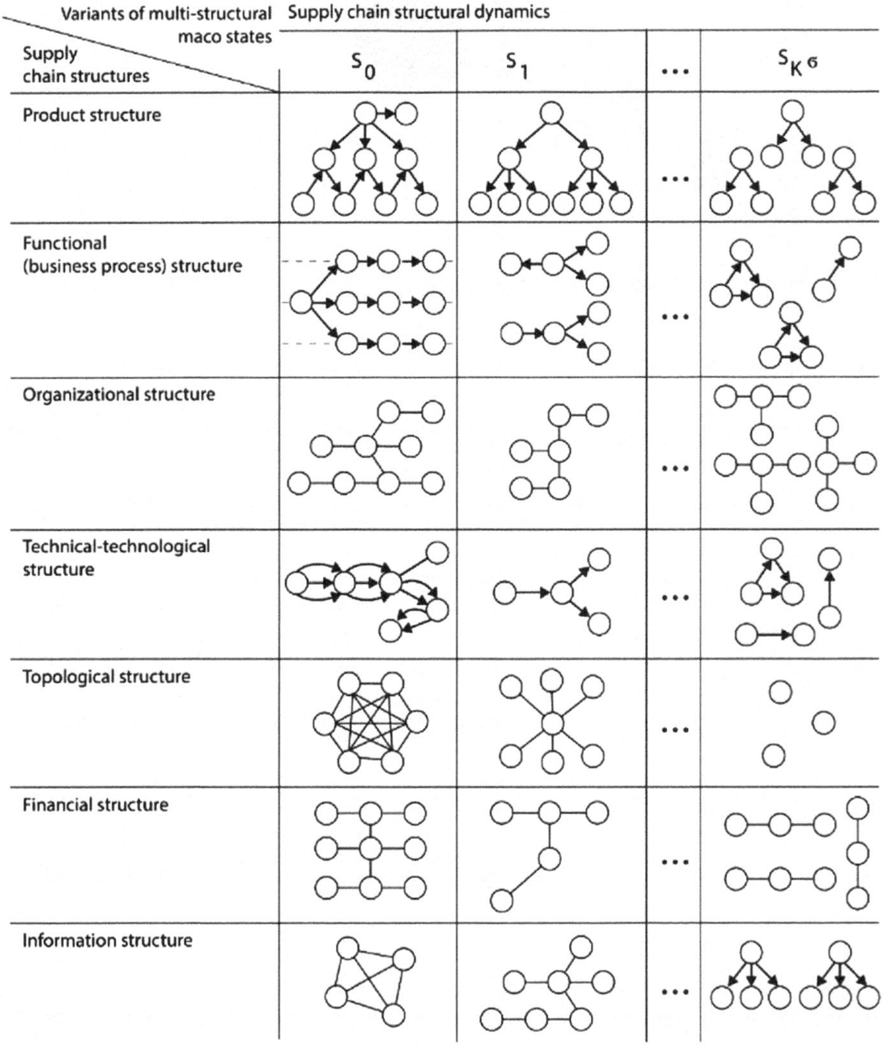

Fig. 4 Supply chain multi-structural composition and structural dynamics (based on Ivanov et al., 2010)

structures and their changes in dynamics. The composition of different structures at different points in time results in supply chain multi-structural macrostates *S*. Multi-structural macrostates describe supply chain design evolution over time due to planned (controllable) and uncertain factors.

The multi-dimensional dynamic space, along with coordinated and distributed decision-making, guides us in understanding modern supply chains as *multi-structural dynamic systems* (Ivanov et al., 2010).

1.4 Supply Chain Performance, Resilience and Ripple Effect Control

One of the main objectives of supply chain management is to increase total supply chain performance, which is basically referred to as supply chain effectiveness (i.e. sales and service level) and efficiency (supply chain costs). At the same time, the achievement of planned performance can involve the impact of disruptions in a real-time execution environment. Supply chain execution is subject to uncertainty at the planning stage and disruption at the execution stage. Cost efficiency comes with a huge hidden expense should a major disruption (i.e. a more severe impact than a routine disturbance) occur. This requires supply chain protection against and efficient reaction to disturbances and disruptions. Therefore, supply chains need to be planned to be *stable, robust and resilient* enough to (1) maintain their basic properties and ensure execution; and (2) be able to adapt their behaviour in the case of disturbances in order to achieve planned performance using recovery actions.

Decisions in supply chain structural dynamics control can be roughly classified into proactive and reactive stages (Fig. 5).

Resilient supply chain design extends traditional supply chain design approaches with regard to the incorporation of redundancies such as backup facilities, inventory and capacity flexibility. These redundancies create, at the proactive planning stage, some flexibility that can be used at the reactive control stage in the case of disruptions in supply chain structures in order to recover system performance and operational processes.

In Fig. 6, we summarize relations of redundancy, robustness, resilience and flexibility (see also Ivanov, 2018b; Ivanov & Sokolov, 2013).

There is strong and growing literature on robustness and resilience as two fundamental concepts to analyse SC performance with severe uncertainty consideration and with regards to scattered disruptive events resulting in SC structural dynamics. An SC is called *robust* if it is able to absorb disturbances and continue execution with minimal impact on performance. The performance of such an SC is insensitive to the negative impacts of disruptions (Han & Shin, 2016; Ivanov & Sokolov, 2013). Robustness is typically guaranteed by some redundancy such as structural diversification, flexible response options and system adaptation condition improvement. At the same time, we may distinguish between *being safe* and

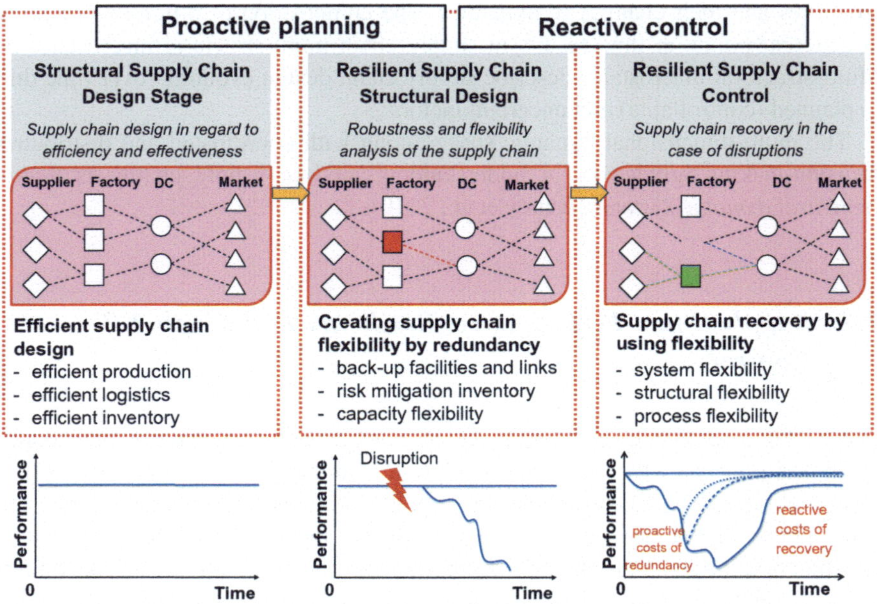

Fig. 5 Supply chain structural dynamics control (Ivanov, 2018b)

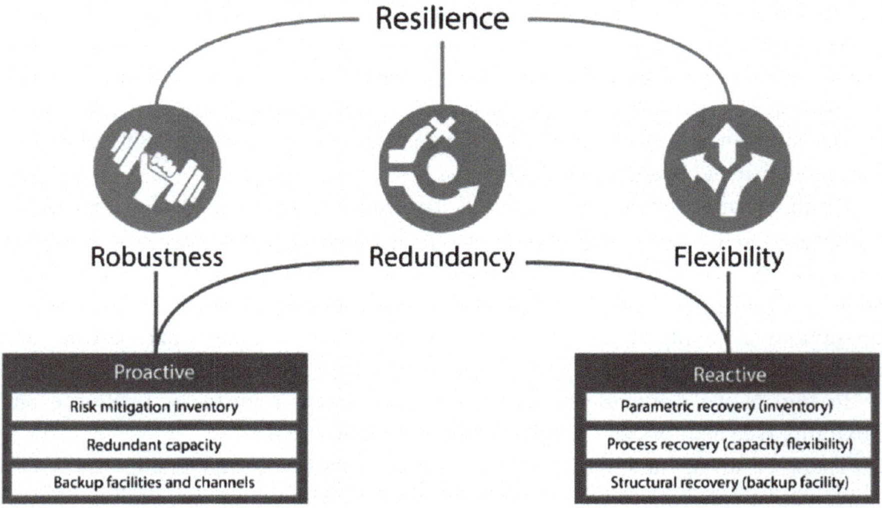

Fig. 6 Resilience control elements (Ivanov, 2018b)

performing safely. In contrast to robustness that considers proactive redundancy (e.g. buffer capacities, backup suppliers, or risk mitigation inventory) at the pre-disruption stage, *resilience* deals with the system's ability to sustain or restore its functionality and performance following a significant change in the system and

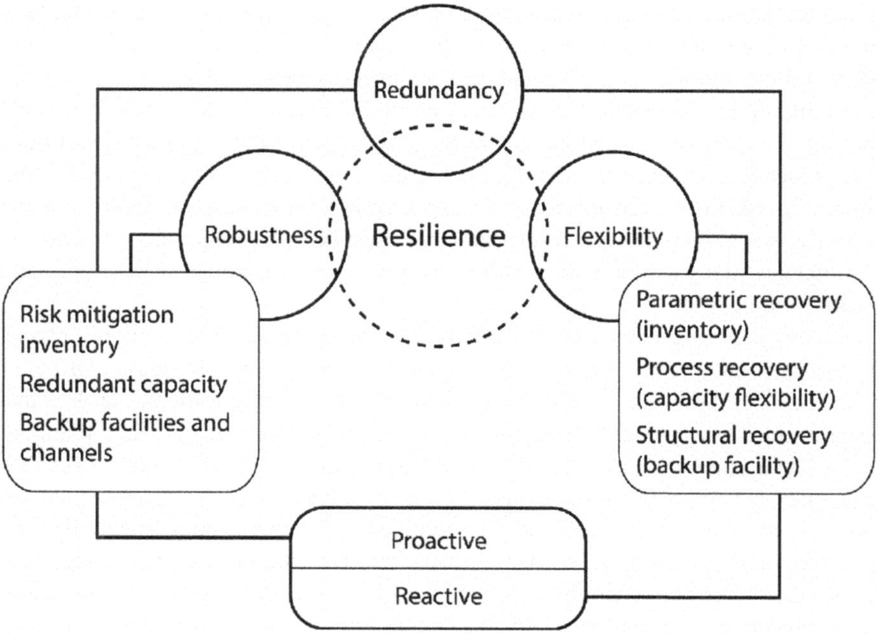

Fig. 7 Ripple effect control elements (Dolgui et al., 2018)

environment conditions (Aven, 2017). SC resilience encompasses both proactive and reactive stages. As such, an integration of pro- and reactive decisions is important for increasing SC resilience by utilizing the synergetic effects between mitigation and contingency policies (Ivanov, 2024c; Kinra et al., 2020; Li et al., 2021).

In Fig. 7, we summarize the relationships between redundancy, robustness, flexibility and resilience.

According to the ripple effect control framework (Dolgui et al., 2018) and other literature on the disruption propagation in the SC (e.g. Scheibe & Blackhurst, 2018; Wang & Zhang, 2018), disruption risks and their propagation in the SC are mainly caused by single sourcing, low risk mitigation inventory, overutilization of capacities, low-level safety technologies and missing contingency plans (Ivanov, 2020).

The width of the ripple effect and how it impacts economic performance is reliant on redundancies such as inventory or capacity buffers, also called robustness reserves, and on the speed and extent of recovery measures. As a result, it is necessary that, in the proactive mode, risk and SC resilience are assessed and incorporated at the design and planning stages. In the reactive mode, operationalization of contingency plans, such as alternative suppliers or shipping routes, must occur quickly in the control stage. This ensures quick stabilization and recovery, which is required to maintain supply continuity and prevent long-term impact. In order to assess the impact of the disruption on the SC, and both the costs and effects of material flow redirection, companies require a tool supported by collaboration and SC visibility solutions to implement these recovery policies.

Ripple effect control in the SC requires two critical capacities: resistance and recovery. For resistance, which is the SC's ability to protect against disruptions and reduce impact once the disruption occurs, some redundancy such as backup sourcing, risk mitigation inventory or capacity flexibility must be built in at the proactive stage. For recovery, this redundancy must be activated jointly with reactive contingency plans with regard to risk mitigation inventory, capacity flexibility and backup sources.

Recent literature has identified different methods to strengthen supply chains to mitigate uncertainty impacts and ensure supply chain robustness. Different robustness reserves can include material inventory, capacity buffers, etc. For this issue, valuable approaches and models for supply chain design and planning under uncertainty were elaborated. An increase in inventory, additional production capacities and alternative transportation methods or backup facilities would increase costs. At the same time, these so-called redundant elements would potentially lead to an increase in sales and service levels. The robustness elements would also reduce the risk of perturbations which may influence schedule execution. Therefore, target objectives (e.g. on-time delivery) can be better achieved. This will positively influence sales and service levels. Redundancy elements may also increase supply chain flexibility and have positive effects on both service levels and costs. The resilient state of a supply chain requires a balanced robustness and flexibility which allows for achieving maximum performance with disruption risk considerations at acceptable redundancy costs.

2 Taxonomies of the Ripple Effect

2.1 Classification of the Ripple Effect Analysis Problems

Analysis of literature allows the identification of several problem classes and datasets; it is recommended to analyse these using optimization, simulation, or hybrid simulation–optimization techniques. The literature has been analysed regarding the modelling techniques used, the problems addressed, the performance measures and the scope of the ripple effect analysis. More specifically, the following characteristics have been analysed to derive the classifications following standard problem classifications in supply chain management at design, planning and control decision-making levels (Ivanov, 2019): (i) supply chain structural and operational parameters at the supply chain design level, (ii) inventory, sourcing, shipment and production control policies at the supply chain planning level and (iii) recovery policies at the supply chain control level. The following classification has been obtained (Fig. 8).

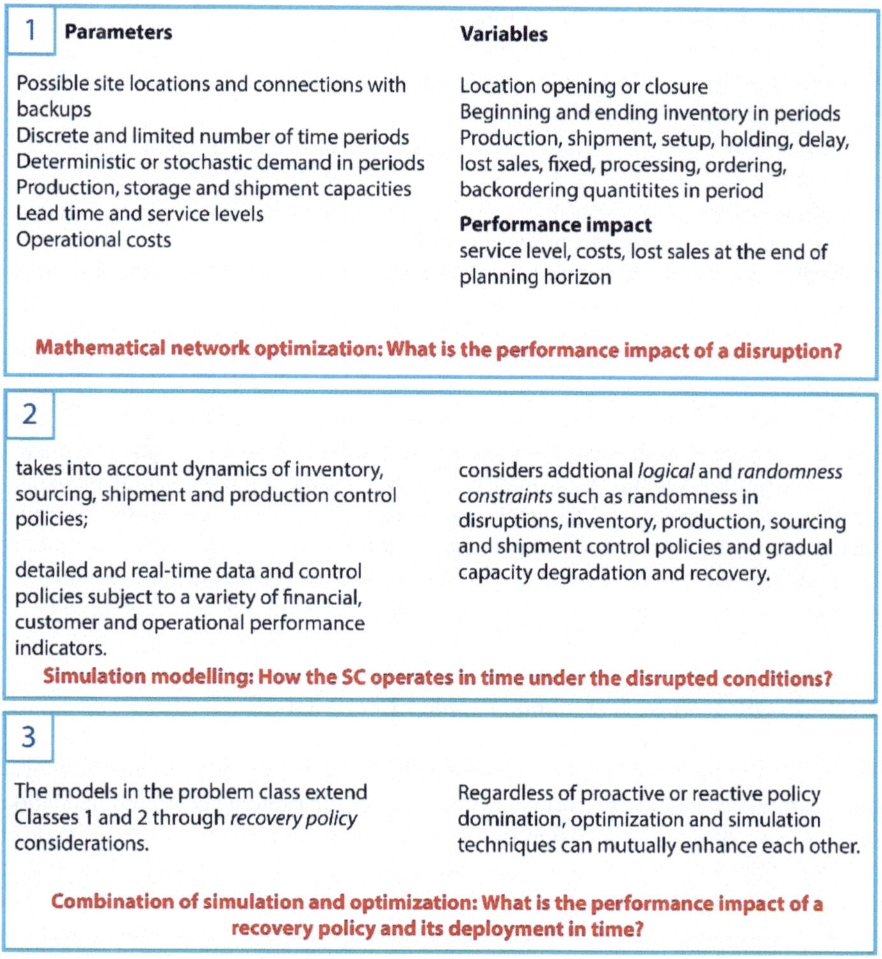

Fig. 8 Three problem classes in the ripple effect analysis

Let us consider these three classes of the ripple effect analysis in detail.

2.1.1 Problem Class 1. Static Ripple Effect Analysis

The models in the problem class allow computation of the performance impact of disruption and recommendation of a resilient supply chain design based on aggregate location and flow data subject to cost minimization or profit maximization. This problem class considers the following dataset:

Parameters:

- Possible site locations and connections (nodes and paths) with backups

- Discrete and limited number of time periods
- Deterministic or stochastic demand in periods
- Production, storage and shipment capacities in periods
- Lead time and service levels
- Operational costs
- Variables
- Location opening or closure
- Beginning and ending inventory in periods
- Production, shipment, setup, holding, delay, lost sales, fixed, processing, ordering, backordering quantities in periods

Performance impact: service level, costs, lost sales at the end of the planning horizon.

Mathematical network optimization has been typically used for this class. Those models are placed at the supply chain design level and help to analyse the impact of the disruption on the performance by deactivating some structural elements, changing some operational parameters (e.g. capacity) and observing the resulting variations in costs or sales. This analysis is helpful at the strategic decision-making level. At the same time, those models do not take into account the dynamics of inventory, sourcing, shipment and production control policies.

2.1.2 Problem Class 2. Dynamic Ripple Effect Analysis

The models in the problem class allow supply chain behaviour to be analysed over time, computation of the performance impact of the disruption and recommendation of a resilient supply chain design based on detailed and real-time data and control policies subject to a variety of financial, customer and operational performance indicators. In addition to the more detailed data from the Class 1 dataset, this problem class considers additional *logical* and *randomness constraints* such as randomness in disruptions, inventory, production, sourcing and shipment control policies, and gradual capacity degradation and recovery. For problems in this class, simulation has been dominantly applied. Since simulation studies on the ripple effect deal with time-dependent parameters, duration of recovery measures and capacity degradation and recovery, they have earned an important role in academic research. Simulation has the advantage that it can extend the handling of omplex problem settings in Class 1 with situational behaviour changes in the system over time.

2.1.3 Problem Class 3. Dynamic Ripple Effect Analysis with Recovery Considerations

The models in the problem class extend Classes 1 and 2 through recovery policy considerations. Independent of proactive or reactive policy domination, optimization and simulation techniques can mutually enhance each other. For problems in this class, a combination of network optimization and simulation (e.g. simulation runs over optimization results) can be recommended. The research considering the recovery stage is still new and requires an extension. We consider problem class 3 as an especially promising future research avenue.

2.2 *Literature Classification Taxonomy*

Ivanov and Dolgui proposed the following literature classification taxonomy (Table 3).

2.2.1 Semantic Level: Structural Properties, Complexity Role and Critical Nodes

Disruptions and the resulting ripple effect cause SC structural changes, and it is also referred to as SC structural dynamics. Structural SC properties have been recognized to have a crucial impact on the ripple effect and SC robustness and resilience. A body of literature has been established that examines the impacts of different structural variations on SC performance for various risk attitudes in a decision-maker, ranging from risk-neutral to risk-averse. This literature at the structural level targets semantic network analysis in order to identify underlying interdependencies between network graph forms and SC robustness, flexibility, adaptability and resilience.

The semantic network analysis literature pertains to the dependencies of SC robustness and resilience on the structural complexity that increases uncertainty and disruption risk propagation. The quantitative methodologies used mostly include mathematical optimization, simulation, graph theory, game theory, control theory, complexity theory, financial analysis and reliability theory. The major findings in this research stream pose the impact of different structural SC designs, e.g., in terms of the critical nodes on disruption-based SC structural and performance dynamics. The issues of segmentation, diversification, backup suppliers, facility fortification, globalization and localization are considered important managerial levers to increase SC resilience at the proactive and reactive stages. In summary, *structural variety and recoverability* can be considered a major SC resilience driver, as identified at the semantic structural analysis level.

2.2.2 Process Level

Flexibility has been mostly analysed at the process level. The literature mostly focuses on product and process flexibility to ensure SC robustness and resilience. The literature recognizes flexibility as a major driver of resilient SCs. The papers in this research stream investigate the use of flexible production and sourcing

Table 3 Literature classification scheme

Analysis levels	Proactive stage			Reactive stage		
Network structure and variety	I	Complexity	A, B, C, D, E, F	I	Complexity	A, B, C, D, E, F
		Centralization			Centralization	
	1	Diversification		7	Diversification	
		Localization			Localization	
	II	Complexity	A, B, C, D, E, F	II	Complexity	A, B, C, D, E, F
		Centralization			Centralization	
	2	Diversification		8	Diversification	
		Localization			Localization	
Process flexibility	I	Backup/dual s.	A, B, C, D, E, F	I	Backup/dual s.	A, B, C, D, E, F
		Postponement			Postponement	
	3	Product subst.		9	Product subst.	
		Coordination			Coordination	
	II	Backup/dual s.	A, B, C, D, E, F	II	Backup/dual s.	A, B, C, D, E, F
		Postponement			Postponement	
	4	Product subst.		10	Product subst.	
		Coordination			Coordination	
Parametric redundancy	I	Inventory	A, B, C, D, E, F	I	Inventory	A, B, C, D, E, F
		Capacity			Capacity	
	5	Lead time		11	Lead time	
	II	Inventory	A, B, C, D, E, F	II	Inventory	A, B, C, D, E, F
		Capacity			Capacity	
	6	Lead time		12	Lead time	

I—Supply Chain Structural Design
II—Supply Chain Process Planning and Control
1–12—Research field numbers
A–F—Methodologies:
A—Mathematical Optimization (deterministic mixed-integer, stochastic, robust, goal and fuzzy optimization)
B—Simulation (discrete-event simulation, agent-based simulation, system dynamics)
C—Game Theory (cooperative/non-cooperative, dynamic differential and symmetric/asymmetric (incomplete information) games)
D—Control Theory (optimal control, model-predictive control, feedback control)
E—Reliability Theory (probabilistic, statistical, logic and graph models)
F—Hybrid Methodology
Coding example for Ivanov et al. (2014b). Optimal distribution (re)planning in a centralized multi-stage network under conditions of ripple effect and structure dynamics. European Journal of Operational Research, 237(2), 758–770
10 – II – Ba – F:AD—this study focuses on the SC planning level and the impact of backup sourcing at the process recovery stage using a hybrid optimization-control theory methodology

processes to achieve SC robustness and resilience under disruptions. The coping strategies, the authors indicate, consider dual and multiple sourcing whereby the focus of analysis includes a tremendous variety of proactive and reactive measures such as backup supplier contracts, pricing policy adjustment, advanced, spot and contingency purchasing, risk mitigation inventory, capacity reservations, product flexibility and postponement and collaboration and visibility.

Flexibility is the central theme of the research conducted at the process level referring to the ability of production, sourcing and transportation systems in the SC to change (adapt) in dynamic environments. The methodologies used include mathematical optimization, discrete-event simulation, game theory and real options. Backup and dual sourcing, postponement, product substitution, production capacity flexibility and coordination have been identified as major elements of the contingency processes and SC resilience drivers are to be addressed at the process management level. Increasing SC resilience is considered in the flexibility framework in light of some process redundancy (e.g. a more expensive backup source) as opposed to process leanness.

2.3 Control Level

The research focus at the control level is directed at process parameters such as inventory, capacity utilization and lead time. High inventory, capacity reservations and lead-time reserves may help to increase SC resilience but might negatively affect efficiency. Parametric redundancy is a central research category at the control level. Insufficient redundancy is risky. Redundancy is costly. This trade-off presents a central issue in the research at the parametric redundancy control level. High inventory, capacity reservations and lead-time reserves may help in increasing SC resilience, but they negatively influence SC efficiency. The methodologies used in this research area include mathematical optimization, discrete-event simulation, system dynamics and control theory.

3 Future Research Perspectives on the Ripple Effect

While ripple effect and disruption risks have attracted considerable research attention, this research domain seems to be at the beginning stage of development. Some future research avenues are summarized in this section. With regard to the current research, we refer the readers to a recent state-of-the-art survey in the given domain for more detailed analysis (Dolgui et al., 2018; Fahimnia et al., 2015; Ho et al., 2015; Ivanov et al. 2017a, b; Klibi et al., 2010; Snyder et al., 2016).

3.1 Supply Chain Risk Analytics, Digitalization and Industry 4.0

Innovations in digital technologies influence the development of new paradigms, principles and models in SCs. The Internet of Things (IoT), cyber-physical systems, additive manufacturing and smart, connected products facilitate the development of Industry 4.0-driven digital SC. Such technology advances are facilitated by the advent of big data analytics and advanced tracking and tracing technologies. Accompanying such technological advances are similar advances in organizational practice and culture, shaped by socio-technical considerations of new technology use. The dynamic nature of digitalization demands research that can help analyse, understand and evaluate its drivers, facilitators and performance outcomes. Such outcomes could range from time competitiveness to risk management and resilience.

The impact of digitalization on resilient operations and the SC can be quite complex. Consider some interplays. Risk in the SC can be mitigated by the descriptive and predictive use of big data analytics to gain visibility and forecast accuracy, reduce information disruption risks and improve contingency plan activation. Reductions in supply and time risks can be achieved by using advanced trace and tracking systems leading to real-time coordinated activation of contingency policies.

SCs typically hedge against disruptions by means of risk mitigation inventory, capacity reservations and backup sources. Such protection is expensive to maintain (in anticipation) and deploy. Blockchain digitalization could help reduce risk and associated preventive costs, if a record of activities and data needed for recovery exists for synchronized contingency plans. Similarly, additive manufacturing can reduce the need for risk mitigation inventory and capacity reservations, as well as diminish the need for expensive backup contingent suppliers. The decentralized control principles in Industry 4.0 systems make it possible to diversify risks and reduce the need for structural SC redundancy with the help of manufacturing flexibility. Big data analytics and advanced trace & tracking systems in general, and Blockchain technology in particular, can help us to trace the roots of disruptions to observe disruption propagation (i.e. the ripple effect) (Dolgui et al., 2018; Ivanov, 2018a), to select short-term stabilization actions based on a clear understanding of what capacities and inventories are available (emergency planning), to develop mid-term recovery policies, and to analyse the long-term performance impact of ripples effects. Additive manufacturing has a potential to reduce disruption propagation in the SC, since the number of SC layers and the resulting complexity would be reduced. Resilience may improve, resultantly.

Initial efforts to understand the impact of digital technologies on the SC risk management are underway. However, both conceptual and granular understandings of the contribution and the interplay of different digital technologies in regard to specific SC and operations resilience and sustainability require further analysis (Ivanov, 2023, 2024b).

The impact of digitalization and Industry 4.0 on the ripple effect and disruption risk control analytics in the SC is therefore a promising research avenue. The purpose of the research in the given area is to investigate the interplay between digitalization, SC resilience and SC risks. The scope synthesizes research from two distinct areas, i.e. the impact of digitalization on logistics, and the impact of supply chain management on risk control. As such, the topics of this domain connect business, information, engineering and quantitative analysis perspectives on digitalization to control and the supply chain risks issues. Such studies would connect business, information, engineering and analytics perspectives on digitalization and SC risks in order to bring the discussion further with the help of a conceptual framework for researching the relationships between digitalization and SC disruption risks. Examples of the questions to be answered are, e.g., (1) what relations exist between big data analytics, Industry 4.0, additive manufacturing, advanced trace & tracking systems and SC disruption risks; (2) how digitalization can contribute to enhancing ripple effect control; and (3) what digital technology-based extensions can trigger the developments towards SC risk analytics.

At the proactive level, optimization and simulation models produce notable insights for managers and can be applied where the probability of disruption can be roughly estimated. On the one hand, big data analytics and advanced trace and tracking systems may help in predicting disruptions and providing more accurate data to build sophisticated disruption scenarios for resilient SC design analysis. Digital technologies open new problems for resilient SC design. For example, additive manufacturing changes SC designs whereby new resilient sourcing problems may arise. This area can further be enhanced using collaborative purchasing platforms.

At the reactive level and with regards to mitigation strategies and identifying disruption impact on finance and operational performance, digital technologies can be extensively used to obtain real-time information on the scope and scale of disruptions, their propagation in the SC and to simulate possible recovery strategies. In addition, at the reactive level, adaptation is necessary for achieving desired output performance by ensuring the possibility of changing SC plans and inventory policies. Adaptation processes in ripple effect control can be supported by feedback and adaptive control methods using decentralized agent techniques with the help of digital technologies. Visualizing these processes through virtual reality-supported simulation has not yet been done extensively to model the ripple effect in the supply chain. For this, simulation models, along with new digital technologies, can improve tools which are already used in developing SC agility and visibility in terms of disruption velocity.

3.2 *Low-Certainty-Need Supply Chains*

Uncertainty and risk predictions are commonly researched in studies of SC disruption management, mostly assuming known disruptive event or disruption scenario probability. The resulting resource allocation and costs have frequently resulted in expensive systems which help businesses cope with uncertainty. Without undermining the importance of further developing this common perspective, new approaches need to be developed that focus on the reduction of SC behaviour dependence on environmental changes.

The unpredictability of the occurrence of disruption and its magnitude suggests that designing SCs with a low need for "certainty" may be as important, if not more so, than predetermined pre-disruption strategies. While the problem of disruption impact investigation with disruption probability estimations has attracted consid-

Table 4 Research gaps at semantic, process and control levels

Analysis levels	Research gaps
Semantic level	The semantic network analysis pertains to the dependencies of SC robustness and resilience on structural network properties. Which structural SC designs, e.g., in terms of the critical nodes, can help to increase SC robustness and reduce the need for disruption-driven process changes? How can segmentation, diversification, backup suppliers, facility fortification, globalization and localization be applied to increase SC resilience whilst remaining lean and efficient? Which SC design patterns can provide quicker and more efficient recoverability?
Process level	Backup and dual sourcing, postponement, product substitution, production capacity flexibility and coordination are major elements of contingency processes and drivers of SC resilience. How can process redundancy be allocated to increase SC robustness and reduce the need for disruption-driven process changes? How can process redundancy (e.g. a backup source) be applied whilst remaining lean and efficient? Which reactive process flexibility policies can help in efficient SC recovery?
Control level	High inventory, capacity reservations and lead-time reserves may help in increasing SC resilience, but they negatively influence SC efficiency. How can parametric redundancy be applied to increase SC robustness and resilience whilst remaining lean and efficient? Which reactive control policies can help in efficient SC recovery?

erable research attention, some fundamental issues in this research stream need to be pointed out, such as fair probability estimation of rare events, consideration of only “known” events and the exclusion of “unknown” events, and the consideration of mainly the direct effects of disruptions in model outputs rather than disruption propagation chains and the resulting indirect effects (Ivanov, 2019; Pavlov et al., 2019; He et al., 2018; Mizgier, 2017). Such new perspectives in SC disruption management can be placed under the umbrella of low-certainty-need (LCN) SCs. The ultimate objective of the LCN SCs is to develop the ability to operate according to planned performance regardless of environmental changes.

In the given research domain, the task is first to identify the characteristics of the LCN framework and its management. For example, structural variety, process flexibility and parametrical redundancy are identified as key LCN SC characteristics that ensure disruption resistance as well as recovery resource allocation and that allow for SC operation in a broad range of environmental states. Two efficiency capabilities of the LCN SC, i.e. low need for uncertainty consideration in planning decisions and low need for recovery coordination efforts, need to be investigated. The LCN SC does not necessarily imply higher costs but rather seeks an efficient combination of lean and resilient elements. The results of this research would allow the identification of an LCN SC framework as well as missing themes and new research questions which contribute to a better understanding of SC disruption risk management and control.

Table 4 summarizes the research gaps identified at semantic, process and control levels with regard to the LCN framework.

As shown in Table 4, a number of research gaps can be identified that motivate the development of the LCN SC framework. First, structural SC design patterns need to be identified that allow for both efficient robustness and recoverability. Second, process flexibility policies need to analysed which enable the reduction of disruption-driven process changes and efficient SC recovery. Finally, at the control level, the efficient usage of parametric redundancy and the development of reactive control policies are also research gaps that drive the pursuit of establishing the LCN SC framework.

3.3 *Proactive Planning, Network Redundancy Optimization and Situational Recovery Control*

The research in ripple effect control needs to be united by three basic principles of system-cybernetic research. The *first* principle is the integrated modelling of resilient network structures. New principles and methods of SC structural dynamics control will be developed using a variety of methodologies for multi-criteria network synthesis and analysis. A particular focus will be directed towards the deployment of post-disruption management and understanding which factors fit the particular dynamics the SC structures confront. The *second* principle is proactive planning and network redundancy optimization. The given paradigm combines both SC robustness (i.e. the ability to absorb disturbances and continue execution with minimal impact on performance), monitoring (i.e. real-time disruption identification and data-driven replanning preparation) and resilience (i.e. the ability to sustain and restore SC functionality using recovery and adaptation policies).

The *third* principle is situational proactive control. A disruptive event, planning of the recovery control policy and implementation of this policy are distributed in time and subject to SC structural and parametrical dynamics. In other words, the environment, SC structures and operational parameters may change during the period between the planning of the recovery control policy and its implementation. As such, situational proactive control with a combined usage of simulation–optimization and analytics are needed to improve the transition processes from a disrupted to a restored SC state. This allows reducing investments in robustness and increasing resilience by obviating the transition process control problems. A combination of these three principles builds a framework of future decision-support systems for SC disruption risk management which utilizes two major ideas, i.e. (i) low-certainty-need SC designs and network redundancy optimization with an optimal combination of robustness and adaptation elements to ensure both efficient and resilient SCs and (ii) integrated SC ripple effect modelling with simulation, optimization and analytics components to support situational forecasting, predictive simulation, prescriptive optimization and adaptive learning (Ivanov, 2022; Ivanov & Dolgui, 2020).

3.4 Empirical Research and Simulation

Even if simulation and optimization studies provide valuable insights on preventing and mitigating the ripple effect in the SC, there is a lack of practical validation. Only a few studies incorporated real company data. At the same time, empirical research in SC management has also developed a variety of valuable approaches and methods to tackle the ripple effect. In this setting, combined empirical simulation, studies are encouraged. An example of an area for such integrated research is coordinated contingency plans. In addition, the identification of information patterns needed to make decisions on ripple effect identification and recovery policies would be in the scope of this research.

3.5 Complexity Theory, Dynamics, Performance Analysis and Control

For dealing with the ripple effect in the SC, complexity management and system modelling might provide a theoretical basis. Based on Ashby's principle of requisite variety, the problem of a system under control and uncertainty implies an area under control and an area under uncertainty, according to the perspective of complexity management. The system control can be adapted by widening one area and narrowing the other (Ivanov, 2010; Ivanov & Sokolov, 2010). Therefore, the connection between the system and the environmental spaces are categorized according to amplification of *control variety* or attenuation of *environmental variety*. A balance of control and disruption impact and maintenance of planned execution processes and a cost-efficient, fast recovery post-disruption can be achieved by amplifying the variety of the control area and reducing the area of uncertainty.

Further research can be initiated in this area as it regards structural network properties and the identification of structural patterns in SC design which cause a greater or lesser ripple effect. In addition, the ripple effect and the impact of recovery and proactive strategies within one feedback framework including planning and adaptation control loops are revealed by applying methods of dynamic control theory.

Analysis of short-term and long-term impacts of the ripple effect on the SC and the creation of a performance measurement system is a promising research avenue. Even though some key performance indicators have been presented in literature episodically, there is a lack of systematic performance management techniques for the ripple effect in the SC.

3.6 Disruptions and Perishable Products

Generally, inventory constitutes an SC resilience drive in literature. In the case of SCs for perishable products, there are limits to inventory holding durations because of the short storage and expiration periods. The resilience of these kinds of SCs may be affected by the risk of goods write-off and customer segmentation by requirements for freshness. Safety stock reductions and an increase in transport frequency result from the constraints inherent in product perishability. However, when disruption risks are considered, this may lead to an increase in safety stock. The bounded capacity of suppliers should also be analysed. Since customer demand tends to be vulnerable, there might be different requirements for the freshness of products and penalties when the product is unavailable or freshness is decreased. Further, in perishable product SCs, issues of batching usually carry more weight.

3.7 Competition and Behavioural Aspects

Since severe disruptions may influence competition in the markets, a research agenda on the ripple effect needs to include this factor. In addition, managerial decisions are of a behavioural nature and subject to individual risk perceptions. Agent-based modelling can be applied to a broader scope of these problems. These principles may include collaboration (trust and information sharing) and an SC risk management culture (e.g. leadership and risk-averse behaviour). In this setting, agent-based modelling would be a suitable method for enhancing the existing simulation impact on SC ripple effect research in regard to non-engineering SC resilience principles.

3.8 Ripple Effect Visualization

Visualizing the ripple effect is an obvious next step for simulation features. Yet, it has not been very frequently used for modelling the ripple effect in the SC. Given this, simulation models would enhance the existing tools in SC agility and visibility concerning disruption velocity.

3.9 Closed-Loop SCs, Sustainability and Humanitarian Logistics

Resilience has a number of intersections with SC sustainability. Since SCs have become more and more global, these network structures build the backbone of the modern economy and directly influence such sustainability issues as employment rates, natural resource consumption, etc. SC sustainability issues include an assessment of SC design resilience and efficient SC structure reconfiguration in the case of disruptions from the perspectives of environmental, political and society impacts.

Disruptions in the reverse part of closed-loop SCs, as well as disruption-based reverse logistics flows have rarely been analysed. Approaches for analysing the disruptions in the reverse part of the closed-loop SCs (e.g. temporary unavailability of a warehouse for collecting the used batteries for electric cars) in regard to (i) their impact on overall SC performance as well as to (ii) proactive and reactive policies with consideration of inventory control policies and sustainable manufacturing concepts are yet to be developed. In addition, disruptions in a region frequently result in both humanitarian catastrophes and industrial disruptions at the same time. In this setting, limited resources need to be fairly allocated to both human life rescue and the stabilization of everyday life and recovery of the industrial sector.

3.10 Human Aspects

Finally, yet importantly—the area of human factors needs to be developed in the future. Our perception (partially derived from the experiments and literature) is that in a short-term perspective, SC adaptability to the disrupted mode is low and recovery actions are at the beginning of their implementation, which causes high coordination efforts. This means a very stressful time for SC recovery teams. It follows that the better the preparation, the less stressful and the more efficient the recovery work will be.

4 Data-Driven Ripple Effect Control: Towards Supply Chain Risk Analytics

Analysis of current and future research trends in ripple effect allows formulating two important insights which lead the discussion further towards SC risk analytics as shown in Sects. 4.1 and 4.2.

4.1 From Competition Between the Supply Chains to Competition Between the Information Services and Analytics Algorithms

The company is as good as the SC behind it. Today and looking at the near future, the SC will be as good as the digital technology behind it. Consider two examples to support this proposition. The first is the logistics service provider UPS. UPS and SAP developed a joint technology which allows UPS to manufacture items using 3D printing directly at the distribution centres (UPS, 2018). The second example is Blockchain technology. Contracts in SCs often involve multi-party agreements with regulatory and logistic constraints. Further complexities may arise from operations in different jurisdictions, as well as dynamic features embedded in the contracts. The flow of information in an SC plays a critical role in the efficiency of the operations. Regulatory processes (e.g. customs) can be expedited by improving confidence in documentation. This, in turn, will reduce waste, risk and insurance premiums. IBM and Maersk are collaborating to create trust and transparency in global SCs (IBM, 2018). They are developing a distributed contract collaboration platform using Blockchain technology. Maersk estimates that shipping a single container of flower from Kenya to Rotterdam requires nearly 200 communications. In their approach, each distinct entity involved in the transaction is allowed to access this system. Shipping from the port of Mombasa requires signatures from three different agencies and six documents: the smart contract will automatically generate after the system receives the signatures. Simultaneously, when the documents about inspection, sealing of refrigerator, pickup by the trucker and approval from customs communicated to the port of Mombasa are uploaded, all the participants can see the data in the meantime, allowing the related entity to prepare for the container.

These and further recent examples of digital technology applications to SCs allow for the new proposition that the competition is not between SCs but rather between SC *services* and the analytics algorithms behind the SCs based on the cyber-physical system approach.

According to Zhuge (2011), the evolution from cyberspace and systems to the cyber-physical-social space and systems can be described by three extensions. It distinguishes two types of cyberspaces: the first one allows users to read information in cyberspace like the web, and the other one allows users to read and write information in cyberspace. Both rely on humans to add information to cyberspace in order to share it with others.

The first extension to this basic concept depicts the extension of cyberspace to the physical space through various sensors. Any significant information in physical space can be automatically sensed, stored and transmitted through cyberspace. Internet of Things can be considered as a kind of cyber-physical space.

The second extension is that user behaviours can be sensed and feedback to cyberspace for analysing the patterns of behaviours, and humans can remotely control the actuators to behave in physical space through cyberspace. This enables cyberspace to adapt its services according to the feedback since behaviour change may indicate some psychological change.

In the third extension, i.e. the cyber-physical system, not only the individual's behaviours, but also social interactions can be feedback into cyberspace for further processing. Users are considered with their social characteristics and relations rather than as isolated individuals. Sensors are limited in their ability to collect all information in the physical space, so users still need to directly collect significant information in the physical space and then put it into cyberspace after analysis (including experiments). Users can also manipulate physical objects in the physical space, which can also be feedback into cyberspace to reflect the real-time situation]. Users' status, interests and knowledge evolve with social interaction and operations in cyberspace.

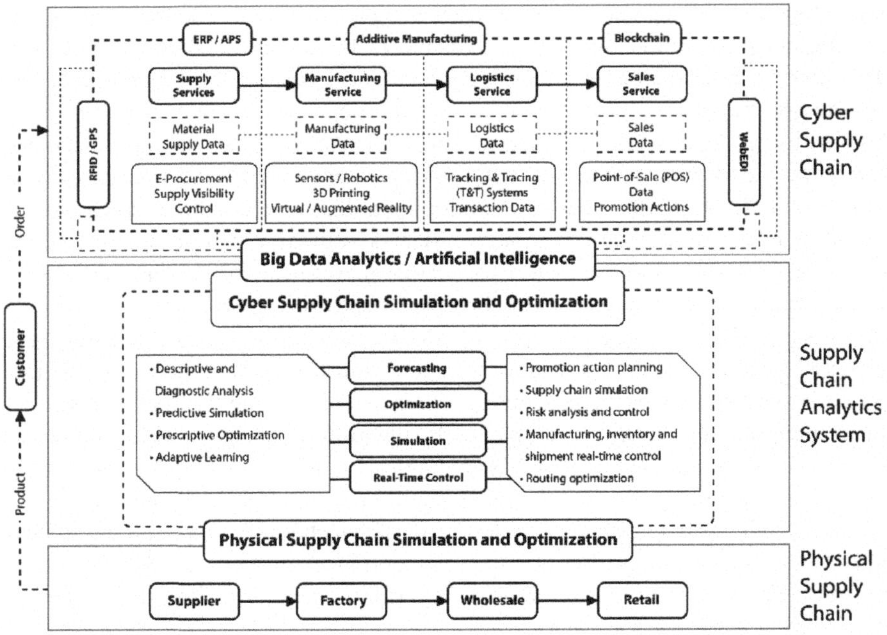

Fig. 9 Service and material flow coordination in the cyber-physical supply chain

The afore-mentioned analysis can be presented as a digital cyber-physical SC framework (Fig. 9).

According to Waller and Fawcett (2013) and KPMG (2017), the application areas of SC analytics can be classified into four areas, i.e.

- Descriptive and diagnostic analysis
- Predictive simulation and prescriptive optimization
- Real-time control
- Adaptive learning

Examples of SC and operations analytics applications include logistics and SC control with real-time data, inventory control and management using sensing data, dynamic resource allocation in Industry 4.0 customized assembly systems, improving forecasting models using Big Data, machine learning techniques for process control, SC visibility and risk control, optimizing systems based on predictive information (e.g. predictive maintenance), combining optimization and machine learning algorithms and simulation-based modelling and optimization for stochastic systems.

Success in SC competition will become more and more dependent on analytics algorithms in combination with optimization and simulation modelling. Initially intended for process automation, business analytics techniques now disrupt markets and business models and have a significant impact on SCM development. As such, new disruptive SC business models will arise where SCs will be understood not as rigid physical systems with a fixed and static allocation of some processes to some firms. Instead, different physical firms will offer services of supply, manufacturing, logistics and sales which will result in a dynamic allocation of processes and dynamic SC structures. Recent literature documented the possibility of modelling such integrated service-material flow SCs (Ivanov et al., 2014; Yang et al., 2017).

In new disruptive SC business models SCs will no longer be understood as a rigid physical system with a fixed and static allocation of some processes to some firms. Instead, different physical firms will offer services of supply, manufacturing, logistics and sales which will result in dynamic allocation of processes and dynamic SC structures. Indeed, this idea is not really new. We can recall the concept of virtual enterprises that was developed about 15–20 years ago (Camarinha-Matos & Macedo, 2010).

The SCs in virtual enterprises were expected to be formed dynamically thru the so-called competence cell or agent networking (Ivanov et al., 2004; Ivanov & Sokolov, 2012a, b; Teich, 2003; Teich & Ivanov, 2012). In essence, the suppliers were integrated into a tool that contained their technological processes and the related operational parameters (e.g. costs and lead time). A customer was able to place an order specification and an automatic algorithm was able to find the suppliers needed to be networked to fulfil this customer order. So while the individual contributors (e.g. robots, sensors, RFID—radio-frequency identification, agents, modular factories, etc.) are not really new, they are becoming more practical and companies are more receptive to using them to stay competitive.

4.2 *Risk Analytics in the Digital SC: Data-Driven Decision Analysis, Modelling, Control and Learning Systems*

With the help of optimization and simulation approaches, current research generates new knowledge about the influence of disruption propagation on SC output performance considering disruption location, duration and propagation and recovery policies. New digital technologies create new challenges for the application of quantitative analysis techniques to SC ripple effect analysis and open new ways and problem statements for these applications.

In the past decades, simulation and optimization have played significant roles in solving complex problems. Successful examples include production planning and scheduling, SC design and routing optimization, to name a few. However, many problems remain challenging because of their complexity and large scale, and/or uncertainty and stochastic nature. In addition, the major application of optimization and simulation methods in the last decades was seen in decision support, meaning that decision-makers were to manually provide the model input and interpret the model output. On the other hand, the rapid rise of business analytics provides exciting opportunities for Operations Research and the reexamination of these hard optimization problems, as well as newly emerging problems.

The modelling stage is devoted to predictive simulation and prescriptive optimization. Disruption scenario simulation, SC design optimization, and recovery optimization are major decisions that need to be supported at this level. A structural dynamics control approach in combination with mathematical optimization can be used. Real-time control area contains supply flow real-time control, disruption identification and real-time performance and recovery control. Feedback control can be applied in this domain with modifications.

It is commonly known that feedback control in socio-organizational differs from technical systems where feedback can be implemented almost immediately. In socio-organizational systems, the feedback information first needs to be evaluated by managers and the adjustment decisions need to be coordinated among different departments in the firms or even cross-organizational. As such, the differences in the system states can be observed between the system state at the moment of starting to prepare the adjustment decisions on the basis of the feedback information and the system state at the moment of decision implementation. In other words, delayed feedback occurs due to system inertia. The correction (adaptation) decisions need to be implemented at the object or system which is different from the object or system that has been considered for the reconfiguration decision planning.

Finally, the learning stage is comprised of risk mitigation learning, disruption recovery learning and disruption pattern recognition. A combination of control algorithms and artificial intelligence can provide a number of new insights in the given area.

Consider some practical examples. Sourcing, manufacturing, logistics and sales data are distributed among very different systems, such as ERP, RFID, sensors and Blockchain. Big data analytics integrates this data into information used by AI algorithms in the cyber SC and managers in the physical SC. As such, a new generation of simulation and optimization models is arising. The pervasive adoption of analytics and its integration with Operations Research shows that simulation and optimization are key, not only in the modelling of physical SC systems but also in the modelling of cyber SC systems and learning from them.

PwC is working with a large car company looking to introduce autonomous vehicles to the public (Wilkinson, 2018). Part of this work employs deep reinforcement learning to develop rules. Together with simulation, deep reinforcement learning is used to determine "optimal" decision rules that allow the vehicles to maximize efficiency while also satisfying customer trip demand. The software environment for the project uses the extensible and practical environment of AnyLogic multimethod simulation software to lever the capabilities of DL4 J for the deep learning environment. Autonomous cars are becoming more common, and the features are already present in many consumer cars. These examples show that artificial intelligence becomes more pervasive in the real world with every project, and necessarily it must be part of the simulation.

With regards to SC risks, Resilience360 at DHL allows comprehensive disruption risk management by mapping end-to-end SC, building risk profiles and identifying critical hotspots in order to initiate mitigation activities and alert in near-real-time mode on incidents that could disrupt the SC (DHL, 2018). RiskMethods

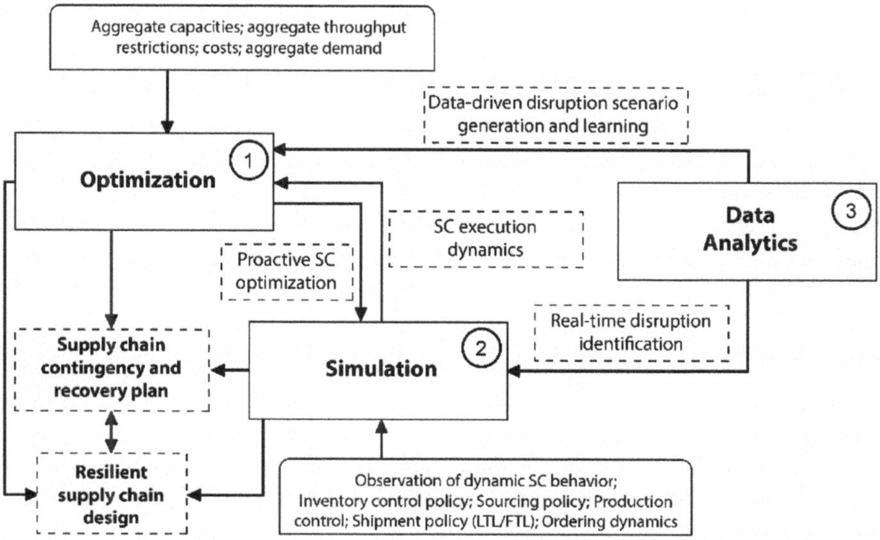

Fig. 10 Concept of a decision-support system for supply chain risk analytics

GmbH developed software for proactive SC risk management that contains modules "Risk radar," "Impact analyzer" and "Action planner" for risk monitoring, impact assessment and planning of the mitigation actions (RM, 2018).

As such, a new generation of simulation and optimization models can be observed that extends the decision-support systems (DSS) towards decision analysis, modelling, control and learning systems (DAMCLS). A DAMCLS example for ripple effect control in the SC that combines simulation, optimization and data analytics is shown in Fig. 10.

The DAMCLS system for SC risk analytics aims at proactive, resilient SC design in anticipation of disruptions and structural–parametrical adaptation in the case of disruptions. The decision-support system is based on a concept that combines simulation, optimization and data analytics. The simulation–optimization part of the system is intended to provide proactive, resilient SC optimization and simulation of SC dynamic behaviour in the event of possible disruptions or disruption scenarios. In addition, this supports reactive, predictive simulation of disruption impacts on SC performance and recovery policies which are subsequently optimized in a prescriptive manner using an analytical model. The data analytics part of the system is applied to disruption identification in real time using process feedback data, e.g., from sensors and RFID. In addition, this aims at automated data input of disruption data into the reactive simulation model for recovery policy simulation and optimization. Finally, data analytics is used as a data-driven learning system at the proactive stage, helping to generate adequate disruption scenarios for resilient SC design and planning.

Future trends in the research on the supply chain ripple effect can be seen in the following areas:

- Digital twins, Metaverse and supply chain visibility (Ivanov, 2023)
- Supply chain viability and Industry 5.0 (Ivanov, 2022; Ivanov & Dolgui, 2020)
- Intertwined supply networks and cross-industry ripple effect (Dolgui et al., 2024)
- Financial ripple effect (Proselkov et al., 2024)
- Ripple effect in the presence of hyper-disruptions like COVID-19 pandemic and global trade system disruptions (Brusset et al., 2023; Ivanov, 2020)
- Ripple effects in closed loop supply chains and circular economy (Park et al., 2022)

The simultaneous existence of different disruptive stressors (e.g., natural catastrophes, epidemics/pandemics, global geopolitical vulnerabilities, critical material shortages, climate change, and financial crises) along with the rapid development of digital technologies, industrial Metaverse and Industry 5.0 raises a number of novel research questions where novel and substantial contributions can be made for further development of the ripple effect theory.

References

Akkermans, H., & van Wassenhove, L. N. (2018). Supply chain tsunamis: Research on low probability high impact disruptions. *Journal of Supply Chain Management, 54*(1), 64–76.

Aven, T. (2017). How some types of risk assessments can support resilience analysis and management. *Reliability Engineering and System Safety, 167*, 536–543.

Brusset, X., Ivanov, D., Jebali, A., La Torre, D., & Repetto, M. (2023). A dynamic approach to supply chain reconfiguration and ripple effect analysis in an epidemic. *International Journal of Production Economics, 263*, 108935.

Camarinha-Matos, L. M., & Macedo, P. (2010). A conceptual model of value systems in collaborative networks. *Journal of Intelligent Manufacturing, 21*(3), 287–299.

Chopra, S., & Sodhi, M. S. (2004). Managing risk to avoid supply-chain breakdown. *MIT Sloan Management Review, 46*, 52–61.

Chopra, S., & Sodhi, M. S. (2014). Reducing the risk of supply chain disruptions. *MIT Sloan Management Review, 55*(3), 73–80.

Deng, X., Yang, X., Zhang, Y., Li, Y., & Lu, Z. (2019). Risk propagation mechanisms and risk management strategies for a sustainable perishable products supply chain. *Computers and Industrial Engineering, 135*, 1175–1187.

DHL. (2018). Retrieved February 4, 2018 from https://resilience360.com/

Dolgui, A., & Ivanov, D. (2021). Ripple effect and supply chain disruption management: New trends and research directions. *International Journal of Production Research, 59*(1), 102–109.

Dolgui, A., Ivanov, D., & Sokolov, B. (2018). Ripple effect in the supply chain: An analysis and recent literature. *International Journal of Production Research, 56*(1-2), 414–430.

Dolgui, A., Ivanov, D., & Rozhkov, M. (2020). Does the ripple effect influence the bullwhip effect? An integrated analysis of structural and operational dynamics in the supply chain. *International Journal of Production Research, 58*(5), 1285–1301.

Dolgui, A., Gusikhin, O., Ivanov, D., Li, X., & Stecke, K. (2024). A network-of-networks adaptation for cross-industry manufacturing repurposing. *IISE Transactions, 56*(6), 666–682.

Fahimnia, B., Tang, C. S., Davarzani, H., & Sarkis, J. (2015). Quantitative models for managing supply chain risks: A review. *European Journal of Operational Research, 247*(1), 1–15.
Ghadge, A., Er, M., Ivanov, D., & Chaudhuri, A. (2022). Visualisation of ripple effect in supply chains under long-term, simultaneous disruptions: A system dynamics approach. *Interna-tional Journal of Production Research, 60*(20), 6173–6186.
Goldbeck, N., Angeloudis, P., & Ochieng, W. (2020). Optimal supply chain resilience with consideration of failure propagation and repair logistics. *Transportation Research: Part E, 133*, 101830.
Gurnani, H., Mehrotra, A., & Ray, S. (2012). *Supply chain disruptions: Theory and practice of managing risk*. Springer.
Han, J., & Shin, K. S. (2016). Evaluation mechanism for structural robustness of supply chain considering disruption propagation. *International Journal of Production Research, 54*(1), 135–151.
Handfield, R. B., & McCormack, K. (2008). *Supply chain risk management: Minimizing disruptions in global sourcing*. Auerbach Publications.
He, J., Alavifard, F., Ivanov, D., & Jahani, H. (2018). A real-option approach to mitigate disruption risk in the supply chain. *Omega*. https://doi.org/10.1016/j.omega.2018.08.008
Heckmann, I. (2016). *Towards supply chain risk analytics*. Springer-Gabler.
Hendricks, K. B., & Singhal, V. R. (2005). Association between supply chain glitches and operating performance. *Management Science, 51*(5), 695–711.
Ho, W., Zheng, T., Yildiz, H., & Talluri, S. (2015). Supply chain risk management: A literature review. *International Journal of Production Research, 53*(16), 5031–5069.
Hosseini, S., & Ivanov, D. (2022). A new resilience measure for supply networks with the ripple effect considerations: A Bayesian network approach. *Annals of Operations Research, 319*, 581–607.
Hosseini, S., Ivanov, D., & Dolgui, A. (2019). Review of quantitative methods for supply chain resilience analysis. *Transportation Research: Part E, 125*, 285–307.
Hosseini, S., Ivanov, D., & Dolgui, A. (2020). Ripple effect modeling of supplier disruption: Inte-grated Markov chain and dynamic Bayesian network approach. *International Journal of Production Research, 58*(11), 3284–3303.
IBM. (2018). Retrieved February 11, 2018 from https://www.youtube.com/watch?v=tdhpYQCWnCw&t=52s
Ivanov, D. (2010). A framework for aligning (re)planning decisions on supply chains strategy, design, tactics, and operations. *International Journal of Production Research, 48*(13), 3999–4017.
Ivanov, D. (2017). Simulation-based ripple effect modelling in the supply chain. *International Journal of Production Research, 55*(7), 2083–2101.
Ivanov, D. (2018a). Revealing interfaces of supply chain resilience and sustainability: A simulation study. *International Journal of Production Research, 56*(10), 3507–3523.
Ivanov, D. (2018b). *Structural dynamics and resilience in supply chain risk management*. Springer.
Ivanov, D. (2019). Disruption tails and revival policies: A simulation analysis of supply chain design and production-ordering systems in the recovery and post-disruption periods. *Computers and Industrial Engineering, 127*, 558–570.
Ivanov, D. (2020). Predicting the impacts of epidemic outbreaks on global supply chains: A simulation-based analysis on the coronavirus outbreak (COVID-19/SARS-CoV-2) case. *Transportation Research Part E: Logistics and Transportation Review, 136*, 101922.
Ivanov, D. (2022). Viable supply chain model: Integrating agility, resilience and sustainability perspectives—Lessons from and thinking beyond the COVID-19 pandemic. *Annals of Operations Research, 319*, 1411–1431.
Ivanov, D. (2023). Intelligent digital twin (iDT) for supply chain stress-testing, resilience, and viability. *International Journal of Production Economics, 263*, 108938.
Ivanov, D. (2024a). Transformation of supply chain resilience research through the COVID-19 pandemic. *International Journal of Production Research, 62*(23), 8217–8238.

Ivanov, D. (2024b). Supply chain resilience: Conceptual and formal models drawing from immune system analogy. *Omega, 127*, 103081.

Ivanov, D. (2024c). Two views of supply chain resilience. *International Journal of Production Research, 62*(11), 4031–4045.

Ivanov, D. (2024d). Cash flow dynamics in the supply chain during and after disruptions. *Transportation Research: Part E, 185*, 103526.

Ivanov, D., & Dolgui, A. (2020). Viability of intertwined supply networks: Extending the supply chain resilience angles towards survivability. A position paper motivated by COVID-19 outbreak. *International Journal of Production Research, 58*(10), 2904–2915.

Ivanov, D., & Dolgui, A. (2021). OR-methods for coping with the ripple effect in supply chains during COVID-19 pandemic: Managerial insights and research implications. *International Journal of Production Economics, 232*, 107921.

Ivanov, D., & Sokolov, B. (2010). *Adaptive supply chain management*. Springer.

Ivanov, D., & Sokolov, B. (2012a). The inter-disciplinary modelling of supply chains in the context of collaborative multi-structural cyber-physical networks. *Journal of Manufacturing Technology Management, 23*(8), 976–997.

Ivanov, D., & Sokolov, B. (2012b). Structure dynamics control-based service scheduling in collaborative cyber-physical supply networks. In L. Camarinha-Matos, L. Xu, & H. Afsarmanesh (Eds.), *Proceedings of the IFIP conference on virtual enterprises PRO-VE 2012 IFIP AICT* (Vol. 380, pp. 280–288). Springer.

Ivanov, D., & Sokolov, B. (2013). Control and system-theoretic identification of the supply chain dynamics domain for planning, analysis, and adaptation of performance under uncertainty. *European Journal of Operational Research, 224*(2), 313–323.

Ivanov, D., Arkhipov, A., & Sokolov, B. (2004). Intelligent supply chain planning in virtual enterprises. In L. Camarihna-Matos (Ed.), *Virtual enterprises and collaborative networks, proceedings of the IFIP conference on virtual enterprises PRO-VE 2004* (pp. 215–223). Kluwer Academic Publishers.

Ivanov, D., Sokolov, B., & Kaeschel, J. (2010). A multi-structural framework for adaptive supply chain planning and operations with structure dynamics considerations. *European Journal of Operational Research, 200*, 409–420.

Ivanov, D., Sokolov, B., & Dilou Raguinia, E. A. (2014). Integrated dynamic scheduling of material flows and distributed information services in collaborative cyber-physical supply networks. *International Journal of Systems Science: Operations & Logistics, 1*(1), 18–26.

Ivanov, D., Sokolov, B., & Pavlov, A. (2014a). Optimal distribution (re)planning in a centralized multi-stage network under conditions of ripple effect and structure dynamics. *European Journal of Operational Research, 237*(2), 758–770.

Ivanov, D., Sokolov, B., & Dolgui, A. (2014b). The ripple effect in supply chains: Trade-off 'efficiency-flexibility-resilience' in disruption management. *International Journal of Produc-tion Research, 52*(7), 2154–2172.

Ivanov, D., Hartl, R., Dolgui, A., Pavlov, A., & Sokolov, B. (2015). Integration of aggregate distribution and dynamic transportation planning in a supply chain with capacity disruption and the ripple effect consideration. *International Journal of Production Research, 53*(23), 6963–6979.

Ivanov, D., Sokolov, B., Pavlov, A., Dolgui, A., & Pavlov, D. (2016). Disruption-driven supply chain (re)-planning and performance impact assessment with consideration of pro-active and recovery policies. *Transportation Research Part E, 90*, 7–24.

Ivanov, D., Dolgui, A., Sokolov, B., & Ivanova, M. (2017a). Literature review on disruption recovery in the supply chain. *International Journal of Production Research, 55*(20), 6158–6174.

Ivanov, D., Pavlov, A., Pavlov, D., & Sokolov, B. (2017b). Minimization of disruption-related return flows in the supply chain. *International Journal of Production Economics, 183*, 503–513.

Ivanov, D., Dolgui, A., Ivanova, M., & Sokolov, B. (2018). Simulation vs. optimization approaches to ripple effect modelling in the supply chain. In M. Freitag, H. Kotzab, & J. Pannek (Eds.),

Dynamics in logistics. LDIC 2018, Bremen 20–22, 2018. Lecture notes in logistics (pp. 34–39). Springer.

Ivanov, D., Dolgui, A., & Sokolov, B. (2019a). The impact of digital technology and industry 4.0 on the ripple effect and supply chain risk analytics. *International Journal of Production Research, 57*(3), 829–846.

Ivanov, D., Tsipoulanidis, A., & Schönberger, J. (2019b). *Global supply chain and operations management* (2nd ed.). Springer.

Khojasteh, Y. (Ed.). (2017). *Supply chain risk management*. Springer Singapore.

Kinra, A., Ivanov, D., Das, A., & Dolgui, A. (2020). Ripple effect quantification by supply risk exposure assessment. *International Journal of Production Research, 58*(19), 5559–5578.

Klibi, W., Martel, A., & Guitouni, A. (2010). The design of robust value-creating supply chain networks: A critical review. *European Journal of Operational Research, 203*(2), 283–293.

Kouvelis, P., Dong, L., Boyabatli, O., & Li, R. (2012). *Handbook of integrated risk management in global supply chains*. Wiley.

KPMG. (2017). *Supply chain big data series*. Retrieved February 17, 2018 from https://assets.kpmg.com/content/dam/kpmg/au/pdf/2017/big-data-shaping-supply-chains-of-tomorrow.pdf

Levner, E., & Ptuskin, A. (2018). Entropy-based model for the ripple effect: Managing environmental risks in supply chains. *International Journal of Production Research, 56*(7), 2539–2551.

Li, Y., Zobel, C. W., Seref, O., & Chatfield, D. C. (2020). Network characteristics and supply chain resilience under conditions of risk propagation. *International Journal of Production Economics, 223*, 107529.

Li, Y., Chen, K., Collignon, S., & Ivanov, D. (2021). Ripple effect in the supply chain network: Forward and backward disruption propagation, network health and firm Vulnera-bility. *European Journal of Operational Research, 291*(3), 1117–1131.

Liberatore, F., Scaparra, M. P., & Daskin, M. S. (2012). Hedging against disruptions with ripple effects in location analysis. *Omega, 40*, 21–30.

Llaguno, A., Mula, J., & Campuzano-Bolarin, F. (2022). State of the art, conceptual framework and simulation analysis of the ripple effect on supply chains. *International Journal of Pro-duction Research, 60*(6), 2044–2066.

Mizgier, K. J. (2017). Global sensitivity analysis and aggregation of risk in multi-product supply chain networks. *International Journal of Production Research, 55*(1), 130–144.

Park, Y. W., Blackhurst, J., Paul, C., & Scheibe, K. P. (2022). An analysis of the ripple effect for disruptions occurring in circular flows of a supply chain network. *International Journal of Production Research, 60*, 4693–4711.

Pavlov, A., Ivanov, D., Dolgui, A., & Sokolov, B. (2018). Hybrid fuzzy-probabilistic approach to supply chain resilience assessment. *IEEE Transactions on Engineering Management, 65*(2), 303–315.

Pavlov, A., Ivanov, D., Pavlov, D., & Slinko, A. (2019). Optimization of network redundancy and contingency planning in sustainable and resilient supply chain resource management under conditions of structural dynamics. *Annals of Operations Research*. https://doi.org/10.1007/s10479-019-03182-6

Pavlov, A., Ivanov, D., Werner, F., Dolgui, A., & Sokolov, B. (2022). Integrated detection of disruption scenarios, the ripple effect dispersal and recovery paths in supply chains. *Annals of Operations Research, 319*, 609–631.

Proselkov, Y., Zhang, J., Xu, L., Hofmann, E., Choi, T., Rogers, D., & Brintrup, A. (2024). Financial ripple effect in complex adaptive supply networks: An agent-based model. *International Journal of Production Research, 62*(3), 823–845.

Quang, H. T., & Hara, Y. (2018). Risks and performance in supply chain: The push effect. *International Journal of Production Research, 56*(4), 1369–1388.

RM. (2018). Retrieved February 10, 2018 from https://www.riskmethods.net/en/software/overview

Sawik, T. (2016). On the risk-averse optimization of service level in a supply chain under disruption risks. *International Journal of Production Research, 54*(1), 98–113.

Sawik, T. (2018). *Supply chain disruption management using stochastic mixed integer programming*. Springer.
Scheibe, K. P., & Blackhurst, J. (2018). Supply chain disruption propagation: A systemic risk and normal accident theory perspective. *International Journal of Production Research, 56*(1–2), 43–59.
Schmitt, T. G., Kumar, S., Stecke, K. E., Glover, F. W., & Ehlen, M. A. (2017). Mitigating disruptions in a multi-echelon supply chain using adaptive ordering. *Omega, 68*, 185–198.
Simangunsong, E., Hendry, L. C., & Stevenson, M. (2012). Supply-chain uncertainty: A review and theoretical foundation for future research. *International Journal of Production Research, 50*(16), 4493–4523.
Simchi-Levi, D., Schmidt, W., & Wei, Y. (2014). From superstorms to factory fires: Managing unpredictable supply chain disruptions. *Harvard Business Review*.
Simchi-Levi, D., Schmidt, W., Wei, Y., Zhang, P. Y., Combs, K., Ge, Y., et al. (2015). Identifying risks and mitigating disruptions in the automotive supply chain. *Interfaces, 45*(5), 375–390.
Snyder, L. V., Atan, Z., Peng, P., Rong, Y., Schmitt, A. J., & Sinsoysal, B. (2016). OR/MS models for supply chain disruptions: A review. *IIE Transactions, 48*(2), 89–109.
Sokolov, B., Ivanov, D., Dolgui, A., & Pavlov, A. (2016). Structural quantification of the ripple effect in the supply chain. *International Journal of Production Research, 54*(1), 152–169.
Tang, O., & Musa, S. N. (2011). Identifying risk issues and research advancements in supply chain risk management. *International Journal of Production Economics, 133*, 25–34.
Teich, T. (2003). *Extended value chain management (EVCM)*. GUC-Verlag.
Teich, T., & Ivanov, D. (2012). Integrated customer-oriented product design and process networking of supply chains in virtual environments. *International Journal of Networking and Virtual Organizations, 11*(1), 48–61.
Tomlin, B. (2006). On the value of mitigation and contingency strategies for managing supply chain disruption risks. *Management Science, 52*, 639–657.
UPS. (2018). Retrieved February 11, 2018 from https://www.youtube.com/watch?v=aYoNd2nQqLg
Waller, M. A., & Fawcett, S. E. (2013). Data science, predictive analytics, and big data: A revolution that will transform supply chain design and management. *Journal of Business Logistics, 34*, 77–84.
Wang, Y., & Zhang, F. (2018). Modeling and analysis of under-load-based cascading failures in supply chain networks. *Nonlinear Dynamics*. https://doi.org/10.1007/s11071-018-4135-z
Waters, D. (2011). *Supply chain risk management: Vulnerability and resilience in logistics* (2nd ed.). Kohan Page.
Wilkinson, G. (2018). *Integrating artificial intelligence with simulation modeling*. Retrieved February 11, 2018 from https://www.anylogic.com/blog/
Yang, Y., et al. (2017). Mitigating supply chain disruptions through interconnected logistics services in the physical internet. *International Journal of Production Research, 55*(14), 3970–3983.
Zhuge, H. (2011). Semantic linking through spaces for cyber-physical-socio intelligence: A methodology. *Artificial Intelligence, 175*(5–6), 988–1019.

A Multi-Portfolio Approach to Integrated Risk-Averse Planning in Supply Chains Under Disruption Risks

Tadeusz Sawik

Abstract This chapter presents a multi-portfolio approach for the time and space integrated decision-making in a supply chain under disruption risks. In the context of supply chain disruptions, the portfolio is defined as the allocation of demand for parts among suppliers or the allocation of demand for products among production facilities. A disruptive event is assumed to impact both primary suppliers of parts and the firm primary assembly plant. Then, the firm selects recovery suppliers and recovery plants, along with the transshipment of parts from disabled primary plants to recovery plants and production and inventory planning in recovery plants. The mitigation and recovery decisions are integrated over time and space: the primary portfolios to be implemented before a disruptive event are optimized simultaneously with recovery portfolios for the aftermath period as well as the portfolios for both part suppliers and product manufacturers in different geographic regions are determined simultaneously. Using conditional cost at risk and conditional service at risk as risk measures, the risk-averse solutions are obtained. The solution results are compared for different demand patterns. The findings indicate that when the objective is to optimize service level with no regard to costs, both supply and demand portfolios are more diversified. The findings also demonstrate that the developed multi-portfolio approach leads to computationally efficient stochastic MIP models with a very strong LP relaxation. The proposed multi-portfolio approach that allows for time and space integrated decision-making may help to better mitigate the impact of disruption propagation on supply chain performance, i.e., the ripple effect.

T. Sawik (✉)
Department of Operations Research, AGH University of Kraków, Kraków, Poland

Indian Institute of Management Mumbai, Maharashtra, India
e-mail: ghsawik@cyf-kr.edu.pl

D. Ivanov et al. (eds.), *Handbook of Ripple Effects in the Supply Chain*,
International Series in Operations Research & Management Science 276,
https://doi.org/10.1007/978-3-031-85508-5_2

1 Introduction

In global supply chains, material flows are more often subject to unexpected disruptions and the resulting losses due to the shortage of material supplies may threaten the financial state of firms. For example, the disruptive events in the automotive and electronics supply chains that occurred in 2011 (the Great East Japan earthquake and tsunami in March and Thailand's floods in October) resulted in huge losses in the automotive and high-tech industry, e.g., Marszewska (2016), Matsuo (2015) and Park et al. (2013). In order to reduce potential losses, different disruption management strategies are applied in practice. Whenever a primary supplier is hit by a disruption, recovery of the disrupted primary supplier can be supported by the firm (e.g., the case of Toyota helping the supplier of automotive semiconductors, Renesas Electronics after the earthquake in 2011, Matsuo, 2015, and after the Kumamoto earthquake in 2016, Marszewska, 2016) or the firm selects an alternate (recovery) supplier, non-disrupted or disrupted less severely than the primary supplier. Similarly, when a disruptive event hits the firm's primary production facility, then a reasonable disruption management strategy would be to move production to alternate production facilities of the firm.

Supply chain disruptions rarely occur as long isolated events. Therefore, they must be contained before they propagate through the supply chain and create greater losses, see Blackhurst et al. (2011). Supply chain disruption propagation in Scheibe and Blackhurst (2018) was defined as the spread of the disruption effects beyond the initial disruption location. In the literature, supply chain disruption propagation is also known as domino effect, snowball effect, and more recently as the ripple effect, e.g., Basole and Bellamy (2014), Dolgui et al. (2018), Ivanov (2018), Ivanov et al. (2014a, b), Liberatore et al. (2012). The ripple effect depends on the supply chain resilience (Ivanov, 2018): multiple sourcing, prepositioning of emergency inventory, suppliers and plants fortification, production, and logistics flexibility are typical factors that might help to mitigate the impact of the ripple effect. In order to contain disruptions and prevent them from spreading through the supply chain, the decision-making on disruption management should be better coordinated and both proactive and reactive decisions integrated over time and space.

This chapter presents a multi-portfolio approach for time and space integrated decision-making in a supply chain under disruption risks to better mitigate the impact of disruption propagation on supply chain performance, i.e., the ripple effect. In the context of supply chain disruptions, the portfolio is defined as the allocation of demand for parts among suppliers or the allocation of demand for products among production facilities. A disruptive event is assumed to impact both primary suppliers of parts and the firm primary assembly plant. Then the firm selects recovery suppliers and recovery plants, along with the transshipment of parts from the disabled primary plant to recovery plants and production and inventory planning in recovery plants. The proactive and reactive decisions are integrated over time and space: the primary portfolios to be implemented before a disruptive event is optimized simultaneously with recovery portfolios for the aftermath period as well as the portfolios for both part suppliers and product manufacturers in different geographic regions are determined simultaneously. The portfolio approach to supply chain disruption management, first proposed for the selection of primary suppliers to mitigate the impact of disruption risks, Sawik (2011b, 2013a), was next enhanced for the selection of primary and recovery suppliers, Sawik (2017), and recently (Sawik, 2018a, 2019a, b) also for the selection of recovery assembly plants. In addition, a dynamic supply portfolio approach was proposed by Sawik (2011c, 2018b) to simultaneously mitigate the impact of the low probability and high-impact supply disruptions and the high-probability and low-impact supply delays.

The chapter is organized as follows. The review of relevant literature is presented in Sect. 2. The problem of selecting primary and recovery suppliers and assembly plants subject to partial disruptions is briefly described in Sect. 3. Stochastic mixed-integer programs for the integrated risk-averse supply, production, and inventory planning in a supply chain under disruption risks are developed in Sect. 4. Numerical examples are provided in Sect. 5 and final conclusions and directions for further research are presented in Sect. 6.

2 Literature Review

Supply chain disruption management utilizes various quantitative methods to optimize disruption mitigation and recovery strategies. Typical methods include scenario-based stochastic programming, mixed-integer programming, fuzzy set theory, game theory, control theory, simulation, genetic algorithms, etc. Reviews of literature on supply chain disruptions and recovery were presented in Snyder et al. (2016), where scholarly works were discussed and organized into six categories: evaluating supply disruptions, strategic decisions, sourcing decisions, contracts and incentives, inventory, and facility location. A systematic literature review and a comprehensive analysis of the decision-making models for supply chain risk were presented in Rajagopal et al. (2017). Stochastic programming and mixed-integer linear programming were found to be the commonly studied modeling methods. The literature on disruption recovery in supply chains was recently reviewed by Ivanov et al. (2017). The paper structured and classified existing research streams and application areas of different quantitative methods subject to different disruption risks and recovery measures.

Typical relevant works are briefly reviewed below. An interesting model for severe disruptions was proposed by MacKenzie et al. (2014) and applied to a simulation based on the Great East Japan earthquake and tsunami in 2011. In the model a disruption simultaneously impacts several suppliers. The model incorporated decisions made by both suppliers and firms during the disruption of random duration. The decisions may include whether or not suppliers move production to an alternate facility and hold parts inventory and whether a firm purchases parts from alternate suppliers that are not impacted, helps a primary supplier recover more quickly, or holds finished product inventory.

In Meena et al. (2014), the problem of determining the number of suppliers at risks of supplier failure due to catastrophic events was studied. A simple heuristic algorithm was proposed to determine the number of suppliers to minimize total costs subject to a target service level and to maximize the service level subject to a total costs constraint.

In Namdar et al. (2018), a scenario-based stochastic mixed-integer programming model was developed for single and multiple sourcing in the presence of disruption and operational risks. To achieve supply chain resilience under disruptions, backup supplier contracts, spot purchasing, and collaboration and visibility were considered.

In Ruiz-Torres et al. (2013), the optimal supply allocation and contingency planning in supply networks made up of multiple suppliers with different cost and reliability characteristics and a set of separate demand points was considered. Suppliers have production flexibility that allows them to deliver a contingency quantity in case other suppliers fail. The problem objective was to minimize the total network costs and was formulated as a mixed-integer program.

A bi-objective mixed possibilistic, two-stage stochastic programming model was developed by Torabi et al. (2015) to address supplier selection and order allocation problems to build a resilient supply base under operational and disruption risks. To enhance the resilience level, the model applies several proactive strategies, suppliers' business continuity plans, fortification of suppliers, and contracting with backup suppliers.

In Yoon et al. (2018), a bi-objective stochastic mixed-integer programming model that integrates supplier selection and risk mitigation strategy selection was proposed. The authors suggested that a combination of upstream and downstream risk mitigation strategies should be jointly considered with supplier selection rather than considering these decisions separately. A similar integrated approach was also proposed by Sawik (2013b, 2014, 2015, 2016).

Most approaches, including the abovementioned, used in supply chain disruption management assume that some estimation of disruptive events probability and potential losses are available. At the same time, a fair probability estimation of rare events is a complicated problem and even small errors in those estimations may significantly impact the modeling results. In Simchi-Levi et al. (2015), a novel risk-exposure model was proposed for analyzing operational-disruption risk with no need to estimate the probability of any specific disruptive events. The model is capable of assessing the impact of a disruption originating anywhere in a supply chain and has been applied by Ford Motor Company to identify risk exposures, evaluate risk mitigation actions, and develop optimal contingency plans.

3 Selection of Supply and Demand Portfolios

In this section the problem of integrated selection of primary and recovery supply, demand, and transshipment portfolios under disruption risks is presented.

Table 1 Notation

Indices
$i =$ supplier, $i \in I$
$j =$ assembly plant, $j \in J$
$k =$ region, $k \in K$
$l =$ disruption level, $l \in L_i, i \in I, l \in L^j, j \in J$
$s =$ disruption scenario, $s \in S$
$t =$ planning period, $t \in T$
Input parameters
$\xi =$ per unit product inventory holding cost
$\zeta =$ per unit and per period penalty cost of delayed demand for products
$\eta =$ per unit penalty cost of unfulfilled demand for products
$c_j =$ per period capacity of non-disrupted plant j
$c^s_{jt} =$ capacity in period t of plant j under disruption scenario s
$d_t =$ demand for products in period t
$e_i =$ per unit price of parts purchased from supplier i
$f_i =$ fixed ordering cost for supplier i
$\varepsilon_j =$ additional per unit production cost at recovery plant $j > 1$
$\varphi_j =$ fixed production setup cost at recovery plant $j > 1$
$g_j =$ per unit transshipment cost from primary plant $j = 1$ to plant $j > 1$
$p_{il} =$ probability of disruption level l for supplier i
$\pi_{jl} =$ probability of disruption level l for plant j
$p^k =$ regional disruption probability for region k
$t_s =$ start time period of disruption event s
$\gamma_{il} =$ fraction of an order delivered by supplier i under disruption level l
$\gamma^s_i =$ fraction of an order delivered by supplier i under scenario s
$\delta_{jl} =$ fraction of capacity of plant j available under disruption level l
$\delta^s_j =$ fraction of capacity of plant j available under scenario s
$\tau_{ij} =$ delivery lead time from supplier i to plant j
$\sigma_j =$ transshipment time from primary plant $j = 1$ to plant $j > 1$
$\theta_{is} =$ time to recover of supplier i from disruption under scenario s
$\vartheta_{js} =$ time to recover of plant j from disruption under scenario s
$\rho_{is} =$ cost to recover of supplier i from disruption under scenario s
$\varrho_{js} =$ cost to recover of plant j from disruption under scenario s

Consider a supply chain in which a single producer assembles one product type in several assembly plants to meet customer demand, using a critical part type that can be manufactured and provided by several suppliers. (for the notation used, see Table 1).

Let $I = \{1, \ldots, m\}$ be the set of m suppliers, $J = \{1, \ldots, n\}$, the set of n assembly plants, and $T = \{1, \ldots, h\}$, the set of h planning periods.

If we denote by d_t the demand for product in period t, then $D_t = \sum_{t' \in T: t' \le t} d_{t'}$ is the cumulative demand for products by period t and $D = \sum_{t \in T} d_t$ is the total demand for the entire planning horizon. If we assume that one part is required to produce one product, then the demand for parts is identical to the demand for products.

Let $j = 1$ be the primary plant, where the total demand for products, D, is initially assigned.

The suppliers of parts and assembly plants are located in different geographic regions, subject to potential regional disasters that may result in a complete shutdown of all suppliers and plants in the same region simultaneously. Denote by I^k and J^k, respectively, the subsets of suppliers and plants in region $k \in K$, and by p^k, the regional disruption probability for region k.

In addition, each supplier $i \in I$ is subject to random local disruptions of different levels, $l \in L_i = \{0, \dots, \overline{L}_i\}$, where disruption level refers to the fraction of an order that can be delivered, e.g., Sawik (2015). Level $l = 0$ represents complete shutdown of a supplier, i.e., no order delivery, while level $l = \overline{L}_i$ represents normal conditions with no disruption, i.e., full order delivery. Denote by $p_{il}$the probability of disruption level l for supplier i and by γ_{il} the fraction of an order that can be delivered by supplier i under disruption level l (fulfillment rate)

$$\gamma_{il} = \begin{cases} 0 & \text{if } l = 0 \\ \in (0, 1) & \text{if } l = 1, \dots, \overline{L}_i - 1 \\ 1 & \text{if } l = \overline{L}_i. \end{cases} \tag{1}$$

Similarly to suppliers, each plant $j \in J$ is subject to random local disruptions of different levels, $l \in L^j = \left\{0, \dots, \overline{L}^j\right\}$, where disruption level refers to the available fraction of full capacity, c_j, available per period under normal conditions. Level $l = 0$ represents complete shutdown of an assembly plant, while level $l = \overline{L}^j$ represents normal conditions, i.e., full capacity, c_j, available. Denote by π_{jl}, the probability of disruption level l for plant j, and by δ_{jl}, the fraction of available capacity of plant j under disruption level l.

$$\delta_{jl} = \begin{cases} 0 & \text{if } l = 0 \\ \in (0, 1) & \text{if } l = 1, \dots, \overline{L}^j - 1 \\ 1 & \text{if } l = \overline{L}^j. \end{cases} \tag{2}$$

The total number of all potential scenarios is $\prod_{i \in I} (\overline{L}_i + 1) \prod_{j \in J} \left(\overline{L}^j + 1\right)$. Each scenario $s \in S$ is represented by an $(m + n)$- dimensional vector $\lambda_s = \{\lambda_{1s}, \dots, \lambda_{ms}, \lambda_{m+1,s}, \dots, \lambda_{m+n,s}\}$, where $\lambda_{is} \in L_i$ is the disruption level of supplier $i \in I$ and $\lambda_{m+j,s} \in L^j$ is the disruption level of plant $j \in J$, under scenario $s \in S$. Disruption s is assumed to occur in period t_s, and the corresponding suppliers $i \in I$ (such that $\lambda_{is} < \overline{L}_i$) and plants $j \in J$ (such that $\lambda_{m+j,s} < \overline{L}^j$) are assumed to be simultaneously hit by the disruption.

The probability P_s of disruption scenario $s \in S$ is $P_s = \prod_{k \in K} P_s^k$, where P_s^k is the probability of realizing disruption scenario s in region k (see Sawik, 2015)

$$P_s^k = \begin{cases} \left(1-p^k\right)\left(\prod_{i\in I^k}\prod_{l\in L_i:\lambda_{is}=l} p_{il}\right)\left(\prod_{j\in J^k}\prod_{l\in L^j:\lambda_{m+j,s}=l}\pi_{jl}\right), \\ if \quad \sum_{i\in I^k}\lambda_{is} + \sum_{j\in J^k}\lambda_{m+j,s} > 0 \\ p^k + \left(1-p^k\right)\left(\prod_{i\in I^k} p_{i0}\right)\left(\prod_{j\in J^k}\pi_{j0}\right), \\ if \quad \sum\limits_{i\in I^k}\lambda_{is} + \sum\limits_{j\in J^k}\lambda_{m+j,s} = 0. \end{cases} \tag{3}$$

When supplier i is hit by disruption at level l, its recovery process to normal conditions takes *TTR*(i, l) time periods (time to recover) and let *CTR*(i, l) be the firm's portion of cost to recover. For each supplier i, denote by θ_{is} and ρ_{is}, respectively, time to recover and firm's portion of cost to recover from disruption under scenario s

$$\theta_{is} = TTR\,(i,l)\,; \quad i \in I, s \in S : l = \lambda_{is} \tag{4}$$

$$\rho_{is} = CTR\,(i,l)\,; \quad i \in I, s \in S : l = \lambda_{is}. \tag{5}$$

Similarly, when plant j is hit by disruption at level l, its recovery process to normal conditions takes *PRT*(j, l) time periods (plant recovery time) and cost *PRC*(j, l) (plant recovery cost). For each plant j, denote by ϑ_{js} and ρ_{js}, respectively, time to recover and cost to recover from disruption under scenario s

$$\vartheta_{js} = PRT\,(j,l)\,; \quad j \in J, s \in S : l = \lambda_{m+j,s} \tag{6}$$

$$\varrho_{js} = PRC\,(j,l)\,; \quad j \in J, s \in S : l = \lambda_{m+j,s}. \tag{7}$$

The orders for parts are assumed to be placed at the beginning of the planning horizon, and under normal conditions, the parts ordered from supplier i are delivered to assembly plant j in period τ_{ij}, where τ_{ij} is the total of manufacturing lead time and transportation time. Denote by σ_j transshipment time from the primary plant $j = 1$ to recovery plant j.

The firm that moves production to an alternate assembly plant j incurs a fixed cost φ_j and encounters additional per unit cost of production ε_j, and per unit cost, g_j, of transshipment of parts from the primary plant, where $\varphi_1 = 0$, $\varepsilon_1 = 0$ and $g_1 = 0$. A recovery plant can be a disrupted primary plant $j = 1$ with reduced capacity during the recovery process and then with its full capacity or a new plant, non-disrupted or disrupted less severely than the primary plant.

The following assumptions are made to formulate the problem.

- Each supplier has sufficient capacity to meet the total demand for parts.
- One unit of a critical part is needed to assemble one unit of product.
- If supplier $i \in I$ (assembly plant $j \in J$) is hit by disruption s in period t_s, its recovery process starts in period $t_s + 1$, so that the disrupted supplier (disrupted plant) returns to its full capacity in period $t = t_s + \theta_{is}$ $(t = t_s + \vartheta_{js})$.

- A single disruption scenario is assumed to be realized over the entire planning horizon. Multiple disruptions, one after the other in a series, during the recovery process are not considered.
- Period-dependent demand for products is considered to be satisfied during the planning horizon using a buildup inventory of products and a penalty cost is charged for demand fulfilled with delay and not fulfilled at all.
- A recovery supplier can be a disrupted primary supplier after its recovery to full capacity or a new supplier.
- A recovery assembly plant can be a disrupted primary assembly plant during and after its recovery to full capacity or a new assembly plant.

4 Risk-Averse Decision-Making

In this section, two stochastic MIP models **Support_CV(c)** and **Support_CV(sl)** are proposed for integrated, risk-averse selection of primary and recovery supply portfolios, recovery demand and transshipment portfolios, and production and inventory planning. The objective is to reduce the risk of worst-case cost by minimizing conditional cost at risk, CVaR^c, and reduce the risk of worst-case service level by maximizing conditional service at risk, CVaR^{sl}, respectively. Problems **Support_CV(c)** and **Support_CV(sl)** are stochastic multi-portfolio selection problems in which primary and recovery supply, demand and transshipment portfolios are simultaneously selected for all potential disruption scenarios (for definitions of first- and second-stage variables, see Table 2). The portfolios are defined below.

Models **Support_CV(c)** and **Support_CV(sl)** presented below are deterministic equivalent mixed-integer programs of the two-stage stochastic mixed-integer programs with recourse (e.g., Birge & Louveaux, 2011). The primary supply portfolio selection variables, u_i, v_i, are referred to as first-stage decisions, and the scenario-dependent recovery supply, demand and transshipment portfolio selection variables, U_i^s, V_{ij}^s, q_j^s, r_j^s and w_j^s, and production and inventory planning variables, x_{jt}^s, y_t^s, z_t^s, a_t^s, b_t^s, are referred to as recourse or second-stage decisions (cf. Table 2).

Support_CV(c): *Selection of* **Sup***ply, transshipment and demand* **port***folios with production and inventory planning to minimize* **CV** *aR of* **c***ost.*

Minimize

$$CVaR^c = VaR^c + (1-\alpha)^{-1} \sum_{s \in S} P_s C_s \tag{8}$$

subject to

Primary supply portfolio selection constraints:

$$\sum_{i \in I} v_i = 1 \tag{9}$$

Table 2 Two-stage problem variables

First-stage variables	
$u_i \in \{0, 1\}$, such that the value 1 means that supplier i is selected as a primary supplier; otherwise $u_i = 0$ (primary supplier selection)	
$v_i \in [0, 1]$, fraction of total demand for parts, ordered from the primary supplier i, to be delivered to primary plant $j = 1$ (primary supply portfolio)	
Second-stage variables	
$q_j^s \in \{0, 1\}$, such that the value 1 means that assembly plant j is selected as a recovery plant under disruption scenario s; otherwise $q_j^s = 0$ (recovery plant selection)	
$r_j^s \in [0, 1]$, fraction of total demand for products to be completed by recovery plant j under disruption scenario s (recovery demand portfolio)	
$U_i^s \in \{0, 1\}$, such that the value 1 means that supplier i is selected as a recovery supplier under disruption scenario s; otherwise $U_i^s = 0$ (recovery supplier selection)	
$V_{ij}^s \in [0, 1]$, fraction of total demand for parts, ordered from recovery supplier i to recovery plant j, under disruption scenario s (recovery supply portfolio)	
$w_j^s \in [0, 1]$, fraction of total demand for parts, transshipped from the primary plant $j = 1$ to recovery plant, j, under scenario s, where w_1^s represents parts that remain in the primary plant $j = 1$ (recovery transshipment portfolio)	
$x_{jt}^s \geq 0$, production in plant j in period t under disruption scenario s (production planning)	
$y_t^s \geq 0$, inventory of products at the beginning of period t under disruption scenario s (inventory planning)	
$z_t^s \geq 0$, shortage of products at the beginning of period t under disruption scenario s (inventory planning)	
Auxiliary variable	
$a_t^s \in \{0, 1\}$, such that value 1 means that there exists an inventory of products at the beginning of period t under disruption scenario s	
$b_t^s \in \{0, 1\}$, such that value 1 means that there exists a shortage of products at the beginning of period t under disruption scenario s	
$\mu_i^s \in \{0, 1\}$, such that the value 1 means that $u_i = U_i^s = 1$; otherwise $\mu_i^s = 0$ (elimination of double fixed ordering costs)	
The primary supply portfolio	$(v_1, \ldots, v_m)$, specifies supplies of parts from the primary suppliers to primary assembly plant $j = 1$, where $v_i \in [0, 1]; i \in I, \quad \sum_{i \in I} v_i = 1.$
The recovery supply portfolio for scenario$s$	$\left(V_{1j}^s, \ldots, V_{mj}^s\right); \quad j \in J,$ Specifies supplies of parts from recovery suppliers to recovery assembly plants, where $V_{ij}^s \in [0, 1]; i \in I, j \in J, \quad \sum_{i \in I} \left(\gamma_i^s v_i + \sum_{j \in J} V_{ij}^s\right) = 1,$ $\sum_{i \in I} \gamma_i^s v_i$ denotes delivery of parts from the primary suppliers to primary plant

(continued)

Table 2 (continued)

The recovery transshipment portfolio for scenario s	$(w_1^s, \ldots, w_n^s)$, Specifies transshipment of parts from the primary assembly plant to recovery plants, where $w_j^s \in [0, 1]; j \in J, \quad \sum_{j \in J} w_j^s = \sum_{i \in I} \gamma_i^s v_i - \sum_{t \in T: t < t_s} x_{1t}^s / D,$ And $\sum_{t \in T: t < t_s} x_{1t}^s$ denotes production at primary assembly plant $j = 1$, before disruption.
The recovery demand portfolio for scenario s	$(r_1^s, \ldots, r_n^s)$, Specifies an allocation of unfulfilled demand for products among recovery plants, where $r_j^s \in [0, 1]; j \in J, \quad \sum_{t \in T: t < t_s} x_{1t}^s / D + \sum_{j \in J} r_j^s = 1.$

$$v_i \leq u_i; \quad i \in I \tag{10}$$

Recovery supply and demand portfolio selection constraints:

$$V_{ij}^s \leq U_i^s; \quad i \in I, j \in J, s \in S \tag{11}$$

$$V_{ij}^s \leq q_j^s; \quad i \in I, j \in J, s \in S \tag{12}$$

$$r_j^s \leq q_j^s; \quad j \in J, s \in S \tag{13}$$

$$w_j^s \leq q_j^s; \quad j \in J, s \in S \tag{14}$$

$$\sum_{i \in I} \left(\gamma_i^s v_i + \sum_{j \in J} V_{ij}^s \right) = 1; \quad s \in S \tag{15}$$

$$\sum_{t \in T: t < t_s} x_{1t}^s / D + \sum_{j \in J} r_j^s = 1; \quad s \in S \tag{16}$$

$$\sum_{i \in I} V_{ij}^s + w_j^s = r_j^s; \quad j \in J, s \in S \tag{17}$$

$$\sum_{j \in J} w_j^s = \sum_{i \in I} \gamma_i^s v_i - \sum_{t \in T: t < t_s} x_{1t}^s / D; \quad s \in S \tag{18}$$

$$\mu_i^s \leq \left(u_i + U_i^s \right) / 2; \quad i \in I, s \in S. \tag{19}$$

Production capacity constraints:

$$x^s_{1t} \leq c_1; \quad t \in T, s \in S : t < t_s \tag{20}$$

$$x^s_{jt} \leq c^s_{jt} q^s_j; \quad j \in J, t \in T, s \in S : t \geq t_s \tag{21}$$

$$\sum_{t \in T : t \geq t_s} x^s_{jt}/D \leq r^s_j; \quad j \in J, s \in S. \tag{22}$$

Supply-transshipment production coordinating constraints:

$$\sum_{t' \in T : t' \leq t} x^s_{1t'} \leq \sum_{i \in I : \tau_{i1} \leq t-1} \gamma^s_i v_i + \sum_{i \in I : t_s + \theta_{is} + \tau_{i1} \leq t-1} V^s_{i1} - \left(\text{if } t \leq \overline{t}_s \text{ then } 0 \text{ else } \sum_{j \in J : j > 1} w^s_j \right); \quad t \in T, s \in S \tag{23}$$

$$\sum_{t' \in T : t' \leq t} x^s_{jt'} \leq \sum_{i \in I : t_s + \theta_{is} + \tau_{ij} \leq t-1} V^s_{ij} + \left(\text{if } t \leq \overline{t}_s + \sigma_j \text{ then } 0 \text{ else } w^s_j \right); \quad j \in J, t \in T, s \in S : j > 1, \tag{24}$$

where $\overline{t}_s = \max\{t_s, \max_{i \in I} \tau_{i1}\}$ and, $\overline{t}_s + 1$, is the start period of transshipment of parts to recovery plants.

Product inventory constraints:

$$y^s_{t+1} - z^s_{t+1} = \sum_{j \in J} \sum_{t' \in T : t' \leq t} x^s_{jt'} - D_t; \quad t \in T, s \in S \tag{25}$$

$$y^s_{t+1} \geq \sum_{j \in J} \sum_{t' \in T : t' \leq t} x^s_{jt'} - D_t; \quad t \in T, s \in S \tag{26}$$

$$z^s_{t+1} \geq D_t - \sum_{j \in J} \sum_{t' \in T : t' \leq t} x^s_{jt'}; \quad t \in T, s \in S \tag{27}$$

$$a^s_t + b^s_t \leq 1; \quad t \in T, s \in S \tag{28}$$

$$y^s_t/D \leq a^s_t; \quad t \in T, s \in S \tag{29}$$

$$z_t^s / D \le b_t^s; \quad t \in T, s \in S \tag{30}$$

$$ {}_{h+1}^{s} = 0; \quad s \in S. \tag{31}$$

Risk constraints:

- the tail cost for scenario s, $\mathcal{C}_s$, is defined as the nonnegative amount by which cost in scenario s exceeds VaR^c,

$$\begin{aligned} \mathcal{C}_s \ge & \sum_{i \in I} \left(f_i \left(u_i + U_i^s - \mu_i^s\right) + \rho_{is} U_i^s + e_i \left(\gamma_i^s v_i + \sum_{j \in J} V_{ij}^s \right) \right) / D \\ & + \sum_{j \in J} \left(g_j w_j^s + \left(\varphi_j + \varrho_{js}\right) q_j^s + \sum_{t \in T} \varepsilon_j x_{jt}^s \right) / D \\ & + \sum_{t \in T} \left(\xi y_t^s + \zeta z_t^s \right) / D + \eta z_{h+1}^s / D - VaR^c; \quad s \in S \end{aligned} \tag{32}$$

Non-negativity and integrality conditions:

$$\mathcal{C}_s \ge 0; \quad s \in S \tag{33}$$

$$a_t^s \in \{0, 1\}; \quad t \in T, s \in S \tag{34}$$

$$b_t^s \in \{0, 1\}; \quad t \in T, s \in S \tag{35}$$

$$q_j^s \in \{0, 1\}; \quad j \in J, s \in S \tag{36}$$

$$r_j^s \in [0, 1]; \quad j \in J, s \in S \tag{37}$$

$$u_i \in \{0, 1\}; \quad i \in I \tag{38}$$

$$v_i \in [0, 1]; \quad i \in I \tag{39}$$

$$U_i^s \in \{0, 1\}; \quad i \in I, s \in S \tag{40}$$

$$V_{ij}^s \in [0, 1]; \quad i \in I, j \in J, s \in S \tag{41}$$

$$w_j^s \in [0, 1]; \quad j \in J, s \in S \tag{42}$$

$$x_{jt}^s \ge 0; \quad j \in J, t \in T, s \in S \tag{43}$$

$$y_t^s \geq 0; \quad t \in T, s \in S \tag{44}$$

$$z_t^s \geq 0; \quad t \in T, s \in S \tag{45}$$

$$\mu_i^s \in \{0, 1\}; \quad i \in I, s \in S, \tag{46}$$

where

$$c_{jt}^s = \begin{cases} \delta_j^s c_j, \ if & t_s \leq t \leq t_s + \vartheta_{js} - 1 \\ c_j, \quad if & t \leq t_s - 1, t \geq t_s + \vartheta_{js}. \end{cases} \tag{47}$$

$\delta_j^s = \delta_{j,\lambda_{m+j,s}}$, and $\lambda_{m+j,s}$ is the disruption level of plant j under scenario s.

Equation (9) is the primary supply portfolio selection constraint and Eq. (10) ensures that demand for parts cannot be assigned to nonselected primary suppliers. Equations (11) and (12) ensure that the unfulfilled demand for parts cannot be assigned to nonselected recovery suppliers and cannot be ordered for nonselected recovery plants, respectively. Equation (13) ensures that the unfulfilled demand for products cannot be assigned to nonselected recovery plants, and Eq. (14) ensures that parts from the primary plant cannot be transshipped to nonselected recovery plants. Equations (15) and (16) are the recovery supply portfolio and the recovery demand portfolio selection constraints, respectively. The supply and demand portfolio flow conservation constraints (17) ensure that for each recovery plant, the recovery supplies and transshipment of parts are in balance with the demand for products to be fulfilled by that plant. The balance constraints for parts (18), ensure that the total transshipments of parts from the primary supplier are equal to the partially fulfilled supplies from the primary suppliers less the usage of parts for production before disruption. Equation (19) ensures that each supplier selected for both primary and recovery portfolios is charged exactly once with fixed ordering cost in the objective function. Equations (20) and (21) ensure that before a disruptive event, the production at a primary plant in every period cannot exceed its capacity, and after a disruptive event, the production at each selected recovery plant in every period cannot exceed the plant available capacity, respectively. Equation (22) guarantees that the total production at each recovery plant cannot exceed the assigned portion of the total demand for products.

The supply-transshipment-production coordinating constraints (23)–(24) ensure that

In the primary assembly plant, the cumulative demand for parts cannot exceed the cumulative deliveries from the primary and recovery suppliers, less transshipment of parts to recovery plants, if $t > \bar{t}_s$, Eq. (23),

In recovery assembly plants, the cumulative demand for parts cannot exceed the cumulative deliveries from recovery suppliers, plus the transshipment of parts from the primary plant, if $t > \bar{t}_s + \sigma_j$, Eq. (24).

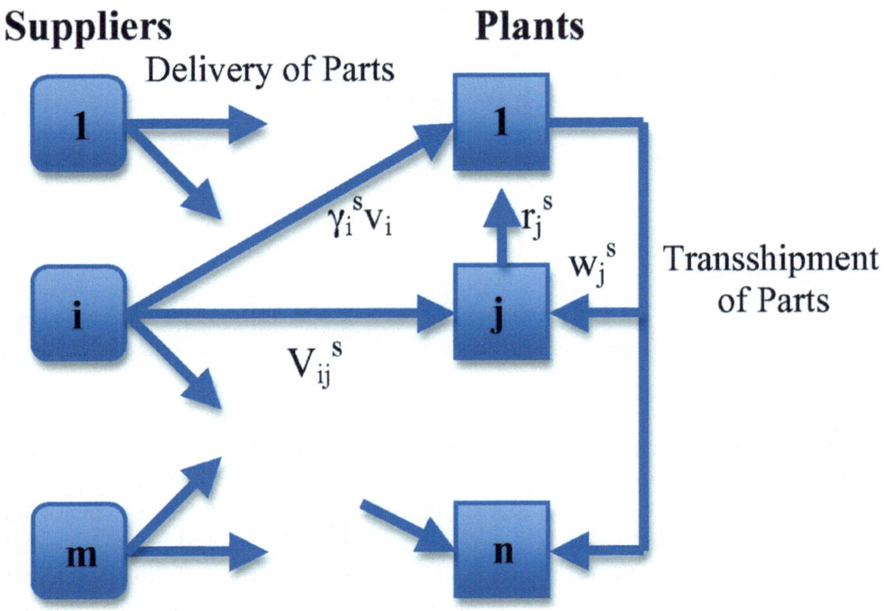

Fig. 1 Primary ($\gamma_i^s v_i$) and recovery (V_{ij}^s, w_j^s) flow of parts, and products (r_j^s)

The product inventory constraints (25)–(27) ensure that inventory or shortage of products in period $t + 1$ are, respectively, positive or negative differences between cumulative production and demand by period t. Equations (28)–(30) indicate that in each period, there is either inventory or shortage or production and demand are in balance. Finally, Eq. (31) ensures that no inventory of products remains at the end of the planning horizon.

The cost per product (see (32)) consists of different fixed and variable costs per product. The fixed cost includes the cost of ordering parts from the primary and recovery suppliers, the cost of recovery processes of suppliers and plants, and the cost of moving production from the primary to recovery plants. The variable cost includes the cost of purchasing parts from the primary and recovery suppliers, the cost of transshipment of parts, the cost of additional production cost in recovery plants, the cost of holding inventory of products, the cost of penalty for demand fulfilled with delay, and cost of penalty for unfulfilled demand.

Notice that the supply and demand portfolio selection constraints form an embedded network flow problem, (see Fig. 1). In particular, Eqs. (17) and (18) are flow conservation constraints under scenario s for each node j (assembly plant) and node $j = 1$ (primary assembly plant), respectively. $\gamma_i^s v_i$ represents the flow of parts from supplier i (source node) to primary plant $j = 1$ (sink/transshipment node), V_{ij}^s represents the flow of parts from recovery supplier i (source node) to recovery plant j (sink node) and w_j^s represents the flow of parts from the primary plant

$j = 1$ (transshipment node) to recovery plant j (sink node). Finally, r_j^s represents the outflow of products from recovery plant j.

Support_CV(sl): *Selection of* **Sup***ply, transshipment and demand* **port***folios with production and inventory planning to maximize* **CV** *aR of* s*ervice* **l***evel.*

Maximize

$$CVaR^{sl} = VaR^{sl} - (1-\alpha)^{-1}\sum_{s\in S} P_s \mathcal{S}_s \tag{48}$$

subject to (9)–(31), (34)–(47) and

Risk constraints:

– the tail service level for scenario s, $\mathcal{S}_s$, is defined as the nonnegative amount by which VaR^{sl} exceeds the service level in scenario s,

$$\mathcal{S}_s \geq VaR^{sl} + z_{max}^s/D - 1; \quad s \in S \tag{49}$$

$$\mathcal{S}_s \geq 0; \quad s \in S, \tag{50}$$

where $z_{max}^s = \max_{t\in T} z_t^s$ is the maximum shortage of products under disruption scenario s.

In the proposed models, CVaR is represented by auxiliary functions (8) and (48), where $\alpha \in (0, 1)$ is the confidence level, $\mathrm{VaR}^{(c)}$ (VaR^{sl}) is the targeted cost (targeted service level) such that for a given confidence level α, for $100\alpha\%$ of the scenarios, the outcome is below $\mathrm{VaR}^{(c)}$ (above VaR^{sl}).

In risk-averse decision-making, the confidence level α is fixed by the decision-maker to control the risk of losses due to supply disruptions. The higher the confidence level α, the more risk-aversive the decision-maker is and the smaller percent of the highest cost (lowest service) outcomes are focused on. The decision maker is willing to accept only solutions for which the total probability of scenarios with a cost greater than VaR^c (service smaller than VaR^{sl}) is not greater than $1-\alpha$.

Worst-Case Service Level

Proposition 1

The lowest service level, $\underline{SL}$, can be calculated as below.

$$\underline{SL} = \min\left\{\underline{SL1}/D, \underline{SL2}/D\right\}. \tag{51}$$

$\underline{SL1} = D - \sum_{t\in T: t\leq t^{SL}} d_t$, is the greatest unfulfilled demand due to nonavailable production capacity in disabled plants, where

$$t^{SL} = \max_{s\in S}\left\{t_s + \max_{j\in J}\left(\max\left(\max_{i\in I}\left\{\theta_{is} + \tau_{ij}\right\}, \vartheta_{js}, \sigma_j\right)\right)\right\},$$

is the longest time required to resume production after disruption.

$\underline{SL2} = \min_{j \in J} \sum_{t \in T: t > \max_{s \in S}\{t_s + \max(\max_{i \in I}\{\theta_{is} + \tau_{ij}\}, \vartheta_{js}, \sigma_j)\}} c_j$, is minimum production capacity available after recovery.

Proof 1

The lowest service level is associated with worst-case disruption scenario $s \in S$ for which primary supplier and primary assembly plant are both shutdown by disruption at time t_s before any production was started. Then, production resumes at plant j with the latest recovery time (after $t_s + \vartheta_{js}$ periods), after the latest delivery of parts by a recovery supplier (i.e., after $t_s + \max_{i \in I}\{\theta_{is} + \tau_{ij}\}$ periods) and after transshippment of parts from the primary plant (i.e., after $t_s + \sigma_j$ periods), whichever occurs later. Thus, $\sum_{t \in T: t \le t^{SL}} d_t$, is the maximum shortage due to disruption, i.e., shortage of products under worst-case scenario. On the other hand, the lowest service level is associated with minimum recovery capacity. The smallest recovery capacity of plant j is determined by the number of periods, $\Sigma_{t \in T: t > \max_{s \in S}\{t_s + \max(\max_{i \in I}\{\theta_{is} + \tau_{ij}\}, \vartheta_{js}, \sigma_j)\}}$, remaining for production after its full recovery.

5 Computational Examples

The basic input data for the computational examples presented in this section are available in Sawik (2018a, b, 2019). Although the input data are hypothetical, their relations to each other are real, and in part, they have been taken from a real case study. In particular, the case studies of Toyota supply chain disruption and recovery after the Great East Japan earthquake and tsunami of March 11, 2011 (e.g., Marszewska, 2016; Matsuo, 2015; Park et al., 2013) have been analyzed. Moreover, demand patterns are based on real-world data from a high-tech manufacturer (Sawik, 2011a). The computational experiments were performed using the AMPL programming language and the Gurobi 7.5.0 solver on a MacBookPro laptop with Intel Core i7 processor running at 2.8 GHz and with 16 GB RAM. The basic input parameters are shown in Table 3.

In order to emphasize the impact of disruptions on the primary assembly plant $j = 1$, the plant has greater capacity and is modeled to be less reliable than plant $j = 2$ ($c_1 = 10000 > c_2 = 5000, \pi_1 = 0.75 < \pi_2 = 0.85$).

The total number of all potential disruptive events was $\left(\overline{L} + 1\right)^m \left(\overline{L^1} + 1\right) \left(\overline{L^2} + 1\right) = \left(4^4\right)\left(2^2\right) = 1024$. In the computational examples, a disruption scenario was defined as a combination of a disruptive event and its start time. The start time t_s of each disruptive event was assumed to be not greater than the maximum delivery lead time to primary assembly plant $j = 1$, $\max_{i \in I}(\tau_{i1}) = 4$. Thus $t_s \in \{1, 2, 3, 4\}$, and the total number of all potential scenarios to be considered was $1024 \times 4 = 4096$, since each disruptive event $s = 1, \ldots, 1024$ may occur at four different start times.

Notice that $\gamma_i^s = 1$, if $t_s > \tau_{i1}$, i.e., full order is delivered by supplier i, if disruption occurs after delivery lead time to the primary assembly plant.

Table 3 Input parameters

$m = 4$ suppliers, $n = 2$ plants, $h = 30$ planning periods	
$K = \{1, 2\}$	Two geographic regions
$I^1 = \{1, 2\}, J^1 = \{1, 2\}$	Two suppliers and two plants in region $k = 1$
$I^2 = \{3, 4\}$	Two suppliers in region $k = 2$
Total demand for parts/products	$D = 300{,}000$
Demand per period for parts/products	
Constant demand pattern:	$d_t = 10{,}000, \forall t \in T$
Increasing demand pattern:	
$d = (0, 2, 7, 0, 10, 8, 6, 0, 9, 8, 9, 11, 2, 5, 5, 10, 11, 13, 17, 7, 17, 13, 18, 10, 7, 19, 28, 18, 15, 15) \times 1000$	
Decreasing demand pattern:	
$d = (0, 23, 21, 14, 15, 20, 13, 9, 18, 18, 16, 7, 10, 12, 11, 10, 5, 5, 2, 13, 9, 10, 9, 0, 6, 8, 0, 3, 7, 6) \times 1000$	
Disruption levels	
$\overline{L}_i = 3$	Four disruption levels for each supplier $i \in I$
$\overline{L}^j = 1$	Two disruption levels for each plant $j \in J$
Local disruption probability for suppliers:	
$p_{i0} = 0.1\ (1—p_{i3})$	Complete shutdown (level $l = 0$)
$p_{i1} = 0.3(1—p_{i3})$	Major disruption (level $l = 1$)
$p_{i2} = 0.6(1—p_{i3})$	Minor disruption (level $l = 2$)
$p_{i3} \in [0.75, 0.95]; i \in I^1, p_{i3} \in [0.65, 0.75]; i \in I^2$	Non disruption (level $l = 3$)
Local nondisruption probability for plants:	$\pi_1 = 0.75, \pi_2 = 0.85$
Regional disruption probability:	$p^1 = 0.001, p^2 = 0.01$
Fulfillment rates of suppliers:	
$\gamma_{i0} = 0\ \forall\ i \in I, \gamma_{i3} = 1\ \forall\ i \in I$	
$\gamma_{i1} \in [0.01, 0.50]\ \forall\ i \in I^1\ \gamma_{i1} \in [0.01, 0.30]\ \forall\ i \in I^2$, $\gamma_{i2} \in [0, 51, 0.99]\ \forall\ i \in I^1\ \gamma_{i2} \in [0, 31, 0.99]\ \forall\ i \in I^2$	
Plant capacity:	
$c_1 = 10{,}000, c_2 = 5000$	
Cost parameters:	
$e = (14, 12, 8, 9), f = (8000, 6000, 12{,}000, 13{,}000)$	
$\varepsilon_2 = 1, \varphi_2 = 100, g_2 = 0.1$, $\xi = \max_{i \in I} e_i/10 = 1.4.\ \zeta = 10, \eta = 100$	
$CTR(i, l) =$ if $l = 0$ then 100,000 *if*; if $l = 1$ then 10,000 *if*; if $l = 2$ then 1000 *if* $\forall i \in I$	
$PRC\ (1, 0) = PRC(2, 0) = 10{,}000$	
Time parameters:	
$\tau_{1j} = \tau_{2j} = 2, \tau_{3j} = \tau_{4j} = 4, \forall j \in J, \sigma_2 = 2$	
$TTR(i, l) =$ if $l = 0$ then 12; if $l = 1$ then 10; if $l = 2$ then 8 $\forall i \in I$	
$PRT(1, 0) = 10, PRT(2, 0) = 5$	

The probability P_s of realizing each disruption scenario $s \in S$ was calculated as follows:

$P_s = 0.1 P_s^1 P_s^2$ for $s \le 1024$

$P_s = 0.2 P_{s-1024}^1 P_{s-1024}^2$ for $1025 \le s \le 2048$

$P_s = 0.3 P_{s-2048}^1 P_{s-2048}^2$ for $2049 \le s \le 3072$

$P_s = 0.4 P_{s-3072}^1 P_{s-3072}^2$ for $3073 \le s \le 4096$

where probabilities $P_s^k;\ k = 1, 2,\ s \le 1024$ were obtained using Formula (3).

Notice that disruptive events with later start times were modeled to be more likely.

The solution results are summarized in Tables 4 and 5, respectively, for model **Support_CV(c)** and **Support_CV(sl)**. The tables show primary supply portfolio and expected demand portfolios for different values of confidence level $\alpha = 0.5$, 0.75, 0.9, 0.95, 0.99. The solution values, CVaR and VaR, are presented along with the associated expected values of cost per product,

$$E(c) = \sum_{s \in S} P_s \sum_{i \in I} \left(f_i \left(u_i + U_i^s - \mu_i^s\right) + \rho_{is} U_i^s + e_i \left(\gamma_i^s v_i + \sum_{j \in J} V_{ij}^s \right) \right) /D$$
$$+ \sum_{s \in S} P_s \sum_{j \in J} \left(g_j w_j^s + (\varphi_j + \varrho_{js}) q_j^s + \sum_{t \in T} \varepsilon_j x_{jt}^s \right) /D$$
$$+ \sum_{s \in S} \sum_{t \in T} P_s \left(\xi y_t^s + \zeta z_t^s \right) /D +$$
$$\eta \sum_{s \in S} P_s z_{h+1}^s /D, \tag{52}$$

and expected service level,

$$E(sl) = 1 - \sum_{s \in S} P_s z_{max}^s /D, \tag{53}$$

as well as with the solution value of LP relaxation of stochastic MIP models **Support_CV(c)** and **Support_CV(sl)**. Table 4 shows that while CVaR, VaR, and E(c) increase with the confidence level α, the expected service level, E(sl), decreases as the decision maker becomes more risk aversive with respect to cost. Similar results for model **Support_CV(sl)** are shown in Table 5. While CVaR, VaR, and E(sl) decrease with the confidence level α, E(c) increases as the decision maker becomes more risk aversive with respect to the service level.

Table 4 indicates that the primary supply portfolio for model **Support_CV(c)** becomes less diversified as the confidence level increases and for the highest $\alpha = 0.99$ a single supplier $i = 2$ is selected only, the cheapest among the most reliable suppliers. In contrast, for model **Support_CV(sl)**, a diversified supply portfolio with all suppliers is selected for all confidence levels, see Table 5. The expected recovery demand portfolios are similar for both models and all confidence levels, with the primary plant $j = 1$ selected as a major recovery supplier.

Table 4 Solution results for model **Support_CV(c)**

Confidence level α	0.50	0.75	0.90	0.95	0.99
Constant demand pattern					
Var. = 365,531, bin. = 56,064, cons. = 693,807, Nonz. = 7804322[a]					
CVaR	64.20	92.25	104.80	120.01	136.71
VaR	34.53	69.26	87.49	90.87	134.92
Exp.Cost E(c), (52)	49.32	59.67	71.63	72.67	90.40
Exp.Service E(sl), (53) × 100%	87.90	85.02	81.69	81.84	80.22
Primary supply portfolio:	1(5)				
Supplier(% of total demand)[b]	2(8)	2(5)	2(4)	2(13)	2(100)
	3(78)	3(82)	3(83)	3(74)	
	4(9)	4(13)	4(13)	4(13)	
Exp.Recovery demand portfolio:	1(90)				
Plant (% of total demand)[c]	2(10)				
LP[d]	63.95	92.06	104.46	119.34	134.96
Increasing demand pattern					
Var. = 596,787, Bin. = 207,504, Cons. = 1,003,583, Nonz. = 9671394[a]					
CVaR	39.00	57.45	67.71	83.54	96.41
VaR	19.39	46.61	51.57	53.43	94.58
Exp.Cost E(c), (52)	29.19	44.27	47.78	50.77	56.00
Exp.Service E(sl), (53) × 100%	93.42	82.47	80.90	79.45	79.83
Primary supply portfolio:	2(9)	2(3)	2(4)	2(16)	2(100)
Supplier(% of total demand)[b]	3(83)	3(88)	3(87)	3(75)	
	4(8)	4(9)	4(9)	4(9)	
Exp.Recovery demand portfolio:	1(89)				
Plant (% of total demand)[c]	2(10)				
LP[d]	38.85	56.85	67.37	82.08	94.80
Decreasing demand pattern					
Var. = 375,267, Bin. = 59,824, Cons. = 699,143, Nonz. = 7778870[a]					
CVaR	96.97	125.04	137.69	152.87	169.64
VaR	67.46	100.42	120.41	123.80	167.86
Exp.Cost E(c), (52)	82.21	98.35	104.28	106.33	128.97
Exp.Service E(sl), (53) × 100%	69.12	64.52	62.99	62.59	60.58
Primary supply portfolio:	1(5)				
Supplier(% of total demand)[b]	2(8)	2(5)	2(3)	2(14)	2(100)
	3(78)	3(82)	3(84)	c(76)	
	4(9)	4(13)	4(13)	4(10)	
Exp.Recovery demand portfolio:	1(89)	1(90)	1(89)	1(90)	1(87)
Plant (% of total demand)[c]	2(10)	2(10)	2(10)	2(10)	2(13)
LP[d]	96.80	124.85	137.36	152.20	167.90

[a]Var. = total number of variables, Bin. = number of binary variables, Cons. = number of constraints, Nonz. = number of nonzero coefficients

[b]1(v_1 × 100), 2(v_2 × 100), 3(v_3 × 100), 4(v_4 × 100)

[c]$1\left(\sum_{s\in S} P_s r_1^s \times 100\right), 2\left(\sum_{s\in S} P_s r_2^s \times 100\right)$

[d]LP relaxation solution value of (8)

Table 5 Solution results for model **Support_CV(sl)**

Confidence level α	0.50	0.75	0.90	0.95	0.99
Constant demand pattern					
Var. = 365,531, Bin. = 56,064, Cons. = 693,807, Nonz. = 7544834[a]					
CVaR	82.55	71.83	70.17	69.10	68.32
VaR	93.33	78.33	71.67	70.00	68.33
Exp.Service E(sl), (53) × 100%	87.94	82.70	82.67	82.34	82.16
Exp.Cost E(c), (52)	534.65	554.09	553.47	556.06	556.38
Primary supply portfolio:	1(33)	1(28)	1(30)	1(25)	1(28)
Supplier(% of total demand)[b]	2(27)	2(25)	2(21)	2(27)	2(27)
	3(13)	3(21)	3(24)	3(21)	3(21)
	4(27)	4(26)	4(25)	4(27)	4(24)
Exp.Recovery demand portfolio:	1(89)	1(92)		1(90)	
Plant (% of total demand)[c]	2(10)	2(8)		2(10)	
LP[d]	82.55	71.83	70.17	69.10	68.32
Increasing demand pattem					
Var. = 596,787, Bin. = 207,504, Cons. = 1,003,583, Nonz. = 9332090[a]					
CVaR	91.24	89.14	87.12	86.27	84.96
VaR	93.33	93.33	88.67	86.67	85.00
Exp.Service E(sl), (53) × 100%	93.55	93.55	90.66	88.78	87.10
Exp.Cost E(c), (52)	513.63	513.62	516.70	519.23	521.70
Primary supply portfolio:	1(35)	1(35)	1(31)	1(32)	1(33)
Supplier(% of total demand)[b]			2(32)		
	3(11)	3(11)	3(17)	3(16)	3(12)
	4(22)	4(22)	4(20)	4(20)	4(23)
Exp.Recovery demand portfolio:	1(88)		1(89)		
Plant (% of total demand)[c]	2(10)		2(11)		
LP[d]	91.24	89.14	87.12	86.27	84.96
Decreasing demand pattern					
Var. = 375,267, Bin. = 59,824, Cons. = 699,143, Nonz. = 7513406[a]					
CVaR	63.99	53.69	51.92	51.10	50.31
VaR	74.33	61.00	53.67	52.00	50.33
Exp.Service E(sl), (53) × 100%	69.16	63.92	63.87	63.62	63.32
Exp.Cost E(c), (52)	569.41	588.41	588.60	590.24	593.00
Primary supply portfolio:	1(33)	1(26)	1(30)	1(27)	1(26)
Supplier(% of total demand)[b]	2(28)	2(26)	2(21)	2(22)	2(28)
	3(12)	3(23)	3(24)	3(23)	3(21)
	4(27)	4(25)	4(25)	4(28)	4(25)
Exp.Recovery demand portfolio:	1(88)			1(90)	
Plant (% of total demand)[c]	2(10)			2(10)	
LP[d]	63.99	53.69	51.92	51.10	50.31

[a] Var. = total number of variables, Bin. = number of binary variables, Cons. = number of constraints, Nonz. = number of nonzero coefficients

[b] $1(v_1 \times 100), 2(v_2 \times 100), 3(v_3 \times 100), 4(v_4 \times 100)$

[c] $1\left(\sum_{s\in S} P_s r_1^s \times 100\right), 2\left(\sum_{s\in S} P_s r_2^s \times 100\right)$

[d] LP relaxation solution value of (48)

For model **Support_CV(c)** and three different demand patterns, Fig. 2 presents expected cumulative production at each plant $j \in J$, by each period,

$$\sum_{s \in S} P_s \sum_{t' \in T: t' \leq t} x^s_{jt'}; \quad t \in T,$$

and expected shortage of products at the end of each period,

$$\sum_{s \in S} P_s z^s_{t+1}; \quad t \in T,$$

associated with the risk-averse solution for $\alpha = 0.99$. In addition, Fig. 2 shows the cumulative demand for products by each period $t \in T$,

$$D_t = \sum_{t' \in T: t' \leq t} d_{t'}.$$

For the top $100(1 - \alpha)\ \% = 1\%$ of the worst-case scenarios $s \in S$ with cost,

$$\begin{aligned}
&\sum_{i \in I} \left(f_i \left(u_i + U^s_i - \mu^s_i\right) + \rho_{is} U^s_i + e_i \left(\gamma^s_i v_i + \sum_{j \in J} V^s_{ij} \right) \right) / D \\
&+ \sum_{j \in J} \left(g_j w^s_j + \left(\varphi_j + \varrho_{js}\right) q^s_j + \sum_{t \in T} \varepsilon_j x^s_{jt} \right) / D \\
&+ \sum_{t \in T} \left(\xi y^s_t + \zeta z^s_t \right) / D + \eta z^s_{h+1} / D,
\end{aligned}$$

exceeding VaR^c, Fig. 3 presents expected worst-case cumulative production at each plant $j \in J$, by each period,

$$\frac{\sum_{s \in S: C_s > 0} P_s \sum_{t' \in T: t' \leq t} x^s_{jt'}}{\sum_{s \in S: C_s > 0} P_s}; \quad t \in T,$$

and expected worst-case shortage of products, at the end of each period,

$$\frac{\sum_{s \in S: C_s > 0} P_s z^s_{t+1}}{\sum_{s \in S: C_s > 0} P_s}; \quad t \in T.$$

Similar diagrams for model **Support_CV(sl)**, three different demand patterns and $\alpha = 0.99$ are shown in Figs. 4 and 5. Figure 4 presents expected cumulative production at each plant and expected shortage of products and Fig. 5 shows expected worst-case cumulative production,

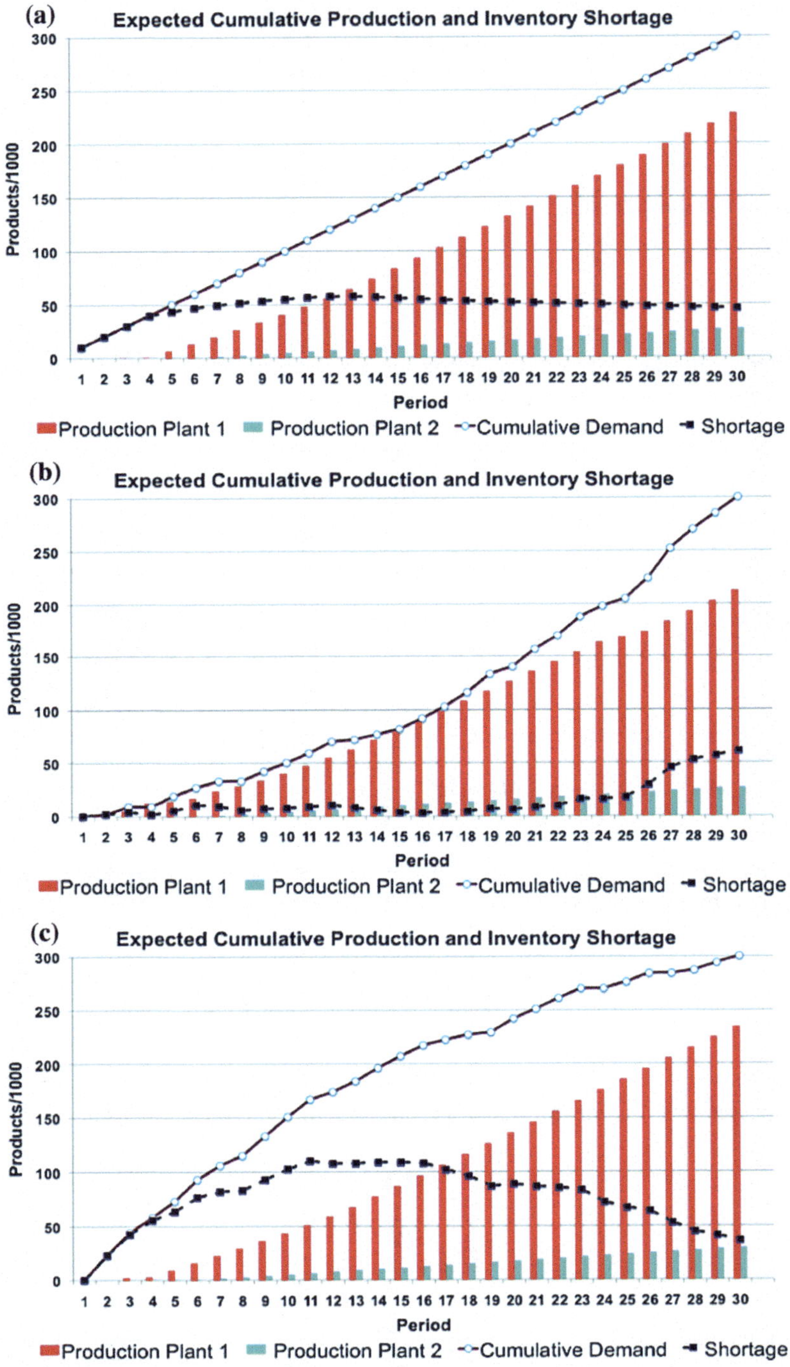

Fig. 2 Expected cumulative production and shortage of products for model **Support_CV(c)**, $\alpha = 0.99$: (**a**) constant demand pattern, (**b**) increasing demand pattern, (**c**) decreasing demand pattern

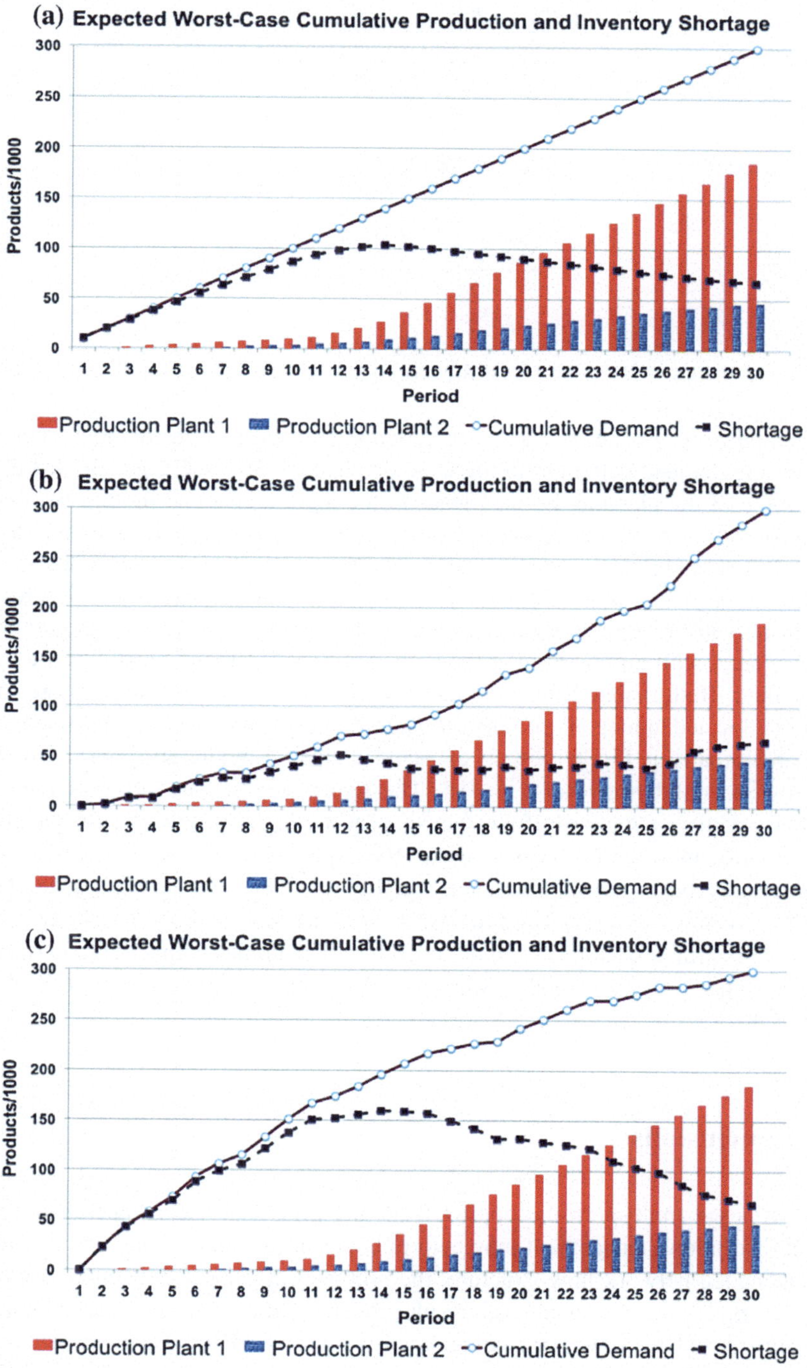

Fig. 3 Expected worst-case cumulative production and shortage of products for model **Support_CV(c)**, $\alpha = 0.99$: (**a**) constant demand pattern, (**b**) increasing demand pattern, and (**c**) decreasing demand pattern

$$\frac{\sum_{s\in S:S_s>0} P_s \sum_{t'\in T:t'\leq t} x^s_{jt'}}{\sum_{s\in S:S_s>0} P_s}; \quad t\in T,$$

and expected worst-case shortage of products,

$$\frac{\sum_{s\in S:S_s>0} P_s z^s_{t+1}}{\sum_{s\in S:S_s>0} P_s}; \quad t\in T,$$

for the top $100(1-\alpha)\% = 1\%$ of the worst-case scenarios $s \in S$ with service level, $1 - z^s_{max}/D$, below VaR^{sl}.

Comparison of solution results for different demand patterns demonstrates that decreasing demand pattern with the most demand for products concentrated at the beginning of the planning horizon leads to the highest cost and the lowest service level for both expected and expected worst-case scenarios, see Tables 4 and 5 and Figs. 2, 3, 4, and 5. In contrast to increasing demand pattern, for which the impact of disruption risks can be mitigated much better. A comparison of solution results for the two different models indicates that when the primary objective is to optimize service level with no regard to costs, both supply and demand portfolios are more diversified. Both demand for parts and unfulfilled demand for products are more evenly allocated among all primary suppliers and recovery plants, respectively.

For comparison, Fig. 6 presents expected results for model **Support_CV(c)** and the lowest confidence level $\alpha = 0.5$. The figure demonstrates that for the risk-aversive primary supply portfolio focusing on the highest 50% of cost outcomes, i.e., oriented more on business as usual, the expected results are better than for a more risk-aversive portfolio, i.e., for a higher confidence level. Clearly, the best-expected results can be achieved for risk-neutral decision-making, i.e., for $\alpha = 0$.

Proven optimal solutions were obtained for all examples with CPU time ranging from around 100 s for model **Support_CV(sl)** to over 600 s for model **Support_CV(c)**.

6 Conclusions

The proposed multi-portfolio approach for the time and space integrated supply chain disruption management may help to better mitigate the ripple effect. Due to an embedded network flow structure, the proposed stochastic MIP models have a very strong LP relaxation and as a result proven optimal solutions can be obtained in a very short CPU time using commercially available MIP solvers.

Future research should concentrate on relaxations of the various simplified assumptions used to formulate the problem. A straightforward enhancement is a multiple part type and product type setting with subsets of suppliers available for

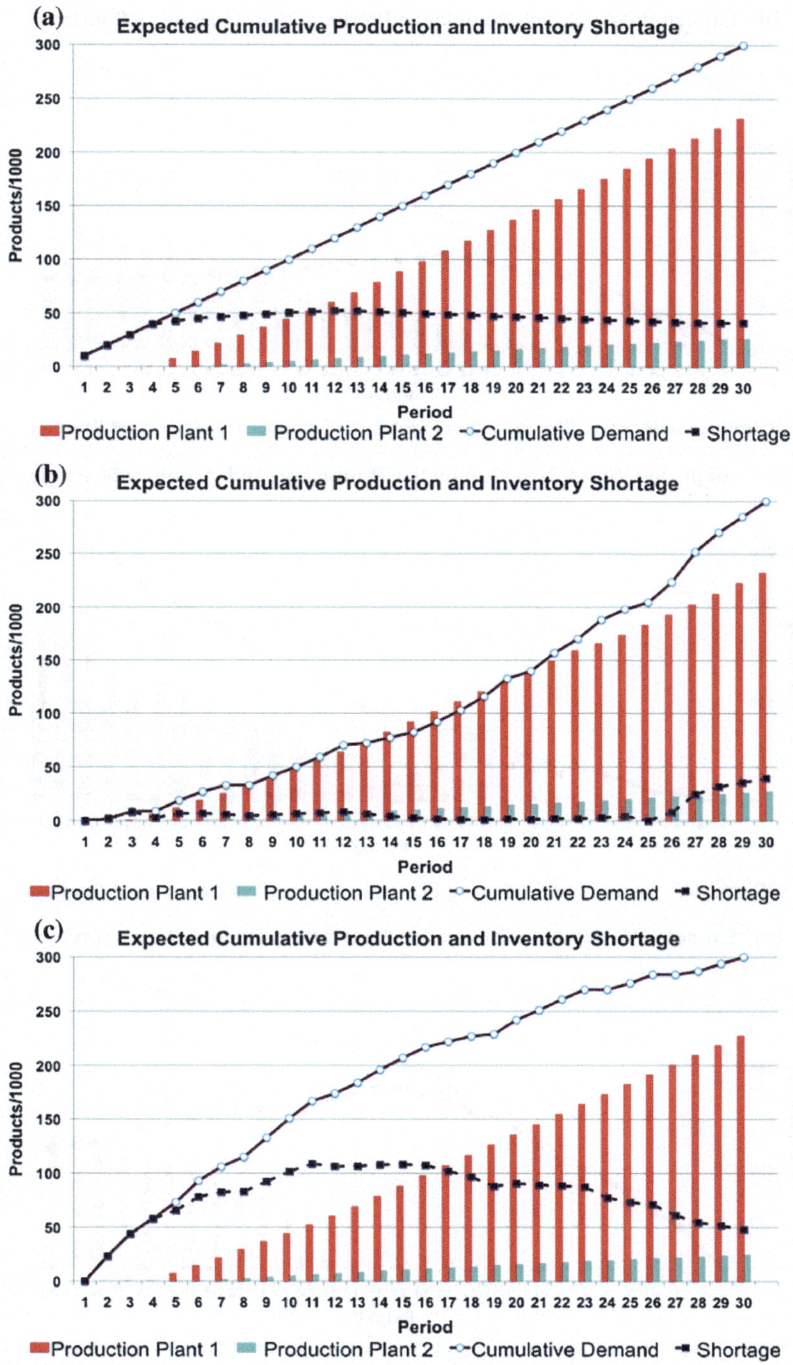

Fig. 4 Expected cumulative production and shortage of products for model **Support_CV(sl)**, $\alpha = 0.99$: (**a**) constant demand pattern, (**b**) increasing demand pattern, and (**c**) decreasing demand pattern

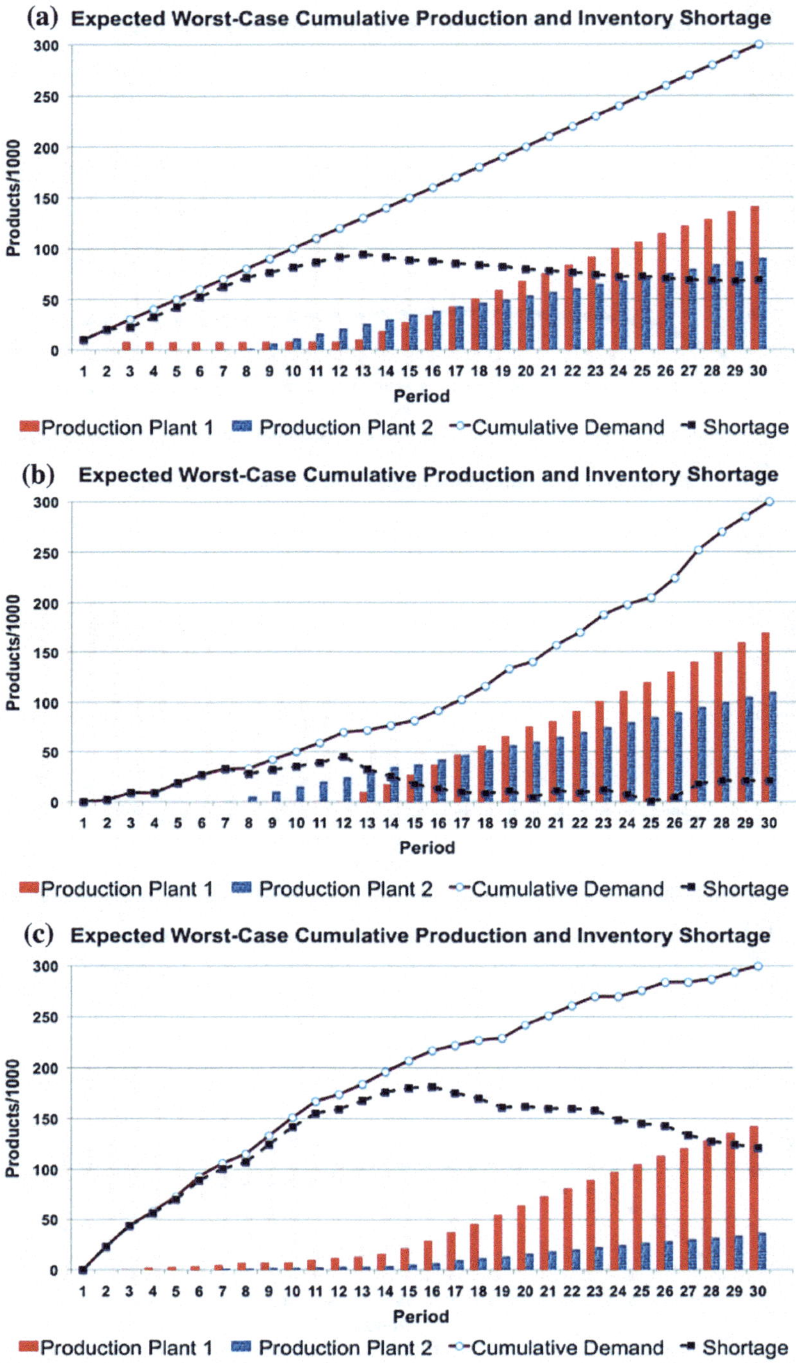

Fig. 5 Expected worst-case cumulative production and shortage of products for model **Support_CV(sl)**, $\alpha = 0.99$: (**a**) constant demand pattern, (**b**) increasing demand pattern, and (**c**) decreasing demand pattern

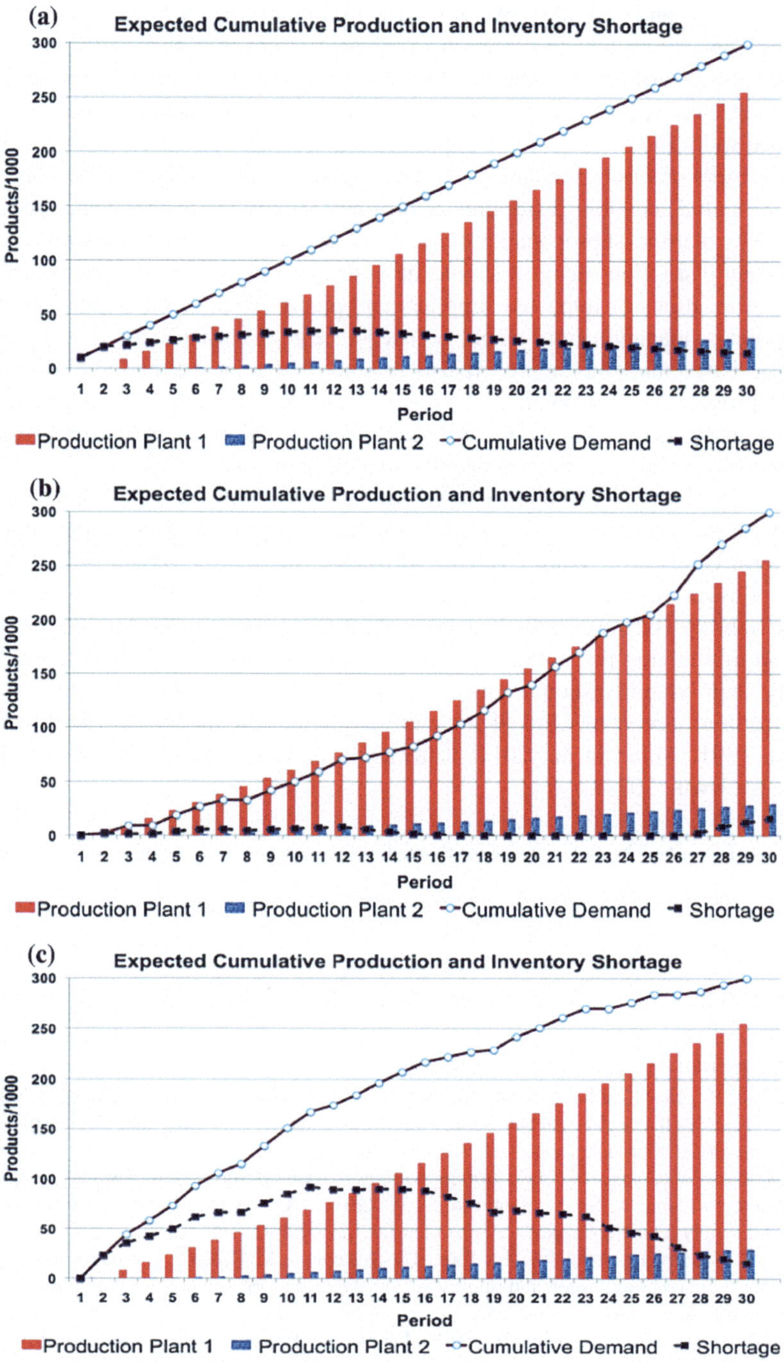

Fig. 6 Expected cumulative production and shortage of products for model **Support_CV(c)**, $\alpha = 0.5$: **a** constant demand pattern, **b** increasing demand pattern, and **c** decreasing demand pattern

each part type and assembly plants for each product type. A partial recovery can also be considered with the fraction of capacity recovered and available in each period. Moreover, different recovery modes with different recovery time and cost can be selected by the decision maker. In the scenario analysis, a disruptive event is assumed to hit the corresponding suppliers and assembly plants simultaneously. In order to more precisely account for the ripple effect, i.e., the disruption propagation from its initial location, disruption of different supply chain members should be delayed by disruption propagation time. Thus, instead of a common disruption start time, each disruption scenario should be associated with different start times, for different locations of impacted suppliers and plants. Then some estimation of the disruption propagation time would be needed, e.g., Ivanov (2020). The proposed scenario-based modeling approach also assumes that some estimation of disruptive events probability and potential losses is available. However, a fair probability estimation of rare events is a complicated problem and even small errors in those estimations may significantly impact the modeling results.

Now, in the post-COVID era, the researchers on supply chain risk management are aware of the ripple effect impact on the recovery process at the times of long-term crises, e.g., Ivanov (2020), Brusset et al. (2023), Sawik (2022, 2024). The optimization objectives under long-term crises more often focus on supply chain viability (Ivanov, 2022; Ivanov et al., 2023; Ivanov & Keskin, 2023) rather than on minimizing performance deviations after a disruption. Future research on supply chain viability should concentrate on developing computationally efficient optimization approaches linked with the mathematical theory of viability (e.g., Aubin, 1991; Aubin et al., 2011; Karacaoglu & Krawczyk, 2021; Sawik, 2023; Sawik & Sawik, 2024).

Acknowledgments The author wishes to thank the editors for their invitation to contribute to this handbook.

References

Aubin, J. P. (1991). *Viability theory*. Birkhauser.

Aubin, J. P., Bayen, M., & Saint-, P. P. (2011). *Viability theory: New directions* (2nd ed.). Springer.

Basole, R., & Bellamy, M. (2014). Supply network structure, visibility, and risk diffusion: A computational approach. *Decision Sciences, 45*(4), 753–789.

Birge, J. R., & Louveaux, F. (2011). *Introduction to stochastic programming*. Springer.

Blackhurst, J., Dunn, K. S., & Craighead, C. W. (2011). An empirically derived framework of global supply resiliency. *Journal of Business Logistics, 32*(4), 374–391.

Brusset, X., Ivanov, D., Jebali, A., La Torre, D., & Repetto, M. (2023). A dynamic approach to supply chain reconfiguration and ripple effect analysis in an epidemic. *International Journal of Production Economics, 263*, 108935.

Dolgui, A., Ivanov, D., & Sokolov, B. (2018). Ripple effect in the supply chain: An analysis and recent literature. *International Journal of Production Research, 56*(1–2), 414–430.

Ivanov, D. (2018). *Structural dynamics and resilience in supply chain risk management*. Springer.

Ivanov, D. (2020). Predicting the impacts of epidemic outbreaks on global supply chains: A simulation-based analysis on the coronavirus outbreak (COVID-19/SARSCoV-2) case. *Transportation Research Part E, 136*, 101–922.

Ivanov, D. (2022). Viable supply chain model: Integrating agility, resilience and sustainability perspectives– Lessons from and thinking beyond the COVID-19 pandemic. *Annals of Operations Research, 319*, 1411–1431.

Ivanov, D., & Keskin, B. B. (2023). Post-pandemic adaptation and development of supply chain viability theory. *Omega, 116*, 102806.

Ivanov, D., Sokolov, B., & Dolgui, A. (2014a). The ripple effect in supply chains: Trade-off 'efficiency-flexibility-resilience' in disruption management. *International Journal of Production Research, 52*(7), 2154–2172.

Ivanov, D., Sokolov, B., & Pavlov, A. (2014b). Optimal distribution (re)planning in a centralized multi-stage supply network under conditions of the ripple effect and structure dynamics. *European Journal of Operational Research, 237*(2), 758–770.

Ivanov, D., Dolgui, A., Sokolov, B., & Ivanova, M. (2017). Literature review on disruption recovery in the supply chain. *International Journal of Production Research, 55*(20), 6158–6174.

Ivanov, D., Dolgui, A., Blackhurst, J., & Choi, T. M. (2023). Toward supply chain viability theory: From lessons learned through COVID-19 pandemic to viable ecosystems. *International Journal of Production Research, 61*(8), 2402–2415.

Karacaoglu, G., & Krawczyk, J. B. (2021). Public policy, systemic resilience and viability theory. *Metroeconomica, 72*(4), 826–848.

Liberatore, F., Scaparra, M. P., & Daskin, M. S. (2012). Hedging against disruptions with ripple effects in location analysis. *Omega, 40*(1), 21–30.

MacKenzie, C. A., Barker, K., & Santos, J. R. (2014). Modeling a severe supply chain disruption and post-disaster decision making with application to the Japanese earthquake and tsunami. *IIE Transactions, 46*(12), 1243–1260.

Marszewska, J. R. (2016). Implications of seismic hazard in Japan on Toyota supply chain disruption risks. In *Proceedings of 13th international conference on industrial logistics, Zakopane, Poland* (pp. 178–185).

Matsuo, H. (2015). Implications of the Tohoku earthquake for Toyota's coordination mechanism: Supply chain disruption of automotive semiconductors. *International Journal of Production Economics, 161*, 217–227.

Meena, P. L., Sarmah, S. P., & Sarkar, A. (2014). Sourcing decisions under risks of catastrophic event disruptions. *Transportation Research Part E, 47*, 1058–1074.

Namdar, J., Li, X., Sawhney, R., & Pradhan, N. (2018). Supply chain resilience for single and multiple sourcing in the presence of disruption risks. *International Journal of Production Research, 56*(6), 2339–2360.

Park, Y., Hong, P., & Roh, J. J. (2013). Supply chain lessons from the catastrophic natural disaster in Japan. *Business Horizon, 56*(1), 75–85.

Rajagopal, V., Venkatesan, S. P., & Goh, M. (2017). Decision-making models for supply chain risk mitigation: A review. *Computers & Industrial Engineering, 113*, 646–682.

Ruiz-Torres, A. J., Mahmoodi, F., & Zeng, A. Z. (2013). Supplier selection model with contingency planning for supplier failures. *Computers & Industrial Engineering, 66*, 374–382.

Sawik, T. (2011a). *Scheduling in supply chains using mixed integer programming*. Wiley.

Sawik, T. (2011b). Selection of supply portfolio under disruption risks. *Omega, 39*, 194–208.

Sawik, T. (2011c). Selection of a dynamic supply portfolio in make-to-order environment with risks. *Computers and Operations Research, 38*(4), 782–796.

Sawik, T. (2013a). Selection of resilient supply portfolio under disruption risks. *Omega, 41*, 259–269.

Sawik, T. (2013b). Integrated selection of suppliers and scheduling of customer orders in the presence of supply chain disruption risks. *International Journal of Production Research, 51*(23–24), 7006–7022.

Sawik, T. (2014). Joint supplier selection and scheduling of customer orders under disruption risks: Single vs. dual sourcing. *Omega, 43*(2), 83–95.

Sawik, T. (2015). Integrated supply chain scheduling under multi-level disruptions. *IFAC-Papers On Line, 48*(3), 1515–1520.
Sawik, T. (2016). Integrated supply, production and distribution scheduling under disruption risks. *Omega, 62*, 131–144.
Sawik, T. (2017). A portfolio approach to supply chain disruption management. *International Journal of Production Research, 55*(7), 1970–1991.
Sawik, T. (2018a). *Supply chain disruption management using stochastic mixed integer programming*. Springer.
Sawik, T. (2018b). Selection of a dynamic supply portfolio under delay and disruption risks. *International Journal of Production Research, 56*(1–2), 760–782.
Sawik, T. (2019a). Two-period vs. multi-period model for supply chain disruption management. *International Journal of Production Research, 57*(14), 4502–4518.
Sawik, T. (2019b). Disruption mitigation and recovery in supply chains using portfolio approach. *Omega, 84*, 232–248.
Sawik, T. (2022). Stochastic optimization of supply chain resilience under ripple effect: A COVID-19 pandemic related study. *Omega, 109*, 102596.
Sawik, T. (2023). A stochastic optimisation approach to maintain supply chain viability under the ripple effect. *International Journal of Production Research, 61*(8), 2452–2469.
Sawik, T. (2024). *Stochastic programming in supply chain risk management: Resilience, viability, and Cybersecurity*. Springer.
Sawik, T., & Sawik, B. (2024). Risk-averse decision-making to maintain supply chain viability under propagated disruptions. *International Journal of Production Research, 62*(8), 2853–2867.
Scheibe, K. P., & Blackhurst, J. (2018). Supply chain disruption propagation: A systemic risk and normal accident theory perspective. *International Journal of Production Research, 56*(1–2), 43–59.
Simchi-Levi, D., Schmidt, W., Wei, Y., Zhang, P. J., Combs, K., Ge, Y., et al. (2015). Identifying risks and mitigating disruptions in the automotive supply chain. *Interfaces, 45*(5), 375–390.
Snyder, L. V., Atan, Z., Peng, P., Rong, Y., Schmitt, A., & Sinsoysal, B. (2016). OR/MS models for supply chain disruptions: A review. *IIE Transactions, 48*(2), 89–109.
Torabi, S. A., Baghersad, M., & Mansouri, S. A. (2015). Resilient supplier selection and order allocation under operational and disruption risks. *Transportation Research Part E, 79*, 22–48.
Yoon, J., Talluri, S., Yildiz, H., & Ho, W. (2018). Models for supplier selection and risk mitigation: A holistic approach. *International Journal of Production Research, 56*(10), 3636–3661.

The Rippling Effect of Supply Chains As Complex Adaptive Non-linear Systems

Virginia L. M. Spiegler, Mohamed M. Naim, and Junyi Lin

Abstract Non-linearities can lead to unexpected dynamic behaviours in supply chain systems that could then either trigger disruptions or make the response and recovery process more difficult. In this chapter, we take a control-theoretic perspective to discuss the impact of non-linearities on the ripple effect. This chapter is particularly relevant for researchers wanting to learn more about the different types of non-linearities that can be found in supply chain systems, the existing analytical methods to deal with each type of non-linearity and the future scope for research based on the current knowledge in this field.

1 Introduction

As a result of globalisation and increasing competitive pressures, modern supply chains have gone through a leaning and lengthening process (Christopher & Peck, 2004) and now back to reshoring (Gray et al., 2013). The managers have attempted to optimise supply chains by reducing holding inventory, outsourcing noncore activities, cutting the number of suppliers and sourcing globally, forgetting that the world market is an erratic and unpredictable place (Kearney, 2003). In addition to this, current trade restrictions resulting from the protectionist political environment in North America and Europe (Manners-Bell, 2018), as well as other recent supply

---

V. L. M. Spiegler (✉)
Kent Business School, University of Kent, Canterbury, UK
e-mail: V.L.Spiegler@kent.ac.uk

M. M. Naim
Cardiff Business School, Cardiff University, Cardiff, UK
e-mail: NaimMM@cardiff.ac.uk

J. Lin
International Business School Suzhou (IBSS), Xi'an Jiaotong-Liverpool University, Suzhou, People's Republic of China
e-mail: Junyi.Lin@xjtlu.edu.cn

D. Ivanov et al. (eds.), *Handbook of Ripple Effects in the Supply Chain*,
International Series in Operations Research & Management Science 276,
https://doi.org/10.1007/978-3-031-85508-5_3

and demand shocks (COVID-19 pandemic (Ivanov, 2020), Russia-Ukraine war (Tsang et al., 2024), extreme weather events (Shu & Fan, 2024)), have significantly influenced global supply chains, introducing additional uncertainty and complexity, making supply chains more vulnerable to disruptions than ever before.

The resulting complex business environment has increased the importance of handling risks which can emerge from the customers, suppliers, manufacturing processes and control systems (Spiegler et al., 2012) and of designing ripple effect mitigation strategies through agile, resilient and viable practices (Dolgui et al., 2018; Ivanov & Dolgui, 2020). Increased complexity also means that supply chain researchers can no longer disregard capacity limitations, restrictive policies and other system constraints, i.e. that the real world is non-linear. Non-linearities can introduce unexpected behaviour in a system causing instability and uncertainty (Spiegler et al., 2016; Wang & Disney, 2012); therefore, it is important to understand how control systems can be designed to influence dynamic behaviours and how non-linearities impact the performance of supply chains.

When looking at supply chain problems, the researchers have created a number of production and inventory models to represent the flows of information and material between different supply chain players. There are a number of research streams that deal with such problems such as Markov demand process, Bayesian approach, moving average or ARIMA process (Zhao et al., 2016), mixed-integer programming, stochastic programming, simulation (via system dynamics, agent-based modelling, discrete event), graph theory (Dolgui et al., 2018) and control theory (via feedback control and optimal control mechanism) (Dolgui et al., 2018; Zhao et al., 2016). The latter approach concerns determining transient responses and systems stability, i.e. understanding and controlling supply chain dynamics. These dynamics are normally driven by the application of different control system policies and can be considered as a source of disruption depending on the control system design (Colicchia et al., 2010). Moreover, a number of frameworks exist for tackling the ripple effect in the supply chain dynamics, control and disruption management domain (Ivanov et al., 2014).

In this chapter, we will discuss the impact of non-linearities on the ripple effect from a control structure perspective, by revisiting the literature on non-linear control theory application in supply chain management. As Ivanov and Sokolov (2013) pointed out 'useful tools for quantitative analysis of control and systems theory for a wide supply chain management research community remain undiscovered'. This chapter reviews new research techniques and recent progress in the analytical understanding of how non-linearities influence dynamic behaviours and affect the performance of supply chains. We start by introducing different types of non-linearities and their typical effect on system transient output response. Then, we suggest the existing methods to analyse each type of non-linearity and detail a selected number of mathematical approaches that can be used alongside simulation methods to explore the hidden dynamics caused by such non-linearities. Next, we discuss the applications of these methods and compile key findings on the rippling effect of non-linearities. Finally, the chapter concludes with a future research agenda.

2 Types of Non-linearities

A non-linear system is one whose performance does not obey the principle of superposition. This means that the output of a non-linear system is not directly proportional to the input and the variables to be solved cannot be expressed as a linear combination of the independent parts (Rugh, 2002). In supply chain systems, non-linearities can naturally occur due to the existence of physical and economic constraints, for instance, fixed and variable capacity constraints in the manufacturing and shipping processes, resource availability, variable delays and variable control parameters, trade and infrastructure constraints (Spiegler et al., 2016).

Since the variety of possible non-linearities in supply chain systems is extremely wide, it may be worthwhile to classify them into different categories, for which appropriate analytical methods will be suggested. The first research found on categorisation of non-linearities in business system dynamics research was done by Mohapatra (1980) who identified three types of non-linearities: limiting functions, table functions and product operators. He also recommends some techniques to deal with such properties, including the omission of redundant functions, linearisation through averaging, best-fit line approximations and small perturbation theory. In the control systems literature, non-linearities are more extensively classified as inherent or intentional, continuous or discontinuous and single- or multiple-valued (Cook, 1986; Vukic et al., 2003), as in Fig. 1.

Inherent non-linearities are intrinsic to the nature of the system and arise from the system's hardware and motion. They are normally undesirable and need to be compensated for by the system designer. Intentional non-linearities are artificial and deliberately introduced by the designer in order to improve system performance (Cook, 1986). Normally in supply chain systems, non-linearities are intrinsic to the system due to physical and economic constraints. These non-linearities may or may not be considered in the system modelling depending on the degree of accuracy and complexity necessary for the supply chain design. On the other hand, supply chain designers may want to include non-linearities that do not exist in reality for the sake of improving certain performance measures. This type of research has been considered by Spiegler et al., 2018 but is yet to be duly explored. Other studies have shown that while the presence of non-linearities may worsen some performance measures, they may improve others. For example, Evans and

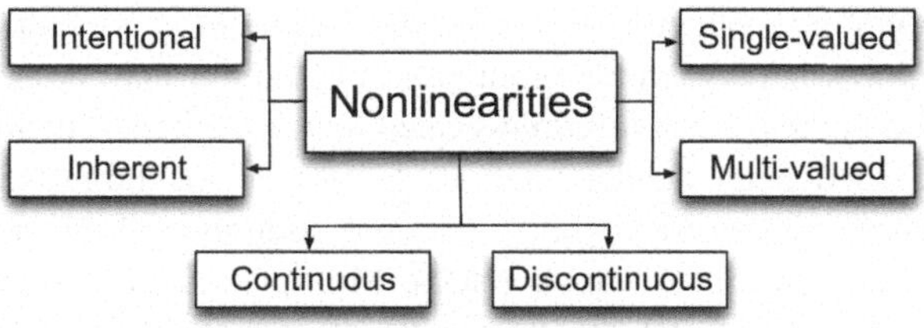

Fig. 1 Types of non-linearities

Table 1 Non-linearities in production and inventory control systems

Non-linearity	Block diagram symbol	Typical output response	Input–output profile
Fixed capacity constraint (discontinuous, single-valued)	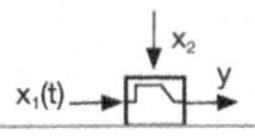	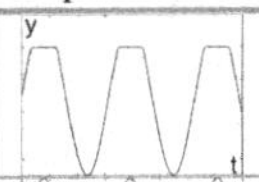	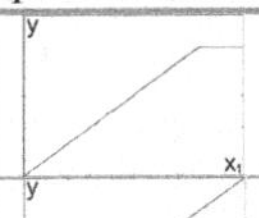
Non-negativity constraint (discontinuous, single-valued)	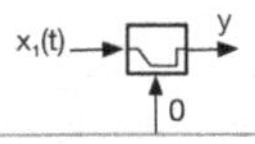	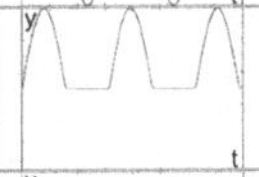	
Variable capacity constraint (discontinuous, multiple-valued)	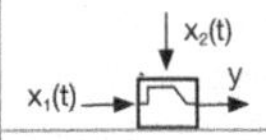	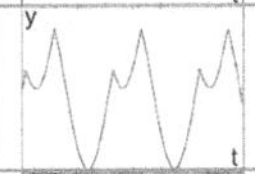	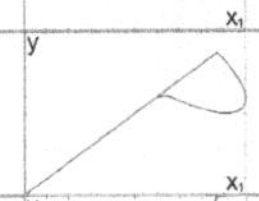
Rounding (discontinuous, single-valued)	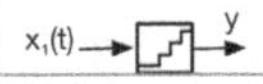	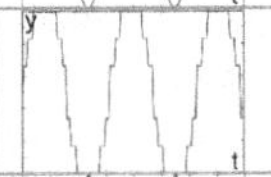	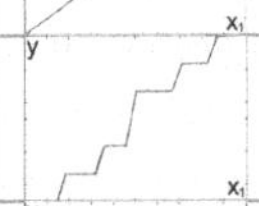
Time-varying parameter (continuous, single-valued)	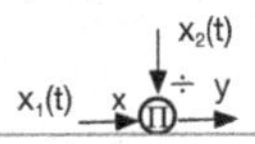	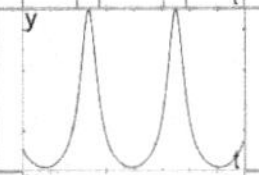	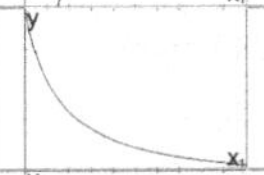
Time-varying parameter (continuous, multiple-valued)	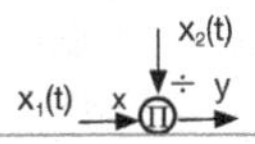	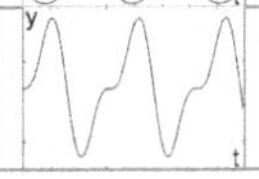	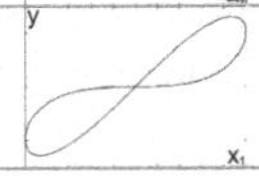

Naim (1994)—demand amplification versus service level Grübbstrom and Wang (2000)—complexity of the production plan versus production cost, Wikner et al. (2007)—lead-time expectations versus dynamic behaviour in the system.

Continuous and discontinuous non-linearities are associated with the rate of change between input and output. Table 1 contains examples of discontinuous (the first four rows) and continuous (the last two rows) non-linearities found in production and inventory control models for supply chain management and their block diagram symbol, typical output response given a sinusoidal input and the rate of change between output and input. A feature of the outputs in continuous functions is that they are smooth enough to possess convergent expansions at all points and therefore can be linearised. Examples include any adaptive control system, where certain control parameters, instead of being fixed, vary depending on the state of other variables (Vukic et al., 2003). Sharp changes in output values or gradients indicate discontinuities. The most common type of discontinuous non-linearity is the piecewise linear functions, which consist of a set of linear relations for different regions.

In the case of single-valued non-linearities, the output is a result of the current value of the input, whereas two or more values of output may be possible for the same input value in the case of multiple-valued non-linearities. The multiple values of the output will depend on the previous history of the input; thus such non-linearities are said to possess memory. The last column of Table 1 demonstrates the difference between these characteristics. Multiple-valued functions are often used in engineering to model the hysteresis of magnetic and elastic materials and mechanical backlash of friction gears (Cook, 1986). In business studies, this kind of non-linear behaviour has been described in economics (Göcke, 2002), for instance between buying/selling states and price (Cross et al., 2009) and unemployment and economy growth rate (Lang & de Peretti, 2009). In supply chain management research, multi-valued non-linearities are not so commonly reported. They have been used to model the switching of certain operation strategies depending on cost directions. Examples include investigations on changes in global sourcing (Kouvelis, 1998) and manufacturing strategies (Kogut & Kulatilaka, 1994) depending on foreign exchange rate directions. From a purely production-inventory control system perspective, this kind of effect has been identified in outbound shipments which depend on relational fluctuations between inventory levels and current demand (Spiegler et al., 2016). The normal thinking is that independent of demand growing direction, the order quantities placed to suppliers or shipped to customers will always match demand. However, when a variable capacity is put in place, these outputs can result in a complex multiple-valued non-linear behaviour.

3 Methods for the Analysis of Non-linearities

When confronted with a non-linear system, the first approach is to linearise it. The rationale for this is that techniques to analyse linear systems are much more established and better understood than non-linear control theory methods (Vukic et al., 2003). Linearisation is generally considered an appropriate choice when the solution can be obtained in this manner. While linear system theory is well acknowledged, the literature on non-linear theory is less conclusive when it comes to generality and applicability (Rugh, 2002). Because of a lack of common terminology and lack of detailed research methods in the non-linear control systems literature, the complete catalogue of all the existing methods and their applicability in the analysis of non-linear feedback systems are laborious. Table 2 presents a list of the methods that have been sufficiently acknowledged in the literature and whose full details were accessible.

There are a number of methods for system linearisation such as small perturbation theory, describing function and averaging or best-fit line approximations. The former allows the system with continuous non-linearities to be analysed through successive approximations in the form of power series around a specific operating point (Cook, 1986). If the system can be represented by the Taylor series or Volterra series, then it can be approximated using perturbation theory (Odame & Hasler, 2010). The Volterra series is often compared with Taylor series, but it is also suitable to approximate outputs with memory, which means that the Volterra series can mimic systems where the output depends on past inputs so they are suitable for multi-valued non-linearities (Rugh, 2002). The describing function method is attributed to a quasi-linearisation, since the approximation process of the non-linear system is for specific inputs. For instance, sinusoidal inputs are more often used since the frequency response approach is a powerful tool for the analysis and design of systems (Gelb & Velde, 1968). Averaging and best-fit line techniques produce

Table 2 Summary of methods used to analyse non-linear systems

	Method of analysis	Applications	Considerations
Linearisation methods	Small perturbation theory with Taylor series expansion	Continuous Single-valued	The assumption that the amplitude of the excitation signal is small. Local stability analysis only
	Describing function	Continuous, discontinuous single-valued, Multi-valued	Less accurate when non-linearities contain higher harmonics. Analysis of systems with periodic or Gaussian random input only
	Small perturbation theory with Volterra/Wiener series expansion	Continuous Multi-valued	The assumption that the amplitude of the excitation signal is small. Difficulty in calculating the kernels and operators of the system, making it impractical for high-order systems
	Averaging and best-fit line approximations	Continuous, discontinuous single-valued, multi-valued	Gross approximation of real responses. Only when better estimates are not possible
Purely control design method	Sliding mode control	Continuous, discontinuous (less effective) Single-valued, Multi-valued (more challenging)	Design of controllers for non-linear systems without requiring linearization, therefore being more robust to model uncertainties, parameter variations, and external disturbances. However, it does not provide a deep understanding of the non-linear behaviour, as it only mitigates and manages the system's complexities
Graphical and simple methods	Phase plane and graphical solutions	Continuous, discontinuous single-valued, multi-valued	Limited to 1st and 2nd order systems only
	Point transformation method	Discontinuous single-valued, multi-valued	Piecewise linear systems only. For high order systems, automated numerical methods must be employed
Exact solutions	Direct solution	Continuous single-valued	Limited to a finite number of equations
Stability method	Lyapunov-based stability analysis for piecewise linear systems	Discontinuous only single-valued examples were found	Piecewise linear systems only. Computation can be complex depending on the system
Simulation	Numerical and simulation solution	Continuous, discontinuous single-valued, multi-valued	Can be time-consuming. Dependent on computer and software calculations capacity

rough estimations and can be a simpler alternative for comprehending more complex systems in a qualitative manner (Mohapatra, 1980). However, whenever precision and reliability are needed, these methods should be avoided (Vukic et al., 2003).

In Sliding Mode Control, the system state is forced onto a predefined sliding surface and kept there using a high-frequency switching control law. When the system reaches the sliding surface, it behaves according to the reduced-order dynamics defined by that surface. While the dynamics on the surface are well-defined, the switching nature of the control input means that the system states fluctuate around the surface (Slotine & Li, 1991). The main advantages of sliding mode control include robustness, finite-time convergence and reduced-order compensated dynamics (Shtessel et al. (2014). However, its disadvantage include chattering (oscillation around the sliding surface,) complex implementation, and higher control effort (Edwards & Spurgeon, 2018).

Then there are graphical techniques, such as the phase plane analysis. However, this technique is limited to second-order systems (Vukic et al., 2003). The point transformation method allows periodicity and stability of piecewise linear systems to be investigated by studying the behaviour of trajectories that repeatedly cross from one region to another (Cook, 1986), but it can be complicated for high-order systems. There are also exact solutions for a finite number of non-linear control systems with low order (Vukic et al., 2003), making its application very limited. More complex and sophisticated techniques such as the one developed by Johansson (2003) are used for stability analysis of piecewise linear systems by combining Lyapunov functions and the convex optimisation process.

Finally, there is a simulation, which, although it is a very helpful technique, should be, in principle, used as a complementary tool to the above analytical methods. Simulation has many advantages, offering a 'middle ground between pure mathematical modelling, empirical observation and experiments for strategic issues in supply chain research' Größler and Schieritz (2005). Because simulation is a numerical technique that allows the analysis of complex models, it does not require specific mathematical forms that are analytically solvable.

In the next subsection, we provide instructions on how to adopt the following linearisation methods: describing functions and small perturbation theory with Taylor and Volterra series expansion. These methods were chosen given their wide applicability, versatility and power in uncovering hidden dynamics caused by different types of non-linearities and in tracing the transient behaviour, which is necessary to estimate the system's performance. In supply chain systems, the understanding of transient responses can elucidate the occurrence of disruptions and how to mitigate their cascading effects on other supply chain members.

3.1 Describing Function

The describing function method is a quasi-linear representation of a non-linear element subjected to a specific input. This is a method that attempts to estimate the output properties, such as frequency, amplitude and stability, after being affected by a non-linear component (Gelb & Velde, 1968). This method is also used to predict limit cycles or sustained oscillations (Spiegler & Naim, 2017).

The basic idea of the describing function is to express a non-linear element in the form of a transfer function, or a gain, determined from its effects on a particular input signal. For asymmetric non-linearities, or symmetric non-linearities subjected to biased inputs, at least two terms of the describing function are needed: one that expresses the change in the output amplitude (N_A) and another that considers the change in the output mean (N_B). This leads to the so-called dual-input describing function (Cook, 1986; Vukic et al., 2003). Another effect caused by this type of non-linearity is the possible change in phase angle (ϕ) of the output response in relation to its input. Next, we give an example of how sinusoidal describing functions can be determined.

Consider the input to the non-linearity:

$$x_1(t) = A.\cos(\omega t) + B \tag{1}$$

where ω is the angular frequency and $\omega = 2\pi/T$. The output y can be approximated to

$$y(t) = N_A.A.\cos(\omega t + \phi) + N_B.B \tag{2}$$

In order to determine the terms of the describing function (N_A, N_B and ϕ) the series has to be expanded and its first harmonic coefficients must be determined. The Fourier series expansion method is used to represent the output y such as

$$\begin{aligned} y(t) &\approx b_0 + a_1\cos(\omega t) + b_1\sin(\omega t) + a_2\cos(2\omega t) + b_2\sin(2\omega t) + \cdots \\ y(t) &\approx b_0 + \sum_{K=1}^{\infty} [a_k\cos(\omega t) + b_2\sin(2\omega t)] \end{aligned} \tag{3}$$

where the Fourier coefficients are given by

$$a_k = \frac{1}{\pi}\int_{-\pi}^{\pi} y(t)\cos(k.\omega t)\, d\omega t \tag{4}$$

$$b_k = \frac{1}{\pi}\int_{-\pi}^{\pi} y(t)\sin(k.\omega t)\, d\omega t \tag{5}$$

$$b_0 = \frac{1}{2\pi}\int_{-\pi}^{\pi} y(t) d\omega t \tag{6}$$

The non-linear function y is then approximated to the first harmonic, resulting in:

$$y(t) = b_0 + a_1 \cos(\omega t) + b_1 \sin(\omega t) = b_0 + \sqrt{a_1^2 + b_1^2}.\cos(\omega t + \phi) \tag{7}$$

where $\phi = \arctan\left(\frac{b_1}{a_1}\right)$

In this way, the two terms of the describing function can be determined as

$$N_A = \frac{\sqrt{a_1^2 + b_1^2}}{A} \tag{8}$$

$$N_B = \frac{b_0}{B} \tag{9}$$

For single-valued non-linearities, the coefficient b_1 will be equal to zero and therefore, the phase angle ϕ will be also zero. In the case of dynamic multi-valued non-linearities, the describing function will be in the form of $N_A(A, \omega)$. Normally, a plot of $N_A(A, \omega)$ versus $\phi(A, \omega)$ for various values of A and ω are used to understand such complex non-linearities (Gelb & Velde, 1968).

Examples of supply chain applications of such methods can be found in Spiegler et al. (2016a, b), Spiegler and Naim (2017), Wang et al. (2015), Lin et al. (2018, 2022, 2023). By replacing the different describing function values in the system transfer functions, these studies were able to determine the effect of non-linearities on the system's natural frequency, damping ratio and stability.

3.2 *Small Perturbation Theory with Taylor Series Expansion*

The Taylor series can be used to approximate the response of a non-linear system to a given input if the output of this system depends strictly on the input at that particular time.

Given a system with single-valued continuous non-linearity

$$\begin{aligned} \dot{x} &= f(x, u) \\ y &= h(x, u) \end{aligned} \tag{10}$$

where x is the state vector, $\dot{x}$ is the time derivative of the state vector, y is the output vector and u is the input vector, we can derive an approximate linear system about a nominal operating state space x^* and for a given input u^* by using small perturbation theory with Taylor series expansion. The linearisation process involved in this approach is such that departures from a steady-state point are small enough to produce transfer function coefficients. Hence, by assuming a small amplitude of the excitation signal, the non-linear differential equations are replaced by a set of linearised differential equations with coefficients dependent upon the steady-state operating point.

The first-order Taylor series approximation of the non-linear state derivatives leads to the following linearised function:

$$\Delta \dot{x} = A\Delta x + B\Delta u \tag{11}$$

$$\Delta y = C\Delta x + D\Delta u \tag{12}$$

where $\Delta x = x - x^*$, $\Delta \dot{x} = \frac{d\Delta x}{dt}$, $\Delta y = y - y^*$, $\Delta u = u - u^*$ and A, B, C, D can be found through the following partial derivatives:

$$\begin{pmatrix} A & B \\ C & D \end{pmatrix} = \begin{pmatrix} \frac{\partial f_1(x^*,u^*)}{\partial x_1} & \cdots & \frac{\partial f_1(x^*,u^*)}{\partial x_n} & \frac{\partial f_1(x^*,u^*)}{\partial u_1} & \cdots & \frac{\partial f_1(x^*,u^*)}{\partial u_k} \\ \vdots & \ddots & \vdots & \vdots & \ddots & \vdots \\ \frac{\partial f_n(x^*,u^*)}{\partial x_1} & \cdots & \frac{\partial f_n(x^*,u^*)}{\partial x_n} & \frac{\partial f_n(x^*,u^*)}{\partial u_1} & \cdots & \frac{\partial f_n(x^*,u^*)}{\partial u_k} \\ \frac{\partial h_1(x^*,u^*)}{\partial x_1} & \cdots & \frac{\partial h_1(x^*,u^*)}{\partial x_n} & \frac{\partial h_1(x^*,u^*)}{\partial u_1} & \cdots & \frac{\partial h_1(x^*,u^*)}{\partial u_k} \\ \vdots & \ddots & \ddots & \vdots & \ddots & \vdots \\ \frac{\partial h_m(x^*,u^*)}{\partial x_1} & \cdots & \frac{\partial h_m(x^*,u^*)}{\partial x_n} & \frac{\partial h_m(x^*,u^*)}{\partial u_1} & \cdots & \frac{\partial h_m(x^*,u^*)}{\partial u_k} \end{pmatrix} \tag{13}$$

where n is the number of state variables, m is the number of outputs and k is the number of inputs.

The nominal operating point (x^*, u^*) normally corresponds to the equilibrium or resting points where all state derivatives are equal to zero and they can be found by applying the final value theorem:

$$x^* = \lim_{t\to\infty} x(t) = \lim_{s\to 0} s.X(s) \tag{14}$$

for a constant input u.

The reader can refer to the works of Spiegler et al. (2016b), Jeong et al. (2000), Wang and Gunasekaran (2017), Lin et al. (2018a) for the application of this method in supply chain models. After linearisation is performed, it is possible to find the system transfer function and the system design will follow linear control system theory.

3.3 *Small Perturbation Theory with Volterra–Wiener Series Expansion*

The Volterra series has the ability to deal with multi-valued non-linearities by capturing memory effects. The output from the Volterra series depends on the previous history of the input to the system. Hence, a continuous multi-valued non-linear output can be approximated to

$$y(t) = h_0 + \int_{\mathbb{R}} h_1(\tau_1)\, x(t-\tau_1)\, d\tau_1 + \int_{\mathbb{R}} h_2(\tau_1, \tau_2)\, x(t-\tau_1)\, x(t-\tau_2)\, d\tau_1 d\tau_2 + \ldots$$
$$= h_0 + \sum_{n=1}^{N} \int_{\mathbb{R}} h_n(\tau_1, \ldots, \tau_n) \prod_{j=1}^{N} x\left(t - \tau_j\right) d\tau_j \tag{15}$$

where h_0 is a constant and for n = 1, 2...N, the function $h_n(\tau_1, \ldots, \tau_n)$ is referred as nth-order Volterra kernel. Note that when $h_0 = 0$ and $N = 1$, the formula describes the system's impulse response by a convolution of $x(t)$. Volterra extended the linear system representation to non-linear systems by adding a series of non-linear integral operators.

The convergence of the Volterra series is comparable to the convergence of the Taylor series expansion of a function which often allows only for small deviations from the starting point. However, the type of convergence required is very rigorous since not only the error has to approach zero with an increasing number of terms, but also the derivatives of the error. Hence, the estimation of Volterra coefficients is generally performed by estimating the coefficients of an orthogonalised series, e.g. the Wiener series, and then recomputing the coefficients of the original Volterra series. The readers can refer to Rugh (2002) for more details. No application of this method was found in the supply chain management literature.

4 The Effects of Non-linearities

In recent years, researchers have put effort into shaping stability regions of non-linear supply chain systems and understanding the factors which will lead to chaotic behaviours. These studies made contributions in explaining and tackling uncertainties, dynamics and disruptions in complex supply chain systems, since they elucidated how non-linearities can change a system's transient response by monitoring the variation in terms of output's amplitude, mean and phase and consequently its repercussion on the system's natural frequency and damping ratio. Figure 2 illustrates a few examples of the effect of some non-linearities (maximum manufacturing capacity and variable delay in filling order from the model in Spiegler et al., 2016b) on the system's step response. The figure demonstrates how simplistic linear assumptions can be and that indeed non-linearities can make significant changes to responses' amplitude and settling time. Another observation is that non-linearities affect different performances in different ways. For instance, while the maximum production capacity seems to diminish inventory levels, it helps to decrease the amplification in manufacturing order rate (bullwhip). The variable delay in filling orders will have a more negative impact on the outbound shipment rate than on the inventory response. When both non-linearities are considered at the same time, outbound shipments' amplitude and recovery time further worsen. Response and recovery time as well as over/undershoot are good indicative of the

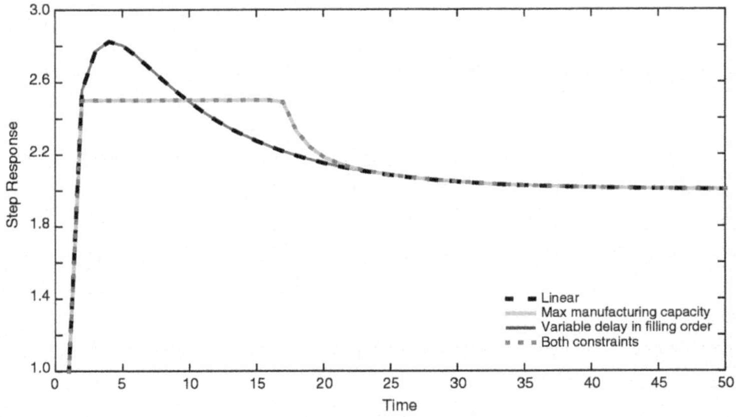

(a) Manufacturing order rate

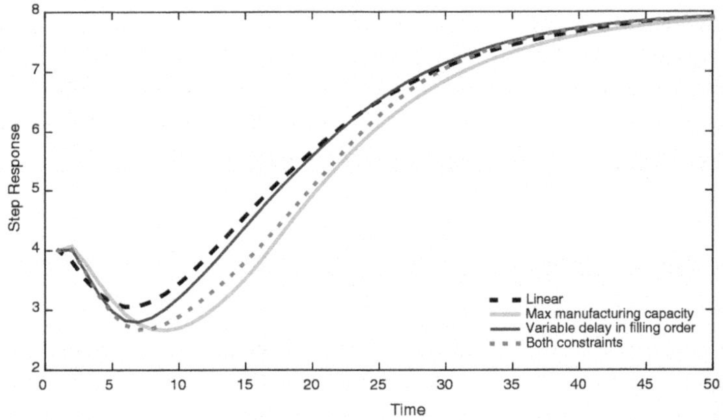

(b) Inventory level

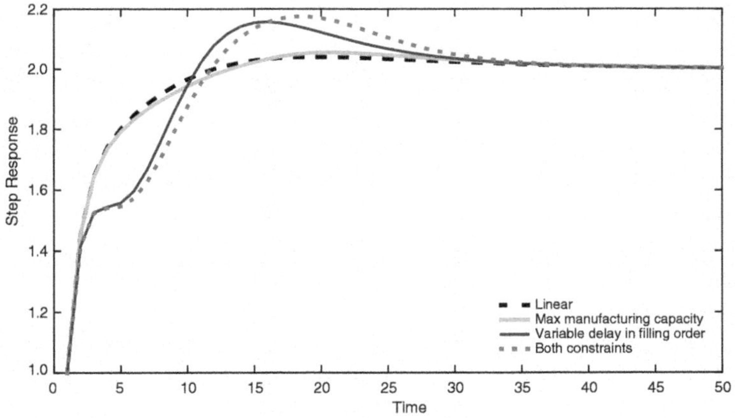

(c) Outbound shipment rate

Fig. 2 Example of non-linearity effects on transient responses system's resilience (Spiegler et al., 2012) and the diminishment in this performance can affect other players in the supply chain.

Table 3 summarises the current understanding of some types of non-linearities and their impact on the ripple effect. Discontinuous, single-valued non-linearities such as maximum capacity constraints, buying quantity constraints and non-negativities have been predominantly studied by describing function methods (Lin et al., 2018, 2023; Spiegler & Naim, 2017; Spiegler et al., 2016a, b; Wang et al., 2015). This method enabled understanding of the impact of such non-linearities on system output responses, for example manufacturing and supplier orders and production rates. Although this non-linearity does not provoke a shift in the output phase, it will change the output's mean and amplitude. These distortions increase the complexity of supply chain dynamics, making it difficult for supply chains to respond and recover from disruptions, therefore potentially aggravating the ripple effect. For instance, studies on fixed manufacturing capacity suggest that this non-linearity decreases the amplification of manufacturing orders, consequently decreasing the Bullwhip effect. However, its impact on the manufacturing output mean can slow down the ripple effect mitigation process. In the case of asymmetrical non-linearities, such as in Lin et al. (2018), the output will be relative depending on the relationship between the minimum (non-negative) and maximum capacity constraints. If the mean of the orders received is less than half of the maximum capacity, then the non-negative order boundary dominates. This leads to an increase in average inventory level and orders, therefore increasing costs. If the mean of the desired orders exceeds half of the maximum capacity, then the dominant impact on system dynamics will be the capacity constraint rather than the non-negative order low boundary. Under such condition, mean gain will increase with demand amplitude, leading to a decrease in average inventory level and orders, therefore increasing the risk of disruption.

Describing functions have also been used to analyse discontinuous multi-valued non-linearities such as variable shipment constraints due to changes in customer orders and inventory levels (Spiegler & Naim, 2017; Spiegler et al., 2016a, b). For low-frequency orders, this dynamic capacity constraint can decrease the output's amplitude and mean and shift the output's phase, making the output response lag behind the input. Hence, disruptions are less likely to affect supply chains with high- and medium-frequency demands. Ripple effect mitigation strategies would include encouraging high-frequency purchasing by developing resilient demand management strategies. In Lin et al. (2018a, b, 2022), a similar non-linearity is applied to switch between 'push' and 'pull' production modes, but the authors decided to evaluate both modes separately through transfer function analysis. This analysis was not able to capture the effect of the switch non-linearity, but the authors were able to conclude that when the upstream operates in make-to-stock mode, other capacity constraints can reduce the bullwhip effect at the expense of increased inventory variability at the expense of decreased resilience.

Only single-valued continuous non-linearities have been identified in the production and inventory control literature, therefore only Taylor series expansion has been applied as a small perturbation method to predict non-linear responses of

Table 3 Summary of rippling effects of non-linearities in production and inventory control systems

Source	Non-linearity	Non-linearity type	Method used	Non-linearity output	Effect on output			Potential consequences to Ripple effect
					Amp.	Mean	Phase	
Spiegler et al. (2016b)	Manufacturing constraint	Discontinuous, single-valued	Describing function	Manufacturing order	↓	↓	0	Although manufacturing constraints decrease the amplification of manufacturing orders, forcing the production manager to prioritise a level production strategy, it substantially decreases the mean of the output, suggesting difficulties in responding to sudden changes in demand and lead time. Hence, this non-linearity can cause disruptions that may cascade downstream affecting all other companies in the supply chain
Spiegler et al. (2016a, b) and Spiegler and Naim (2017)	Shipment constraint	Discontinuous, multi-valued	Describing function	Shipment	↓[a]	↓[a]	←[a]	This dynamic capacity constraint in the outbound shipment occurs due to variation in inventory levels and customer orders. A decrease in the output's amplitude and mean is observed and a phase shift makes the output response lag behind the input, only for low-frequency orders. Hence, disruptions are less likely to affect supply chains with high and medium frequency demands

Jeong et al. (2000) and Spiegler et al. (2016b)	Time-varying delay in filling orders	Continuous, single-valued	Small perturbation with Taylor series	Shipment	↓↑	↓↑	0	This continuous non-linearity can cause different impacts on the system output response depending on the parameter choice. Linearisation with Taylor series expansion enables to determine parameter settings that minimises disruptions
Spiegler et al. (2016a)	Buying quantity constraint	Discontinuous, single-valued	Describing function	Supplier order	↓	↑	0	A combination of non-negativity and batching constraints made this non-linearity very complex to analyse. Insights obtained in the work suggest that products with the same demand pattern should be grouped together to determine order quantities that minimises disruptions
Wang et al. (2015), Spiegler and Naim (2017) Lin et al. (2023)	Forbidden return of orders	Discontinuous, single-valued	Describing function	Supplier order	↓	↑	0	Non-negative constraints in the ordering rule, known as forbidden return constraint, can cause limit cycles, which are oscillations intrinsic to the non-linear production and inventory control system itself and not imposed by the demand. This problem may be exacerbated as this information signal propagates upstream in the supply chain. Distorted orders sent upstream are more likely to cause disruptions through their backlash downstream

(continued)

Table 3 (continued)

Source	Non-linearity	Non-linearity type	Method used	Non-linearity output	Effect on output			Potential consequences to Ripple effect
					Amp.	Mean	Phase	
Wikner et al. (1992) and Naim et al. (2017)	Time-varying delay in filling orders	Continuous, single-valued	Averaging technique + simulation	Shipment	Not able to capture			Although the method was unable to understand the impact of this non-linearity on system response, averaging technique enabled the researcher to propose a design that minimises disruptions and cost
Wang and Gunasekaran (2017)	Resource availability	Continuous, single-valued	Small perturbation with Taylor series	Production rate	↓	↓	0	The effective production rate is related to the amount of available resources in the current environment. As available resources decrease, the production rate decreases in a non-linear fashion, increasing the risk of disruptions
Wang and Disney (2012) and Wang et al. (2014)	Forbidden return of orders	Discontinuous, single-valued	Lyapunov-based stability analysis for piecewise-linear systems	Supplier order	Not able to capture			Although the method was not able to provide an understanding of the impact of this non-linearity on system's responses, it enables to undertake an in-depth stability analysis of the system and recommendations for parameter settings
Lin et al. (2018b)	Manufacturing constraint (non-neg + max capacity)	Discontinuous, single-valued	Averaging technique + simulation	Production rate	↓	↓ ↑ 0[b]	0	The method enables to compare the results between linear and non-linear assumptions under the same control policy settings. A slower recovery speed of assembly work in process and production rate is observed in the non-linear environment due to the effect of non-linearity on the output's amplitude

Lin et al. (2018a, b, 2022)	Push–pull decision	Discontinuous, multi-valued	Transfer function analysis of different modes	Assembly rate	Not able to capture			Very complex non-linearity that was analysed only by transfer function analysis of different operating modes. Insights suggest that the non-linearity would behave similarly to the shipment constraint in Spiegler et al. (2016a, b) and Spiegler and Naim (2017) due to limited recovery inventory. However, further research needs to be done. Other insights suggest that when upstream operates in make-to-stock mode, non-linearities can reduce the bullwhip effect at the expense of increased inventory variability
Lin et al. (2018a)	Rate change between backlog level and shipment rate	Continuous, single-valued	Small perturbation with Taylor series	Lead time	↓↑	↓↑	0	This continuous non-linearity can cause different impacts on the system output response depending on the parameter choice. Linearisation with Taylor series expansion enables to determine parameter settings that minimises disruptions

↓ decrease; ↑ increase; ← lag;
[a]only for low frequencies;
[b]if symmetrical

continuous constraints, such as time-varying delays, resource depletion and real lead-time estimation (Jeong et al., 2000; Lin et al., 2018a; Spiegler et al., 2016b; Wang & Gunasekaran, 2017). In Spiegler et al. (2016b); Jeong et al. (2000); Lin et al. (2018a), the non-linear differential equation involved more than one variable; hence, the impact on the output's amplitude and mean will depend on the input and parameter settings. However, linearisation with Taylor series expansion enabled the determination of parameter settings that minimise operational disruptions and increase supply chain resilience. In the case of Wang and Gunasekaran (2017), the non-linear differential equation is used to represent the depletion of resources that will certainly have a negative impact on the production rate and therefore on the ripple effect. The analysis of this non-linearity can shed light on how to best allocate scarce resources to ensure seamless flows of information and material.

It is worth mentioning that other authors have used analytical methods to analyse non-linearities such as averaging (Naim et al., 2017; Wikner et al., 1992) and stability methods (Wang et al., 2014; Wang & Disney, 2012), which even though was not able to capture the effect of non-linearities on system's responses, it allowed establishing parameter settings that minimise cost and disruptions and meet stability requirements indispensable for supply chain resilience. Similarly, Cuong et al. (2020) combined stability methods with phase plane plots and simulation to perform a dynamical analysis of a complex maritime supply chain system, gaining insights into periodicity, stability and chaotic behaviours caused by non-linearities. Finally, they used sliding mode control to design controllers to overcome parameter uncertainties and external disturbances.

5 Conclusion and Future Scope

This chapter has revisited the literature on non-linear control theory application in supply chain management. The chapter instructed readers on the different types of non-linearities that can commonly appear in supply chain systems and provoke undesirable dynamic behaviours. A number of methods and references to their application have been introduced and key findings on the rippling effect of non-linearities have been discussed. From the main points discussed in this chapter, we outline the following directions for future research:

1. **Supply chain structural development**: There is an opportunity to explore the effect of different capacity constraints to devise tactical and strategic plans regarding potential adjustments in supply chain structure such as investment in infrastructure and resource efficiency and flexibility. Previous research suggests that adequate capacity levels can help attain desired supply chain performance; therefore a performance-based structure plan can help companies to make right investments depending on their focus: customer service, cost efficiency, risk management and so on. For ripple effect mitigation, this performance-based structure plan should include all members of the supply chain.

2. **Supply chain design development**: There is an important avenue for future research regarding the deliberate employment of non-linearities for improved system's design. Computer simulation enabled the researchers to increase model accuracy and validation to better represent reality. However, the analytical methods presented here bring us one step forward in unravelling the mechanisms of non-linear supply chain dynamics. This knowledge can be used as an advantage in the improvement of the supply chain performance, from operational, economic and environmental viewpoints.
3. **Continuous, multi-valued non-linearities**: Despite not being referenced in the supply chain literature, continuous non-linearities with memory can appear in supply chain models, for example, in circular economy supply chains where there is the uncertainty of the volume, timing and quality of both demand and returns; therefore, multiple inputs should be considered. Hence, future research can build on previous efforts and discoveries to identify new non-linearities in supply chain systems to investigate which limitations and constraints they represent and their effect on system's dynamics.
4. **Discontinuous, multi-valued non-linearities**: A limited study has explored the multi-valued discontinuous non-linearities analytically, even though some insights are obtained by the simulation approach. Discontinuous non-linearities with memory are very common in supply chain systems. For example, the shipment constraint in assemble-to-order systems due to the limited customer order decoupling point (CODP) inventory. Also, the constrained remanufacturing order rate in hybrid manufacturing/remanufacturing systems, driven by the availability of recoverable inventory (limited returned products), is also a multi-valued non-linearity. Future research should analytically predict its impact on the ripple effect and suggest the corresponding control strategies in mitigating the unwanted dynamic behaviour.
5. **Effect of lead-time disturbances**: Most research in the application of both linear and non-linear control theories in supply chain management focuses on understanding the impact of demand uncertainty and on improving demand forecasting methods. Lead-time fluctuations can lead to performance degradation and increased production costs, just as demand uncertainties can Dolgui et al. (2013) and disturbances and uncertainties in production and supply lead times are reported to be the main sources of supply chain risk Colicchia et al. (2010). Supply chain control theorists have avoided tackling lead-time disturbance under the assumption that models become non-linear.

References

Christopher, M., & Peck, H. (2004). Building the resilient supply chain. *International Journal of Logistics Management, 15*(2), 1–14.

Colicchia, C., Dallari, F., & Melacini, M. (2010). Increasing supply chain resilience in a global sourcing context. *Production Planning & Control: The Management of Operations, 21*(7), 680–694.

Cook, P. A. (1986). *Nonlinear dynamical systems*. Prentice.

Cross, R., Grinfeld, M., & Lamba, H. (2009). Hysteresis and economics. *IEEE Control Systems, 29*(1), 30–43.

Cuong, T. N., Xu, X., Lee, SD. And You, SS. (2020) Dynamic analysis and management optimization for maritime supply chains using nonlinear control theory. Journal of International Maritime Safety, Environmental Affairs and Shipping, 4 (2), 48–55.

Dolgui, A., Ben-Ammar, O., Hnaien, F., & Louly, M. A. (2013). A state of the art on supply planning and inventory control under lead time uncertainty. *Studies in Informatics and Control, 22*(3), 255–268.

Dolgui, A., Ivanov, D., & Sokolov, B. (2018). Ripple effect in the supply chain: An analysis and recent literature. *International Journal of Production Research, 56*(1–2), 414–430.

Edwards, C., & Spurgeon, S. K. (2018). *Sliding mode control: Theory and applications*. CRC Press.

Evans, G. N., & Naim, M. M. (1994). The dynamics of capacity constrained supply chain. In *Proceedings of international system dynamics conference* (pp. 28–39). Stirling.

Gelb, A., & Velde, W. V. (1968). *Multiple-input describing functions and nonlinear system design*. McGraw-Hill.

Göcke, M. (2002). Various concepts of hysteresis applied in economics. *Journal of Economic Surveys, 16*(2), 167–188.

Gray, J. V., Skowronski, K., Esenduran, G., & Rungtusanatham, M. J. (2013). The reshoring phenomenon: What supply chain academics ought to know and should do. *Journal of Supply Chain Management, 49*(2), 27–33.

Größler, A., & Schieritz, N. (2005). Of stocks, flows, agents and rules-"strategic" simulations in supply chain research. In H. Kotzab, S. Seuring, M. Müller, & G. Reiner (Eds.), *Research methodologies in supply chain management*. Physica-Verlag Heiedlberg.

Grübbstrom, R., & Wang, Z. (2000). Introducing capacity limitations into multi-level, multi-stage production-inventory systems applying the input-output: Laplace transform approach. *International Journal of Production Research, 38*(17), 4227–4234.

Ivanov, D. (2020). Predicting the impacts of epidemic outbreaks on global supply chains: A simulation-based analysis on the coronavirus outbreak (Covid-19/Sars-CoV-2) case. *Transportation Research Part E, 136*, 101922.

Ivanov, D., & Dolgui, A. (2020). Viability of intertwined supply networks: Extending the supply chain resilience angles towards survivability. A position paper motivated by Covid-19 outbreak. *International Journal of Production Research, 58*(10), 2904–2915.

Ivanov, D., & Sokolov, B. (2013). Control and system-theoretic identification of the supply chain dynamics domain for planning, analysis, and adaptation of performance under uncertainty. *European Journal of Operational Research, 224*(2), 313–323.

Ivanov, D., Sokolov, B., & Dolgui, A. (2014). The ripple effect in supply chains: Trade-off 'efficiency-flexibility- resilience' in disruption management. *International Journal of Production Research, 52*(7), 2154–2172.

Jeong, S., Oh, Y., & Kim, S. (2000). Robust control of multi-echelon production-distribution systems with limited decision policy. *KSME International Journal, 14*(4), 380–392.

Johansson, M. (2003). *Piecewise linear control systems*. Springer.

Kearney, A. T. (2003). Supply chains in a vulnerable, volatile world. *Executive Agenda, 6*(3), 5–16.

Kogut, B., & Kulatilaka, N. (1994). Operating flexibility, global manufacturing, and the option value of a multinational network. *Management Science, 10*, 123–139.

Kouvelis, P. (1998). Global sourcing strategies under exchange rate uncertainty. In S. Tayur, R. Ganeshan, & M. Magazine (Eds.), *Quantitative models for supply chain management*. Springer.

Lang, D., & de Peretti, C. (2009). A strong hysteretic model of Okun's law: Theory and a preliminary investigation. *International Review of Applied Economics, 23*(4), 445–462.

Lin, J., Huang, H., Li, S., & Naim, M. (2023). On the dynamics of order pipeline inventory in a nonlinear order-up-to system. *International Journal of Production Economics, 266*, 109061.

Lin, J., Naim, M., & Spiegler, V. (2018a). Delivery time dynamics in an assemble-to-order inventory and order based production control system. In *Presented at: 20th International Working Seminar on Production Economics, Innsbruck, Austria, 19–23 February*, 2018.

Lin, J., Spiegler, V., & Naim, M. (2018b). Dynamic analysis and design of a semiconductor supply chain: A control engineering approach. *International Journal of Production Research, 56*, 4585–4611.

Lin, J., Zhou, L., Spiegler, V., Naim, M., & Syntetos, A. (2022). Push or pull? The impact of ordering policy choice on the dynamics of a hybrid closed-loop supply chain. *European Journal of Operational Research, 300*(1), 282–295.

Manners-Bell, J. (2018). *Supply chain risk management: Understanding emerging threats to global supply chains*. Kogan Page.

Mohapatra, P. K. J. (1980). Nonlinearity in system dynamics models. *Dynamica, 6*(1), 36–52.

Naim, M. M., Spiegler, V. L., Wikner, J., & Towill, D. R. (2017). Identifying the causes of the bullwhip effect by exploiting control block diagram manipulation with analogical reasoning. *European Journal of Operational Research, 263*, 240–246.

Odame, K., & Hasler, P. E. (2010). Nonlinear circuit analysis via perturbation methods and hardware prototyping. *VLSI Design, 2010*, 1–10.

Rugh, W. J. (2002). *Nonlinear system theory: The Volterra/wiener approach*. The Johns Hopkins University Press.

Shtessel, Y., Edwards, C., Fridman, L., & Levant, A. (2014). *Sliding mode control and observation*. Birkhauser, Springer.

Shu, W., & Fan, D. (2024). How do firms perceive and react to extreme weather risk in their supply bases. *International Journal of Production Economics, 268*, 109125.

Slotine, J. J. E., & Li, W. (1991). *Applied nonlinear control*. Prentice Hall.

Spiegler, V. L. M., Naim, M. M., & Wikner, J. (2012). A control engineering approach to the assessment of supply chain resilience. *International Journal of Production Research, 50*(21), 6162–6187.

Spiegler, V. L. M., Potter, A. T., Naim, M. M., & Towill, D. R. (2016a). The value of nonlinear control theory in investigating the underlying dynamics and resilience of a grocery supply chain. *International Journal of Production Research, 54*(1), 265–286.

Spiegler, V. L., Naim, M. M., Towill, D. R., & Wikner, J. (2016b). A technique to develop simplified and linearised models of complex dynamic supply chain systems. *European Journal of Operational Research, 251*(3), 888–903.

Spiegler, V., & Naim, M. (2017). Investigating sustained oscillations in nonlinear production and inventory control models. *European Journal of Operational Research, 261*, 572–583.

Spiegler, V., Naim, M., Syntetos, A., & Zhou, L. (2018). Understanding the effects of lead-time disturbance in supply chains: on the design of an adaptive model to cope with lead-time changes. In *The 60th Conference of Operational Research (OR60), 11–13 September*. UK: Lancaster.

Tsang, Y. P., Fan, Y., Feng, Z. P., & Li, Y. (2024). Examining supply chain vulnerability via an analysis of ESG-prioritized firms amid the Russian-Ukrainian conflict. *Journal of Cleaner Production, 434*, 139754.

Vukic, Z., Kuljaca, L., Donlagic, D., & Tesaknjak, S. (2003). *Nonlinear control systems*. Marcel Dekker.

Wang, G., & Gunasekaran, A. (2017). Modeling and analysis of sustainable supply chain dynamics. *Annals of Operations Research, 250*, 521–536.

Wang, X., & Disney, S. M. (2012). Stability analysis of constrained inventory systems. *European Journal of Operational Research, 223*, 86–95.

Wang, X., Disney, S., & Wang, J. (2014). Exploring the oscilatory dynamics of a forbidden returns inventory system. *International Journal of Production Economics, 147*(A), 3–12.

Wang, Z., Wang, X., & Ouyang, Y. (2015). Bounded growth of the bullwhip effect under a class of nonlinear ordering policies. *European Journal of Operational Research, 247*(1), 72–82.

Wikner, J., Naim, M. M., & Rudberg, M. (2007). Exploiting the order book for mass customized manufacturing control systems with capacity limitations. *IEEE Transactions on Engineering Management, 54*(1), 145–155.

Wikner, J., Naim, M. M., & Towill, D. R. (1992). The system simplification approach in understanding the dynamic behaviour of a manufacturing supply chain. *Journal of Systems Engineering, 2*, 164–178.

Zhao, Y., Zhao, C., He, M., & Yang, C. (2016). A state-feedback approach to inventory control: Analytical and empirical studies. *Production and Operations Management Society, 25*(3), 535–547.

Bullwhip Effect of Multiple Products with Interdependent Product Demands

Srinivasan Raghunathan, Christopher S. Tang, and Xiaohang Yue

Abstract The bullwhip effect has been studied extensively by researchers using analytical and empirical models based on a single product. We extend the current theory to provide insights for a firm that manufactures multiple products in a single product category with interdependent demand streams. We find that interdependency between demand streams plays a critical role in determining the existence and magnitude of the bullwhip effect. More importantly, we show that interdependency impacts whether the firm should manage ordering and inventory decisions at the category level or at the product level and whether the bullwhip effect measure computed at the category level is informative or not.

1 Introduction

For decades, many researchers have examined different ways to develop processes and strategies to mitigate supply chain risks using mathematical models and data analytical models. Several books (Gurnani et al., 2012; Handfield & McCormack, 2008; Heckmann, 2016; Ivanov, 2018; Khojasteh, 2017; Kouvelis & Dong, 2011; Sodhi & Tang, 2012; Waters, 2011; Zsidisin & Henke, 2019) and review articles (Fahimnia et al., 2015; Gupta et al., 2016; Ho et al., 2015; Klibi et al., 2010; Simangunsong et al., 2012) provide insightful overview and introduction to different aspects of this important field.

S. Raghunathan (✉)
School of Management, The University of Texas at Dallas, Richardson, TX, USA
e-mail: sraghu@utdallas.edu

C. S. Tang
UCLA Anderson School, Los Angeles, CA, USA
e-mail: chris.tang@anderson.ucla.edu

X. Yue
Lubar School of Business, The University of Wisconsin-Milwaukee, Milwaukee, WI, USA
e-mail: xyue@uwm.edu

D. Ivanov et al. (eds.), *Handbook of Ripple Effects in the Supply Chain*,
International Series in Operations Research & Management Science 276,
https://doi.org/10.1007/978-3-031-85508-5_4

By and large, supply chain risks can be categorized into two types: operational risk and disruption risk. Extensive research over the last few decades has shown that operational risks associated with stable variations (such as changes in consumer preferences, yield fluctuations, or delivery delays) can be managed effectively. However, disruption risks, which arise from high-impact events with significant uncertainties, continue to pose a major challenge for researchers and practitioners.

As supply chains become more vulnerable, firms are paying close attention to both operational risks and disruption risks (Chopra et al., 2007). These risks significantly impact supply chain performance, including sales, stock return, service level, and costs. The disruptions caused by the COVID-19 pandemic in 2020, the Houthi attacks in the Red Sea in 2024, and the export bans on various products between the US and China since 2021 serve as evidence of the challenges posed by these risks (Ivanov, 2021).

Some researchers (e.g., Hendricks & Singhal, 2005) have quantified the negative effects of supply chain disruption through empirical analysis and found 33–40% lower stock returns relative to their benchmarks over a 3-year time period that started 1 year before and ended 2 years after a disruption. Disruption risk represents a new challenge for supply chain managers who face the "ripple effect" (Dolgui et al., 2018; Ivanov, 2017; Ivanov et al. 2014a, b, 2017). The ripple effect deals with structural disruptions in the supply chain, while the bullwhip effect deals with parametrical deviations in the supply chain.

The bullwhip effect examines weeks/daily demand and lead time fluctuations as primary drivers of the changes in the supply chain. It occurs at the parametric level, and it can be mitigated so that the operations can be restored on a short-term basis. In recent years, the research community has started to investigate severe supply chain disruptions caused by natural disasters, political conflicts, terrorism, maritime piracy, economic crises, the destruction of information systems or transportation infrastructure failures. Some researchers (Ivanov, 2018) refer to these severe natural and man-made disasters as the ripple effect in the supply chain where changes in the supply chain occur at the structural level and recovery may take mid and long-term periods with significant impact on output performance such as annual revenues.

Despite the difference between the ripple effect and the bullwhip effect, there is a linkage between them. Although we focus on how to mitigate the impact of the bullwhip effect of multiple products in a single product category with interdependent product demand streams in this chapter, our model and analysis can be used to examine the ripple effect if we view different products as country-specific products. Then, as the demand for a country-specific product is disrupted due to an economic crisis occurring in one country, the implication of this disruption can affect the demand for other products associated with other countries, especially when the world economies are connected. From this vantage point, our bullwhip effect can be interpreted at a "higher" level so that the net effect is akin to the ripple effect. For this reason, we shall focus our discussion on the bullwhip effect in the remainder of this chapter, even though the result can be interpreted as the ripple effect in the context of a supply chain that produces country-specific products where each of the products is subject to uncertain disruptions arising from different counties.

In the bullwhip effect literature, the seminal work of Lee et al. (1997) examined different drivers of the bullwhip effect—i.e., the amplification of variance of demand as it propagates upstream through the supply chain. Lee et al. employed the autoregressive model of order one (AR (1)) in their analysis of the bullwhip effect of a single product. Subsequent papers examined the bullwhip effect using a variety of demand models, forecasting methods, and inventory policies, but they focused primarily on the single product setting (e.g., Aviv, 2003; Chen et al., 2000; Chen & Lee, 2009; Gaur et al., 2005; Gilbert, 2005; Graves, 1999; Miyaoka & Hausman, 2004; Zhang, 2004a; Zhang, 2004b).[1] Empirical results have been mixed about the prevalence of the bullwhip effect. Using firm-level data, Bray and Mendelson (2012) found that two-thirds of the publicly traded firms exhibit a bullwhip effect, but using industry-level data, Cachon et al. (2007) found no significant bullwhip effect in many industries. Chen and Lee (2012) analytically showed that time and product aggregation can mask the bullwhip effect, while Bray and Mendelson reached similar conclusions using empirical data. The current theory of supply chain bullwhip effect has been developed using single product models. Chen and Lee considered multiple products to examine the impact of product aggregation but did not model or analyze interdependency among individual product demand streams. In this work, we extend the current theory to provide new insights for a firm that manufactures multiple products with interdependent demands in a product category.

The motivation for this work stems from product proliferation commonly observed in most industries. Product proliferation occurs when a firm introduces multiple variants or brands of the same core product. The dramatic increase in product variety over the last three decades in numerous product categories has been well documented by Cox and Alm (1998) and Aichner and Coletti (2013). A large product variety is also observed among contract manufacturers in industries such as electronics and automobiles, who supply components to multiple original equipment manufacturers. Product proliferation poses operational challenges to firms for a variety of reasons. For instance, because the different product variants within a product category are (imperfect) substitutes of each other, their demands tend to be interdependent in the sense they are not only contemporaneously correlated in a given time period but also serially correlated across time periods. The contemporaneous correlation is natural because these are variants of the same product. The serial correlation across different products arises because of consumer-driven factors and firm-driven factors. For example, the well-known variety-seeking behavior of some consumers results in those consumers switching the product variant they buy over time (for example, switching from Kellogg's Corn Flakes in 1 week to Rice Krispies in the following week). Furthermore, firms may promote different product variants at different time periods, which can also contribute to serial correlation (for example, distributing coupons for Corn Flakes in 1 month and

[1] See Wang and Disney (2016) for a comprehensive review of the bullwhip effect literature.

for Rice Krispies in the following month). On the operational side, while some firms manage forecasting, production, ordering, and inventory of these products at the individual product level, others manage them at the product category level. Though these aspects capture the essence of many current real-world scenarios, they have not been adequately addressed in the extant bullwhip effect literature.[2]

We employ a vector autoregressive model of order one (VAR(1)) that captures the interdependency among product demands across periods using cross-correlation coefficient and interdependency within a period using contemporaneous (or spatial) correlation coefficient, in addition to the within-product demand dependency using autocorrelation coefficient.[3] VAR(1) model is a natural extension of the widely used AR(1) model to model demands of multiple products and has been studied and applied in a variety of multiproduct demand forecasting contexts (Box & Tiao, 1977; Chen & Blue, 2010; Sbrana & Silvestrini, 2013; Tiao & Box, 1981; Tiao & Tsay, 1983).[4] In this chapter, we first derive the bullwhip effect formula by assuming that the firm manages the ordering and inventory decisions at the product level and use the formula to obtain the impacts of key model parameters. The well-known result that an increase in the autocorrelation coefficient amplifies the bullwhip effect (Lee et al., 1997) continued to hold in the VAR(1) model, but more importantly, the result extends to the cross-correlation coefficient. The impact of contemporaneous correlation and the number of products on the bullwhip effect depend critically on whether the cross-correlation is positive or negative. When the cross-correlation is negative (positive), we show that an increase in contemporaneous correlation decreases (increases) the bullwhip effect, and an increase in the number of products increases (decreases) the bullwhip effect. That is, demand pooling and negative contemporaneous correlation do not necessarily mitigate the product-level bullwhip effect.

Subsequently, we derive the bullwhip effect formula when the firm manages the ordering and inventory decision at the product category level. We show that category-level bullwhip effect is increasing in the autocorrelation, cross-correlation, contemporaneous correlation, and the number of firms. Therefore, the conventional intuitions about the impact of these factors on the bullwhip effect persist at the category-level bullwhip measure. Furthermore, we also show that the bullwhip effect

[2] There is extensive literature on aggregate and individual product-level forecasting, commonly known as top-down and bottom-up forecasting, in the operations management literature (e.g., Sbrana & Silvestrini, 2013; Chen & Blue, 2010 are some recent studies that compare the forecast accuracy of top-down and bottom-up approaches).

[3] This work is based on Raghunathan et al. (2017). All proofs of propositions are available in Raghunathan et al. (2017).

[4] While we do not have firm-level data to test whether the VAR(1) model fits brand demand streams, an examination of the data contained in the Compustat database shows that the VAR(1) model provides a statistically significant fit for the cost of goods sold (a widely used proxy for demand in the bullwhip effect literature) of firms within an industry. Furthermore, Chen and Blue (2010) gave an example of a semiconductor manufacturer where the VAR(1) model is appropriate to model the demand data of two product variants.

when the firm manages at the category level is identical to the bullwhip effect that is computed (or measured) by aggregating the demand and order data for all products even though the ordering and inventory decisions are made at the individual product level. This result demonstrates that correct implications cannot be derived based on the aggregate bullwhip effect measure without knowledge about how the products are actually managed within the firm.

Finally, we compare the product-level and category-level bullwhip effects. Whether the category-level bullwhip effect is greater than or smaller than the product-level bullwhip effect depends critically on the cross-correlation coefficient—the product-level bullwhip effect is smaller (larger) than the category-level bullwhip effect when the cross-correlation coefficient is highly positive (negative). Therefore, for a multiproduct firm, from the bullwhip effect point of view, the cross-correlation plays a vital role in the firm's choice regarding category-level versus product-level management of ordering and inventory decisions. Furthermore, the cross-correlation also plays a role in determining whether the bullwhip effect measured using aggregate data masks or exaggerates the underlying bullwhip effect if the firm adopts product-level management.

2 The Model

We consider a manufacturer who produces and sells $n \geq 2$ products in a product category. The demand for product i in period t takes on the following form:

$$D_{it} = d + \rho D_{i(t-1)} + \gamma \sum_{j=1}^{n} \frac{D_{j(t-1)}}{n} + \varepsilon_{it} \tag{1}$$

where D_{it} is product i's demand in period t, $\overline{D}_{(t-1)} = \sum_{j=1}^{n} D_{j(t-1)}$ is the product category's demand in period t-1, ρ is the autocorrelation coefficient, γ is the cross-correlation coefficient, ε_{it} is the error term which is identically normally distributed with mean zero and variance σ^2 for all i and t, ε_{it} for a given i is independent across time, and the contemporaneous correlation coefficient between ε_{it} and ε_{jt} is ω, $\forall\, i \neq j$. Of course, $-1/(n\text{-}1) < \omega < 1$ to ensure variance positivity. Thus, the demand streams for the products can be expressed as the following VAR(1) process:

$$\begin{bmatrix} D_{1t} \\ D_{2t} \\ \vdots \\ D_{nt} \end{bmatrix} = \begin{bmatrix} d \\ d \\ \vdots \\ d \end{bmatrix} + \begin{bmatrix} \rho+(\gamma/n) & \gamma/n & \cdots & \gamma/n \\ \gamma/n & \rho+(\gamma/n) & \cdots & \gamma/n \\ \vdots & \gamma/n & \ddots & \gamma/n \\ \gamma/n & \gamma/n & \cdots & \rho+(\gamma/n) \end{bmatrix} \begin{bmatrix} D_{1(t-1)} \\ D_{2(t-1)} \\ \vdots \\ D_{n(t-1)} \end{bmatrix} + \begin{bmatrix} \varepsilon_{1t} \\ \varepsilon_{2t} \\ \vdots \\ \varepsilon_{nt} \end{bmatrix}. \tag{2}$$

The manufacturer procures the raw materials required for production from a supplier. Without loss of generality, we assume that each unit of a product requires one unit of raw material. We consider two potential replenishment strategies that the manufacturer could adopt. Under the *product-level* replenishment strategy, the ordering, replenishment, and inventory decisions are managed separately for each product, as if there is a separate upstream supply chain for each product. Under the *category-level* replenishment strategy, the decisions are made for the whole product category, as if there is a single upstream supply chain for all products combined. Regardless of the replenishment strategy used, we assume that the manufacturer adopts the optimal state-dependent base-stock policy to determine the order quantity (Chen et al., 2000; Lee et al., 1997).

The key aspects of the replenishment process are the following. Let L denote the replenishment lead time of raw materials, which we assume to be the same for all products. At the beginning of period t, the demands for all products in (t-1) are observed, and all orders made in period t-L are received from the supplier. Then, either an order z_{it} for product i (for a total of n orders) or a single order of $\overline{z}_t$ for the entire product category is placed with the supplier depending on the replenishment strategy used by the manufacturer. The production of products occurs during the period and demands for all products are realized and satisfied at the end of t. There is no system capacity constraint, no order batching, and free return of excess inventory. Inventory-related costs are proportional and are identical across products. Since the focus of this study is demand interdependency, we abstract away many operational details. For instance, we assume symmetric products to eliminate the impact of product-specific characteristics from our findings.

Finally, we impose the following *stationarity condition* to make the demand process covariance stationary.

Stationarity Assumption: $|\rho| < 1$ *and* $|\rho + \gamma| < 1$.

3 Derivation and Analysis of the Bullwhip Effect

Suppose the manufacturer uses the product-level replenishment strategy. Then, to compute the optimal base-stock level for a product, we need the expected value and variance of its demand over lead time L. This requires the forecasts, over the lead time, for not only the product's own demand but also the category demand. The joint demand process is used to perform minimum mean square error (MMSE) forecasting updates.

The joint demand process is derived as follows. Using (1), we compute the category demand $\overline{D}_t$ to be the following:

$$\overline{D}_t = \sum_{i=1}^{n} D_{it} == nd + (\rho + \gamma)\,\overline{D}_{(t-1)} + \sum_{i=1}^{n} \varepsilon_{it} \tag{3}$$

Note that the category demand follows the AR(1) process. Now, we can express the firm demand and the category demand jointly as the following 2 × 2 VAR(1) process:

$$\begin{bmatrix} D_{it}-\mu_D \\ \overline{D}_t-n\mu_D \end{bmatrix}=\begin{bmatrix} \rho & \gamma/n \\ 0 & \rho+\gamma \end{bmatrix}\begin{bmatrix} D_{i(t-1)}-\mu_D \\ \overline{D}_{t-1}-n\mu_D \end{bmatrix}+\begin{bmatrix} \varepsilon_{it} \\ \sum_{i=1}^{n}\varepsilon_{it} \end{bmatrix}, \tag{4}$$

where $\mu_D = d/(1-\rho-\gamma)$..

Let $\tau \geq 0$ be an integer, $\pi_k = \frac{(\rho+\gamma)^{k+1}-\rho^{k+1}}{\gamma}$ and $\Pi_\tau = \sum\limits_{k=0}^{\tau}\pi_k$. The future demand for product i at period $(t+\tau)$, $D_{i(t+\tau)}$, can be expressed as a function of last observed product demand $D_{i(t-1)}$ and category demand $\overline{D}_{t-1}$, and future errors such as:

$$\begin{bmatrix} D_{i(t+\tau)}-\mu_D \\ \overline{D}_{t+\tau}-n\mu_D \end{bmatrix}=\begin{bmatrix} \rho^{\tau+1} & \rho\pi_\tau \\ 0 & (\rho+\gamma)^{\tau+1} \end{bmatrix}\begin{bmatrix} D_{i(t-1)}-\mu_D \\ \overline{D}_{t-1}-n\mu_D \end{bmatrix}$$
$$+\sum_{k=0}^{\tau}\begin{bmatrix} \rho^{\tau-k} & (\gamma/n)\pi_{\tau-k-1} \\ 0 & (\rho+\gamma)^{\tau-k} \end{bmatrix}\begin{bmatrix} \varepsilon_{i(t+k)} \\ \sum_{i=1}^{n}\varepsilon_{i(t+k)} \end{bmatrix}$$

The optimal MMSE τ-period-ahead forecast is given by the conditional expectation of future demand $D_{t+\tau}$ given $D_{i(t-1)}$ and $\overline{D}_{t-1}$. The above joint demand system yields the following MMSE forecasting equations:

$$E\left(D_{i(t+\tau)}|D_{i(t-1)},\overline{D}_{(t-1)}\right)=\mu_D+\rho^{\tau+1}\left(D_{i(t-1)}-\mu_D\right)+(\gamma/n)\pi_\tau\left(\overline{D}_{t-1}-n\mu_D\right) \tag{5}$$

The lead time demand forecast can then be obtained by summing the τ-period-ahead forecasts such as

$$E\left[\left(\sum_{\tau=0}^{L}D_{i(t+\tau)}\right)\middle| D_{i(t-1)},\overline{D}_{(t-1)}\right]=\sum_{\tau=0}^{L}E\left(D_{i(t+\tau)}\,|D_{i(t-1)},\overline{D}_{(t-1)}\right)$$
$$=n\mu_D+\rho\left(\frac{1-\rho^{L+1}}{1-\rho}\right)\left(D_{i(t-1)}-\mu_D\right)$$
$$+(\gamma/n)\Pi_L\left(\overline{D}_{t-1}-n\mu_D\right).$$

So, the optimal base-stock level for product i for period t, S_{it}^*, is given by

$$S_{it}^*=E\left[\left(\sum_{\tau=0}^{L}D_{i(t+\tau)}\right)\middle| D_{i(t-1)},\overline{D}_{(t-1)}\right]+K\sqrt{\operatorname{Var}\left(\sum_{\tau=0}^{L}D_{i(t+\tau)}|D_{i(t-1)},\overline{D}_{(t-1)}\right)}$$
$$=n\mu_D+\rho\left(\frac{1-\rho^{L+1}}{1-\rho}\right)\left(D_{i(t-1)}-\mu_D\right)+(\gamma/n)\Pi_L\left(\overline{D}_{t-1}-n\mu_D\right)$$
$$+K\sqrt{\operatorname{Var}\left(\sum_{\tau=0}^{L}D_{i(t+\tau)}|D_{i(t-1)},\overline{D}_{(t-1)}\right)}$$

where K represents the service-level coefficient. Therefore, the optimal order quantity z_{it}^* is given by

$$\begin{aligned} z_{it}^* &= S_{it}^* - S_{i(t-1)}^* + D_{i(t-1)} \\ &= \rho\left(\frac{1-\rho^{L+1}}{1-\rho}\right)\left(D_{i(t-1)} - D_{i(t-2)}\right) + (\gamma/n)\,\Pi_L\left(\overline{D}_{t-1} - \overline{D}_{t-2}\right) + D_{i(t-1)} \end{aligned} \tag{6}$$

Using the variances of z_{it}^* and D_{it}, we can compute the formula for the bullwhip effect associated with product *i*. Lemmas 1 and 2 provided in the appendix derive these variances. The bullwhip effect is commonly measured according to either the (unconditional) variance of order quantity minus the variance of demand, which we refer to as the *bullwhip difference*, or according to the (unconditional) variance of order quantity divided by variance of demand, which we refer to as *bullwhip ratio*. Since the upstream supply chain costs (e.g., inventory holding costs) are proportional to the square root of the order variance for a given demand model (Chen & Lee, 2012), bullwhip difference measure is useful when absolute costs associated with the bullwhip are desired. On the other hand, the ratio measure is useful when percentage costs associated with the bullwhip effect are desired. We examine the bullwhip difference in detail first, followed by an examination of the bullwhip ratio.

3.1 *Product-Level Bullwhip Effect: Bullwhip Difference*

Define the following for notational convenience and expositional clarity.

$$\theta = \left(1 - \frac{1}{n}\right)(1 - \omega);$$

$$M(\lambda) = \frac{2\lambda\left(1 - \lambda^{L+1}\right)\left(1 - \lambda^{L+2}\right)}{(1 - \lambda)\left(1 - \lambda^2\right)}\sigma^2$$

Proposition 1 *The product-level bullwhip difference is given by*

$$BWE_{diff}^{product} = Var\left(z_{it}^*\right) - Var\left(D_{it}\right) = \theta M(\rho) + (1 - \theta)\, M(\rho + \gamma) \tag{7}$$

Equation (7) is a generalization of the bullwhip difference derived by Lee et al. (Eq. (3.5)) for a single product AR(1) demand model. If $\gamma = 0$, Eq. (7) reduces to Lee et al.'s equation. We note that the product-level bullwhip difference is a convex linear combination of $M(\rho)$ and $M(\rho + \gamma)$. $M(\lambda)$ is the bullwhip difference in the case of an AR(1) demand stream with autocorrelation coefficient λ. Since the VAR(1) demand model we examine can be viewed as a linear combination of two AR(1) models—one with an autocorrelation coefficient of $\rho + \gamma$ representing the category demand and the other with an autocorrelation coefficient of γ representing the own demand—one could interpret $M(\rho)$ and $M(\rho + \gamma)$ as the bullwhip contribution of the product's own demand stream and that of the category demand stream, respectively. The parameter θ indicates the relative influence of the product's own demand stream vis-à-vis that of the category demand stream in determining a product's bullwhip, reflecting the differential effects of these two streams on a product's demand. Clearly, an increase in n or a decrease in ω increases the relative influence of the firm's own demand process on the bullwhip difference. When n increases or ω decreases, the variance of the category demand per product decreases because of the increase in the demand pooling effect, reducing the relative influence of the category demand on the product's bullwhip effect. When n approaches infinity, the influence of category demand stream and interdependency on a product's bullwhip vanishes.

Equation (7) reveals that a positive autocorrelation coefficient is not sufficient to ensure that the bullwhip effect exists; bullwhip exists only when $\frac{M(\rho)}{M(\rho+\gamma)} > \left(1 - \frac{1}{\theta}\right)$. A set of sufficient conditions for the product-level bullwhip effect to exist are $\rho > 0$ and $\rho + \gamma > 0$. These conditions represent an extension of Lee et al. who show that a positive autocorrelation coefficient is sufficient for the existence of the bullwhip effect in the single product case.

Proposition 2 *Suppose $\rho, \rho + \gamma > 0$. Then, (i) $BWE_{diff}^{product}$ is increasing in ρ, (ii) $BWE_{diff}^{product}$ is increasing in γ, (iii) $BWE_{diff}^{product}$ is decreasing in n if and only if $\gamma > 0$, and (iv) $BWE_{diff}^{product}$ is increasing in ω if and only if $\gamma > 0$.*

Proposition 2

(i) generalizes the well-known impact of autocorrelation on the bullwhip difference in the single product scenario to the multiproduct scenario with interdependent demands—an increase in autocorrelation amplifies the firm-level bullwhip difference in both models, assuming bullwhip effect exists. The result that an increase in cross-correlation also amplifies the product-level bullwhip difference is new to the literature. Interestingly, an increase in the number of products does not always decrease the product-level bullwhip difference. One effect of an increase in the number of products is that it decreases the variance of the category demand because of demand pooling, and one would expect this effect to decrease the product's bullwhip difference. However, we find that an increase in the number of products has an opposing effect also on a product's bullwhip difference in our context of interdependent demands—an increase in the number of products decreases the influence of category demand and increases the influence of product's own demand in the product's bullwhip. Whether the former effect or the latter effect dominates depends on whether $M(\rho) < M(\rho + \gamma)$. Only when $\gamma > 0$, $M(\rho) < M(\rho + \gamma)$, implying that the sign of cross-correlation can fundamentally alter the impact of the number of products on a product's bullwhip effect. The explanation for the impact of

ω on the product-level bullwhip difference stated as Proposition 2(iv) is analogous to the explanation regarding the impact of n. In essence, the conventional intuition that a decrease in contemporaneous correlation or an increase in the number of products enhances the demand pooling effect and reduces the variance of the category demand, and hence has a mitigating effect on the bullwhip difference does not hold if the cross-correlation is negative.

3.2 Category-Level Bullwhip Difference: Impact of Product Aggregation

Recent research on the bullwhip effect has shown that measuring the bullwhip effect using aggregate data instead of individual product-level data generally masks the underlying bullwhip effect (Chen & Lee, 2012). This study assumes that ordering decisions are made separately for each product, but the bullwhip effect measurement is done using aggregate data. In our context, product aggregation can be viewed from two perspectives. The first perspective is the one used by the prior literature, i.e., the manufacturer uses product-level replenishment strategy but the measurement of the bullwhip effect is made using aggregate data. The other perspective is that the manufacturer uses a category-level replenishment strategy and the measurement is also done using category-level data. The distinction between these two perspectives is important because the implications that can be drawn from the measured bullwhip are very different under the two perspectives. Under the first perspective, any incorrect implication we draw about the underlying bullwhip effect is attributed solely to the measurement problem. On the contrary, under the second perspective, the implications relate to the underlying replenishment strategy used by the manufacturer.

If the manufacturer uses a product-level replenishment strategy, using (6), we obtain the total order quantity as

$$\begin{aligned}\sum_{i=1}^{n} z_{it}^{*} &= \sum_{i=1}^{n}\Big[\rho\left(\tfrac{1-\rho^{L+1}}{1-\rho}\right)\left(D_{i(t-1)} - D_{i(t-2)}\right)\\ &\quad + (\gamma/n)\,\Pi_L\left(\overline{D}_{t-1} - \overline{D}_{t-2}\right) + D_{i(t-1)}\Big]\\ &= \tfrac{(\rho+\gamma)\left(1-(\rho+\gamma)^{L+1}\right)}{1-(\rho+\gamma)}\left(\overline{D}_{t-1} - \overline{D}_{t-2}\right) + \overline{D}_{t-1}.\end{aligned}$$

Hence, the bullwhip difference measured using the aggregate data is computed as $Var\left(\sum_{i=1}^{n} z_{it}^{*}\right) - Var\left(\overline{D}_{it}\right)$.

On the other hand, if the manufacturer uses a category-level replenishment strategy, the bullwhip difference is computed using the variance of the order quantity for the category, which is now determined using the optimal base-stock level for the whole category, and the variance of the category demand. Since the category demand follows an AR(1) process (recall Eq. (3)), we can compute the bullwhip difference using Eq. (3.5) of Lee et al. In the following result, we show that regardless of the replenishment strategy used, the category-level bullwhip difference is given by the following result.

Proposition 3 *Under both product-level replenishment and category-level replenishment strategies, the category-level bullwhip difference is given by*

$$BWE_{diff}^{catgeory} = (1-\theta)\, n^2 M\,(\rho + \gamma) \tag{8}$$

The finding that category-level bullwhip difference is the same regardless of the replenishment strategy used demonstrates that appropriate implications cannot be derived based on the category-level bullwhip difference without knowledge about how the products are managed within the firm. Specifically, if the manufacturer uses a category-level replenishment strategy, then the category-level bullwhip difference is informative in the sense it reveals the true underlying bullwhip effect; on the other hand, if the manufacturer uses a product-level replenishment strategy, then the category-level bullwhip difference is uninformative in the sense it may not reveal the true underlying bullwhip effect. It is easy to verify that if $\rho + \gamma > 0$, then, the category-level bullwhip effect exists.

Proposition 4 *Suppose $\rho + \gamma > 0$. Then, (i) $BWE_{diff}^{catgeory}$ is increasing in ρ, (ii) $BWE_{diff}^{catgeory}$ is increasing in γ, (iii) $BWE_{diff}^{catgeory}$ is increasing in n, and (iv) $BWE_{diff}^{catgeory}$ is increasing in ω.*

Proposition 4 is not surprising when we recognize that the variance of the category demand is increasing in all the parameters stated in the proposition. More importantly, contrasting Proposition 4 with Proposition 2 reveals that the counterintuitive impacts of n and ω on the product-level bullwhip effect when cross-correlation is negative are reversed at the category level. This result suggests that the impact of interdependency on the bullwhip difference is nontrivial and depends critically on the replenishment strategy used.

A key question is how product aggregation affects the bullwhip difference. To see the impact of aggregation, we compare the total bullwhip difference for all the products combined under the two replenishment strategies. Thus, we compare $BWE_{diff}^{catgeory}$ with $n^{*}BWE_{diff}^{product}$ since $BWE_{diff}^{product}$ is identical for each of the n products.

Proposition 5 *$BWE_{diff}^{catgeory} < n^{*}BWE_{diff}^{product}$ if and only if $\frac{M(\rho)}{M(\rho+\gamma)} > 1 + \frac{n\omega}{1-\omega}$.*

Clearly, the category-level bullwhip difference can be higher or lower than the product-level bullwhip difference. If the category-level bullwhip difference is smaller (larger) than the product-level bullwhip difference, and the manufacturer uses the product-level replenishment strategy, then aggregation masks (exaggerates) the underlying bullwhip difference. If the category-level bullwhip difference is

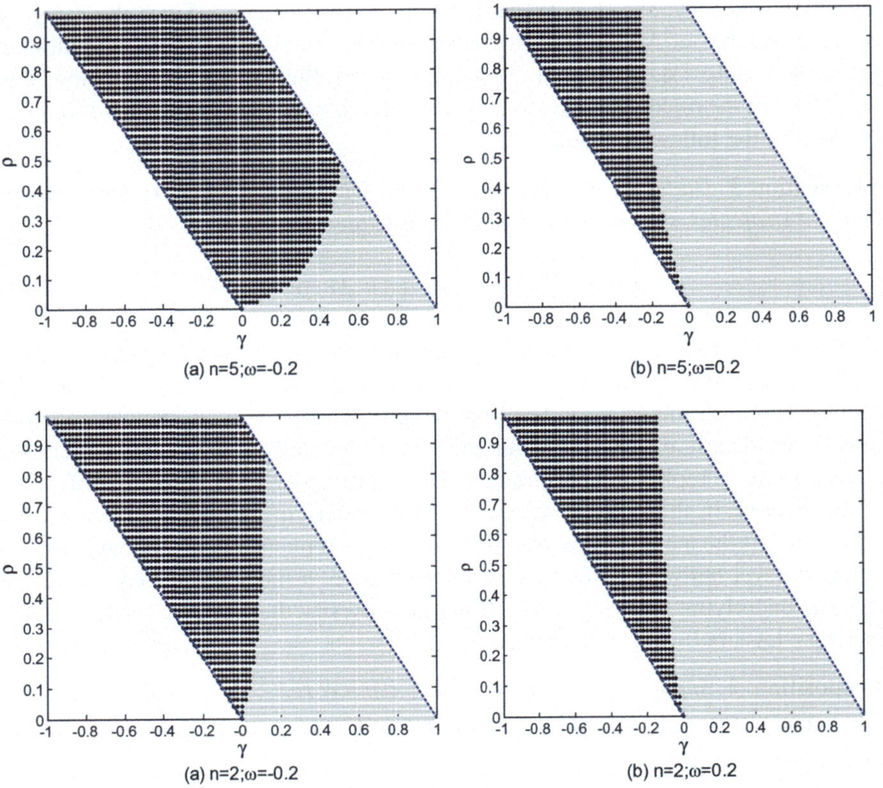

Fig. 1 Comparison of Product-level and Category-level Bullwhip Difference (Notes: $L = 2$. $\sigma^2 = 1$. Shaded region is the relevant and feasible region of comparison (i.e., $0 < \rho < 1$ and $0 < \rho + \gamma < 1$). $BWE_{diff}^{catgeory} \leq n^* BWE_{diff}^{product}$ in the darkly shaded region. $BWE_{diff}^{catgeory} > n^* BWE_{diff}^{product}$ in the lightly shaded region)

smaller (larger) than the product-level bullwhip difference, then using the category-level replenishment strategy rather than the product-level replenishment strategy mitigates (amplifies) the true bullwhip effect.

Figure 1 illustrates Proposition 5 assuming that bullwhip exists at both the product and category levels. In the figure, the unshaded regions represent the parameter space where either the stationarity assumption does not hold or the bullwhip does not exist at the category level. Figure 1 reveals that the total bullwhip difference is likely to be smaller at the category level than product level when n is high, ω is negative, and γ is not too high. On the other hand, the total bullwhip difference is likely to be smaller at the product level than the category level when n is high, ω is positive, and γ is not too low.

3.3 Bullwhip Ratio

The bullwhip ratio is useful when cost implications of the bullwhip effect are analyzed on a percentage basis, as mentioned in the previous section. Moreover, the bullwhip ratio is independent of the demand error variance σ^2, thus normalizing the effect of inherent demand variability on the bullwhip effect.

Proposition 6 *(i) The product-level bullwhip ratio is given by* $BWE_{ratio}^{product} = \frac{Var(z_{it}^*)}{Var(D_{it})} = 1 + \frac{\theta M(\rho)/\sigma^2 + (1-\theta)M(\rho+\gamma)/\sigma^2}{\left(\frac{\theta}{(1-\rho^2)} + \frac{(1-\theta)}{(1-(\rho+\gamma)^2)}\right)}$. *(ii) The category-level bullwhip ratio is given by* $BWE_{ratio}^{category} = \frac{Var\left(\sum_{i=1}^{n} z_{it}^*\right)}{Var(\overline{D}_{it})} = 1 + \left(1 - (\rho+\gamma)^2\right) M(\rho+\gamma)/\sigma^2$.

Note that $M(.)/\sigma^2$ is independent of σ^2. We find that the impacts of model parameters on the bullwhip ratio, both at product level and at category level, are non-monotonic. Specifically, we find that for $x \in \{\rho, \gamma, n, \omega\}$, $\frac{\partial BWE_{ratio}^{product}}{\partial x} > 0$ if and only if $BWE_{ratio}^{product} < \left(\frac{\partial Var(z_{it}^*)}{\partial x}\right) / \left(\frac{\partial Var(D_{it})}{\partial x}\right)$. Similarly, for $x \in \{\rho, \gamma\}$, $\frac{\partial BWE_{ratio}^{category}}{\partial x} > 0$ if and only if $BWE_{ratio}^{category} < \left(\frac{\partial Var\left(\sum_{i=1}^{n} z_{it}^*\right)}{\partial x}\right) / \left(\frac{\partial Var(\overline{D}_{it})}{\partial x}\right)$.

This finding suggests that if the rate of increase in demand variance is significantly higher than the rate of increase in order variance with respect to a parameter, then the bullwhip ratio is likely to be decreasing in that parameter. This result is somewhat similar to Proposition 9 of Chen and Lee (2012) which shows that bullwhip ratio approaches one (i.e., bullwhip ratio is masked) as number of products aggregated approaches infinity if the ratio of the order variance to the demand variance approaches zero.

Proposition 7 $BWE_{ratio}^{catgeory} < BWE_{ratio}^{product}$ *if and only if* $\frac{M(\rho)}{M(\rho+\gamma)} > \frac{1-(\rho+\gamma)^2}{1-\rho^2}$.

Proposition 7 is qualitatively similar to Proposition 5. Specifically, it reveals that aggregation does not necessarily mask (or mitigate) the bullwhip ratio nor exaggerate (or amplify) the bullwhip ratio, as found for the case of bullwhip difference. Furthermore, $\frac{M(\rho)}{M(\rho+\gamma)}$ needs to be sufficiently large for masking or mitigation to occur under both ratio and difference measures of the bullwhip effect. In contrast to bullwhip difference, n and ω do not affect the impact of aggregation on bullwhip ratio. Figure 2 illustrates Proposition 7. We find that the category-level bullwhip ratio is smaller (larger) than the product-level bullwhip ratio in most of the parameter space where γ is negative (positive).

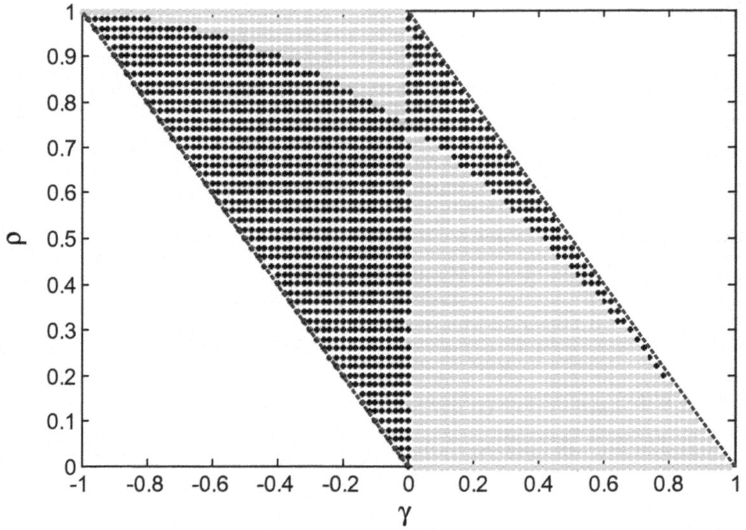

Fig. 2 Comparison of product-level and category-level bullwhip ratio

(Notes: $L = 2$. $\sigma^2 = 1$. Shaded region is the relevant and feasible region of comparison (i.e., $0 < \rho < 1$ and $0 < \rho + \gamma < 1$). $BWE_{ratio}^{catgeory} < n^* BWE_{ratio}^{product}$ in the darkly shaded region. $BWE_{ratio}^{catgeory} > n^* BWE_{ratio}^{product}$ in the lightly shaded region.)

4 Managerial Implications and Conclusion

Our findings have significant implications for academics as well as practitioners. From an academic research perspective, our findings reveal that the impact of the demand interdependency on the bullwhip effect is nontrivial and it can be counterintuitive. The findings are important because, on the one hand, they cannot be obtained or inferred through an analysis of single-product demand models or multiproduct models with independent demand streams, but, on the other hand, many real-world supply chains manage interdependent products. Furthermore, managing products with interdependent demands as one category is a natural and widely practiced strategy in modern supply chains. However, the extant bullwhip effect literature has examined aggregation solely from a measurement perspective. The multiproduct demand model presented in this chapter provides preliminary insights into the impacts of category management practice on the bullwhip effect. Further examination of multiproduct supply chains seems warranted to obtain a much richer and more complete understanding of the bullwhip phenomenon in modern supply chains.

From a practitioner perspective, our findings imply that category-level replenishment is not always better than product-level replenishment from a bullwhip effect standpoint, despite the benefits demand pooling associated with category management may offer. The nature of the demand interdependency across product demands (i.e., the sign and the magnitude of the cross-correlation coefficient) determines whether category-level management is better than product-level management. In general, a negative cross-correlation tends to favor the category-level management. Furthermore, the implications practitioners should draw and bullwhip management strategies they should choose to adopt based on bullwhip estimates depend not only on the level at which the bullwhip is measured but also on how the replenishment decisions are made within the firm. Measuring the bullwhip effect at the product category level is informative only if the firm manages the ordering process at the category level and is uninformative if the firm manages the replenishment process at the product level.

The findings also have implications for firms' other strategies related to the different brands or product variants to the extent the nature of cross-correlation is the result of these strategies. For instance, suppose brand-level promotion induces a negative cross-correlation between demand streams. Then, our results suggest that the category-level replenishment process is likely to be better than the brand-level replenishment process from a bullwhip perspective. In this case, applying uniform marketing and replenishment strategies regarding the level at which products are managed could be sub-optimal.

Finally, this chapter provides potentially new insights into the reasons for the observation of mixed results regarding the presence of the bullwhip effect at the firm and industry levels. For instance, Cachon et al. (2007) reported that the bullwhip effect is noticeably absent in the US economy as a whole. Subsequently, many researchers have attributed data aggregation as an explanation for this finding (Chen & Lee, 2012). We suspect that products of firms within the same industry will likely have interdependent demand streams because they are likely to compete with or complement products. Therefore, we hypothesize that such interdependencies could be an important contributing factor to the mixed results. An empirical study that tests this hypothesis will be a valuable complement to our study.

References

Aichner, T., & Coletti, P. (2013). Customers' online shopping preferences in mass customization. *Journal of Direct, Data and Digital Marketing Practice, 15*(1), 20–35.

Aviv, Y. (2003). A time-series framework for supply chain inventory management. *Operations Research, 51*(2), 210–227.

Box, G. E. P., & Tiao, G. C. (1977). A canonical analysis of multiple time series. *Biometrika, 64*(2), 355–365.

Bray, R., & Mendelson, H. (2012). Information transmission and the bullwhip effect: An empirical investigation. *Management Science, 58*(5), 860–875.

Cachon, G., Randall, T., & Schmidt, G. (2007). In search of the bullwhip effect. *Manufacturing and Service Operations Management, 9*(4), 457–479.
Chen, A., & Blue, J. (2010). Performance analysis of demand planning approaches for aggregating, forecasting and disaggregating interrelated demands. *International Journal of Production Economics, 128*, 586–602.
Chen, L., & Lee, H. L. (2009). Information sharing and order variability control under a generalized demand model. *Management Science, 55*(5), 781–797.
Chen, L., & Lee, H. L. (2012). Bullwhip effect measurement and its implications. *Operations Research, 60*(4), 771–784.
Chen, F., Drezner, Z., Ryan, J. K., & Simchi-Levi, D. (2000). Quantifying the bullwhip effect in a simple supply chain: The impact of forecasting, lead times, and information. *Management Science, 46*(3), 436–443.
Chopra, S., Reinhardt, G., & Mohan, U. (2007). The importance of decoupling recurrent and disruption risks in a supply chain. *Naval Research Logistics, 54*(5), 44–555.
Cox, M. W., & Alm, R. (1998). *The right stuff. America's move to mass customization*. Federal Reserve Bank of Dallas.
Dolgui, A., Ivanov, D., & Sokolov, B. (2018). Ripple effect in the supply chain: An analysis and recent literature. *International Journal of Production Research*. (Published online).
Fahimnia, B., Tang, C. S., Davarzani, H., & Sarkis, J. (2015). Quantitative models for managing supply chain risks: A review. *European Journal of Operational Research, 247*(1), 1–15.
Gaur, V., Giloni, A., & Seshadri, S. (2005). Information sharing in a supply chain under ARMA demand. *Management Science, 51*(6), 961–969.
Gilbert, K. (2005). An ARIMA supply chain model. *Management Science, 51*(2), 305–310.
Graves, S. C. (1999). A single-item inventory model for a nonstationary demand process. *Manufacturing & Service Operations Management, 1*(1), 50–61.
Gupta, S., Starr, M. K., Farahani, R. Z., & Matinrad, N. (2016). Disaster management from a POM perspective: Mapping a new domain. *Production and Operations Management, 25*, 1611–1637.
Gurnani, H., Mehrotra, A., & Ray, S. (2012). *Supply chain disruptions: Theory and practice of managing risk*. Springer.
Handfield, R. B., & McCormack, K. (2008). *Supply chain risk management: Minimizing disruptions in global sourcing*. Auerbach Publications.
Heckmann, I. (2016). *Towards supply chain risk analytics*. Springer-Gabler.
Hendricks, K. B., & Singhal, V. R. (2005). An empirical analysis of the effect of supply chain disruptions on long-run stock price performance and equity risk of the firm. *Production and Operations Management, 14*(1), 35–52.
Ho, W., Zheng, T., Yildiz, H., & Talluri, S. (2015). Supply chain risk management: A literature review. *International Journal of Production Research, 53*(16), 5031–5069.
Ivanov, D. (2017). Simulation-based ripple effect modelling in the supply chain. *International Journal of Production Research, 55*(7), 2083–2101.
Ivanov, D. (2018). *Structural dynamics and resilience in supply chain risk management*. Springer.
Ivanov, D. (2021). Supply chain risks, disruptions, and ripple effect. In *Introduction to supply chain resilience*. Springer.
Ivanov, D., Sokolov, B., & Dolgui, A. (2014a). The ripple effect in supply chains: Trade-off 'efficiencyflexibility-resilience' in disruption management. *International Journal of Production Research, 52*(7), 2154–2172.
Ivanov, D., Sokolov, B., & Pavlov, A. (2014b). Optimal distribution (re)planning in a centralized multistage network under conditions of ripple effect and structure dynamics. *European Journal of Operational Research, 237*(2), 758–770.
Ivanov, D., Dolgui, A., Sokolov, B., & Ivanova, M. (2017). Literature review on disruption recovery in the supply chain. *International Journal of Production Research, 55*(20), 6158–6174.
Khojasteh, Y. (Ed.). (2017). *Supply chain risk management*. Springer.
Klibi, W., Martel, A., & Guitouni, A. (2010). The design of robust value-creating supply chain networks: A critical review. *European Journal of Operational Research, 203*(2), 283–293.

Kouvelis, P., & Dong, L. (2011). *Handbook of integrated risk management in global supply chains*. Wiley.

Lee, H. L., Padmanabhan, P., & Whang, S. (1997). Information distortion in a supply chain: The bullwhip effect. *Management Science, 43*, 546–558.

Miyaoka, J., & Hausman, W. H. (2004). How a base stock policy using stale forecasts provides supply chain benefits. *Manufacturing & Service Operations Management, 6*(2), 149–162.

Raghunathan, S., Tang, C., & Yue, X. (2017). Analysis of the bullwhip effect in a multi-product setting with interdependent demands. *Operations Research, 65*(2), 424–432.

Sbrana, G., & Silvestrini, A. (2013). Forecasting aggregate demand: Analytical comparison of top-down and bottom-up approaches in a multivariate exponential smoothing framework. *International Journal of Production Economics, 146*, 185–198.

Simangunsong, E., Hendry, L. C., & Stevenson, M. (2012). Supply-chain uncertainty: A review and theoretical foundation for future research. *International Journal of Production Research, 50*(16), 4493–4523.

Sodhi, M. M. S., & Tang, C. S. (2012). *Managing Supply Chain Risk*. Springer.

Tiao, G. C., & Box, G. E. P. (1981). Modeling multiple times series with applications. *Journal of the American Statistical Association, 76*(376), 802–816.

Tiao, G. C., & Tsay, R. S. (1983). Multiple time series modeling and extended sample cross-correlations. *Journal of Business and Economic Statistics, 1*(1), 43–59.

Wang, X., & Disney, S. (2016). The bullwhip effect: Progress, trends and directions. *European Journal of Operations Research, 250*(3), 691–701.

Waters, D. (2011). *Supply chain risk management: Vulnerability and resilience in logistics* (2nd ed.). Kogan Page.

Zhang, X. (2004a). The impact of forecasting methods on the bullwhip effect. *International Journal of Production Economics, 88*(1), 15–27.

Zhang, X. (2004b). Evolution of ARMA demand in supply chains. *Manufacturing and Services Operations Management, 6*(2), 195–198.

Zsidisin, G. A., & Henke, M. (2019). *Research in supply chain risk: Historical roots and future perspectives: Revisiting supply chain risk* (Vol. 7). Springer Publishers.

Demand Shocks in Supply Networks: Ripple Effects

Nikolay Osadchiy

Abstract This chapter explores the ripple effects of demand shocks in supply networks, focusing on the mechanisms that drive propagation of these shocks and the factors that can either amplify or mitigate their impact. The analysis is based on three key research areas: idiosyncratic shocks and the bullwhip effect, correlated shocks and systematic risks, and the role of community structure in shock propagation. Since the focus is on demand shock propagation, it is important to establish a measure of upstreamness for firms in the network from the most downstream to upstream. We call this measure layer and describe its construction and variations. We then proceed with the analysis of idiosyncratic shock propagation, where small fluctuations in consumer demand lead to increasingly large variations in order upstream in the supply chain. We find that in complex supply networks, the propagation of these idiosyncratic shocks is often mitigated by statistical aggregation and strategic customer base management. We then continue with the analysis of correlated shocks, such as those caused by economic downturns or global crises, which impact multiple nodes simultaneously, leading to systematic risks, and characterize conditions for upstream amplification of systematic shocks. Finally, we investigate how community structures within supply networks influence the propagation of shocks. We argue that tightly knit communities can localize demand or supply disruptions with implications for network resilience and preparedness for ripple effects.

1 Introduction

To function effectively, supply chains need to match supply with demand. Shocks to either supply or demand can throw supply chains into disarray and cascade down supply chains, a phenomenon known as a ripple effect (Ivanov et al., 2014;

N. Osadchiy (✉)
Goizueta Business School, Emory University, Atlanta, GA, USA
e-mail: nikolay.osadchiy@emory.edu

D. Ivanov et al. (eds.), *Handbook of Ripple Effects in the Supply Chain*,
International Series in Operations Research & Management Science 276,
https://doi.org/10.1007/978-3-031-85508-5_5

Dolgui et al., 2018; Li et al., 2021). A substantial body of literature focused on propagation and mitigation of supply shocks treating demand as given (cf Tang, 2006; Craighead et al., 2007; Pavlov et al., 2022). Such supply shocks are typically thought of as due to natural or man-made disasters or operational disruptions. In these situations, the shortage of supply capacity can lead to downstream disruptions when customers exhaust their inventory buffers. However, the inertia and limited flexibility of supply process can expose suppliers to demand shocks, when demand changes unexpectedly leaving upstream suppliers with backlogged demand or excess inventory. Demand shocks can be due to idiosyncratic reasons, such as local changes of consumer preferences or due to systematic factors such as business cycles, or macroeconomic events. Moreover, supply disruptions can lead to demand shocks when customers search for alternative supply options, as, for example, in the recent COVID-19 pandemic (Ivanov, 2024). Similarly to supply, demand shocks can result in backlogs or excess inventory, negatively affecting performance of supply chains. The goal of this chapter is to build on the current literature on demand shock propagation and characterize the changing demand variability structure as it propagates upstream in a supply network.

2 Vertical Propagation of Demand Shocks: Upstreamness

Before discussing propagation of shocks in a network, we need to establish the measure of a vertical position for a firm in the network. The main challenge lies in determining a reference point with respect to which vertical positions of other firms will be determined. Antràs et al. (2012) propose a solution to that challenge by measuring upstreamness position as the distance to the final demand. They reason that for each industry i, its output Y_i is either used as an intermediate input by other industries or consumed as final demand F_i. Thus, upstreamness is

$$U = 1 \times \frac{F_i}{Y_i} + 2 \times \frac{\sum_{j=1}^{N} d_{ij} F_j}{Y_i} + 3 \times \frac{\sum_{j=1}^{N} \sum_{k=1}^{N} d_{ik} d_{kj} F_j}{Y_i} + \dots, \tag{1}$$

where $d_i j$ is the value of input from industry i required to produce one dollar of output of industry j, and N is the number of industries in the economy. Practically, upstreamness can be computed from the industry-level census data that reports fraction of each industry's output allocated to the final demand and to the intermediate inputs used in other industries' production.

At the firm level, Gofman et al. (2020) propose using network distance to the set of firms serving final demand as a measure of upstreamness. They call it *layer*. The set of firms serving final demand is determined exogenously by using the Global Industry Classification System (GICS): Firms in GICS segments 25 (consumer staples) and 30 (consumer discretionary) are assigned upstreamness of one, most downstream. For any other firm upstreamness is determined as the minimum shortest path from that firm to the set of firms with upstreamness one. This measure is elegant and ordinal, in the sense that a firm with upstreamness $l + 1$ is never going to have a customer with upstreamness $l - 1$, which firms with upstreamness l have. However, the measure does not reflect other paths to final demand (not necessarily the shortest) that firms may have and does not reflect the volume of trade occurring across these paths. The latter is an especially difficult challenge, given that the volume of trade in a buyer-supplier pair is rarely available.

A solution to that challenge was proposed by Agrawal and Osadchiy (2024). That measure is called *weighted layer*, and it blends approaches by Antràs et al. (2012) and Gofman et al. (2020): Any firm i in segments GICS 25, 30 serves final demand F_i and their output $Y_i = F_i$, while all others have $F_i = 0$ and their output is used as intermediate input by its customers such that $d_{ij} = \frac{Y_i}{\sum_{k \in C(i)} Y_k}$, where $C(i)$ is a set including all customers of i. Constructed in that way, weighted layer captures the multitude of paths to the final demand and their relative importance.

3 Ripple Effects of Idiosyncratic Demand Shocks

Idiosyncratic shocks arrive independently across firms and time. Under what conditions do they amplify or dampen? A potential mechanism for their amplification is the bullwhip effect—one of the most extensively studied phenomena in supply chain management. It describes how small fluctuations in demand can lead to larger fluctuations in orders placed to upstream suppliers. In principle, a small change in consumer demand at the downstream end of a supply chain can lead to increasingly larger variances in orders as they propagate upstream. This effect results in inefficiencies such as excess inventory, higher costs, and stockouts, ultimately reducing overall supply chain performance (Lee et al., 1997).

Idiosyncratic shocks refer to disturbances that affect individual nodes within a supply network, such as a sudden spike in customer orders or an unexpected delay in production. The bullwhip effect represents a classic example of how such idiosyncratic shocks can lead to significant variability in supply chain operations. However, the propagation of these shocks is not uniform across all supply chains. In complex supply networks, the interaction between multiple firms, each with its own customer and supplier relationships, creates a dynamic environment where the impact of these shocks can be either amplified or dampened depending on network characteristics.

3.1 *Mechanisms of the Bullwhip Effect*

The bullwhip effect can be viewed as a within firm or as a supply chain-wide phenomenon. It is driven by two interrelated mechanisms: demand distortion

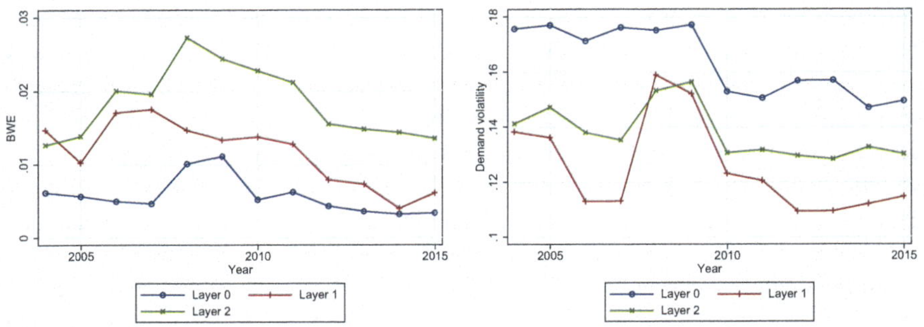

Fig. 1 Demand distortion and variance amplification in supply networks

and variance amplification. Demand distortion occurs when firms misinterpret or overreact to fluctuations in customer orders, leading to exaggerated order quantities upstream. This distortion is often a result of forecasting errors, order batching, and extended lead times. Variance amplification, on the other hand, refers to the tendency for order variability to increase as it moves upstream through the supply chain. In traditional serial supply chains, these mechanisms are closely intertwined, in fact, demand distortion implies upstream variance amplification, leading to a compounding effect as demand shocks propagate through multiple tiers of suppliers.

Osadchiy et al. (2021a) offer a novel perspective by examining the bullwhip effect within supply networks rather than linear, serial supply chains. Their research emphasizes the importance of understanding both intrafirm processes, which lead to demand distortion, and interfirm dynamics, which determine how variance amplification occurs across a network. The authors argue that while demand distortion remains a critical factor, the structure of the supply network and the strategic choices made by firms within it can significantly influence the upstream variance amplification of the bullwhip effect (Fig. 1).

3.2 Empirical Evidence and Network Perspective

The study draws on a comprehensive dataset that tracks the evolution of supply relationships over time. The dataset provides detailed information on buyer-supplier relationships, allowing the authors to construct a dynamic model of a supply network. Their analysis reveals that while demand variability at the individual firm level is substantial (with a median increase in order variability of 1.6 percentage points, Fig. 2, left), this variability does not lead to a corresponding amplification in aggregate demand variability as the shocks propagate upstream (Fig. 2). Indeed, across from layer 0 to layer 2 demand variability actually decreases from 16 to roughly 13 percentage points (Fig. 2, right).

This finding challenges the traditional assumption that variance amplification is inevitable in supply chains. The authors attribute this dampening effect to two

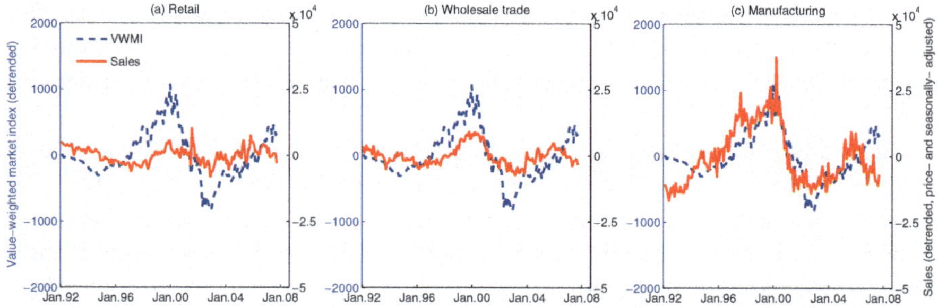

Fig. 2 The detrended time series of sales and the level of value-weighted market index for the US retailers, wholesalers, and manufacturers, reproduced from Osadchiy et al. (2015)

key factors: statistical aggregation and deliberate customer portfolio management. Statistical aggregation occurs when firms serve multiple customers whose demand patterns are not perfectly correlated. As a result, fluctuations in one customer's demand can be offset by opposite fluctuations in another customer's demand, leading to a smoother aggregate demand for the supplier. However, the authors argue that statistical aggregation alone is insufficient to explain the observed reduction in variability. Instead, they point to customer base management as a critical factor.

3.3 Strategic Customer Base Management

Customer base management involves the strategic selection and management of customers to achieve operational objectives such as demand smoothing. The selection is nontrivial and must consider the demand patterns of current customers to make the best addition to the customer portfolio (Schmidt et al., 2021). For example, a customer with even demand over time may turn out to be less preferred than a customer with variable demand that fills capacity in periods with low utilization. Moreover, that customer can be distorting demand or bullwhipping.

The study finds that firms actively manage their customer portfolios to reduce the impact of idiosyncratic shocks and demand distortion. For example, suppliers may seek to add customers whose demand patterns complement those of their existing customers, thereby reducing overall variability. Conversely, suppliers may choose to terminate relationships with customers whose demand is highly volatile or negatively correlated with the rest of their customer base or swap them for a customer with more favorable demand pattern. Note that the latter two strategies go contrary to the statistical economies of scale, as the customer base reduces or remains the same.

The empirical evidence supports this hypothesis. The addition of a new customer is associated with a median reduction in demand variability of 1.3 percentage points, while dropping a customer results in a reduction of 1.7 percentage points. Interestingly, these reductions occur even when the added or dropped customers exhibit high levels of intrafirm variability (i.e., their own orders are highly volatile). This suggests that suppliers are not simply aggregating demand to achieve smoothing but are instead making deliberate choices based on the overall impact on their demand portfolio.

The study also explores the role of customer base events (such as adding or dropping a customer) in driving long-term changes in demand variability. The authors find that firms that experience a high number of customer base events within a given period tend to see significant reductions in demand variability in subsequent periods. This finding underscores the importance of continuous customer base management as a strategy for mitigating the bullwhip effect.

3.4 Implications for Supply Chain Management

The insights from this study have important implications for supply chain management. Traditionally, strategies for mitigating the bullwhip effect have focused on intrafirm processes, such as reducing order lead times, improving demand forecasting, and minimizing order batching. While these strategies remain relevant, the findings suggest that a broader network perspective is essential for managing idiosyncratic shocks. In effect, the network can shield a supplier from a variety of idiosyncratic shocks, in a way that is complementary to traditional efforts to smooth demand variability and reduce demand distortion.

By strategically managing their customer portfolios and leveraging the interconnected nature of supply networks, firms can reduce the impact of demand variability and limit the propagation of the bullwhip effect. This approach highlights the importance of collaboration and coordination across the supply network and the need for real-time visibility into demand patterns and customer behavior.

In conclusion, the ripple effects of idiosyncratic shocks in supply networks are shaped by a combination of intrafirm processes and interfirm relationships. By adopting a network perspective and focusing on strategic customer base management, firms can mitigate the adverse effects of the bullwhip effect and enhance the resilience of their supply chains.

4 Ripple Effects of Correlated Shocks: Systematic Risk in Supply Chains

In this section we explore how demand shocks that are correlated between each other through a common driving factor propagate upstream through the network. Because of the common driving factor, those shocks are systematic. The associated systematic risks represent a different type of challenge for supply networks compared to idiosyncratic risks. While the latter affects individual firms or nodes within the network, systematic risks are characterized by correlated shocks that impact multiple nodes simultaneously. Examples of such risks include economic downturns, global health crises like pandemics, and sector-wide disruptions. The simultaneous nature of these shocks can lead to synchronized failures across the network, making them particularly difficult to contain and manage. Moreover, systematic shocks are difficult to diversify; therefore managing systematic risk emerging from such shocks is particularly important.

4.1 *Introduction to Systematic Risks in Supply Networks*

Conceptually, the systematic risk is the undiversifiable risk in demand which is defined in Osadchiy et al. (2015) as the correlation between changes in demand and changes in the state of the economy, the latter being proxied by a return on a broad financial portfolio. For firms in the supply network, it is therefore important to understand how such undiversifiable shocks propagate through the network and what role the network's topology—its structure and the relationships between its nodes—plays in determining firms' vulnerability to systematic risk.

Unlike idiosyncratic shocks for which the network serves as a dampener, systematic shocks tend to amplify as they propagate upstream supply networks. Osadchiy et al. (2015) report that systematic risk ρ is the lowest downstream at retailers ($\rho = 0.33$, average layer 1) and increases for wholesalers ($\rho = 0.42$, average layer 2.2) to manufacturers ($\rho = 0.50$, average layer 2.9). This is evident even from visual examination of the industry wide sales and the trajectory of the value-weighted market index. The same aggregation mechanism that attenuates idiosyncratic shocks amplifies systematic risk. That result is echoed by Elliott et al. (2022) who find that networks endogenously form and achieve robustness to idiosyncratic shocks, yet shocks that systemically affect the network amplify massively.

One last observation is that systematic risk is correlated with financial beta of industry portfolios (Osadchiy et al., 2015). Because systematic risk is grounded in the characteristics of sales revenue and financial market index time series, those characteristics can be viewed as revenue-based microfoundations of financial beta.

4.2 *Network Analysis of Systematic Risk*

Key network characteristic, such as degree, is identified as a critical factor in determining how systematic risks spread. A greater number of customers generally are associated with a greater systematic risk. The magnitude of demands placed by the customers also plays a role: All things being equal a portfolio of customers with equally sized orders will have a greater systematic risk, compared to a portfolio of customers varying in size.

While degree captures the effect of immediate customers, the effect of systematic risk aggregation from more distant customers can be captured via betweenness centrality. Indeed, a high betweenness centrality firm belongs to more paths between any pair of firms in the network. This is consistent with results from empirical finance where high centrality industries have a higher financial systematic risk and a higher return (Ahern, 2013).

Finally, upstreamness has a critical role in determining the degree of systematic risk: In real-world networks upstream firms have more systematic risk. For this mechanism to work, there should be relatively fewer firms upstream compared to downstream. This observation does hold in real-world networks: Both Gofman et al. (2020) and Osadchiy et al. (2021a) report that networks tend to become thinner upstream. Thus systematic risk upstream amplification would typically be observed in distribution type of networks. In assembly type of network, where there are more upstream components than downstream products, such amplification is less likely.

4.3 Strategies for Mitigating Systematic Risks

Diversifying systematic risk is difficult, yet some strategies can be employed for mitigating the impact of systematic risk in supply networks. We take stock of them below.

Redundancy and Buffers One of the most effective ways to mitigate systematic risks is by building redundancy into the supply network. This can be achieved by maintaining additional inventory, diversifying suppliers, or establishing backup production facilities. Redundancy provides a buffer against correlated shocks, yet it comes at cost, and the investment in redundancy at locally optimal levels may not be sufficient to protect from correlated shocks, leading to supply network fragility (Elliott et al., 2022). Buffers also give the network a temporal advantage, and they can be used to break down simultaneity in orders across customers and give the system more time to react and adapt.

Reducing Delays and Lead Times Systematic risks aggregate over time. Thus, orders with long lead times and scheduling delays are more likely to be revised, and revisions are going to be correlated with each other, amplifying systematic risk.

Decentralizing Decision-Making While centralized and coordinated decision-making across supply chains can be efficient under normal conditions, it creates a bottleneck during crises, as it may struggle to respond quickly to widespread disruptions. Decentralizing decision-making allows for more rapid and localized responses, which can prevent the spread of shocks across the network. Moreover, local responses may help breaking the correlation to the original shock.

Scenario Planning and Stress Testing Regular stress tests and scenario planning exercises can help firms identify potential vulnerabilities and develop contingency plans for systematic risks. By employing scenario analysis or digital twins (Ivanov,

2023), firms can better understand how their network would respond and what measures (e.g., buffers or lead time reductions) would be most effective in mitigating disruptions.

Collaborative Risk Management Given the interconnected nature of supply networks, risk management cannot be solely the responsibility of individual firms. Collaborative approaches, such as joint risk assessments and coordinated contingency planning, can enhance the overall resilience of the network. Firms can work together to share information, align strategies, and pool resources in preparation for potential correlated shocks.

4.4 Implications for Supply Chain Resilience

The findings from this study underscore the complexity of managing systematic risks in supply networks. Unlike idiosyncratic shocks, which can often be contained at the firm level, systematic risks require a more holistic approach that considers the entire network's structure and dynamics. Firms that fail to account for these broader risks may find themselves unprepared for the cascading failures that can result from correlated shocks. In that sense, firms should be aware of their position in a broader supply network and coordinate their strategy accordingly.

The study also highlights the importance of proactive risk management. Rather than reacting to crises after they occur, firms should invest in building resilient network structures and implementing strategies that can mitigate the impact of systematic risks. This may involve trade-offs between efficiency and resilience, as building redundancy and decentralizing decision-making can introduce additional costs. However, these investments are often justified by the significant reduction in risk exposure.

In conclusion, the ripple effects of correlated shocks in supply networks are driven by the interplay between network topology and strategic decision-making. By understanding the specific vulnerabilities of their network structure and adopting targeted risk management strategies, firms can enhance their ability to withstand systematic risks and maintain operational continuity during crises.

5 Propagation of Shocks: Community Structure in Supply Networks

One main implication of the above analysis is that demand shocks change their structure when they propagate upstream: Idiosyncratic downstream shocks tend to aggregate out, while systematic shocks tend to amplify. As a result, a greater component of demand variability upstream is attributable to systematic, economy-wide, shocks. The next question we are addressing is along which paths such shocks are more likely to propagate. For that, we turn to the recent empirical work on network structure.

5.1 Understanding Community Structure in Supply Networks

The challenge of understanding the structure of evolving production networks starts with mapping of supply chains into the network, a nontrivial and data demanding task (MacCarthy et al., 2022; Gaur et al., 2023). Mapping a global supply network involves a combination of data sources from various countries, agencies, and data aggregators (Pichler et al., 2023). Real-world networks are complex because they are formed by often competing mechanisms such as cost minimization, geographical proximity, cultural compatibility, etc. For those reasons, it is likely that the production networks exhibit community structure. Community structure refers to the presence of densely interconnected groups of nodes, or communities, within a network. In the context of supply networks, these communities typically consist of groups of firms that are more closely connected to each other than to firms outside their group. The existence of such communities has significant implications for how shocks propagate through the network.

In a recent study (Osadchiy et al., 2021b) the authors focus on community structure and evolution of the global production network and explore the role of communities in propagation of shocks. They find that from 2003 to 2018 production networks have become more fragmented, that is, they consist of more communities. Moreover, whereas in 2003-2009 the network was largely dominated by a few large communities, such dominance disappears over time with the network comprising of a larger number of relatively equal-sized communities.

5.2 The Role of Communities in Shock Containment

Communities within a supply network can act as buffers that localize the impact of disruptions. They also isolate a relatively small set of firms linking distinct communities through which shocks in one community propagate to another. When a shock occurs within a community, its effects may be contained if the community is relatively isolated from the rest of the network. For example, a localized disruption in a small group of suppliers might have minimal impact on the broader supply chain if those suppliers primarily serve each other and have limited connections to other communities.

However, if the community has strong links to other communities, the shock can spread more easily, potentially leading to cascading failures. The study finds that the extent to which a community can contain a shock depends on several factors, including the density of connections within the community, the strength of inter-community links, and the overall modularity of the network.

Highly modular networks, where communities are tightly clustered but weakly connected to each other, are more effective at containing shocks within a single community. In contrast, networks with strong inter-community links are more vulnerable to widespread disruptions, as shocks can quickly propagate from one community to another.

5.3 Implications for Supply Network Design

The findings from this study have important implications for the design and management of supply networks. Firms can enhance network resilience by strategically managing community structures and intercommunity connections. Key strategies include:

Modular Network Design Encouraging modularity within the network can help contain shocks and prevent them from spreading across communities. This approach involves creating clusters of firms that are closely connected internally but have limited external connections. By isolating communities, firms can reduce the risk of cascading failures and ensure that disruptions are contained within a specific area of the network.

Strengthening Critical Links In networks with important intercommunity links, firms should focus on strengthening these links and ensuring they are robust enough to withstand disruptions. This might involve investing in inventory, diversifying suppliers, or enhancing communication channels to improve coordination and response times. Importantly, Osadchiy et al. (2021b) find that a large fraction of firms linking communities persists in that function over time, even though about one quarter of the rest of the network is rewired. Thus, investments in redundancy are justified from the time perspective.

Dynamic Network Reconfiguration Supply networks are not static; their community structures can evolve over time as firms enter or exit the network, new relationships are formed, or existing ones are dissolved. Firms should regularly assess and, if necessary, reconfigure their networks to maintain an optimal balance between efficiency and resilience. This might involve adding alternative suppliers that belong to a different community, geography, or one that serves a distinct set of customers and re-evaluating the strength of intercommunity links based on changing risk profiles.

Cross-Community Collaboration In networks with strong intercommunity connections, firms should invest in building collaborative relationships that enable rapid and coordinated responses to disruptions. By aligning risk management strategies across communities, firms can reduce the likelihood of shocks spreading uncontrollably and ensure that recovery efforts are well-coordinated.

5.4 *Conclusion: The Dual Role of Community Structure*

Community structures within supply networks play a dual role in shock propagation. On the one hand, they can act as barriers that localize disruptions and prevent them from spreading across the network. On the other hand, if intercommunity connections are strong, these same structures can facilitate the rapid spread of disruptions, leading to widespread cascading failures. Understanding and managing this dynamic is crucial for building resilient supply networks capable of withstanding both local and systemic shocks.

Community structure can be beneficial for firms. The main benefit comes from the fact that once the community structure is known, firms linking communities are easily identifiable, and those critical links can receive the needed attention. For that, the knowledge of the broad supply network is critical, and firms should adopt a tailored approach that takes into account the specific characteristics of their network, the nature of the risks they face, and the strategic objectives they seek to achieve. By doing so, they can strike the right balance between efficiency, resilience, and adaptability in their supply networks.

6 Conclusion

The ripple effects of demand shocks in supply networks are complex and multifaceted, influenced by a combination of idiosyncratic, correlated, and structural factors. This chapter has explored three key areas of research, highlighting different aspects of how demand shocks propagate through supply networks and the strategies that firms can use to mitigate these effects.

We began with the focus on idiosyncratic shocks and the bullwhip effect, emphasizing the importance of strategic customer base management in controlling the propagation of these shocks. While traditional approaches to managing the bullwhip effect have centered on intrafirm processes such as demand forecasting and lead time reduction, we highlight the value of the customer network management. By carefully managing customer relationships and leveraging the interconnected nature of supply networks, firms can achieve significant reductions in demand variability and enhance overall resilience.

The second section examined the impact of correlated shocks and systematic risks, which present a more significant challenge due to their simultaneous impact on multiple nodes within the network. We underscore the critical role of network position in determining vulnerability to systematic risks. By understanding the structural characteristics of their supply networks—such as centrality, redundancy, and connectivity—firms can develop targeted risk management strategies that enhance their ability to withstand correlated shocks.

The third section explored how community structures within supply networks influence the propagation of shocks. The findings revealed that while tightly knit communities can act as buffers that localize disruptions, strong intercommunity links can rapidly spread the shocks across the network. This dual role of community structures highlights the need for a nuanced approach to supply network design and management, the one that considers both modularity (i.e., community structure) and connectivity.

Supply chain resilience cannot be achieved through a one-size-fits-all approach. The complex interplay between different types of shocks and the structural characteristics of supply networks requires a holistic strategy that integrates insights from dynamic network analysis and risk management. Firms that invest in understanding and managing these factors are better positioned to build supply networks that are not only efficient but also resilient in the face of diverse disruptions. By adopting a multifaceted approach that addresses both local idiosyncratic disturbances and systematic risks, supply chain managers can enhance the robustness of their operations and ensure continuity as supply chains become more intertwined and grow in complexity.

References

Agrawal, D., & Osadchiy, N. (2024). Inventory productivity and stock returns in manufacturing networks. *Manufacturing & Service Operations Management, 26*(2), 573–593.

Ahern, K. R. (2013). Network centrality and the cross section of stock returns. Available at SSRN 2197370.

Antràs, P., Chor, D., Fally, T., and Hillberry, R. (2012). Measuring the upstreamness of production and trade flows. *American Economic Review, 102*(3), 412–416.

Craighead, C. W., Blackhurst, J., Rungtusanatham, M. J., Handfield, R. B. (2007). The severity of supply chain disruptions: design characteristics and mitigation capabilities. *Decision Sciences, 38*(1), 131–156.

Dolgui, A., Ivanov, D., & Sokolov, B. (2018). Ripple effect in the supply chain: an analysis and recent literature. *International Journal of Production Research, 56*(1–2), 414–430.

Elliott, M., Golub, B., & Leduc, M. V. (2022). Supply network formation and fragility. *American Economic Review, 112*(8), 2701–2747.

Gaur, V., Osadchiy, N., & Udenio, M. (2023). Why it's so hard to map global supply chains. In *Supply chain: The insights you need from harvard business review*. Harvard Business Press.

Gofman, M., Segal, G., Wu, Y. (2020). Production networks and stock returns: The role of vertical creative destruction. *The Review of Financial Studies, 33*(12), 5856–5905.

Ivanov, D. (2023). Intelligent digital twin (IDT) for supply chain stress-testing, resilience, and viability. *International Journal of Production Economics, 263*, 108938.

Ivanov, D. (2024). Exiting the covid-19 pandemic: After-shock risks and avoidance of disruption tails in supply chains. *Annals of Operations Research, 335*(3), 1627–1644.

Ivanov, D., Sokolov, B., Dolgui, A. (2014). The ripple effect in supply chains: trade-off 'efficiency-flexibility-resilience' in disruption management. *International Journal of Production Research, 52*(7), 2154–2172.

Lee, H. L., Padmanabhan, V., & Whang, S. (1997). Information distortion in a supply chain: The bullwhip effect. *Management Science, 43*(4), 546–558.

Li, Y., Chen, K., Collignon, S., & Ivanov, D. (2021). Ripple effect in the supply chain network: Forward and backward disruption propagation, network health and firm vulnerability. *European Journal of Operational Research, 291*(3), 1117–1131.

MacCarthy, B. L., Ahmed, W. A., & Demirel, G. (2022). Mapping the supply chain: Why, what and how? *International Journal of Production Economics, 250*, 108688.

Osadchiy, N., Gaur, V., & Seshadri, S. (2015). Systematic risk in supply chain networks. *Management Science, 62*(6), 1755–1777.

Osadchiy, N., Schmidt, W., & Wu, J. (2021a). The bullwhip effect in supply networks. *Management Science, 67*(10), 6153–6173.

Osadchiy, N., Udenio, M., & Gaur, V. (2021b). Have supply networks become more fragmented over time? Available at SSRN https://ssrn.com/abstract=3867369V

Pavlov, A., Ivanov, D., Werner, F., Dolgui, A., & Sokolov, B. (2022). Integrated detection of disruption scenarios, the ripple effect dispersal and recovery paths in supply chains. *Annals of Operations Research, 319*(1), 609–631.

Pichler, A., Diem, C., Brintrup, A., Lafond, F., Magerman, G., Buiten, G., Choi, T. Y., Carvalho, V. M., Farmer, J. D., & Thurner, S. (2023). Building an alliance to map global supply networks. *Science, 382*(6668), 270–272.

Schmidt, W., Osadchiy, N., & Wu, J. (2021). The right way to mix and match your customers. *MIT Sloan Management Review, 62*(4), 12–14.

Tang, C. S. (2006). Robust strategies for mitigating supply chain disruptions. *International Journal of Logistics: Research and Applications, 9*(1), 33–45.

Performance Impact Analysis of Disruption Propagations in the Supply Chain

Dmitry Ivanov, Alexander Pavlov, and Boris Sokolov

Abstract Despite a wealth of literature on disruption considerations in the supply chain (SC), a method for quantification of the ripple effect that describes disruption propagation in the SC has not yet been developed. In addition, there are only a few studies that incorporate recovery into the performance impact assessment. This chapter develops a method to quantify the ripple effect in the SC with recovery policy considerations. We study a four-stage SC over time and consider both performance impact assessment and recovery decisions. The performance impact index developed is used to compare sales (revenue) in different SC designs to measure the estimated annual magnitude of the ripple effect. First, we compute optimal SC replanning for two disruption scenarios. Second, we estimate the performance impact of disruptions for six proactive SC designs. Finally, we compare the performance impact index of different SC designs and draw conclusions about the ripple effect in these SC designs along with recommendations for the selection of a proactive strategy. The performance impact index developed can be used to analyze how different markets are exposed to the ripple effect and how to compare different SC designs according to their resilience to severe disruptions.

1 Introduction

Systemic approaches to risk management in the supply chain (SC) have become a visible research topic over the past decade. Fires, explosions, tsunamis, and strikes

---

D. Ivanov (✉)
Department of Business and Economics, Berlin School of Economics and Law, Berlin, Germany
e-mail: divanov@hwr-berlin.de

A. Pavlov · B. Sokolov
St. Petersburg Federal Research Center of the Russian Academy of Sciences (SPC RAS), St. Petersburg, Russia
e-mail: sokol@iias.spb.su

D. Ivanov et al. (eds.), *Handbook of Ripple Effects in the Supply Chain*,
International Series in Operations Research & Management Science 276,
https://doi.org/10.1007/978-3-031-85508-5_6

at production plants, distribution centers, and transportation channels are typical disruptions in supply chains (SC). The ripple effect in the SC occurs if a disruption cannot be localized and cascades downstream impacting SC performance (Ivanov et al., 2014a). Methodical elaborations on the evaluation and understanding of low-frequency-high-impact disruptions are, therefore, vital for understanding and further development of network-based supply concepts (Aldrighetti et al., 2023a; Iftikhar et al., 2021; Ivanov & Rozhkov, 2020; Liberatore et al., 2012; Sawik, 2016; Stecke & Kumar, 2009; Tomlin, 2006).

It has been extensively documented in literature that severe disruptions may ripple quickly through global SCs and cause losses in SC performance that can be measured by such KPIs (key performance indicators) as revenues, sales, service level, and total profits (Ivanov, 2018; Ivanov & Keskin, 2023; Schmitt et al., 2017; Schmitt & Singh, 2012; Simchi-Levi et al., 2015; Snyder et al., 2016). Such risks are a new challenge for research and industry who face the *ripple effect* that arises from vulnerability, instability, and disruptions in SCs (Ho et al., 2015; Ivanov, 2024a, c; Ivanov et al., 2014a, b, 2017a, b, c, 2023; Liberatore et al., 2012). As an opposite to the well-known bullwhip effect that considers high-frequency-low-impact *operational risks*, the ripple effect describes low-frequency-high-impact *disruptive risks* (Alikhani et al., 2023; Cavalcantea et al., 2019; Han & Shin, 2016; Hosseini et al., 2019a, b; Ivanov et al., 2014a, 2017b; Ramani et al., 2022; Rolf et al., 2023; Sawik, 2017; Simchi-Levi et al., 2015; Snyder et al., 2016; Sokolov et al., 2016).

Recent studies have extensively considered disruption risks in light of the impact of disruption propagation (Ivanov, 2017; Ivanov et al., 2013, 2014b; Lim et al., 2013; Paul et al., 2014; Sawik, 2022, 2023; Wilson, 2007). Previous studies also suggested several measures for quantifying disruption risks (Basole & Bellamy, 2014; Choi & Shi, 2022; Han & Shin, 2016; Ivanov, 2025a, b; Lin et al., 2017; Liu et al., 2024; Sardesai & Klingebiel, 2023; Zobel, 2011). However, single-stage disruptions have mostly been considered. Disruption propagation has been ignored, and a method for ripple effect quantification has not yet been developed. In addition, there are only a few studies that incorporate the recovery stage into the performance impact assessment. We are not aware of any published research that considers ripple effect quantification with SC recovery considerations.

The objective of this study is to quantify the ripple effect in the SC with recovery considerations. The remainder of this paper is organized as follows. Section 2 analyzes recent literature. Section 3 considers the methodology and modeling approach. Section 4 is devoted to the model presentation and experimental calculation. Section 5 considers the performance impact assessment and managerial insights. The paper concludes by summarizing the most important findings and outlining future research needs.

2 Literature Review

We structure the literature review according to performance impact assessment techniques and engineering methodologies to model severe SC disruptions. Two engineering methodologies—optimization and simulation—dominate studies on SC performance impact assessment in light of severe disruptions (Aldrighetti et al., 2023b; Aron et al., 2024; Blackhurst et al., 2018; Dolgui et al., 2020; He et al., 2019; Hosseini & Barker, 2016; Ivanov, 2019; Ivanov, 2023a, b, c, 2024b; Käki et al., 2015; Ojha et al., 2018; Pavlov et al., 2018, 2019a; Talluri et al., 2013). Petri nets have been applied to analyze disruption propagation through the SC and to evaluate the performance impact of disruptions (Wu et al., 2007). They allow the study of how changes disseminate through a SC and calculate the impact of the attributes by determining the states that are reachable from a given initial marking in a SC. Wilson (2007) considered transportation disruptions in a multistage SC in order to reveal the ripple effect's impact on fulfilment rate and inventory fluctuations. The findings suggest that transportation disruptions between the Tier-1 supplier and the warehouse have the greatest performance impact.

Tuncel and Alpan (2010) extended the body of knowledge by incorporating multiple disruption scenarios (disruptions in demand, transportation, and quality). In addition, this study also considers recovery actions and the performance impact of such mitigation strategies. Carvalho et al. (2012) presented a simulation study for a four-stage automotive SC. Focusing on the research question of how different recovery strategies influence SC performance in the event of disruptions, the authors analyzed two recovery strategies and six disruption scenarios. The scenarios differ in terms of the presence or absence of a disturbance and the presence or absence of a mitigation strategy. The performance impact was analyzed according to the lead time ratio and total SC costs.

Ivanov et al. (2014b) used a hybrid optimization-control model for the simulation of SC recovery policies for multiple disruptions in different periods in a multistage SC. The developed approach allows simultaneous performance impact analysis of SC disruptions and recovery policies simulation. Basole and Bellamy (2014) used the measure "number of healthy nodes" to quantify the level of risk diffusion as the relation of the number of healthy nodes at time t to the network size.

With regards to robustness and resilience analysis, Nair and Vidal (2011) studied SC robustness against disruptions using graph-theoretical topology analysis. They studied 20 SCs which were subjected to random demand. The performances of the SCs were evaluated by considering varying probabilities of the random failures of nodes and targeted attacks on nodes. Furthermore, the analysis of the study also included the severity of these disruptions by considering the downtime of affected nodes. The results were obtained using multiagent simulation. Furthermore, the authors went into detail on the impact of SC structural design on robustness in the presence of both demand and disruption uncertainty.

Zhao et al. (2011) quantified network connectivity and accessibility with the help of the largest connect component and average and maximum path length. Zobel

(2011) computed predicted resilience assuming a sudden onset disruption and linear recovery behavior. Zobel (2014) extended to a more generalized case considering nonlinearity and average loss per time unit.

Simchi-Levi et al. (2015) developed a risk-exposure index for the case of an automotive SC. The index computation is based on two models—time-to-recovery and time-to-survive—to assess the performance impact of a disruption in the SC. Raj et al. (2015) analyzed SC resilience based on a survival model to represent a time period from the system's failure to operate to the time the system returns to its function (i.e., recovery). The input to the model is a failure event; the output of the model is the recovery time. The model allows a quantitative measurement of SC resilience in terms of recovery time.

Sokolov et al. (2016) quantified the ripple effect in the SC with the help of selected indicators from graph theory and developed a static model for performance impact assessment of disruption propagation in a distribution network. Han and Shin (2016) evaluated the structural robustness of the SC in random networks and compared this with the likelihood of network disruption resulting from random risk.

Ivanov et al. (2016) extended the performance impact assessment and SC plan reconfiguration with consideration of the duration of disruptions and the costs of recovery. They analyzed seven proactive SC structures, computed recovery policies to re-direct material flows in the case of two disruption scenarios, and assessed the performance impact for both service level and costs with the help of a SC (re)planning model containing elements of control theory and linear programming. This study reveals the impact of different parametrical and structural resilience measures on SC service level and efficiency. Ivanov (2025c) shows that performance impact assessment can inform managers about both network connectivity and delivery reliability. In the current paper, we use the basic model and the proactive strategies for SC design in Sect. 4 in order to analyze the performance impact of the ripple effect in Sect. 5.

The reliability of a multistage SC is evaluated by Lin et al. (2017) as the probability that market demand, and therefore sufficient commodity delivery, can be met by the SC through multiple stations of transit within the appropriate time frame. In this study, system reliability acts as the delivery performance index and is assessed by the number of minimal paths. Ivanov et al. (2018) developed a control-theoretic method to assess SC resilience with the consideration of recovery policies using attainable sets.

From the given literature, it can be observed that the scope of SC rippling and its impact on economic performance depends both on robustness reserves (e.g., redundancies like inventory or capacity buffers), flexibility in products and processes, disruption duration, and the speed and scale of recovery measures. Despite significant advancements in this research field, ripple effect assessment with SC recovery considerations is still an under-researched area and presents a promising field for future research.

3 Problem Statement and Research Methodology

We study a four-stage SC over time and consider both performance impact assessment and recovery decisions. The SCD comprises Tier 2 suppliers, Tier 1 suppliers, assembly plants, and markets. The disruptions and recovery policies are considered as given scenarios, the production and transportation quantities are the decision variables, and SC sales (revenue) is the KPI for measuring the estimated annual magnitude of the ripple effect.

The optimization model is developed to replan SC flows to maximize SC sales. The model aims to find the aggregate product flows to be moved from suppliers through the intermediate stages to the markets subject to revenue maximization (i.e., lost sales minimization) under (i) constrained capacities and processing rates, (ii) SC disruptions, and (iii) SC recovery for a multi-period case.

The modeling approach is the optimization-based simulation that allows simultaneous re-computing of the material flows in a multistage SC after a disruption and a comparison of the performance impact of different SCDs. Based on the optimization results, the performance impact of disruptions is computed.

For the computation of performance impact, we suggest introducing an index of performance impact (PI) that represents the relation between the planned KPI in a disruption-free mode and the real KPI in the disruption case (Eq. 1):

$$PI = \frac{KPI_{plan}}{KPI_{disruption}} \tag{1}$$

Such an index can be computed for each i-node in the SC, $i = 1, \ldots. N$. Subsequently, we can compute a product of the i-PIs in order to calculate the overall PI in the SC (Eq. 2):

$$PI_{general} = \prod_{i=1}^{N} PI \tag{2}$$

The organization of the rest of this manuscript is as follows. First, we compute optimal SC replanning for two disruption scenarios and KPI. In order to make the PI analysis more depictive, we restrict ourselves to the analysis of estimated annual magnitude in terms of the revenue at markets. Second, we perform the first step for six proactive strategies in SC design. Finally, we compare the PI of seven SC designs (i.e., the initial SC design and six proactive strategies) and draw conclusions on the ripple effect in these SC designs along with recommendations on the proactive strategy.

4 Mathematical Model

The SC design model can be represented as follows (Table 1).

Objective function

Table 1 Supply chain design formalization

Notation	Meaning
Structure	
$X_\chi(t) = \{A_{\chi i}(t), i \in N_\chi\}$	Set of nodes in the SC design in the χ disruption scenario at the point of time t, where N is the set of node numbers
$E_\chi(t) = \{e_{\chi ij}(t) \in \{0, 1\}, i, j \in N_\chi\}$	Set of arcs in the SC design in the χ disruption scenario
Parameters	
$W_\chi(t) = \{w_{\chi ij\rho}(t), i, j \in N_\chi, \rho \in P\}$	Set of operational characteristics for the production, transportation (if $i \neq j$) or processing at warehouse (if $i = j$) in the χ disruption scenario
$V_{\chi i}(t)$	Maximum warehouse throughput capacity at the node $A_{\chi i}$
$\omega_{\chi ij\rho}(t)$	Maximum transportation channel throughput capacity for the commodity ρ between $A_{\chi i}$ and $A_{\chi j}$
$\psi_{\chi i\rho}(t)$	Maximum inbound throughput capacity for the commodity ρ in $A_{\chi i}$
$\phi_{\chi i\rho}(t)$	Maximum outbound throughput capacity for the commodity ρ in $A_{\chi i}$
$I_{\chi k}$	Total ordered quantity from all suppliers in the period number k
γ_ρ	Importance of the commodity ρ
λ_ρ	Urgency of the commodity ρ
$\varepsilon_{ij}(t)$	Preset matrix time function of time-spatial constraints; we have $\varepsilon_{ij}(t) = 1$, if the channel between A_i and A_j is available and not disrupted within the given period of time, and $\varepsilon_{ij}(t) = 0$, otherwise
Indexes	
$k = \{1, 2, \ldots, L_\chi\}$	A number of the planning period in the planning horizon $T = (t_0, t_f]$
$T = (t_0, t_f]$	Planning horizon
$\rho \in P = \{1, 2, \ldots, p\}$	A number of the commodity in the SC
Decision variables	
$x_{\chi ij\rho k}$	Amount of commodity ρ transmitted from $A_{\chi i}$ to $A_{\chi j}$ and received at $A_{\chi j}$ at time interval number k
$y_{\chi j\rho k}$	Amount of commodity ρ to be stored at the warehouse $A_{\chi i}$ at time interval number k
$g_{\chi j\rho k}$	Amount of commodity ρ to be delivered to $A_{\chi j}$ at time interval k
$z_{\chi j\rho k}$	Amount of commodity ρ at $A_{\chi j}$ to be returned (as caused by the missing capacity of SC nodes and channels) at time interval number k

$$J_{\chi 2} = \sum_{\rho=1}^{p} \lambda_\rho \sum_{i=1}^{n_\chi} \sum_{k=1}^{L_\chi} g^{-}_{\chi i\rho k}; \tag{3}$$

The objective function describes throughput maximization (i.e., sales indicator).
Constraints:

$$\begin{aligned} & I_{\chi i\rho k} + y_{\chi i\rho(k-1)} + \sum_{j\in N^{-}_{\chi ik}} \omega_{\chi ji\rho k} \cdot u_{\chi ji\rho k} + \phi^{+}_{\chi i\rho k} \cdot \vartheta^{+}_{\chi i\rho k} = \\ & = \phi^{-}_{\chi i\rho k} \cdot \vartheta^{-}_{\chi i\rho k} + \sum_{j\in N^{+}_{\chi ik}} \omega_{\chi ij\rho k} \cdot u_{\chi ij\rho k} + y_{\chi i\rho k}\Big) + z_{\chi i\rho k} \end{aligned} \tag{4}$$

$$\left(\sum_{j\in N^{+}_{\chi ik}} x_{\chi ij\rho k} - \sum_{j\in N^{-}_{\chi ik}} x_{\chi ji\rho k}\right) + \left(y_{\chi i\rho k}\Big) - y_{\chi i\rho(k-1)}\right) + \left(g^{-}_{\chi i\rho k} - g^{+}_{\chi i\rho k}\right) + z_{\chi i\rho k} = I_{\chi i\rho k} \tag{5}$$

$$\begin{aligned} & 0 \le x_{\chi ij\rho k} \le \omega_{\chi ij\rho k} \cdot (t_k - t_{k-1}),\ 0 \le \sum_{\rho=1}^{p} y_{\chi i\rho k}\Big) \le V_{\chi i}, \\ & 0 \le g^{-}_{\chi i\rho k} \le \phi^{-}_{\chi i\rho k} \cdot (t_k - t_{k-1}), \\ & 0 \le g^{+}_{\chi i\rho k} \le \phi^{+}_{\chi i\rho k} \cdot (t_k - t_{k-1}),\ z_{\chi i\rho k} \ge 0. \end{aligned} \tag{6}$$

Equations (4) and (5) describe the flow balances of the product ρ subject to the node $A_{\chi i}$ and ensure that the sum of the outgoing flow (subscript (−)), inventory, and return flow should equal the incoming flow (subscript (+)). Equation (6) contains capacity and nonnegativity constraints.

5 Experimental Results

We consider a four-stage SC in the automotive industry. Two Tier 2 suppliers deliver speedometers to a Tier 1supplier that supplies two assembly plants with cockpits. The assembly plants deliver cars to one of two markets. SC structural elements can become fully or partially unavailable for a certain period of time. This capacity reduction may have a performance impact on sales in the markets.

For computational experiments, the following data set was used (Table 2).

In the first scenario, which we call "optimistic," a disruption at assembly plant #2 happens in the second period and destroys the capacity of this plant 100%. This disruption lasts two periods. In period #4, the capacity of this production plant is recovered 100%. In addition, in period #3 a fire at the Tier 2 supplier #1 happens that makes deliveries from this supplier to the Tier 1 supplier in period #3 impossible. In the next period, deliveries can run in normal mode again. Finally, due to strikes

Table 2 Input data

Parameter	Value
Number of periods	6
Demand distribution over six periods	*Market 1:* 250-240-230-240-250-240 *Market 2*: 220-210-200-210-220-210
Delivery quantity to tier 1 supplier	*From* supplier #1: 400 units in each period *From* supplier #2: 100 units in each period
Maximum processing throughput capacities at plants	*Tier 1 supplier:* 550 units a period Assembly plants: 300 units a period each
Maximum transportation throughput capacity	Channel tier 2 supplier #1 to the tier 1 supplier: 500 units a period Channel tier 2 supplier #2 to the tier 1 supplier: 150 units a period Channel tier 1 supplier to the assembly plant #1: 300 units a period, Channel tier 1 supplier to the assembly plant #2: 250 units a period Channel assembly plant #1 to the market #1: 280 units a period Channel assembly plant #2 to the market #2: 240 units a period
Maximum warehouse storage capacities	Tier 2 supplier #1: 150 units per period Tier 2 supplier #2: 70 units per period Tier 1 supplier: 250 units per period Assembly plant #1: 100 units a period Assembly plant #2: 100 units a period Market #1: 50 units a period Market #2: 50 units a period
Price of the final product. $	*65*
Bill-of-material factor	*1:1*, i.e., *one speedometer is needed for one cockpit, and one cockpit is needed for one car*

at a railway company in periods #4 and #6, the transportation channels between the assembly plant #1 and market #1 and between the Tier 2 supplier #2 and the Tier 1 supplier become unavailable, respectively (see Fig. 1).

In Fig. 1, the following parameters are represented:

- Tier 2 supplier #1: node #1
- Tier 2 supplier #2: node #2
- Tier 1 supplier: node #3
- Assembly plant #1: node #5
- Assembly plant #2: node #6
- Market #1: node #8
- Market #2: node #9

Maximum processing throughput capacities are marked in rectangles, maximum transportation throughput capacities are presented on the arcs, and maximum storage capacities are depicted in triangles. The disruptions are marked red. In Fig. 2, the so-called “pessimistic scenario” is shown.

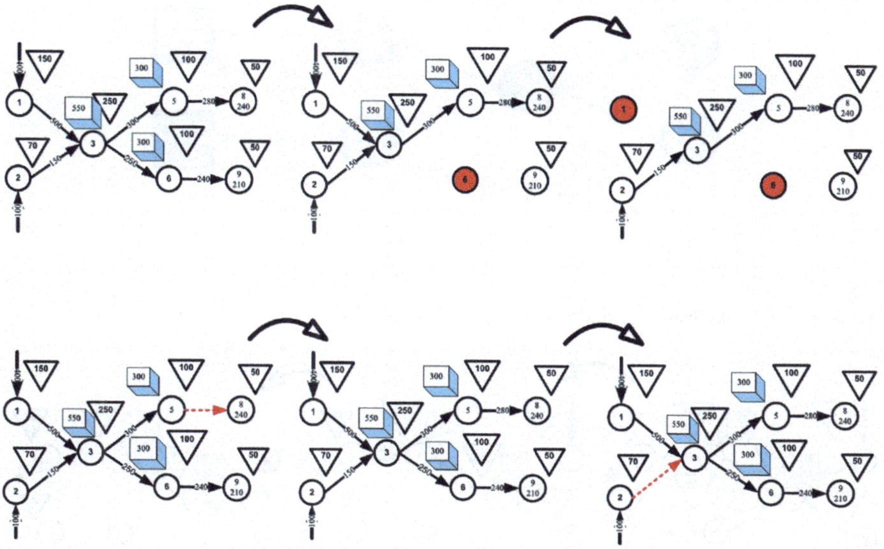

Fig. 1 "Optimistic" disruption scenario

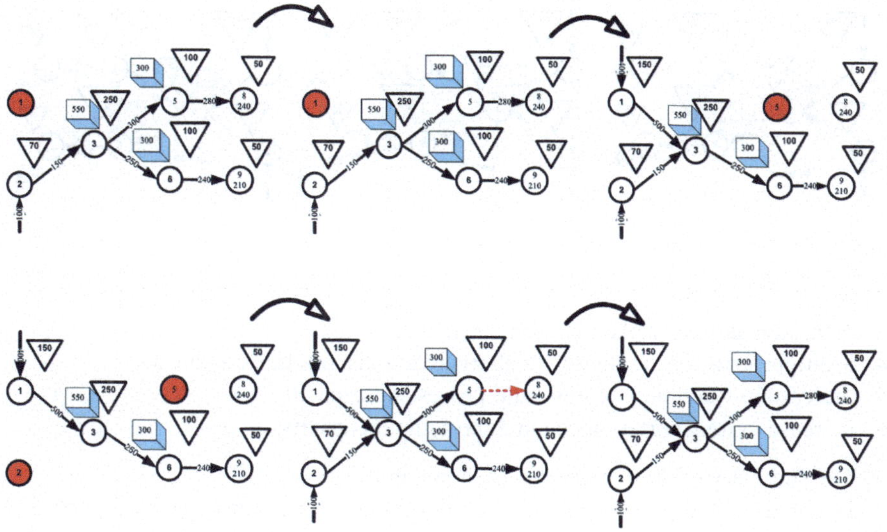

Fig. 2 "Pessimistic" disruption scenario

The following alternative proactive strategies for SC design can be considered:

1. Increase in the flexibility of supplier #2 so that it delivers, under normal conditions, 100 units in each period and can extend the quantity to 400 units if needed.
2. New backup supplier is introduced at the Tier 1 stage.

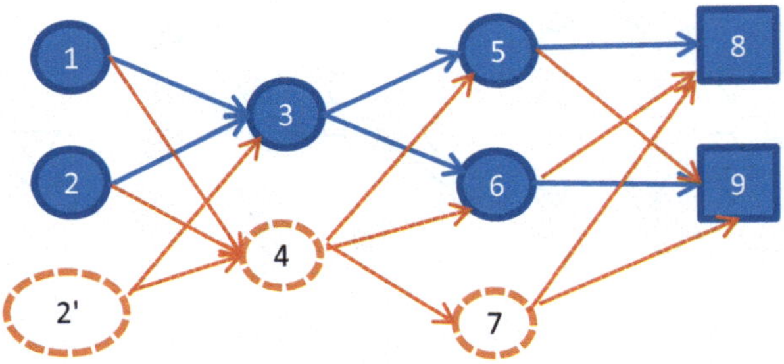

Fig. 3 Possible extensions to SC structure (Ivanov et al., 2016)

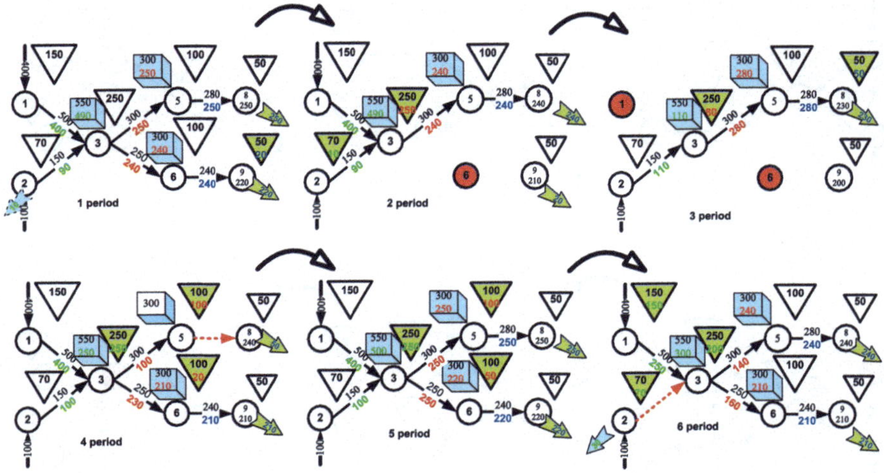

Fig. 4 Replanning results for optimistic scenario

3. A backup assembly plant is introduced.
4. Multiple sourcing strategy with alternative transportation channels.
5. Increase in warehouse storage and processing capacity.
6. Increase in transportation channel throughput capacity.

The summary of these measures is shown in Fig. 3.

In regard to the processing and transportation throughput capacity of new SC design elements, we assume the following: throughput capacity of new transportation channels is 120% of the processing throughput capacity of the outgoing node; processing capacity at node #4 is 250 units in each period, and processing capacity at node #7 is 150 units in each period.

In Fig. 4, the result of optimal replanning for the initial SC design (cf. Figs. 1 and 2) subject to the given data set (cf. Table 2) is presented.

In Fig. 3, the production and shipment quantities of the speedometers, cockpits, and cars are depicted and marked green, red, and blue, respectively. The yellow triangles show the storage capacities and their actual utilization. The grey rectangles

Table 3 Performance impact of different proactive policies in the optimistic scenario

№	Performance indicators	SC designs in the optimistic scenario						
		Initial	Proactive SC designs					
			1	2	3	4	5	6
1	Revenue in the disruption scenario	139,100	140,075	140,530	139,100	139,100	140,985	142,610
2	Maximum revenue in disruption-free scenario	176,800	176,800	176,800	176,800	176,800	176,800	176,800
3	Value of lost sales	*37,700*	*36,725*	*36,270*	*37,700*	*37,700*	*35,815*	*34,190*

Table 4 Performance impact of different proactive policies in the pessimistic scenario

№	Performance indicators	SC designs in the optimistic scenario						
		Initial	Proactive SC designs					
			1	2	3	4	5	6
1	Revenue in the disruption scenario	83,200	83,200	83,200	85,800	88,400	85,800	88,400
2	Maximum revenue in disruption-free scenario	176,800	176,800	176,800	176,800	176,800	176,800	176,800
3	Value of lost sales	*93,600*	*93,600*	*93,600*	*91,000*	*88,400*	*91,000*	*88,400*

depict the manufacturing maximum and used processing capacity, respectively. The numbers on the arcs represent the maximum transportation capacities and their actual utilization. The red nodes and channels are disrupted. The yellow arrows at nodes #8 and #9 depict the delivered quantity of goods at the markets.

In Tables 3 and 4 and Figs. 5 and 6, the planning results are summarized for two scenarios.

It can be observed from Tables 3 and 4 and Figs. 5 and 6 that structural changes significantly impact SC performance. The highest revenue of $140,985 in the optimistic scenario can be achieved if we apply the SC design according to proactive strategy #5, i.e., an increase in warehouse storage and processing capacity is needed.

The highest revenue of $85,800 in the pessimistic scenario can be achieved if we apply the SC design according to proactive strategies #4 and #6, i.e., introduction of alternative transportation channels or/and increase in transportation capacity of the

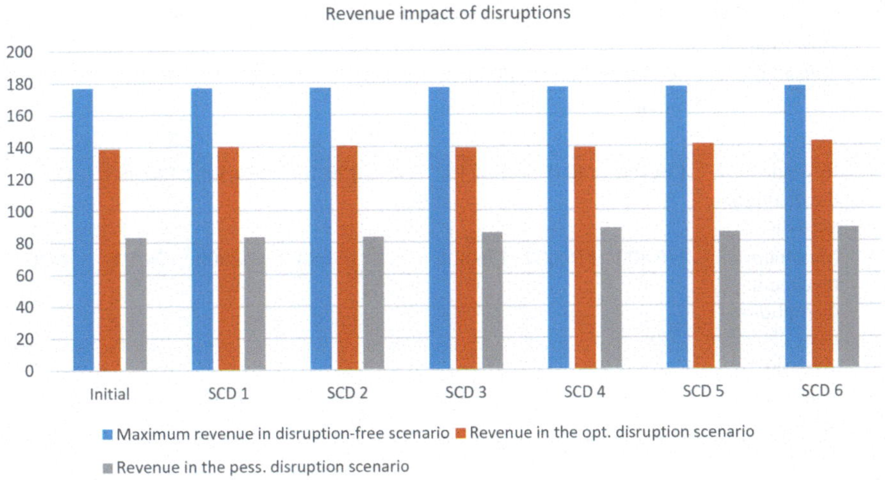

Fig. 5 Revenue impact of disruptions

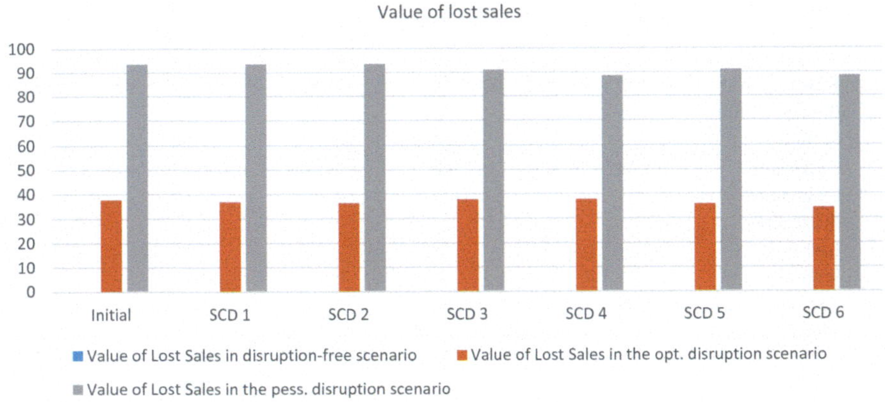

Fig. 6 Lost sales impact of disruptions

existing channels. To make a final decision, the costs of these proactive measures need to be analyzed.

Further, it can be observed that lost sales are much higher in the "pessimistic" scenario. Such a scenario comparison allows the drawing of conclusions on the performance impact of disruptions at different parts in the SC. For example, the Tier 2 supplier #1 (node #1), the production plant #1 (node #5), and the transportation channel between the node #5 and node #8 need to be considered as critical elements in the SC design. According to the study of Simchi-Levi et al. (2015), dual-sourcing policies, risk-sharing contracts, and continuous capacity monitoring can be recommended for these SC elements.

6 Performance Impact of the Ripple Effect and Managerial Insights

Now, we use the computation results from Sect. 4 and compute the PI index using Eqs. 1 and 2. The diagrams with the computational results for seven SC designs for both scenarios can be found in an online supplement. First, we compute PI for two markets individually using Eq. 1. In the next step, we aggregate these partial PI to general PI using Eq. 2. We consider SC sales (revenue) as the KPI for measuring the estimated annual magnitude of the ripple effect. To make the analysis more depictive, we restrict ourselves to the PI consideration at the markets without PI analysis at the intermediate stage in the SC. This is an admissible restriction since the revenue KPI is directly related to the market stage in the SC. The results of PI computation are shown in Table 5.

Let us present these results graphically and analyze their managerial implications (Figs. 7, 8 and 9).

Table 5 and Figs. 7, 8 and 9 can be used as a dashboard for SC design comparison in regard to the ripple effect. The results can also help to analyze the disruption impact at different markets individually in order to derive recommendations for securing supplier and customer satisfaction. For the given data set it can be observed that the SC for market #2 (node #9) is much more resilient and was less exposed to the ripple effect compared to market #1 (node #8). In both scenarios, the PI in regard to market #2 (node #9) does not exceed 1.51 while the maximum PI in regard to market #1 (node #8) is 6.04.

In the case of the optimistic scenario, the lowest ripple effect can be observed in SC design #6 both at the market #1 (node #8) and at the market #2 (node #9), PI = 1.14284 and PI = 1.3744, respectively. In the case of the pessimistic scenario, the lowest ripple effect can be observed in the initial SC design at market #1 (node #8, PI = 3.295455), while at market #2 (node #9), minimum PI = 1.133929 for SC designs #4 and #6.

In total cumulative PI in the optimistic scenario, a large gap between the minimum PI (i.e., initial SCD) and PI for other SC designs can be observed in Fig. 9. At the same time, the total PI values in the pessimistic scenario are very close to each other with a small advantage for SC design #6.

Therefore, PI computation supports the results of the optimization model described in Sect. 4. For the considered example, it becomes obvious that market #1 (node #8) is highly exposed to the ripple effect in the negative scenario while market #2 (node #9) shows disruption resistance and performs in the pessimistic scenario even better than in the optimistic one.

In the joint analysis of two markets, the initial SC design can be recommended since it exhibits the lowest ripple effect. At market #8, risk-sharing contracts can be recommended with the customers. In addition, storage capacity extensions and higher safety stocks can be applied in this market.

Table 5 PI computation results

	Initial	SCD 1	SCD 2	SCD 3	SCD 4	SCD 5	SCD 6
Maximum revenue market #8 in the disruption-free scenario	94,250	94,250	94,250	94,250	94,250	94,250	94,250
Revenue market #8 in the opt. Disruption scenario	81,900	82,225	82,470	81,900	81,900	82,225	82,470
PI market #8 in the opt. disruption scenario	*1,150,794*	*1,146,245*	*1,150,794*	*1,150,794*	*1,150,794*	*1,146,245*	*114,284*
Revenue market #8 in the pess. disruption scenario	28,600	15,600	15,600	15,600	15,600	15,600	15,600
PI market #8 in the pess. disruption scenario	*3,295,455*	*6,041,667*	*6,041,667*	*6,041,667*	*6,041,667*	*6,041,667*	*6,041,667*
Maximum revenue market #9 in the disruption-free scenario	82,550	82,550	82,550	82,550	82,550	82,550	82,550
Revenue market #9 in the opt. Disruption scenario	57,200	57,850	57,850	57,200	57,200	58,760	59,930
PI market #9 in the opt. disruption scenario	*1,443,182*	*1,426,966*	*1,426,966*	*1,443,182*	*1,443,182*	*1,404,867*	*137,744*
Revenue market #9 in the pess. disruption scenario	54,600	67,600	67,600	70,200	72,800	70,200	72,800
PI market #9 in the pess. disruption scenario	*1,511,905*	*1,221,154*	*1,175,926*	*1,175,926*	*1,133,929*	*1,175,926*	*1,133,929*

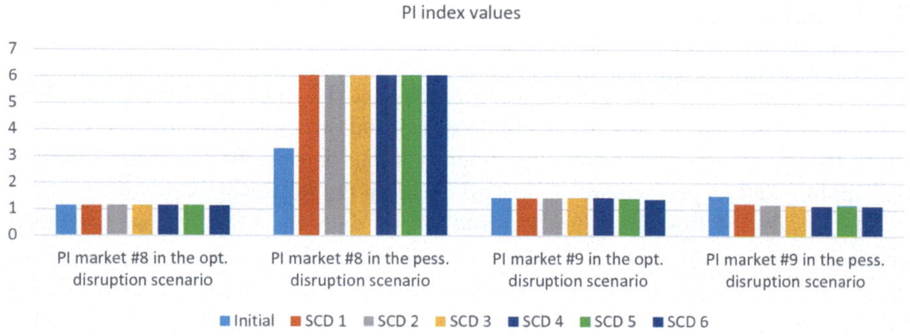

Fig. 7 PI computation for individual markets

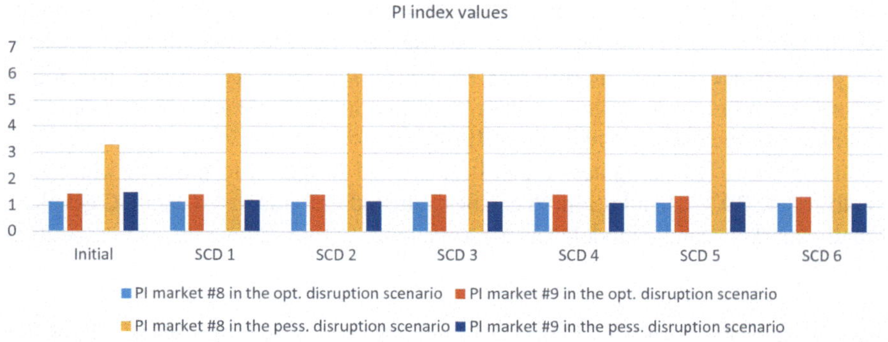

Fig. 8 PI computation for individual markets and different SC designs

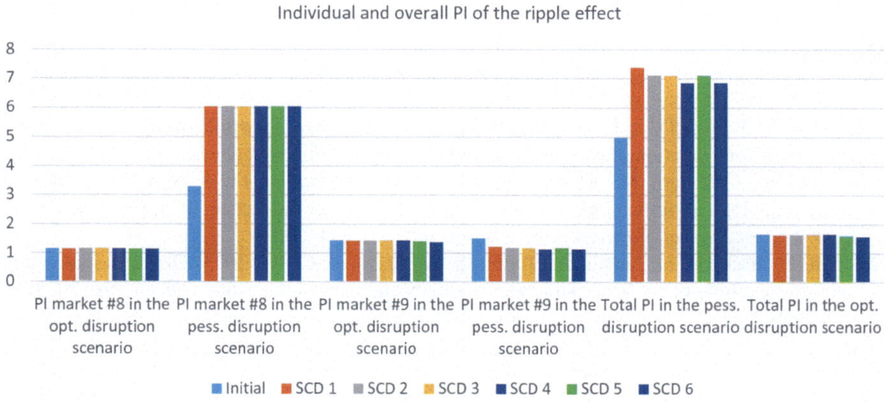

Fig. 9 Comparison of individual and total PI computation

7 Conclusions

The experiments with the optimization model and PI index analysis depict how disruption risks may result in the ripple effect and structure dynamics in the SC. From the developed model and experiments, it can be observed how the scope of the rippling and its performance impact depend on the SC design structure.

The results of this study are twofold. First, with the help of the developed approach, severe disruptions in the SC can be modeled subject to temporary unavailability of some SC elements and their recovery. Second, a method to compare SC designs with a performance impact assessment of the ripple effect has been developed.

The developed method of ripple effect evaluation helps to analyze effective ways to recover and reallocate resources and flows in the SC and to select a resilient SC design. As such, the model can be used by SC risk specialists to analyze the performance impact of different resilience and recovery actions and adjust mitigation and recovery policies with regard to critical SC design elements and SC planning parameters (Pavlov et al., 2019b).

In the future, PI computation can be extended in regard to multiple objectives Pavlov et al., 2024). In addition, PI can be evaluated individually at each SC echelon. Finally, markets can be modeled as heterogeneous entities, allowing for competitive elements.

Acknowledgments The research of Boris Sokolov related to this chapter was supported by project No. 24-29-00706.

References

Aldrighetti, R., Battini, D., & Ivanov, D. (2023a). Efficient resilience portfolio design in the supply chain with consideration of preparedness and recovery investments. *Omega, 117*, 102841.

Aldrighetti, R., Calzavara, M., Martignago, M., Zennaro, I., Battini, D., & Ivanov, D. (2023b). A methodological framework for the design of efficient resilience in supply networks. *International Journal of Production Research, 62*(1–2), 271–290.

Alikhani, R., Ranjbar, A., Jamali, A., Torabi, S. A., & Zobel, C. W. (2023). Towards increasing synergistic effects of resilience strategies in supply chain network design. *Omega, 116*, 102819.

Aron, C., Sgarbossa, F., Ballot, E., & Ivanov, D. (2024). Cloud material handling systems: A cyber-physical system to enable dynamic resource allocation and digital Interoperability. *Journal of Intelligent Manufacturing, 35*, 3815–3836.

Basole, R. C., & Bellamy, M. A. (2014). Supply network structure, visibility, and risk diffusion: A computational approach. *Decision Sciences, 45*(4), 1–49.

Blackhurst, J., Rungtusanatham, M. J., Scheibe, K., & Ambulkar, S. (2018). Supply chain vulnerability assessment: A network based visualization and clustering analysis approach. *Journal of Purchasing and Supply Management, 24*(1), 21–30.

Carvalho, H., Barroso, A. P., Machado, V. H., Azevedo, S., & Cruz-Machado, V. (2012). Supply chain redesign for resilience using simulation. *Computers & Industrial Engineering, 62*(1), 329–341.

Cavalcantea, I. M., Frazzon, E. M., Forcellinia, F. A., & Ivanov, D. (2019). A supervised machine learning approach to data-driven simulation of resilient supplier selection in digital manufacturing. *International Journal of Information Management, 49*, 86–97.

Choi, T. M., & Shi, X. (2022). Reducing supply risks by supply guarantee deposit payments in the fashion industry in the "new normal after COVID-19". *Omega, 109*, 102605.

Dolgui, A., Ivanov, D., & Rozhkov, M. (2020). Does the ripple effect influence the bullwhip effect? An integrated analysis of structural and operational dynamics in the supply chain. *International Journal of Production Research, 58*, 1285–1301.

Han, J., & Shin, K. S. (2016). Evaluation mechanism for structural robustness of supply chain considering disruption propagation. *International Journal of Production Research, 54*(1), 135–151.

He, J., Alavifard, F., Ivanov, D., & Jahani, H. (2019). A real-option approach to mitigate disruption risk in the supply chain. *Omega, 88*, 133–149.

Ho, W., Zheng, T., Yildiz, H., & Talluri, S. (2015). Supply chain risk management: A literature review. *International Journal of Production Research, 53*(16), 5031–5069.

Hosseini, S., & Barker, K. (2016). A Bayesian network for resilience-based supplier selection. *International Journal of Production Economics, 180*, 68–87.

Hosseini, S., Ivanov, D., & Dolgui, A. (2019a). Review of quantitative methods for supply chain resilience analysis. *Transportation Research: Part E, 125*, 285–307.

Hosseini, S., Morshedlou, N., Ivanov, D., Sarder, M. D., Barker, K., & Al Khaled, A. (2019b). Resilient supplier selection and optimal order allocation under disruption risks. *International Journal of Production Economics, 213*, 124–137.

Iftikhar, A., Purvis, L., & Giannoccaro, I. (2021). A meta-analytical review of antecedents and outcomes of firm resilience. *Journal of Business Research, 135*, 408–425.

Ivanov, D. (2017). Simulation-based ripple effect modelling in the supply chain. *International Journal of Production Research, 55*(7), 2083–2101.

Ivanov, D. (2018). *Structural dynamics in supply chain risk management*. Springer.

Ivanov, D. (2019). Disruption tails and revival policies: A simulation analysis of supply chain design and production-ordering systems in the recovery and post-disruption periods. *Computers and Industrial Engineering, 127*, 558–570.

Ivanov, D. (2023a). Collaborative emergency adaptation for ripple effect mitigation in intertwined supply networks. *Annals of Operations Research*. https://doi.org/10.1007/s10479-023-05408-0

Ivanov, D. (2023b). Intelligent digital twin (iDT) for supply chain stress-testing, resilience, and viability. *International Journal of Production Economics, 263*, 108938.

Ivanov, D. (2023c). Design and deployment of sustainable recovery strategies in the supply chain. *Computer & Industrial Engineering, 183*, 109444.

Ivanov, D. (2024a). Transformation of supply chain resilience research through the COVID-19 pandemic. *International Journal of Production Research, 62*(23), 8217–8238.

Ivanov, D. (2024b). Exiting the COVID-19 pandemic: After-shock risks and avoidance of disruption tails in supply chains. *Annals of Operations Research, 335*, 1627–1644.

Ivanov, D. (2024c). Conceptualisation of a 7-element digital twin framework in supply chain and operations management. *International Journal of Production Research, 62*(6), 2220–2232.

Ivanov, D. (2025a). When is the supply chain resilient? Customer and operational perspectives. *International Journal of Production Research*, 1–16. https://doi.org/10.1080/00207543.2025.2454331

Ivanov, D. (2025b). Comparative analysis of product and network supply chain resilience. *International Transactions in Operational Research*. https://doi.org/10.1111/itor.13612

Ivanov, D. (2025c). What network and performance indicators can tell us about supply chain and sourcing resilience (and what they cannot). *Journal of Purchasing and Supply Management, 101014*.

Ivanov, D., & Keskin, B. B. (2023). Post-pandemic adaption and development of supply chain viability theory. *Omega, 116*, 102806.

Ivanov, D., & Rozhkov, M. (2020). Coordination of production and ordering policies under capacity disruption and product write-off risk: An analytical study with real-data based simulations of a fast moving consumer goods company. *Annals of Operations Research, 291*, 387–407.

Ivanov, D., Sokolov, B., & Pavlov, A. (2013). Dual problem formulation and its application to optimal re-design of an integrated production–distribution network with structure dynamics and ripple effect considerations. *International Journal of Production Research, 51*(18), 5386–5403.

Ivanov, D., Sokolov, B., & Dolgui, A. (2014a). The ripple effect in supply chains: Trade-off 'efficiency-flexibility-resilience' in disruption management. *International Journal of Production Research, 52*(7), 2154–2172.

Ivanov, D., Sokolov, B., & Pavlov, A. (2014b). Optimal distribution (re)planning in a centralized multi-stage network under conditions of ripple effect and structure dynamics. *European Journal of Operational Research, 237*(2), 758–770.

Ivanov, D., Sokolov, B., Pavlov, A., Dolgui, A., & Pavlov, D. (2016). Disruption-driven supply chain (re)-planning and performance impact assessment with consideration of pro-active and recovery policies. *Transportation Research Part E, 90*, 7–24.

Ivanov, D., Pavlov, A., Pavlov, D., & Sokolov, B. (2017a). Minimization of disruption-related return flows in the supply chain. *International Journal of Production Economics, 183*, 503–513.

Ivanov, D., Tsipoulanidis, A., & Schönberger, J. (2017b). *Global supply chain and operations management* (1st ed.). Springer.

Ivanov, D., Dolgui, A., Sokolov, B., & Ivanova, M. (2017c). Literature review on disruption recovery in the supply chain. *International Journal of Production Research, 55*(20), 6158–6174.

Ivanov, D., Dolgui, A., & Sokolov, B. (2018). Scheduling of recovery actions in the supply chain with resilience analysis considerations. *International Journal of Production Research, 56*(19), 6473–6490.

Ivanov, D., Dolgui, A., Blackhurst, J. V., & Choi, T. M. (2023). Toward supply chain viability theory: From lessons learned through COVID-19 pandemic to viable ecosystems. *International Journal of Production Research, 61*(8), 2402–2415.

Käki, A., Salo, A., & Talluri, S. (2015). Disruptions in supply networks: A probabilistic risk assessment approach. *Journal of Business Logistics, 36*(3), 273–287.

Liberatore, F., Scaparra, M. P., & Daskin, M. S. (2012). Hedging against disruptions with ripple effects in location analysis. *Omega, 40*(1), 21–30.

Lim, M. K., Bassamboo, A., Chopra, S., & Daskin, M. S. (2013). Facility location decisions with random disruptions and imperfect estimation. *Manufacturing and Service Operations Management, 15*(2), 239–249.

Lin, Y. K., Huang, C. F., Liao, Y.-C., & Yeh, C. T. (2017). System reliability for a multistate intermodal logistics network with time windows. *International Journal of Production Research, 55*(7), 1957–1969.

Liu, M., Ding, Y., Chu, F., Dolgui, A., & Zheng, F. (2024). Robust actions for improving supply chain resilience and viability. *Omega, 123*, 102972.

Nair, A., & Vidal, J. M. (2011). Supply network topology and robustness against disruptions: An investigation using multiagent model. *International Journal of Production Research, 49*(5), 1391–1404.

Ojha, R., Ghadge, A., Tiwari, M. K., & Bititci, U. S. (2018). Bayesian network modelling for supply chain risk propagation. *International Journal of Production Research, 56*(17), 5795–5819.

Paul, S. K., Sarker, R., & Essam, D. (2014). Real time disruption management for a two-stage batch production–inventory system with reliability considerations. *European Journal of Operational Research, 237*, 113–128.

Pavlov, A., Ivanov, D., Dolgui, A., & Sokolov, B. (2018). Hybrid fuzzy-probabilistic approach to supply chain resilience assessment. *IEEE Transactions on Engineering Management, 65*(2), 303–315.

Pavlov, A., Ivanov, D., Pavlov, D., & Slinko, A. (2019a). Optimization of network redundancy and contingency planning in sustainable and resilient supply chain resource management under conditions of structural dynamics. *Annals of Operations Research.* https://doi.org/10.1007/s10479-019-03182-6

Pavlov, A., Pavlov, D., & Zakharov, V. (2019b). Possible ways of assessing the resilience of supply chain networks in conditions of unpredictable disruptions. *IFAC-PapersOnLine, 52*(13), 1283–1288. https://doi.org/10.1016/j.ifacol.2019.11.375

Pavlov, A. N., Pavlov, D. A., Kulakov, A. Y., & Zakharov, V. V. (2024). Study of technology for the reliability and survivability modelling of onboard control system of small spacecraft operating in complex modes. *Journal of Applied Engineering Science, 22*(3), 612–620. https://doi.org/10.5937/jaes0-50149

Raj, R., Wang, J. W., Nayak, A., Tiwari, M. K., Han, B., Liu, C. L., et al. (2015). Measuring the resilience of supply chain systems using a survival model. *IEEE Systems Journal, 9*(2), 377–381.

Ramani, V., Ghosh, D., & Sodhi, M. S. (2022). Understanding systemic disruption from the Covid-19-induced semiconductor shortage for the auto industry. *Omega, 113*, 102720.

Rolf, B., Jackson, I., Müller, M., Lang, S., Reggelin, T., & Ivanov, D. (2023). A review on reinforcement learning alogrithms and applications in supply chain management. *International Journal of Production Research, 61*(20), 7151–7179.

Sardesai, S., & Klingebiel, K. (2023). Maintaining viability by rapid supply chain adaptation by using a process capability index. *Omega, 115*, 102778.

Sawik, T. (2016). On the risk-averse optimization of service level in a supply chain under disruption risks. *International Journal of Production Research, 54*(1), 98–113.

Sawik, T. (2017). A portfolio approach to supply chain disruption management. *International Journal of Production Research, 55*(7), 1970–1991.

Sawik, T. (2022). Stochastic optimization of supply chain resilience under ripple effect: A COVID-19 pandemic related study. *Omega, 109*, 102596.

Sawik, T. (2023). A stochastic optimisation approach to maintain supply chain viability under the ripple effect. *International Journal of Production Research, 61*(8), 2452–2469.

Schmitt, A. J., & Singh, M. (2012). A quantitative analysis of disruption risk in a multi-echelon supply chain. *International Journal of Production Economics, 139*(1), 23–32.

Schmitt, T. G., Kumar, S., Stecke, K. E., Glover, F. W., & Ehlen, M. A. (2017). Mitigating disruptions in a multi-echelon supply chain using adaptive ordering. *Omega, 68*, 185–198.

Simchi-Levi, D., Schmidt, W., Wei, Y., Zhang, P. Y., Combs, K., Ge, Y., et al. (2015). Identifying risks and mitigating disruptions in the automotive supply chain. *Interfaces, 45*(5), 375–390.

Snyder, L. V., Zümbül, A., Peng, P., Ying, R., Schmitt, A. J., & Sinsoysal, B. (2016). OR/MS models for supply chain disruptions: A review. *IIE Transactions, 48*(2), 89–109.

Sokolov, B., Ivanov, D., Dolgui, A., & Pavlov, A. (2016). Structural analysis of the ripple effect in the supply chain. *International Journal of Production Research, 54*(1), 152–169.

Stecke, K. E., & Kumar, S. (2009). Sources of supply chain disruptions, factors that breed vulnerability, and mitigating strategies. *Journal of Marketing Channels, 16*, 193–226.

Talluri, S., Kull, T. J., Yildiz, H., & Yoon, J. (2013). Assessing the efficiency of risk mitigation strategies in supply chains. *Journal of Business Logistics, 34*(4), 253–269.

Tomlin, B. (2006). On the value of mitigation and contingency strategies for managing supply chain disruption risks. *Management Science, 52*, 639–657.

Tuncel, G., & Alpan, G. (2010). Risk assessment and management for supply chain networks–a case study. *Computers in Industry, 61*(3), 250–259.

Wilson, M. C. (2007). The impact of transportation disruptions on supply chain performance. *Transportation Research Part E: Logistics and Transportation Review, 43*, 295–320.

Wu, T., Blackhurst, J., & O'Grady, P. (2007). Methodology for supply chain disruption analysis. *International Journal of Production Research, 45*(7), 1665–1682.

Zhao, K., Kumar, A., Harrison, T. P., & Yen, J. (2011). Analyzing the resilience of complex supply network topologies against random and targeted disruptions. *IEEE Systems Journal, 5*(1), 28–39.

Zobel, C. W. (2011). Representing perceived tradeoffs in defining disaster resilience. *Decision Support Systems, 50*(2), 394–403.

Zobel, C. W. (2014). Quantitatively representing nonlinear disaster recovery. *Decision Sciences, 45*(6), 1053–1082.

AI-Powered Supply Chain and Operations Management (SCOM): Capabilities and Challenges

Yuhong Li, Kedong Chen, and Jiawei Zhang

Abstract Artificial intelligence (AI) has emerged as one of the most transformative technologies that revolutionize various sectors including supply chain and operations management (SCOM). AI's ability to process large datasets, generate predictions, and optimize decision-making processes offers significant operational advantages. However, the integration of AI also introduces challenges, particularly from the aspects of data management and integration, operational and technical challenges in decision-making, and relational and ethical considerations. The enhanced interconnectedness, when combined with AI-powered automation, also significantly amplifies the ripple effect throughout the supply chain when disruptions occur. This chapter explores the capabilities and challenges of AI in SCOM, focusing on data management, decision-making under uncertainties, and buyer-supplier relationship management in the context of resilience and ripple effect. Through a systematic review of the literature, we aim to provide a comprehensive understanding of AI's role in SCOM and present frameworks for managing the complexities and risks associated with AI-powered operations.

1 Introduction

Artificial intelligence (AI) has been recognized as one of the most transformative technologies in recent years (Senoner et al., 2023; Jackson et al., 2024; Rolf et al., 2023). The public release of ChatGPT in November 2022 especially showcased

Y. Li (✉)
Strome College of Business, Old Dominion University, Norfolk, VA, USA
e-mail: y9li@odu.edu

K. Chen
Lally School of Management, Rensselaer Polytechnic Institute, Troy, NY, USA
e-mail: chenk16@rpi.edu

J. Zhang
PRA Group, Norfolk, VA, USA

D. Ivanov et al. (eds.), *Handbook of Ripple Effects in the Supply Chain*,
International Series in Operations Research & Management Science 276,
https://doi.org/10.1007/978-3-031-85508-5_7

the remarkable capabilities of AI (Fosso Wamba et al., 2024) on communication, data processing, and problem-solving across various domains. The capabilities demonstrated by ChatGPT have gained widespread global attention, redefining the scope of what AI can accomplish and setting a new standard for machine-human interaction. This breakthrough has ignited further exploration of AI's potential across various business domains (Jackson et al., 2024; Wu et al., 2022; Dolgui & Ivanov, 2023; Jackson & Ivanov, 2023), expanding the horizons of how AI can be utilized to enhance operational efficiency, improve decision-making processes, and drive innovation in areas such as supply chain management, finance, healthcare, and beyond.

In response to the growing demands for operational efficiency, cost reduction, and enhanced customer satisfaction, AI has emerged as a critical enabler, offering capabilities that were previously beyond the reach of traditional methods (Simchi-Levi & Wu, 2018; Dwivedi et al., 2021). The transformative potential of AI in supply chain management lies in its ability to process vast amounts of data, generate accurate predictions, and optimize complex decision-making processes (Rodríguez-Espíndola et al., 2020; Cannas et al., 2024; Le & Behl, 2023). For example, Amazon uses AI algorithms to analyze vast data from different sources (AmazonScience, 2021), including customer purchasing patterns, seasonal trends, and external impactors such as economic indicators and social media sentiment. Through in-depth analysis, Amazon has built an AI-driven demand forecasting system, allowing the company to optimize its inventory levels, reduce storage costs, and ensure that products are available when customers want them. This precise forecasting helps Amazon maintain its competitive edge by improving supply chain efficiency and customer satisfaction.

However, while the integration of AI in supply chain offers numerous benefits, it also introduces new challenges and risks that must be carefully managed and prepared (Sodhi et al., 2022; Brintrup et al., 2024; Kumar & Mallipeddi, 2022). For example, Maersk, the global shipping giant, has heavily invested in AI technologies to optimize its shipping routes, predict equipment maintenance needs, and enhance overall operational efficiency. However, during the NotPetya cyberattack in 2017, Maersk's AI-driven systems were severely disrupted. The ransomware attack spread rapidly across Maersk's global network, causing massive operational delays and financial losses estimated at up to $300 million (Jordan, 2017). This incident highlights the vulnerabilities that AI systems can introduce, particularly related to cybersecurity. Despite AI's ability to enhance efficiency, the interconnected nature of AI-driven supply chains also creates potential points of failure that have not been exploited before.

The adoption of AI technologies introduces new complexities and vulnerabilities that necessitate a thorough reevaluation of existing risk management frameworks tailored specifically for AI-powered supply chains. Being motivated, we explore the capabilities and challenges associated with AI-powered supply chain and operations management (SCOM), focusing on three primary aspects: data management, decision-making, and relationship management. Through a systematic review of existing literature, we aim to provide a comprehensive overview of the current state of AI in SCOM, identifying key trends, methodologies, and areas of focus within the research landscape. Additionally, we will present frameworks that can help better understand and manage the complexities introduced by AI in SCOM.

The rest of this chapter is organized as follows: Sect. 2 provides background information on AI technologies commonly applied in SCOM and their capabilities. Section 3 presents an in-depth review of the literature on AI in supply chains and topic analyses of news about AI in supply chains. Section 4 provides frameworks that help better understand the challenges of managing AI-powered supply chains. Section 5 concludes this chapter.

2 Background of AI

AI is a technology that enables computers and machines to learn from experiences, adjust to new inputs, and perform human intelligence and problem-solving capabilities (Zamani et al., 2023; Li & Li, 2022; Cui et al., 2022). To be able to perform human-like decisions, AI generally operates through three major phases: perception, decision-making, and learning (Jackson et al., 2024; Fosso Wamba et al., 2024). In the perception phase, AI systems gather and process various forms of input, including data from sensors, human natural language, images, audio, and more. This diverse range of inputs allows AI to interpret and understand the task environment. During the decision-making phase, AI systems autonomously analyze the processed data to make informed decisions and predictions, enabling them to perform human-like complex tasks. In the learning phase, AI continuously analyzes the inputs to build its knowledge base, identify underlying patterns, and improve its performance over time. This learning process is often iterative, enabling AI to adapt to new data and situations.

This section provides a general introduction to widely employed AI technologies (Fig. 1), particularly those closely related to and commonly applied in supply chain and operations management (SCOM), such as machine learning (ML) and natural language processing. These technologies offer significant potential for enhancing SCOM. We will discuss some of their key capabilities and how they can be effectively leveraged to improve supply chain performance.

2.1 Common AI Technologies

Machine learning is perhaps the most widely used AI technology, which can be broadly categorized into supervised learning, unsupervised learning, and reinforcement learning (James, 2013). Supervised learning is a foundational approach that leverages labeled datasets to train models for outcome prediction and pattern recognition. In contrast, unsupervised learning involves grouping unlabeled data

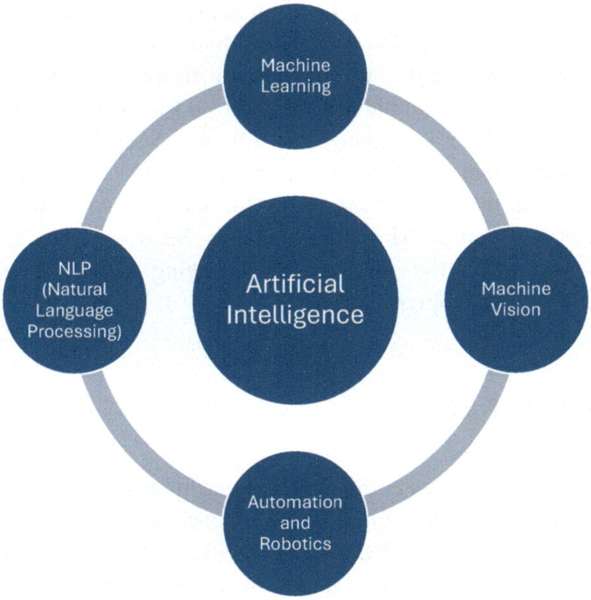

Fig. 1 Top four techniques of AI, adapted from Pushkar (2024)

points into meaningful clusters where data points share similar characteristics (James, 2013). Distinct from both, reinforcement learning trains models to achieve a defined goal by evaluating cumulative rewards across different actions. It is guided by the Markov decision process and systematically trains an agent to explore and evaluate a sequence of possible actions over time.

Each of the three machine learning techniques offers distinct approaches and applications. Supervised learning relies on labeled data to train models, making it highly effective for tasks like spam filtering, image classification, and defect detection, where clear input-output pairs are available. In contrast, unsupervised learning does not require labeled data, allowing it to excel in discovering hidden patterns within data, such as in anomaly detection, recommender systems, and topic modeling. This method is particularly useful when the underlying structure of the data is unknown. Reinforcement learning differs from both by focusing on decision-making in dynamic environments. It is well suited for applications like self-driving vehicles, healthcare treatments, and robotics, where the system must continuously learn and adapt to achieve optimal outcomes. In sum, supervised and unsupervised learning typically deal with static data, and reinforcement learning's strength lies in its ability to handle continuous feedback and evolving scenarios.

Natural language processing (NLP) is also widely used in practices such as language translation, topic modeling, large language models (LLMs), and generative intelligent chatbots (Jackson et al., 2024). At its core, NLP involves the transformation of raw text data into a standardized format. NLP models are designed to tackle a variety of text-related tasks. For instance, topic modeling, an unsupervised learning approach, identifies hidden themes within a large corpus of text by summarizing keywords associated with different topics. This is particularly valuable for categorizing extensive text documents and uncovering underlying semantic structures. Sentiment analysis, on the other hand, is a supervised classification model that predicts the sentiment expressed in a text—whether positive, negative, or neutral—toward a specific subject, such as a policy or a movie. Another significant advancement in NLP is the development of LLMs, which are deep learning models trained on vast amounts of text data using transformer architectures. These models have the capacity to understand and generate human-like text, enabling a wide range of applications, from text recognition to complex generative tasks.

Another AI technology, machine vision, also has wide-ranging applications. Machine vision models can extract, transform, and interpret visual information from images and videos, identify objects, and respond appropriately to visual stimuli. Perhaps the most famous example is autonomous driving. Machine vision models continuously analyze video input from sensors, extracting crucial information such as traffic signals, lane markings, and road conditions, which is then fed into reinforcement learning models to guide decision-making and vehicle control. Machine vision also powers facial recognition for security and authentication and agricultural applications like crop health monitoring. Overall, various AI technologies are transforming industries by enabling advanced data analysis, automation, and decision-making across diverse applications.

2.2 AI Capabilities in SCOM

AI technologies have profoundly affected SCOM in the following three aspects, namely richer data generation and collection, advanced analytics, and strengthened buyer-supplier relationships (Min, 2010). At the core of AI's impact is its ability to collect and process vast amounts of data from diverse sources, such as sensors, market trends, and customer behavior (Choi et al., 2018), thereby laying a robust data foundation that supports advanced analytics and facilitates more accurate decision-making (Min, 2010). Machine learning algorithms and predictive models allow organizations to anticipate demand fluctuations, optimize inventory levels, and streamline operations with unprecedented precision (Wamba et al., 2015). Additionally, AI enhances supplier-buyer relationships by improving data transparency, fostering real-time communication, and enabling collaborative forecasting, which collectively contribute to a more resilient and efficient supply chain (Shahzadi et al., 2024).

AI's ability to generate and analyze vast amounts of data has fundamentally changed how supply chains operate. Utilizing advanced sensors, Internet of Things (IoT) devices, Robotic Process Automation (RPA), and other advanced technologies, AI continuously gathers real-time data from various sources in supply chains, such as inventory levels, shipment statuses, and environmental conditions. The data collection process is automated and more accurate, leading to a richer, more comprehensive dataset with minimal human error and manual effort. Companies like Amazon and Unilever exemplify the practical applications of AI in this context. Amazon's AI-powered systems, particularly in its automated warehouses, continuously update data on inventory levels, item locations, and movement patterns, significantly reducing errors and enhancing operational efficiency (About Amazon Team, 2024). Similarly, Unilever leverages AI-driven natural language processing tools to analyze unstructured data from online platforms, enabling the company to rapidly adapt its product offerings to evolving consumer preferences (Cooper, 2023).

AI also enhances decision-making in supply chains by employing predictive analytics and providing data-driven insights. By analyzing real-time data, AI identifies patterns and trends that inform critical decisions in areas such as inventory management, demand forecasting, and route optimization. AI can also simulate various scenarios, helping managers foresee potential disruptions and proactively develop contingency plans (Ganesh & Kalpana, 2022). This simulation enables supply chain managers to explore "what-if" scenarios, allowing them to assess the potential impact of different strategies and decisions under varying conditions. For instance, Zara has harnessed AI to optimize trend forecasting and inventory management by analyzing social media and sales data in real time, which allows the company to adjust its product offerings swiftly (Quan et al., 2023). Similarly, General Electric (GE) employs AI-powered digital twins to simulate various operational scenarios, enabling predictive maintenance and optimizing asset utilization (Ivanov, 2023; Samuel et al., 2023).

Furthermore, AI strengthens buyer-supplier relationships by facilitating seamless information sharing and real-time communication (Grover et al., 2024). This enhanced communication, transparency, and collaboration allow suppliers and buyers to align their operations more closely. Additionally, AI can monitor supplier performance and compliance with agreed standards, providing both buyers and suppliers with the information needed to maintain high levels of quality and reliability (Pandey et al., 2024). Programs like Siemens' AI-powered supplier collaboration platform and Procter & Gamble's (P&G) "Business Sphere" initiative illustrate how AI can enhance supplier performance and synchronization. These programs enable suppliers and buyers to align their operations strategically, ensuring efficient inventory management and reducing the risk of disruptions (Rathore et al., 2014). The real-time responsiveness provided by AI allows companies to adapt quickly to market changes and helps foster stronger, trust-based relationships between suppliers and buyers (DeNittis, 2024).

3 AI in SCOM

AI application in SCOM has gained substantial attention from both researchers and practitioners, reflecting its potential to revolutionize various aspects of SCOM

(Zamani et al., 2023; Fosso Wamba et al., 2024; Merhi & Harfouche, 2024). In this section, we present a systematic review of literature and news focused on the integration of AI within SCOM, aiming to provide a comprehensive overview of the state of AI in SCOM, identifying key trends, methodologies, and areas of focus within the current research landscape.

3.1 Literature Review and Summary

Our literature review is centered on AI and SCOM, with a particular emphasis on developments in recent years. To ensure a comprehensive selection of relevant articles, we follow a multistep selection process as follows:

- **Keyword-Based Search.** We conduct a comprehensive keyword-based search using the ABI/INFORM database, a widely recognized database that covers a range of business disciplines. To ensure a thorough exploration of relevant literature, we use the following combination of keywords: ("AI" OR "artificial intelligence" OR "machine learning") AND ("supply chain" OR "operations management" OR "operational management"). This approach was intended to capture studies that examine the integration of AI technologies into supply chain processes and their impact on various aspects of operations management.
- **Database Filtering and Journal Quality Assessment.** We focus on peer-reviewed articles over the past 10 years, specifically from 2014 onward till August 22, 2024. To maintain the academic rigor of our study, we limit our selection to research articles published in English and exclude other types of publications, such as editorial notes and biographical items. In order to ensure that our review captures high-quality research in the focused area, we further refine our selection to include only those articles published in top-tier operations management journals. Specifically, we target journals that are ranked as "A" or higher in operations management according to the Australian Business Deans Council (ABDC) journal list. As a result, we have identified a total of 154 journal articles.
- **Content Relevance Evaluation.** We subsequently conduct a thorough examination of each article to assess whether its contents align with our thematic framework. At the end, we select 110 research articles that were published between 2014 and 2024. The temporal distribution of these selected articles, as shown in Fig. 2, illustrates the progression and growing interest in this field over the past decade. Table 1 presents the distribution of the selected articles across various journals.

Our analysis of these articles is conducted through a multifaceted approach, examining them from three key perspectives: topic, methodology, and authorship. This comprehensive analysis allows us to identify the key themes and areas of research and understand the diversity of methodologies employed and the leading authors in the area.

Fig. 2 Number of publications related to AI in SCOM

Table 1 Journal distribution

Journal	No. of articles
International Journal of Production Research	37
Production and Operations Management	14
Annals of Operations Research	14
IEEE Transactions on Engineering Management	8
International Journal of Logistics Management	7
Expert Systems with Applications	7
International Journal of Information Management	6
Journal of Business Logistics	4
Technological Forecasting and Social Change	3
Supply Chain Management	2
International Journal of Operations & Production Management	2
IEEE Transactions on Intelligent Transportation Systems	2
Production Planning & Control	1
International Journal of Physical Distribution & Logistics Management	1
British Journal of Management	1
Benchmarking	1
Grand Total	**110**

3.1.1 Research Topics

These studies cover a wide range of topics about AI in SCOM, and we broadly classify them into four key areas: AI application in SCOM, the impact of AI on SCOM, the influencing factors of AI adoption in SCOM, and comprehensive overview of AI in SCOM. Figure 3 displays the topic distribution for our selected articles.

The most popular topic is related to AI applications in SCOM, which primarily concentrate on the development and deployment of AI technologies and

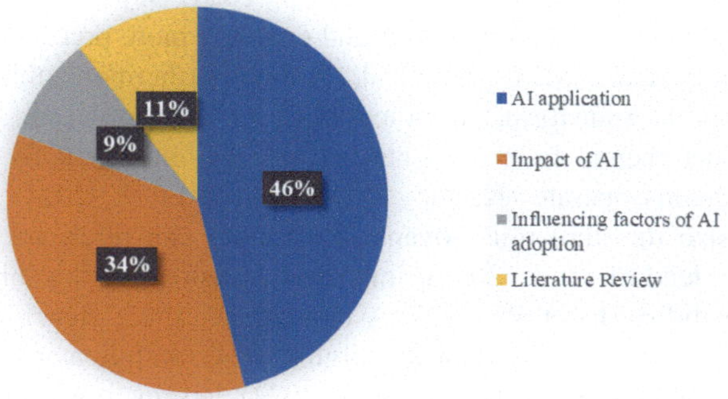

Fig. 3 Study topic distribution

algorithms to tackle specific SCOM challenges (Rolf et al., 2023; Kosasih et al., 2024). Research in this area often focuses on how advanced AI-driven techniques, including machine learning and deep learning, are leveraged to enhance various aspects of SCOM. For example, (Senoner et al., 2023) propose a data-driven methodology grounded in adversarial learning, designed to effectively manage distributional shifts in manufacturing environments, and (Zhu et al., 2021) construct an innovative demand forecasting framework that leverages cross-series training by utilizing time series data. Key areas of focus within this domain include improving sales or demand forecasting (Lau et al., 2018; Zhu et al., 2021; Kharfan et al., 2021; Zhang et al., 2024), strengthening risk management frameworks (Merrick et al., 2022; Bassiouni et al., 2023; Wang et al., 2023; Li et al., 2020), credit forecasting (Wu et al., 2022; Belhadi et al., 2024), price prediction (Ben Ameur et al., 2024), and optimizing supply chain operations (Senoner et al., 2023; Svoboda & Minner, 2022; Raghuram et al., 2022). These studies not only explore the theoretical underpinnings of AI technologies (Kosasih et al., 2024; Rolf et al., 2023) but also emphasize their practical implementation (Wu et al., 2022; Akhtar et al., 2023), offering innovative solutions to the complex and dynamic problems inherent in modern SCOM. Additionally, some research extends beyond individual AI methods (Badakhshan & Ball, 2024; Patel et al., 2024; Tavana et al., 2016), proposing hybrid approaches that integrate AI with traditional SCOM strategies to create more robust and resilient systems. This exploration into AI applications demonstrates a clear trend toward utilizing AI to predict, manage, and mitigate risks in SCOM, thereby enhancing both efficiency and resilience. The growing interest reflects the increasing recognition of AI's potential to revolutionize SCOM.

The second most prominent area of research focuses on the impact of AI on various facets of SCOM, which looks into the broader implications of AI technologies and explores how they reshape decision-making processes, operational efficiency, and SCOM performance (Li et al., 2021; Li & Li, 2022; Weisz et al., 2023; Dolgui & Ivanov, 2024). A significant portion of the literature highlights the critical role of AI-driven systems for enhancing transparency and accountability

(Wong et al., 2024; Kusiak, 2018; Cannas et al., 2024; Dey et al., 2024), providing real-time insights into SCOM activities and enabling more precise monitoring and control (Cannas et al., 2024; Dey et al., 2024). Furthermore, this research area often considers the transformative potential of AI (Jackson et al., 2024), not just in terms of incremental improvements but also in fundamentally redefining the way supply chains operate (Kusiak, 2018; Jackson et al., 2024). These studies often emphasize the numerous advantages that AI can offer, such as enhanced efficiency (Bodendorf et al., 2022), improved decision-making, higher corporate social responsibility (Li et al., 2021; Asokan et al., 2022; Benzidia et al., 2021) and customer satisfaction (Mariani & Borghi, 2024), and greater resilience across supply chains (Mithas et al., 2022; Li et al., 2024). However, it is important to note that relatively few of these papers (Brintrup et al., 2024; Fosso Wamba et al., 2024; Kumar & Mallipeddi, 2022) address the emerging challenges associated with the adoption of AI-powered SCOM. Given that the potential risks and complexities associated with AI integration are less frequently explored, issues such as data privacy, algorithmic bias, the need for new skills, and the ethical implications of AI deployment are critical areas that warrant deeper investigation. The lack of attention to these challenges suggests a gap in the current literature.

Researchers also undertake comprehensive literature reviews from diverse perspectives, each offering unique insights into specific areas of AI's integration into supply chains. Some of these reviews concentrate on particular niches, such as the adoption of explainable AI in SCOM (Kosasih et al., 2024), where the focus is on making AI decisions more transparent and understandable to stakeholders. Others explore the broader benefits and challenges associated with AI adoption (Rolf et al., 2023), shedding light on the potential advantages and obstacles that organizations face while integrating AI into their operations. Moreover, researchers also look into the evolution of retail analytics (Rooderkerk et al., 2022), how AI tools are being employed in supply chain risk management (Baryannis et al., 2019), and AI-augmented healthcare delivery systems (Dai & Tayur, 2022). Finally, the intersection of AI with social responsibilities within supply chains (Asokan et al., 2022) is another critical area of investigation. These studies analyze how AI can be leveraged to ensure ethical practices, sustainability, and social accountability across the supply chain network. These varied perspectives highlight the multidisciplinary nature of the field, where AI intersects with multiple domains to address complex, real-world challenges.

The last major study topic is the influencing factors of AI adoption. Understanding the factors that facilitate or hinder the adoption of AI in SCOM is crucial for realizing its potential benefits (Merhi & Harfouche, 2024; Sodhi et al., 2022). This area of research explores the external and internal factors that impact the effectiveness of AI implementation in SCOM. Topics under this category include the organizational readiness for AI adoption (Virmani et al., 2024), the role of leadership and organizational factors (Dey et al., 2024; Hasija & Esper, 2022), and the influence of institutional pressures and resources (Bag et al., 2021). Additionally, some studies build conceptual frameworks to better understand how these factors interact with AI technologies within the context of SCOM (Merhi & Harfouche, 2024; Leoni et al., 2022).

3.1.2 Research Methods

Researchers utilize a diverse array of methods to investigate this research area, reflecting the complexity and multifaceted nature of the field. These methodologies range from empirical analyses to advanced computational models, each contributing unique insights. Figure 4 illustrates the methodological distribution of the selected studies.

Empirical methods are widely utilized to provide valuable insights into the practical impacts of AI and identify the factors that contribute to the successful adoption of AI (Fosso Wamba et al., 2024; Sodhi et al., 2022). These empirical approaches can be broadly categorized into three major classes: quantitative analysis, case studies, and other qualitative studies. Quantitative studies often rely on statistical methods to examine the relationships between different variables within the supply chain context. By employing robust statistical techniques, researchers can identify significant patterns and correlations, offering a deeper understanding of how AI technologies influence supply chain dynamics. For example, Structural Equation Modeling (SEM) is frequently applied to explore the mediating and moderating effects of organizational factors on AI-driven outcomes (e.g., Fosso Wamba et al., 2024; Belhadi et al., 2024; Le & Behl, 2023; Modgil et al., 2022). The case study offers a qualitative perspective, providing in-depth insights into the implementation and impact of AI within SCOM contexts. These studies typically involve detailed examinations of real-world applications of AI, offering practical examples of challenges and successes (e.g., Cannas et al., 2024; Zheng et al., 2023; Rodríguez-Espíndola et al., 2020; Brintrup et al., 2020). The other

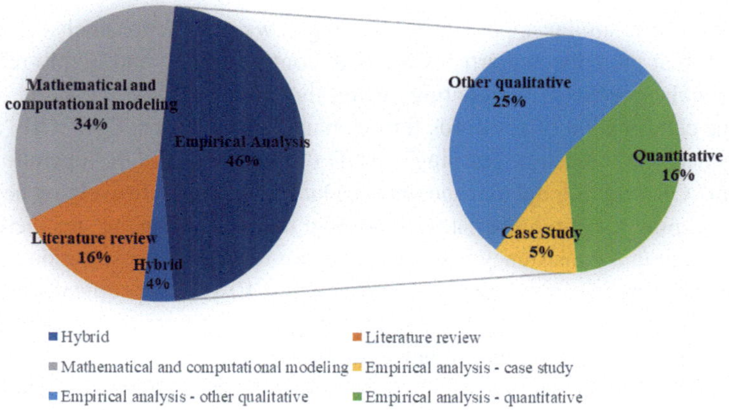

Fig. 4 Methodology distribution

qualitative studies (e.g., Sodhi et al., 2022; Dwivedi et al., 2021; Brau et al., 2024) incorporate a range of methodologies, including compare analysis, interviews and experts opinions, thematic analysis, fuzzy set theoretical approach, and so on.

Mathematical and computational modeling techniques are widely employed to develop and implement sophisticated algorithms aimed at enhancing various decision-making processes within the supply chain (Senoner et al., 2023; Zhu et al., 2021; Merrick et al., 2022). These techniques, particularly those involving deep learning and other advanced machine learning algorithms, are instrumental in developing predictive models that tackle essential challenges, including demand forecasting (Zhang et al., 2024; Kharfan et al., 2021; Zhu et al., 2021; Jackson & Ivanov, 2023), price prediction (Ben Ameur et al., 2024), risk detection (Merrick et al., 2022; Badakhshan & Ball, 2024; Kusiak, 2018; Li et al., 2024) and other aspects of operations (Wang et al., 2023; Tavana et al., 2016; Preil & Krapp, 2022). By leveraging these models, researchers and practitioners are able to significantly improve the efficiency and responsiveness of supply chains, enabling more accurate predictions and better-informed decisions that contribute to the overall effectiveness of supply chain operations.

Some studies combine multiple methods, such as surveys, simulations, and case studies, to provide a more comprehensive analysis of AI in SCOM. These hybrid approaches enable researchers to triangulate findings and offer robust conclusions. For example, Badakhshan and Ball (2024) integrate simulation, machine learning, and optimization to determine the optimal production, storage, and distribution values in order to maximize supply chain service level while minimizing total costs under disruptions. Tavana et al. (2016) propose a hybrid Adaptive Neuro-Fuzzy Inference System (ANFIS)-Neural Network (ANN) model to assist managers in their supplier evaluation process.

3.2 News Analysis

We employed a similar approach in selecting news articles for analysis, designed to ensure that the collected articles are both comprehensive and highly relevant to our research objectives. We continue using the ABI/INFORM database, applying the same combination of keywords for the main subject search: ("AI" OR "artificial intelligence" OR "machine learning") AND ("supply chain" OR "operations management" OR "operational management"). Our focus was exclusively on newspaper articles, aiming to capture the latest trends and insights within the specified content area. To maintain consistency and relevance, we restricted our selection to articles published in English, covering a period of the past 10 years, specifically from 2014 onward till August 22, 2024. As a result, we obtained a total of 973 news articles. Figure 5 illustrates the number of news across the years.

Figure 5 shows a clear upward trend, reflecting the growing interest and attention toward AI and its applications in SCOM. The lower number of articles for 2024 is due to the fact that our selection only includes data up to August 2024, rendering

Fig. 5 Number of news across years

the 2024 data incomplete. The sharp rise in article counts, particularly since 2022, indicates a heightened recognition of the importance of AI in these fields.

To gain a deeper understanding of the content, we conducted a thorough content analysis. Prior to the analysis, we undertook a two-step data preprocessing procedure: content extraction and keyword selection. During the content extraction, we focus on the main text, primarily using the Python Spacy package (Jugran et al., 2021). Initially, we applied tokenization (Jugran et al., 2021), breaking down the text into individual components, including words, punctuation marks, numbers, and spaces. Following tokenization, we removed non-textual elements to ensure the textual data were clean and ready for in-depth analysis. Finally, we performed lemmatization to the remaining textual data (Smelyakov et al., 2020), a process that reduces words to their base or root form, standardizing the text for later analysis.

During the keywords selection stage, we implemented a three-step process to refine the dataset. (1) We began by removing stop words, which are commonly used words such as "yes," "is," and "the" that do not carry significant meaning to the text. (2) We conduct the Tf-idf (term frequency-inverse document frequency) technique to filter out additional noise (Khan et al., 2019). This method helped us remove words that appeared either too frequently (e.g., "like to" and "would be") or too infrequently (e.g., specific human names). Using the TfidfVectorizer package in Python, we set the "min_df" parameter to 0.03, ensuring that words appearing in fewer than 3% of total articles were excluded. This approach effectively removed rarely occurring words while retaining more significant ones that appeared in a substantial number of articles. (3) We conduct N-gram selection (Gaydhani et al., 2018), considering not only individual words but also phrases containing two to five words (e.g., "supply chain" and "artificial intelligence"). This was achieved by setting the ngram_range parameter to (1,5). Ultimately, we selected the top 19,498 words and phrases with the highest Tf-idf values from these articles for further analysis.

Subsequently, we performed topic modeling using the selected words and phrases. We employed the BERTopic package (Grootendorst, 2022; Lee et al., 2023), which utilizes the transformer algorithm as its core machine learning methods and is well suited for conducting complex topic modeling. In our implementation, we evaluate multiple configurations and determine that setting the number of topics to 11 and minimum number of keywords to 7 can yield the most meaningful topic selection. To ensure the reproducibility of the topic modeling results, we incorporated a umap_model within the BERTopic pipeline, setting random_state=100 to control the stochastic nature of the dimensional reduction process. Of the 11 topics generated, one was designated as a buffer for potential additional topics. Considering further modeling within this buffer did not reveal any significant new topics, we reclassified these articles into one of the top ten topics based on the highest topic probability. Figure 6 presents the topic distributions of news.

To observe trends in how attention to each topic has evolved over time, we plot Fig. 7 that illustrates the number of news articles published each year across the top five topics related to AI in supply chains. Clearly from the figure, the "AI in supply chain cooperation" topic shows a consistent rise, especially from 2020 onward, indicating a growing interest in this area. Similarly, the "Healthcare supply chain" topic sees significant attention around 2018, likely due to increasing discussions around AI applications in healthcare during that period. In contrast, topics like "College education in the era of AI" and "AI customer and supply information" display more variable trends, with noticeable peaks and troughs. The topic "AI in the semiconductor supply chain" also gains traction, particularly from 2021, reflecting the growing importance of AI in this critical industry sector.

To analyze the relationships among these topics, we utilized the intertropical distance map, a visualization tool frequently employed in topic modeling. This tool effectively represents the relationships and relative distances between different

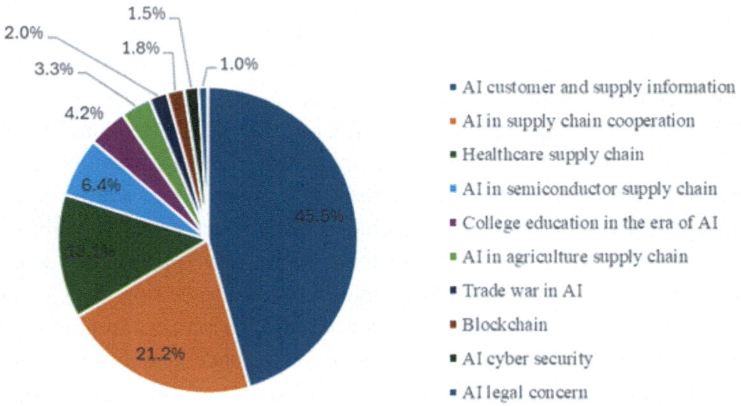

Fig. 6 Topic distribution of news

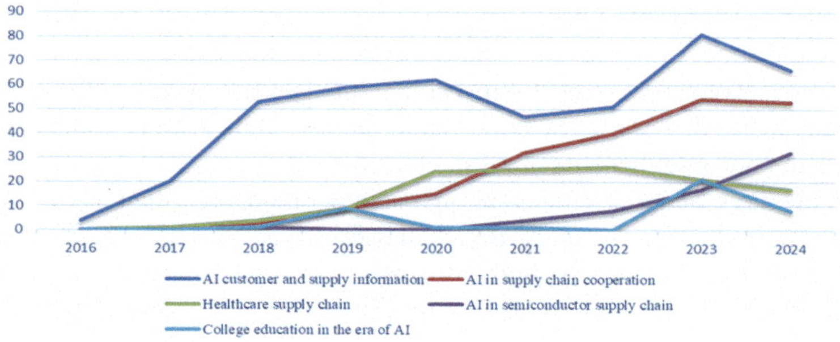

Fig. 7 Number of news (different topics) across years

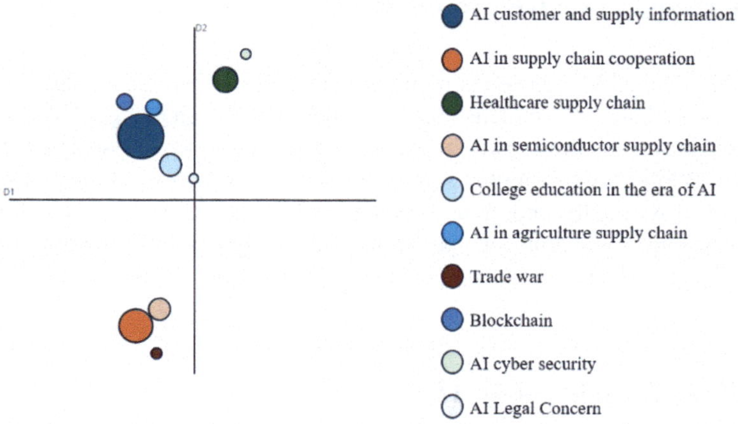

Fig. 8 Intertopic distance map

topics within a text corpus, aiding researchers in understanding how topics are distributed and interconnected within the data. Figure 8 displays the intertopic distance map.

Figure 8 illustrates the relationships among the topics, highlighting those that are closely related and those that are more distinct. This visualization is particularly useful for identifying clusters of related themes and for understanding the overall structure of the topics within the dataset. In our analysis, three major clusters emerge. The first cluster includes "AI customer and supply information," "AI in agricultural supply chain," "College education in the ear of AI," and "AI cyber security." The second cluster includes "AI in supply chain cooperation," "AI in semiconductor supply chain," and "Trade war." The third cluster includes "Healthcare supply chain" and "AI cyber security." Topics that are positioned closer together on the map share more common words and themes, indicating a degree of overlap or relatedness. In contrast, topics that are farther apart represent distinct themes or areas of focus. Additionally, any circles that appear isolated or significantly distant from others, such as "AI Legal Concern," suggest that these topics are relatively unique and distinct from the other topics identified in the dataset.

The analysis of news topics also reveals a limited understanding of the challenges associated with AI-powered SCOM. While there is considerable enthusiasm surrounding the potential benefits of AI in enhancing efficiency and decision-making, the complexities and risks inherent in its integration are often underexplored. The lack of in-depth analysis of the challenges, such as data security concerns, algorithmic biases, and the scalability of AI solutions, underscores the need for more comprehensive research and discussion to fully realize AI's potential in SCOM.

4 Challenges in AI-Powered SCOM

While AI-powered SCOM offers numerous capabilities that can revolutionize how businesses operate, its implementation is not without challenges. The existing literature, however, often overlooks the risks and challenges associated with AI adoption. While recognizing the significant benefits and opportunities AI presents to SCOM, it is equally important to understand the associated risks to effectively manage AI implementation and operations. This section explores the potential risks and challenges AI can introduce, specifically focusing on SCOM.

4.1 Risks Introduced by AI

This subsection examines the following potential risks introduced by AI: (1) data management, including issues of data quality, integration, and security, (2) decision-making, such as the risk of over-reliance on algorithms and the difficulties in ensuring transparency and accountability, and (3) relationship vulnerabilities, including concerns around data sharing, trust, and the equitable distribution of power within partnerships. Addressing these challenges is crucial for companies aiming to fully leverage AI's capabilities while minimizing potential downsides in their supply chain strategies.

4.1.1 Data Management and Integration

Data privacy and security are critical concerns in AI-powered SCOM. AI systems frequently process large volumes of sensitive data, including customer information, supplier details, and proprietary business intelligence (Minelli et al., 2013), making the protection of these data essential to prevent breaches and misuse. A notable example is the 2017 cyberattack on logistics giant Maersk, which caused severe disruptions across its global supply chain operations (Jones & Khan, 2021). This highlights the vulnerability of supply chains to cybersecurity threats. This incident underscored the importance of securing AI-powered systems.

In addition to data privacy and security, the effectiveness of AI in SCOM heavily depends on the quality and integration of data from various sources. AI algorithms require accurate, consistent, and comprehensive data to function correctly (Aldoseri et al., 2023). However, in many supply chains, data are often fragmented across different systems, formats, and departments, leading to challenges in achieving the level of data quality and integration necessary for AI to deliver meaningful insights. Poor data quality and integration can result in inaccurate predictions, flawed decision-making, and ultimately, operational inefficiencies (Hazen et al., 2014). Among the major challenges, a common one is the existence of data silos—isolated pockets of data within an organization that are not easily accessible to other departments or systems (Cromity & De Stricker, 2011). Data silos can significantly hinder the ability of AI systems to provide a unified view of the supply chain, leading to suboptimal decision-making, inefficiencies, and increased risk of errors.

4.1.2 Operational and Technical Challenges in Decision-Making

One of the significant challenges in implementing AI for SCOM decision-making is the risk of algorithmic bias, which can lead to unfair and discriminatory outcomes (Akter et al., 2021). AI systems learn from historical data, and if the data contain biases, AI is likely to replicate or even amplify these biases in its decision-making processes. For example, if an AI system is trained on data that reflects past discriminatory practices in supplier selection, it might continue to favor certain suppliers over others based on biased criteria, such as the demographic characteristics of supplier leadership, rather than performance or quality.

Another critical challenge in using AI for supply chain decision-making is ensuring that AI solutions are scalable and flexible enough to adapt to changing business needs and disruptions (Eck et al., 2020). Many AI solutions are initially developed for specific use cases and may not be easily scalable or adaptable to new conditions (Marr, 2019). AI systems must be able to scale quickly and adapt to dynamic supply chain situations and unexpected changes such as shifts in consumer demand, disruptions due to geopolitical events, or sudden changes in supplier availability, to remain effective.

4.1.3 Relational and Ethical Considerations

AI integration into SCOM represents a profound paradigm shift (Tauchert, 2022), often met with significant resistance from employees. This resistance frequently stems from fear of job displacement, concerns over the erosion of expertise, and uncertainties about the future role of human workers in increasingly automated environments (Koo et al., 2021; Spencer, 2018). As AI technologies advance, they not only transform operational processes but also challenge traditional job roles and skill sets, creating apprehension among staff. Effective change management strategies are therefore crucial in addressing these concerns and facilitating a smoother transition.

AI implementation in SCOM raises ethical and legal concerns that must be addressed to ensure responsible and effective use. One of the primary ethical concerns is the need for transparency in AI systems. As AI algorithms become increasingly complex, understanding and interpreting their decision-making processes can be difficult, which complicates the ability to hold these systems accountable for their outcomes (Stahl & Stahl, 2021; Tiwari, 2023). This lack of transparency can obscure how decisions are made, leading to issues such as discrimination or bias in AI-driven outcomes. For instance, in the case of AI used in hiring or lending, if the underlying algorithms are not transparent, it may be challenging to identify and rectify biased outcomes that disproportionately affect certain groups (Tol, 2019).

4.2 Resilience Practices Against Risks

This subsection discusses best practices adopted by companies to mitigate the preceding risks. These practices correspond to the three categories of AI-related risks. First, to address concerns around data management and integration, compliance with regulations such as the General Data Protection Regulation (GDPR) in the European Union and the California Consumer Privacy Act (CCPA) in the United States is crucial to maintaining data integrity and trust. Cybersecurity plays a central role in safeguarding the data, as AI systems are prime targets for cyberattacks (Chakraborty et al., 2023) due to their strategic values. Companies can implement advanced encryption techniques, multifactor authentication, and continuous monitoring of AI systems to detect and respond to potential threats in real time (Reddy, 2021). Ensuring data privacy and security is not just a technical challenge but a critical component of maintaining trust with customers and partners in an increasingly digital supply chain environment.

To address the challenges related to data quality and integration, particularly data from disparate sources, companies, e.g., Walmart and Ford, have invested heavily in data integration technologies. Walmart's investment ensures that data from suppliers, warehouses, and stores are accurate and consistent (Transights For Training and Consultancy, 2024). Ford implemented an AI-driven platform that integrated data across different departments such as manufacturing, logistics, and sales, breaking down silos and enabling real-time visibility and decision-making (Marr, 2021). Both companies have created a unified data platform, and this move not only improved operational efficiency but also enhanced their ability to respond swiftly to market changes and disruptions. Nonetheless, achieving this level of data integration requires significant investment in data governance, standardization, and cleansing processes, which can be a substantial hurdle for many organizations.

Second, to address operational and technical challenges in decision-making, companies proactively invest in ongoing monitoring and auditing of AI algorithms to ensure they are making fair and unbiased decisions and diversify the datasets used to train these algorithms to reflect a wider range of perspectives and experiences. Companies also spare no efforts recruiting skilled talents although the scarcity of skilled talents remains a big concern, particularly in ML roles (Petuum, Inc., 2022). Some companies attempt to bridge this gap by training existing staff or investing in productivity-enhancing tools. Addressing this challenge requires strategic investments in both workforce development and technology to mitigate the impact of talent shortages on AI implementation and scalability.

Third, to address relational and ethical risks, many companies promote clear communication about AI's benefits, the future of work within the organization, and the role of human supervision in AI systems (Vishwanath & Vaddepalli, 2023). To mitigate resistance and foster acceptance, comprehensive change management practices such as engaging employees early in the AI adoption process, providing training and upskilling opportunities, and demonstrating how AI can complement rather than replace human expertise should be adopted (Chowdhury et al., 2023; Morandini et al., 2023). Additionally, creating a culture of transparency and inclusivity, where employees are involved in decision-making and understand the AI's role in enhancing rather than undermining their work, can alleviate concerns and promote a positive outlook toward AI integration (Benbya et al., 2020). Addressing the human element of AI adoption is essential for leveraging the full potential of AI while maintaining a motivated and engaged workforce.

To address legal concerns related to AI accountability and liability, establishing clear guidelines for responsibility in the event of AI system errors or failures is essential to mitigate risks and ensure fair outcomes. For example, when autonomous vehicles are involved in accidents, determining liability—whether it falls on the manufacturer, software developer, or another party—can be complex and contentious (Alawadhi et al., 2020). Regulatory frameworks are still evolving to address these issues, but as AI systems become more integral to various sectors, it is imperative to develop robust legal standards that outline accountability and address potential harms (Díaz-Rodríguez et al., 2023).

4.3 Ripple Effect in AI-Powered Supply Chain

The potential of AI in SCOM is vast, including several key areas such as automated data input, AI-driven decision-making, and enhanced interconnectivity (Dwivedi et al., 2021; Cannas et al., 2024). The enhanced interconnectedness, when combined with AI-powered automation, may significantly amplify the ripple effect throughout the supply chain when disruptions occur. The high degree of automation and seamless data flow between systems and stakeholders can accelerate both the spread and the impact of disruptions. In this sense, the ripple effect within AI-powered supply chains can be amplified in both severity and scope, as disruptions have the potential to propagate more rapidly and extensively across the network.

The 2024 Microsoft-CrowdStrike outage highlights how highly automated and interconnected system can exacerbate systemic vulnerabilities (Herzlich, 2024). CrowdStrike, a cybersecurity company, and Microsoft, a cloud services provider, both rely heavily on automation, AI, and machine learning for detecting threats, managing systems, and offering real-time responses. This incident, triggered by a faulty software update from CrowdStrike, resulted in the widespread "Blue Screen of Death" on millions of Microsoft Windows devices globally. Various sectors were impacted, leading to an estimated $5.4 billion total financial losses for Fortune 500 firms (Herzlich, 2024). For airlines, the impact was particularly severe. Delta Air Lines, for instance, experienced 7,000 flight cancellations over 5 days, resulting in an estimated $500 million loss (Rogelberg, 2024). The disruption was exacerbated by the reliance on AI-powered systems for flight scheduling, passenger management, and operational logistics. Without these systems functioning properly, airlines struggled to revert to manual processes, further intensifying the operational chaos (Zahn, 2024). Other major airlines, such as American Airlines and United Airlines, also faced significant disruptions, though they recovered more quickly partially by relying on alternative systems or older infrastructure (Ashare, 2024).

As AI systems become increasingly integrated into decision-making processes and operational workflows, even small disturbances in one area of the supply chain may quickly propagate across the entire network. A notable example occurred in 2021, when a ransomware attack targeted the AI-powered systems of Colonial Pipeline (Srinivasan & Ni, 2023), a critical infrastructure responsible for managing fuel supply logistics along the East Coast of the United States. This attack caused the pipeline operator to halt its operation, resulting in widespread fuel shortages, sharp increases in fuel prices, and a surge in panic buying. Colonial Pipeline ended up paying $4.4 million to the hackers to regain control of their systems. This incident highlighted cybersecurity vulnerabilities inherent in AI-driven systems (Kumar & Mallipeddi, 2022), illustrating how a localized disruption in one segment of the supply chain can quickly affect a much larger network. It showcased the cascading effects of compromised AI systems.

When AI-powered SCOM encounters an issue—whether it is a data miscalculation, algorithmic bias, or cybersecurity threat—the effects can quickly spread across multiple areas, from inventory management and production scheduling to logistics and customer service. This heightened sensitivity of AI-powered systems highlights the need for a comprehensive redesign of a robust risk management framework for AI-powered SCOM and contingency planning strategies.

5 Conclusion

This chapter explores the potential and associated challenges of AI integration in supply chain and operations management. Through an in-depth review of the related literature and news articles, we identify key topics and methodologies that have emerged in the papers, reflecting the scholarly community's areas of interest. Additionally, we analyze the topics and intertopic connections as they appear in news articles, which highlight the priorities of industrial practitioners. This evidence suggests that AI technologies offer unparalleled capabilities in enhancing operational efficiency, improving decision-making processes, and fostering more resilient supply chains.

Despite the significant advantages, our analysis reveals that the adoption of AI in supply chains introduces new challenges. We explore these challenges and potential risks in detail, focusing on issues related to data management, algorithmic biases, and the necessity for robust risk management framework specifically tailored to AI-powered systems. These challenges emphasize the importance of developing more sophisticated and adaptable strategies to fully utilize the potential of AI while mitigating its associated risks. In conclusion, although AI has already begun to revolutionize SCOM, its full potential can only be realized through a comprehensive and in-depth understanding of the system, including both its capabilities and challenges.

Acknowledgments This work was supported by COVA CCI (Commonwealth Cyber Initiative) (C-3Q24-ODU-04: A Paradigm Shift: Innovating Supply Chain Security for AI-Assisted Devices)

References

About Amazon Team (2024). Amazon announces over €700 million investment in robotics and AI powered technologies across Europe. Published on April 11, 2024. Accessed on August 18, 2024. https://www.aboutamazon.eu/news/innovation/amazon-announces-over-700-million-investment-in-robotics-and-ai-powered-technologies-across-europe

Akhtar, P., Ghouri, A. M., Khan, H. U. R., Amin ul Haq, M., Awan, U., Zahoor, N., Khan, Z., & Ashraf, A. (2023). Detecting fake news and disinformation using artificial intelligence and machine learning to avoid supply chain disruptions. *Annals of operations research, 327*(2), 633–657.

Akter, S., McCarthy, G., Sajib, S., Michael, K., Dwivedi, Y. K., D'Ambra, J., & Shen, K. N. (2021). *Algorithmic bias in data-driven innovation in the age of AI.*

Alawadhi, M., Almazrouie, J., Kamil, M., & Khalil, K. A. (2020). Review and analysis of the importance of autonomous vehicles liability: A systematic literature review. *International Journal of System Assurance Engineering and Management, 11*, 1227–1249.

Aldoseri, A., Al-Khalifa, K. N., & Hamouda, A. M. (2023). Re-thinking data strategy and integration for artificial intelligence: concepts, opportunities, and challenges. *Applied Sciences, 13*(12), 7082.

AmazonScience (2021). The history of amazon's forecasting algorithm. Published on August 09, 2021. Accessed on August 30, 2024. https://www.amazon.science/latest-news/the-history-of-amazons-forecasting-algorithm

Ashare, M. (2024). American airlines credits it teams with quick recovery from crowdstrike disruption. Published on July 25, 2024. Accessed on September 6, 2024. https://www.ciodive.com/news/american-airlines-crowdstrike-outage-operatioins-recovery/722428/

Asokan, D. R., Huq, F. A., Smith, C. M., & Stevenson, M. (2022). Socially responsible operations in the industry 4.0 era: Post-covid-19 technology adoption and perspectives on future research. *International Journal of Operations & Production Management, 42*(13), 185–217.

Badakhshan, E. & Ball, P. (2024). Deploying hybrid modelling to support the development of a digital twin for supply chain master planning under disruptions. *International Journal of Production Research, 62*(10), 3606–3637.

Bag, S., Pretorius, J. H. C., Gupta, S., & Dwivedi, Y. K. (2021). Role of institutional pressures and resources in the adoption of big data analytics powered artificial intelligence, sustainable manufacturing practices and circular economy capabilities. *Technological Forecasting and Social Change, 163*, 120420.

Baryannis, G., Validi, S., Dani, S., & Antoniou, G. (2019). Supply chain risk management and artificial intelligence: state of the art and future research directions. *International journal of production research, 57*(7), 2179–2202.

Bassiouni, M. M., Chakrabortty, R. K., Hussain, O. K., & Rahman, H. F. (2023). Advanced deep learning approaches to predict supply chain risks under covid-19 restrictions. *Expert Systems with Applications, 211*, 118604.

Belhadi, A., Mani, V., Kamble, S. S., Khan, S. A. R., & Verma, S. (2024). Artificial intelligence-driven innovation for enhancing supply chain resilience and performance under the effect of supply chain dynamism: An empirical investigation. *Annals of Operations Research, 333*(2), 627–652.

Ben Ameur, H., Boubaker, S., Ftiti, Z., Louhichi, W., & Tissaoui, K. (2024). Forecasting commodity prices: empirical evidence using deep learning tools. *Annals of Operations Research, 339*(1), 349–367.

Benbya, H., Davenport, T. H., & Pachidi, S. (2020). Artificial intelligence in organizations: Current state and future opportunities. *MIS Quarterly Executive, 19*(4).

Benzidia, S., Makaoui, N., & Bentahar, O. (2021). The impact of big data analytics and artificial intelligence on green supply chain process integration and hospital environmental performance. *Technological Forecasting and Social Change, 165*, 120557.

Bodendorf, F., Merkl, P., & Franke, J. (2022). Artificial neural networks for intelligent cost estimation–a contribution to strategic cost management in the manufacturing supply chain. *International Journal of Production Research, 60*(21), 6637–6658.

Brau, R. I., Sanders, N. R., Aloysius, J., & Williams, D. (2024). Utilizing people, analytics, and AI for decision making in the digitalized retail supply chain. *Journal of Business Logistics, 45*(1), e12355.

Brintrup, A., Kosasih, E., Schaffer, P., Zheng, G., Demirel, G., & MacCarthy, B. L. (2024). Digital supply chain surveillance using artificial intelligence: Definitions, opportunities and risks. *International Journal of Production Research, 62*(13), 4674–4695.

Brintrup, A., Pak, J., Ratiney, D., Pearce, T., Wichmann, P., Woodall, P., & McFarlane, D. (2020). Supply chain data analytics for predicting supplier disruptions: A case study in complex asset manufacturing. *International Journal of Production Research, 58*(11), 3330–3341.

Cannas, V. G., Ciano, M. P., Saltalamacchia, M., & Secchi, R. (2024). Artificial intelligence in supply chain and operations management: A multiple case study research. *International Journal of Production Research, 62*(9), 3333–3360.

Chakraborty, A., Biswas, A., & Khan, A. K. (2023). Artificial intelligence for cybersecurity: Threats, attacks and mitigation. In *Artificial Intelligence for Societal Issues* (pp. 3–25). Springer.

Choi, T.-M., Wallace, S. W., & Wang, Y. (2018). Big data analytics in operations management. *Production and Operations Management, 27*(10), 1868–1883.

Chowdhury, S., Dey, P., Joel-Edgar, S., Bhattacharya, S., Rodriguez-Espindola, O., Abadie, A., & Truong, L. (2023). Unlocking the value of artificial intelligence in human resource management through AI capability framework. *Human Resource Management Review, 33*(1), 100899.

Cooper, R. G. (2023). The artificial intelligence revolution in new-product development. *IEEE Engineering Management Review, 52*(1), 195–211.

Cromity, J. & U. De Stricker (2011). Silo persistence: It's not the technology, it's the culture! *New Review of Information Networking, 16*(2), 167–184.
Cui, R., Li, M., & Zhang, S. (2022). AI and procurement. *Manufacturing & Service Operations Management, 24*(2), 691–706.
Dai, T. & Tayur, S. (2022). Designing AI-augmented healthcare delivery systems for physician buy-in and patient acceptance. *Production and Operations Management, 31*(12), 4443–4451.
DeNittis, N. (2024). Artificial intelligence at Procter & Gamble. Published on February 2, 2024. Accessed on August 18, 2024. https://emerj.com/ai-sector-overviews/artificial-intelligence-at-procter-gamble/
Dey, P. K., Chowdhury, S., Abadie, A., Vann Yaroson, E., & Sarkar, S. (2024). Artificial intelligence-driven supply chain resilience in Vietnamese manufacturing small-and medium-sized enterprises. *International Journal of Production Research, 62*(15), 5417–5456.
Díaz-Rodríguez, N., Del Ser, J., Coeckelbergh, M., de Prado, M. L., Herrera-Viedma, E., & Herrera, F. (2023). Connecting the dots in trustworthy artificial intelligence: From AI principles, ethics, and key requirements to responsible AI systems and regulation. *Information Fusion, 99*, 101896.
Dolgui, A. & Ivanov, D. (2023). Metaverse supply chain and operations management. *International Journal of Production Research, 61*(23), 8179–8191.
Dolgui, A. & Ivanov, D. (2024). Internet of behaviors: conceptual model, practical and theoretical implications for supply chain and operations management. *International Journal of Production Research, 63*(1), 1–8.
Dwivedi, Y. K., Hughes, L., Ismagilova, E., Aarts, G., Coombs, C., Crick, T., Duan, Y., Dwivedi, R., Edwards, J., Eirug, A., et al. (2021). Artificial intelligence (AI): Multidisciplinary perspectives on emerging challenges, opportunities, and agenda for research, practice and policy. *International Journal of Information Management, 57*, 101994.
Eck, B., Fusco, F., Gormally, R., Purcell, M., & Tirupathi, S. (2020). Scalable deployment of AI time-series models for IoT. arXiv preprint arXiv:2003.12141.
Fosso Wamba, S., Guthrie, C., Queiroz, M. M., & Minner, S. (2024). ChatGPT and generative artificial intelligence: an exploratory study of key benefits and challenges in operations and supply chain management. *International Journal of Production Research, 62*(16), 5676–5696.
Ganesh, A. D. & Kalpana, P. (2022). Future of artificial intelligence and its influence on supply chain risk management–a systematic review. *Computers & Industrial Engineering, 169*, 108206.
Gaydhani, A., Doma, V., Kendre, S., & Bhagwat, L. (2018). Detecting hate speech and offensive language on twitter using machine learning: An n-gram and TFIDF based approach. arXiv preprint arXiv:1809.08651.
Grootendorst, M. (2022). Bertopic: Neural topic modeling with a class-based tf-idf procedure. arXiv preprint arXiv:2203.05794.
Grover, V., Balusamy, B., Milanova, M., & Felix, A. Y. (2024). *Blockchain, IoT, and AI technologies for supply chain management*. Springer.
Hasija, A. & Esper, T. L. (2022). In artificial intelligence (AI) we trust: A qualitative investigation of AI technology acceptance. *Journal of Business Logistics, 43*(3), 388–412.
Hazen, B. T., Boone, C. A., Ezell, J. D., & Jones-Farmer, L. A. (2014). Data quality for data science, predictive analytics, and big data in supply chain management: An introduction to the problem and suggestions for research and applications. *International Journal of Production Economics, 154*, 72–80.
Herzlich, T. (2024). Microsoft to take massive hit as fortune 500 firms suffer $5.4b in losses from crowdstrike outage. Published on July 24, 2024. Accessed on September 6, 2024. https://nypost.com/2024/07/24/business/microsoft-to-take-hit-as-fortune-500-suffers-5-4b-in-crowdstrike-losses-study/
Ivanov, D. (2023). Intelligent digital twin (IDT) for supply chain stress-testing, resilience, and viability. *International Journal of Production Economics, 263*, 108938.

Jackson, I. & Ivanov, D. (2023). A beautiful shock? exploring the impact of pandemic shocks on the accuracy of AI forecasting in the beauty care industry. *Transportation Research Part E: Logistics and Transportation Review, 180*, 103360.

Jackson, I., Ivanov, D., Dolgui, A., & Namdar, J. (2024). Generative artificial intelligence in supply chain and operations management: a capability-based framework for analysis and implementation. *International Journal of Production Research, 62*(17), 6120–6145.

Jackson, I., Jesus Saenz, M., & Ivanov, D. (2024). From natural language to simulations: applying AI to automate simulation modelling of logistics systems. *International Journal of Production Research, 62*(4), 1434–1457.

James, G. (2013). *An introduction to statistical learning*.

Jones, A. & Khan, O. (2021). Surviving NotPetya: Global supply chains in the era of the cyber weapon. In *Cyber security and supply chain management: risks, challenges, and solutions* (pp. 133–146). World Scientific Publishing Company.

Jordan, N. (2017). Shipping company Maersk says June cyberattack could cost it up to $300 million. Published on August 16, 2017. Accessed on August 30, 2024. https://www.cnbc.com/2017/08/16/maersk-says-notpetya-cyberattack-could-cost-300-million.html

Jugran, S., Kumar, A., Tyagi, B. S., & Anand, V. (2021). Extractive automatic text summarization using spacy in Python & NLP. In *2021 International conference on advance computing and innovative technologies in engineering (ICACITE)* (pp. 582–585). IEEE.

Khan, R., Qian, Y., & Naeem, S. (2019). Extractive based text summarization using k-means and tf-idf. *International Journal of Information Engineering and Electronic Business, 12*(3), 33.

Kharfan, M., Chan, V. W. K., & Firdolas Efendigil, T. (2021). A data-driven forecasting approach for newly launched seasonal products by leveraging machine-learning approaches. *Annals of Operations Research, 303*(1), 159–174.

Koo, B., Curtis, C., & Ryan, B. (2021). Examining the impact of artificial intelligence on hotel employees through job insecurity perspectives. *International Journal of Hospitality Management, 95*, 102763.

Kosasih, E. E., Papadakis, E., Baryannis, G., & Brintrup, A. (2024). A review of explainable artificial intelligence in supply chain management using neurosymbolic approaches. *International Journal of Production Research, 62*(4), 1510–1540.

Kumar, S. & Mallipeddi, R. R. (2022). Impact of cybersecurity on operations and supply chain management: Emerging trends and future research directions. *Production and Operations Management, 31*(12), 4488–4500.

Kusiak, A. (2018). Smart manufacturing. *International Journal of Production Research, 56*(1-2), 508–517.

Lau, R. Y. K., Zhang, W., & Xu, W. (2018). Parallel aspect-oriented sentiment analysis for sales forecasting with big data. *Production and Operations Management, 27*(10), 1775–1794.

Le, T. T. & Behl, A. (2023). Linking artificial intelligence and supply chain resilience: Roles of dynamic capabilities mediator and open innovation moderator. *IEEE Transactions on Engineering Management*.

Lee, H., Lee, S. H., Lee, K. R., & Kim, J. H. (2023). ESG discourse analysis through BERTopic: comparing news articles and academic papers. *Computers, Materials & Continua, 75*(3), 6023–6037.

Leoni, L., Ardolino, M., El Baz, J., Gueli, G., & Bacchetti, A. (2022). The mediating role of knowledge management processes in the effective use of artificial intelligence in manufacturing firms. *International Journal of Operations & Production Management, 42*(13), 411–437.

Li, G., Li, N., & Sethi, S. P. (2021). Does CSR reduce idiosyncratic risk? Roles of operational efficiency and AI innovation. *Production and Operations Management, 30*(7), 2027–2045.

Li, M. & Li, T. (2022). AI automation and retailer regret in supply chains. *Production and Operations Management, 31*(1), 83–97.

Li, X., Krivtsov, V., Pan, C., Nassehi, A., Gao, R. X., & Ivanov, D. (2025). End-to-end supply chain resilience management using deep learning, survival analysis, and explainable artificial intelligence. *International Journal of Production Research, 63*(3), 1174–1202.

Li, Z., Guo, H., Barenji, A. V., Wang, W. M., Guan, Y., & Huang, G. Q. (2020). A sustainable production capability evaluation mechanism based on blockchain, LSTM, analytic hierarchy process for supply chain network. *International Journal of Production Research, 58*(24), 7399–7419.

Mariani, M. M. & Borghi, M. (2024). Artificial intelligence in service industries: Customers' assessment of service production and resilient service operations. *International Journal of Production Research, 62*(15), 5400–5416.

Marr, B. (2019). *Artificial intelligence in practice: How 50 successful companies used AI and machine learning to solve problems*. Wiley.

Marr, B. (2021). The amazing ways the ford motor company uses artificial intelligence and machine learning. Published on December 10, 2021. Accessed on August 18, 2024. https://www.forbes.com/sites/bernardmarr/2019/05/17/the-amazing-ways-the-ford-motor-company-uses-artificial-intelligence-and-machine-learning/

Merhi, M. I. & Harfouche, A. (2024). Enablers of artificial intelligence adoption and implementation in production systems. *International Journal of Production Research, 62*(15), 5457–5471.

Merrick, J. R., Dorsey, C. A., Wang, B., Grabowski, M., & Harrald, J. R. (2022). Measuring prediction accuracy in a maritime accident warning system. *Production and Operations Management, 31*(2), 819–827.

Min, H. (2010). Artificial intelligence in supply chain management: Theory and applications. *International Journal of Logistics: Research and Applications, 13*(1), 13–39.

Minelli, M., Chambers, M., & Dhiraj, A. (2013). *Big data, big analytics: Emerging business intelligence and analytic trends for today's businesses* (vol. 578). Wiley.

Mithas, S., Chen, Z.-L., Saldanha, T. J., & De Oliveira Silveira, A. (2022). How will artificial intelligence and industry 4.0 emerging technologies transform operations management? *Production and Operations Management, 31*(12), 4475–4487.

Modgil, S., Singh, R. K., & Hannibal, C. (2022). Artificial intelligence for supply chain resilience: Learning from covid-19. *The International Journal of Logistics Management, 33*(4), 1246–1268.

Morandini, S., Fraboni, F., De Angelis, M., Puzzo, G., Giusino, D., Pietrantoni, L., et al. (2023). The impact of artificial intelligence on workers' skills: Upskilling and reskilling in organisations. *Informing Science, 26*, 39–68.

Pandey, B. K., Kanike, U. K., George, A. S., & Pandey, D. (2024). *AI and machine learning impacts in intelligent supply chain*. IGI Global.

Patel, D., Sahu, C. K., & Rai, R. (2024). Security in modern manufacturing systems: Integrating blockchain in artificial intelligence-assisted manufacturing. *International Journal of Production Research, 62*(3), 1041–1071.

Petuum, Inc. (2022). 5 organizational issues holding back your AI team. Published on June 6, 2022. Accessed on August 28, 2024. https://petuum.medium.com/5-organizational-issues-holding-back-your-ai-team-3f8eabd84319

Preil, D. & Krapp, M. (2022). Artificial intelligence-based inventory management: A Monte Carlo tree search approach. *Annals of Operations Research, 308*(1), 415–439.

Pushkar, A. (2024). Artificial intelligence (AI) techniques. Published on July 18, 2024. Accessed on August 26, 2024. https://intellipaat.com/blog/artificial-intelligence-techniques/

Quan, N. M. et al. (2023). Analyzing the impact of big data analytics on supply chain efficiency in the fashion industry. *Journal of Empirical Social Science Studies, 7*(4), 64–81.

Raghuram, P., Bhupesh, S., Manivannan, R., Anand, P. S. P., & Sreedharan, V. R. (2022). Modeling and analyzing the inventory level for demand uncertainty in the VUCA world: evidence from biomedical manufacturer. *IEEE Transactions on Engineering Management, 70*(8), 2944–2954.

Rathore, S., Panwar, A., & Soral, P. (2014). Critical factors for successful implementation of business analytics: Exploratory findings from select cases. *International Journal of Business Analytics and Intelligence, 2*(2), 11.

Reddy, A. R. P. (2021). The role of artificial intelligence in proactive cyber threat detection in cloud environments. *NeuroQuantology, 19*(12), 764–773.

Rodríguez-Espíndola, O., Chowdhury, S., Beltagui, A., & Albores, P. (2020). The potential of emergent disruptive technologies for humanitarian supply chains: The integration of blockchain, artificial intelligence and 3d printing. *International Journal of Production Research, 58*(15), 4610–4630.

Rogelberg, S. (2024). The crowdstrike outage had an outsized impact on delta, analysts say–and its rivals are soaring because of it. Published on September 5, 2024. Accessed on September 6, 2024. https://fortune.com/2024/09/05/crowdstrike-outage-delta-impact-bernstein-american-airlines-united-microsoft/

Rolf, B., Jackson, I., Müller, M., Lang, S., Reggelin, T., & Ivanov, D. (2023). A review on reinforcement learning algorithms and applications in supply chain management. *International Journal of Production Research, 61*(20), 7151–7179.

Rooderkerk, R. P., DeHoratius, N., & Musalem, A. (2022). The past, present, and future of retail analytics: Insights from a survey of academic research and interviews with practitioners. *Production and Operations Management, 31*(10), 3727–3748.

Samuel, P., Saini, A., Poongodi, T., & Nancy, P. (2023). Artificial intelligence–driven digital twins in industry 4.0. In *Digital Twin for Smart Manufacturing* (pp. 59–88). Elsevier.

Senoner, J., Kratzwald, B., Kuzmanovic, M., Netland, T. H., & Feuerriegel, S. (2023). Addressing distributional shifts in operations management: The case of order fulfillment in customized production. *Production and Operations Management, 32*(10), 3022–3042.

Shahzadi, G., Jia, F., Chen, L., & John, A. (2024). AI adoption in supply chain management: A systematic literature review. *Journal of Manufacturing Technology Management, 35*(6), 1125–1150.

Simchi-Levi, D. & Wu, M. X. (2018). Powering retailers' digitization through analytics and automation. *International Journal of Production Research, 56*(1-2), 809–816.

Smelyakov, K., Karachevtsev, D., Kulemza, D., Samoilenko, Y., Patlan, O., & Chupryna, A. (2020). Effectiveness of preprocessing algorithms for natural language processing applications. In *2020 IEEE International Conference on Problems of Infocommunications. Science and Technology (PIC S&T)* (pp. 187–191). IEEE.

Sodhi, M. S., Seyedghorban, Z., Tahernejad, H., & Samson, D. (2022). Why emerging supply chain technologies initially disappoint: Blockchain, IoT, and AI. *Production and Operations Management, 31*(6), 2517–2537.

Spencer, D. A. (2018). Fear and hope in an age of mass automation: debating the future of work. *New Technology, Work and Employment, 33*(1), 1–12.

Srinivasan, S., & Ni, L.-K. (2023, March). Ransomware attack at colonial pipeline company. *Harvard Business School Teaching Note*, 123–070.

Stahl, B. C. & Stahl, B. C. (2021). Ethical issues of AI. In *Artificial Intelligence for a better future: An ecosystem perspective on the ethics of AI and emerging digital technologies* (pp. 35–53).

Svoboda, J. & Minner, S. (2022). Tailoring inventory classification to industry applications: The benefits of understandable machine learning. *International Journal of Production Research, 60*(1), 388–401.

Tauchert, C. (2022). Adoption of AI-based Information Systems from an Organizational and User Perspective. *Technical University of Darmstadt.*

Tavana, M., Fallahpour, A., Di Caprio, D., & Santos-Arteaga, F. J. (2016). A hybrid intelligent fuzzy predictive model with simulation for supplier evaluation and selection. *Expert Systems with Applications, 61*, 129–144.

Tiwari, R. (2023). Ethical and societal implications of AI and machine learning. *International Journal of Scientific Research in Engineering and Management, 7*(01).

Tol, J. (2019). *Ethical implications of face recognition tasks in law enforcement.* Ph. D. thesis, diploma thesis, University of Amsterdam.

Transights For Training and Consultancy (2024). Walmart's transformation through data analytics. Published on July 3, 2024. Accessed on August 18, 2024. https://www.linkedin.com/pulse/walmarts-transformation-through-data-ogchf/

Virmani, N., Singh, R. K., Agarwal, V., & Aktas, E. (2024). Artificial intelligence applications for responsive healthcare supply chains: A decision-making framework. *IEEE Transactions on Engineering Management, 71*(X), 8591–8605.
Vishwanath, B. & Vaddepalli, S. (2023). The future of work: Implications of artificial intelligence on HR practices. *Tuijin Jishu/Journal of Propulsion Technology, 44*(3), 1711–1724.
Wamba, S. F., Akter, S., Edwards, A., Chopin, G., & Gnanzou, D. (2015). How 'big data' can make big impact: Findings from a systematic review and a longitudinal case study. *International journal of production economics, 165*, 234–246.
Wang, L., Jia, F., Chen, L., & Xu, Q. (2023). Forecasting SMEs' credit risk in supply chain finance with a sampling strategy based on machine learning techniques. *Annals of Operations Research, 331*(1), 1–33.
Weisz, E., Herold, D. M., & Kummer, S. (2023). Revisiting the bullwhip effect: how can AI smoothen the bullwhip phenomenon? *The International Journal of Logistics Management, 34*(7), 98–120.
Wong, L.-W., Tan, G. W.-H., Ooi, K.-B., Lin, B., & Dwivedi, Y. K. (2024). Artificial intelligence-driven risk management for enhancing supply chain agility: A deep-learning-based dual-stage PLS-SEM-ANN analysis. *International Journal of Production Research, 62*(15), 5535–5555.
Wu, J., Zhang, Z., & Zhou, S. X. (2022). Credit rating prediction through supply chains: A machine learning approach. *Production and Operations Management, 31*(4), 1613–1629.
Zahn, M. (2024). Why did delta take days to restore normal service after crowdstrike outage? experts weigh in. Published on July 27, 2024. Accessed on September 6, 2024. https://abcnews.go.com/Business/delta-days-restore-normal-service-after-crowdstrike-outage/story?id=112299966
Zamani, E. D., Smyth, C., Gupta, S., & Dennehy, D. (2023). Artificial intelligence and big data analytics for supply chain resilience: A systematic literature review. *Annals of Operations Research, 327*(2), 605–632.
Zhang, D., Wu, P., Wu, C., & Ngai, E. W. (2024). Forecasting duty-free shopping demand with multisource data: a deep learning approach. *Annals of Operations Research, 339*(1), 861–887.
Zheng, G., Kong, L., & Brintrup, A. (2023). Federated machine learning for privacy preserving, collective supply chain risk prediction. *International Journal of Production Research, 61*(23), 8115–8132.
Zhu, X., Ninh, A., Zhao, H., & Liu, Z. (2021). Demand forecasting with supply-chain information and machine learning: Evidence in the pharmaceutical industry. *Production and Operations Management, 30*(9), 3231–3252.

Supply Chain Plasticity: Reshaping the Future of Risk Management

Ellie C. Falcone, Laharish Guntuka, and Steven Carnovale

Abstract Traditional risk management strategies often focus on restoring normal operations as quickly as possible. However, a new concept—supply chain plasticity—is gaining traction that offers a more dynamic and resilient approach. Supply chain plasticity refers to a firm's capability to rapidly and significantly alter its supply chain network structure to accommodate substantial shifts and disruptions in the business environment. Unlike conventional methods that aim to return to a pre-disruption state, plasticity involves making dramatic changes to how a firm's supply chain operates, often resulting in a new protocol that better aligns with the post-disruption environment. When implemented correctly, supply chain plasticity has the potential to significantly improve recovery, resilience, and responsiveness of supply chains.

1 Introduction

In today's rapidly evolving business landscape, disruptions in the supply chain are inevitable. From natural disasters to geopolitical tensions and global pandemics, firms face myriad challenges that can destabilize their operations. Traditional risk management strategies often focus on restoring normal operations as quickly as

E. C. Falcone
Neeley School of Business, Texas Christian University, Fort Worth, TX, USA
e-mail: e.falcone@tcu.edu

L. Guntuka
Department of Supply Chain Management, Saunders College of Business, Rochester Institute of Technology, Rochester, NY, USA
e-mail: lguntuka@saunders.rit.edu

S. Carnovale (✉)
Information Technology and Operations Management Department, College of Business, Florida Atlantic University, Boca Raton, FL, USA
e-mail: scarnovale@fau.edu

D. Ivanov et al. (eds.), *Handbook of Ripple Effects in the Supply Chain*,
International Series in Operations Research & Management Science 276,
https://doi.org/10.1007/978-3-031-85508-5_8

possible (Boyson et al., 2022). However, a new concept—supply chain plasticity—is gaining traction that offers a more dynamic and resilient approach.

Supply chain plasticity refers to a firm's capability to rapidly and significantly alter its supply chain structure to accommodate substantial shifts and disruptions in the business environment (Zinn & Goldsby, 2019). Unlike conventional methods that aim to return to a pre-disruption state, plasticity involves making dramatic changes to how a firm's supply chain operates, often resulting in a new protocol that better aligns with the post-disruption environment. These changes often result in new protocols that better align with the post-disruption environment, enabling firms to not just recover but thrive in the new conditions.

The importance of supply chain plasticity cannot be overstated in modern supply chains. As global markets become increasingly interconnected and complex, the potential for disruptions has multiplied (Guntuka et al. 2024a, b, c; Bode & Wagner, 2015). Supply chain plasticity allows firms to quickly adapt to these challenges, minimizing the impact of disruptions and maintaining continuity in their operations (Carnovale et al., 2023). This capability is crucial for sustaining competitive advantage and meeting customer expectations in a volatile market.

The concept of supply chain plasticity has evolved as businesses and scholars have recognized the limitations of traditional risk management strategies. Zinn and Goldsby (2019) highlight that conventional approaches often emphasize returning to a 'normal' state as quickly as possible, without necessarily accounting for changes in the external environment. This approach can leave firms vulnerable to future disruptions and fail to capitalize on new opportunities that arise from changing market conditions. The idea of plasticity, therefore, represents a paradigm shift towards a more proactive and flexible strategy in supply chain management. This chapter delves into the fundamental elements of supply chain plasticity, its importance in risk management, and its role in understanding and mitigating the ripple effects of disruptions.

2 Foundations of Supply Chain Plasticity

2.1 What Is Supply Chain Plasticity?

Supply chain plasticity is the ability of a company to quickly and significantly alter its supply chain structure in response to major disruptions. Unlike traditional approaches that focus on returning to a pre-disruption state, supply chain plasticity embraces change and often results in a new, more resilient operational model (Guntuka et al., 2024b). Introduced by Zinn and Goldsby in 2019, the concept emphasizes adapting to environmental impacts and other disruptions. It involves several key elements, including major redesigns, urgency in response, acquiring new skills and resources, and making permanent changes to operations. Studies by Falcone et al. (2022) and Hughes et al. (2023) have underscored the importance of a well-coordinated network of partners and the role of firm and managerial characteristics in enabling a plastic response. The aim is not merely to survive disruptions but to emerge stronger and better suited to new conditions. Recent research by Guntuka et al. (2024b) has explored plasticity as a supply chain network phenomenon that bridges the gap between responsiveness and resilience, finding that as disruptions rise, firms that shift their network structures experience benefits to recovery and performance. Yet, the concept still requires further empirical assessment, indicating a need for further research to fully understand this concept and its broad applications.

2.2 *Fundamental Elements of Supply Chain Plasticity*

Supply chain plasticity is defined by several essential elements that distinguish it from traditional supply chain resilience strategies. These elements include major redesign in a supply chain, urgency in the need to shift, acquisition of new supply chain relationships, and adopting a new permanent way to operate (Hughes et al., 2023). In addition, empirical work has shown that plasticity is a network-rooted phenomenon (Guntuka et al., 2024b).

2.2.1 Major Redesign in a Supply Chain

The foundation of supply chain plasticity is the capacity to implement substantial redesigns of a supply chain, including comprehensive, multi-tier changes that extend throughout the entire supply chain network. This involves more than just minor adjustments; it requires a comprehensive rethinking of the supply chain's relationships, structure, and processes. For example, during the COVID-19 pandemic, many companies had to overhaul their supply chains to cope with sudden changes in demand and supply and network augmentation to achieve superior performance outcomes through operational repurposing (Falcone et al., 2022). Such an example highlighted that a well-coordinated, ready-to-change interorganizational network is essential for successful supply chain plasticity. This capability involves more than merely returning to a pre-disruption state; it includes adopting new ways of operating.

2.2.2 Urgency in the Need to Shift

Another key characteristic of supply chain plasticity is the urgency with which changes need to be made (e.g., Dolgui et al., 2023). Disruptions often require swift action to mitigate immediate impacts (Ivanov, 2022). Firms must be capable of making rapid decisions and implementing changes quickly. The case of Foxconn's immediate shift of production from Zhengzhou City to Shenzhen City in response to COVID-19 mandate lockdowns in China exemplifies the urgency and agility required for supply chain plasticity (Lee, 2022; Guntuka et al., 2024b).

Mini Case Study: The Plasticity of Foxconn's Supply Chain During the COVID-19 Lockdowns

Background

Foxconn, a leading electronics manufacturer, and primary supplier for Apple, operates major production facilities in China. During the COVID-19 pandemic, the city of Zhengzhou, where one of Foxconn's key production sites is located, faced severe lockdowns by the local government to control the spread of the virus. This situation created significant disruptions to Foxconn's operations, including halted production and logistical challenges.

Challenge:

The lockdown in Zhengzhou severely impacted Foxconn's ability to maintain its production schedules. With production lines stalled and workforce availability reduced, the company faced the risk of missing critical deadlines and failing to meet customer demand.

Response:

In response to the operational disruption, Foxconn demonstrated exceptional supply chain plasticity by rapidly shifting its production operations from Zhengzhou to its facility in Shenzhen. This decision involved several key actions:

- *Geographic Relocation*: Foxconn leveraged its diversified network of production facilities by moving operations to Shenzhen, an alternative site less affected by the lockdown.
- *Resource Reallocation*: The company reallocated certain resources, including machinery and personnel, to the Shenzhen facility. This required meticulous planning and coordination to ensure that the new location could quickly ramp up production.
- *Operational Agility*: Foxconn exhibited remarkable agility by adapting its production processes to the new site. This involved overcoming logistical hurdles and streamlining operations to resume normal output levels as swiftly as possible.
- *Effective Communication*: Coordination between the Zhengzhou and Shenzhen sites was crucial. Foxconn maintained clear communication channels to ensure a smooth transition and mitigate any potential disruptions during the shift.

Outcome:

Foxconn's quick adaptation allowed the company to minimize production delays and continue meeting its supply chain commitments despite the lockdowns. The successful relocation underscored the importance of having flexible and adaptable supply chain strategies that can respond effectively to unexpected disruptions.

2.2.3 Acquisition of New Supply Chain Relationships

Implementing supply chain plasticity often necessitates acquiring more up-to-date skills, resources, and technologies through new supply chain partners. Such an acquisition often results in changes in a firm's network position (Guntuka et al., 2024b). This aspect underscores the importance of having a flexible and adaptive supply chain relationships, as well as a strong network of suppliers that can support these changes (Choi & Holmen, 2023).

2.2.4 A Permanent and New Way to Operate

A critical aspect of supply chain plasticity is the establishment of a new and permanent way of operating, rather than merely returning to the pre-disruption state. This involves implementing long-term structural changes that enhance the supply chain's resilience and adaptability to future disruptions. The aim is not just to recover from a crisis but to emerge stronger and more capable of handling unforeseen challenges. This can include adopting new technologies, shifting to more sustainable practices, or reconfiguring the supply chain to better align with market demands and environmental considerations (Mukandwal et al., 2024). For instance, companies may invest in digital transformation initiatives to improve visibility and coordination across the supply chain or develop new strategies for risk management and inventory control. These permanent changes are integral to building a more robust and agile supply chain that can withstand and adapt to future disruptions.

2.3 *How Is Supply Chain Plasticity Different from Other Concepts in Supply Chain Risk Management*

2.3.1 Flexibility vs. Plasticity

Supply chain flexibility refers to the capability of a firm to adapt and respond to changes or disruptions within its existing operational framework (Gligor & Holcomb, 2014). This concept includes making adjustments in production, sourcing, or distribution to accommodate fluctuations in demand or supply. Flexibility can manifest in both operational and strategic dimensions, such as incorporating redundancy in capacity, employing multiple sourcing strategies, and utilizing flexible production systems. The focus of flexibility is on maintaining performance and meeting customer expectations within the established parameters of the supply chain. In contrast, supply chain plasticity goes beyond the adjustments encompassed by flexibility. Plasticity involves a more profound and transformative process that includes structural changes to the supply network. It is not merely about adapting within the current framework but reconfiguring the framework itself to better align with a changing environment. While flexibility is an essential aspect of responding to disruptions, it is a necessary but not sufficient condition for achieving the broader and deeper changes associated with plasticity. Plasticity requires a shift in the overall design and strategy of the supply chain to enhance its capacity to thrive in dynamic conditions.

2.3.2 Agility vs. Plasticity

Agility refers to the ability to quickly adjust tactics and operations in response to changes (Nagel & Dove, 1998). This concept encompasses several attributes, including alertness to market changes, accessibility of resources and information, decisiveness in decision-making, swiftness in execution, and flexibility in operations. Agility emphasizes a proactive approach, where companies continuously monitor market trends, make swift decisions, and leverage digital capabilities to adapt efficiently (Christopher & Peck, 2004). Supply chain plasticity incorporates the characteristics of agility but extends beyond them by emphasizing the need for transformative changes. While agility focuses on the ability to respond quickly and efficiently within the existing supply chain framework, plasticity involves more profound changes, such as redesigning the supply network and making strategic shifts. Thus, while agility is a critical component that contributes to the ability to respond to immediate challenges, plasticity encompasses a broader scope, including long-term, structural changes that fundamentally alter the supply chain's configuration.

2.3.3 Adaptability vs. Plasticity

Both adaptability and plasticity focus on the responsiveness of supply chains to evolving business environments (Richey et al., 2022). Adaptability involves modifying supply chain structures and practices to remain competitive and resilient. Typically, adaptability manifests through incremental adjustments and short-term responses to changes in the market or disruptions in the supply chain. On the other hand, supply chain plasticity emphasizes significant and transformative changes. It involves a strategic rethinking and restructuring of the supply network to align better with future business needs and challenges. While adaptability can lead to meaningful changes within the supply chain, plasticity requires a more comprehensive approach that includes both reactive and proactive measures. It prioritizes long-term strategic goals over immediate responses, aiming for a more resilient and adaptable supply chain through substantial structural changes.

2.3.4 Plasticity As a Link Between Responsiveness and Resilience

Richey et al. (2022) defined responsiveness as an organizational strategy focused on changing behaviors, norms, and policies within a supply chain to adapt post-disruption. Roh et al. (2022) also emphasize responsiveness's strategic impact, highlighting the role of organizational design and leadership in improving adaptability. In contrast, Asamoah et al. (2021) take a more tactical view, identifying responsiveness as a multidimensional construct that includes logistics, operations, and supplier network responsiveness. On the other hand, supply chain resilience, defined by Pettit et al. (2013) and Hosseini et al. (2019), is the ability to survive, adapt, and grow amidst turbulent changes. It requires preparation, collaboration, agility, and a risk management culture.

Supply chain plasticity extends beyond resilience and responsiveness, involving the capability to rapidly change supply network structures and strategies. It enables firms to respond to immediate challenges and proactively redesign their supply networks for greater robustness and adaptability. This concept serves as a bridge between responsiveness and resilience, allowing firms to navigate volatile environments, mitigate risks, and maintain competitiveness through structural transformations and proactive risk management.

3 Importance of Supply Chain Plasticity in Risk Management

3.1 Enhancing Resilience

One of the primary benefits of supply chain plasticity is its ability to enhance resilience. Traditional risk management strategies (e.g., Schoenherr et al., 2023) often focus on short-term recovery, aiming to return to a pre-disruption state. In contrast, plasticity enables firms to adapt and thrive in a new environment, making them more resilient to future disruptions. By redesigning their supply chains, firms can better align with changing market conditions and mitigate the impact of disruptions (Falcone et al., 2022).

3.2 Mitigating Ripple Effects

Disruptions in the supply chain can have far-reaching ripple effects, impacting multiple tiers of suppliers and customers. Supply chain plasticity helps firms to manage these ripple effects more effectively. By altering their supply chain structure, firms can reduce dependencies on vulnerable nodes and create more robust networks that can absorb shocks and continue to operate smoothly (Son et al., 2021).

How to Mitigate Ripple Effects with Supply Chain Plasticity Ripple effects refer to the cascading impact of disruptions that can spread across the supply chain, affecting multiple stakeholders and processes. Supply chain plasticity helps mitigate these ripple effects through:

- Decentralized Risk Management: By decentralizing critical functions and diversifying supplier bases, firms can limit the scope of disruptions. A plastic supply chain reduces the dependency on single suppliers or production sites, thus containing the ripple effects of disruptions.
- Adaptive Buffering: Implementing adaptive buffering strategies, such as safety stock or flexible warehousing, allows firms to manage fluctuations in supply and demand. These buffers can absorb shocks and prevent them from propagating through the supply chain.
- Collaborative Networks: Building collaborative relationships with suppliers and partners enables information sharing and joint problem-solving. This collaboration helps coordinate responses to disruptions and minimizes their spread across the supply chain.
- Scenario Planning and Simulation: Using scenario planning and simulation tools, firms can anticipate potential disruptions and develop contingency plans. These tools allow for testing different responses and identifying the most effective strategies to mitigate ripple effects.

Facilitating Innovation and Competitiveness Supply chain plasticity also fosters innovation and competitiveness. By continuously evolving and adapting their supply chains, firms can leverage new technologies, processes, and business models. This dynamic approach not only helps in managing risks but also opens up new opportunities for growth and competitive advantage. Firms that embrace plasticity are better positioned to respond to market changes and capitalize on emerging trends (Falcone et al., 2022).

4 Building Supply Chain Plasticity

To enhance supply chain plasticity, firms need to implement specific strategies and actionable steps that focus on adaptability, resilience, and long-term innovation. Here are some recommended approaches for building supply chain plasticity:

4.1 Reconfiguring Supply Chain Partnerships

Strategy: Adapt and strengthen partnerships with suppliers and other stakeholders to ensure stability and collaboration during disruptions.

Action items:

- *Supplier diversification*: Expand the supplier base by onboarding alternative suppliers to minimize reliance on single sources and geographic regions. This approach reduces the risk associated with supplier dependency and ensures a more stable supply chain.

- *Strategic alliances:* Develop strategic alliances with key suppliers to share risk management strategies and resources. Collaborative efforts in risk mitigation can enhance overall resilience and provide a unified response to disruptions.
- *Supplier risk assessments:* Conduct regular risk assessments of suppliers to identify potential vulnerabilities. Collaborate with suppliers to create and implement joint mitigation plans that address these risks effectively.

4.2 Strengthening Logistics and Distribution Networks

Strategy: Build resilient and responsive logistics and distribution networks that can adapt to changes induced by disruptions.

Action items:

- *Multi-modal transport:* Employ multiple transportation modes (such as air, sea, rail, and road) to ensure flexibility in shipping options and routes. This diversification helps maintain supply chain continuity in the face of transport disruptions.
- *Decentralized warehousing*: Implement a decentralized warehousing strategy by establishing distribution centers in various locations. This approach mitigates the impact of regional disruptions and enhances overall supply chain stability.
- *Collaborative logistics networks*: Partner with third-party logistics providers to enhance scalability and responsiveness in distribution operations. Leveraging external expertise and resources can improve logistical flexibility and efficiency.

4.3 Enhancing Supplier Collaboration During Disruption

Strategy: Foster robust collaboration with suppliers to ensure coordinated and effective responses to disruptions.

Action items:

- *Regular communication channels*: Set up regular communication channels with suppliers to facilitate information sharing, coordination of actions, and prompt issue resolution. Effective communication is crucial for managing disruptions and maintaining strong partnerships.
- *Joint problem-solving*: Engage in joint problem-solving sessions with suppliers to develop innovative solutions and contingency plans for managing disruptions. Collaborative efforts can lead to more effective and creative strategies.
- *Mutual support agreements*: Establish mutual support agreements with key suppliers, ensuring that resources and support are shared during critical periods. These agreements strengthen the partnership and improve resilience during disruptions.

4.4 Post-disruption Recovery and Improvement

Strategy: Focus on recovery and continuous improvement in operations and partnerships after the disruption.

Action items:

- *Post-mortem analysis*: Perform thorough post-mortem analyses of the disruption to identify weaknesses and areas for improvement in operations and supply chain management. This analysis provides valuable insights for enhancing future resilience.
- *Process reengineering*: Reengineer processes based on lessons learned from the disruption. This approach enhances resilience and adaptability, ensuring that the supply chain is better prepared for future challenges.
- *Supplier performance reviews:* Conduct comprehensive reviews of supplier performance during the disruption. Use these reviews to renegotiate terms or seek new partnerships as necessary to improve overall supply chain performance.

4.5 Investing in Long-Term Supply Chain Innovation

Strategy: Invest in long-term supply chain innovation to enhance overall resilience and competitiveness.

Action items:

- *R&D for supply chain technologies*: Allocate resources to research and develop new supply chain technologies and practices that improve flexibility and responsiveness. Investing in cutting-edge solutions can drive competitive advantage.
- *Pilot innovative solutions*: Implement pilot programs to test and validate innovative solutions. Successful initiatives should be integrated into broader operations to enhance supply chain performance.
- *Continuous learning and improvement*: Promote a culture of continuous learning and improvement within the organization. Encourage teams to stay updated with the latest supply chain trends and technologies to remain competitive and resilient.

5 Conclusion

5.1 Challenges and Barriers

Implementing supply chain plasticity presents several challenges and barriers. The complexity of redesigning supply networks, establishing new partnerships, and integrating advanced technologies can be resource-intensive and require meticulous planning. Internal resistance to change is another significant obstacle, as employees and managers may be reluctant to adopt new practices or technologies. Additionally, the costs associated with these changes can be substantial, placing a financial strain on organizations, particularly smaller firms with tighter budgets. Integrating new strategies with existing systems poses its own set of difficulties, especially when legacy systems are involved. Coordinating with a diverse supplier base to align goals and establish effective communication channels can also be complex and time-consuming. Furthermore, while supply chain plasticity aims to enhance resilience, it introduces new risks that need to be managed effectively to prevent unintended consequences.

5.2 Future Trends in Supply Chain Plasticity

The future of supply chain plasticity is set to be influenced by several emerging trends. Increased digital integration through technologies like AI, machine learning, and IoT will enhance real-time monitoring, predictive analytics, and automation, allowing for more sophisticated and responsive supply chain operations. Sustainability will become a central focus, with firms integrating environmental and social considerations into their supply chain strategies while maintaining efficiency (Guntuka et al., 2024c). Collaborative networks will be enhanced through platforms that facilitate better information sharing and joint problem-solving among stakeholders. Agile and adaptive supply chain models will become more prevalent, focusing on flexibility and rapid reconfiguration in response to market changes. Advanced risk management techniques, including scenario planning and analytics, will improve preparedness and response to disruptions. Blockchain technology will provide greater transparency and security, streamlining transactions and improving data accuracy. Lastly, human-centric approaches will gain importance, emphasizing employee training, innovation, and team collaboration to support the successful implementation and sustainability of plasticity initiatives.

References

Asamoah, D., Nuertey, D., Agyei-Owusu, B., & Akyeh, J. (2021). The effect of supply chain responsiveness on customer development. *The International Journal of Logistics Management, 32*(4), 1190–1213.

Bode, C., & Wagner, S. M. (2015). Structural drivers of upstream supply chain complexity and the frequency of supply chain disruptions. *Journal of Operations Management, 36*, 215–228.

Boyson, S., Gerst, M., Guntuka, L., Linton, T., Muraski, G., Vakil, B., & Vakil, S. (2022). How exposed is your supply chain to climate risks? *Harvard Business Review*.

Carnovale, S., Guntuka, L., & Falcone, E. (2023). Perfection in supply Chain plasticity? *Distribution Business Management Journal, 25*.

Choi, T. Y., & Holmen, E. (2023). Ready for another leap? Making a case for tetrads. *Journal of Purchasing and Supply Management, 29*(2), 100825.

Christopher, M., & Peck, H. (2004). The five principles of supply chain resilience. *Logistics Europe, 12*(1), 16–21.

Dolgui, A., Gusikhin, O., Ivanov, D., Li, X., & Stecke, K. (2023). A network-of-networks adaptation for cross-industry manufacturing repurposing. *IISE Transactions*. https://doi.org/10.1080/24725854.2023.2253881

Falcone, E. C., Fugate, B. S., & Dobrzykowski, D. D. (2022). Supply chain plasticity during a global disruption: Effects of CEO and supply chain networks on operational repurposing. *Journal of Business Logistics, 43*(1), 116–139.

Gligor, D. M., & Holcomb, M. (2014). The road to supply chain agility: An RBV perspective on the role of logistics capabilities. *The International Journal of Logistics Management, 25*(1), 160–179.

Guntuka, L., Corsi, T. M., & Cantor, D. E. (2024a). Recovery from plant-level supply chain disruptions: Supply chain complexity and business continuity management. *International Journal of Operations & Production Management, 44*(1), 1–31.

Guntuka, L., Carnovale, S., & Falcone, E. (2024b). Supply Chain plasticity: A responsive network capability to ensure resilience. *Journal of Business Logistics, Forthcoming*.

Guntuka, L., Mukandwal, P. S., Aktas, E., & Paluvadi, V. S. K. (2024c). From carbon-neutral to climate-neutral supply chains: A multidisciplinary review and research agenda. *The International Journal of Logistics Management, 35*(3), 916–942.

Hosseini, S., Ivanov, D., & Dolgui, A. (2019). Review of quantitative methods for supply chain resilience analysis. *Transportation Research Part E: Logistics and Transportation Review, 125*, 285–307.

Hughes, M. M., Zhou, Z., Zinn, W., & Knemeyer, A. M. (2023). Plastic response to disruptions: Significant redesign of supply chains. *Journal of Business Logistics, 44*(1), 80–108.

Ivanov, D. (2022). Viable supply Chain model: Integrating agility, resilience and sustainability perspectives—Lessons from and thinking beyond the COVID-19 pandemic. *Annals of Operations Research, 319*, 1411–1431.

Lee, Y. (2022). Exclusive output of Apple iPhones at major China plant could fall 30% amid Covid curbs -source. *Reuters*. Retrieved from https://www.reuters.com/technology/foxconn-covid-woes-may-hit-up-30-iphone-nov-shipments-zhengzhou-plant-source-2022-10-31/

Mukandwal, P. S., Guntuka, L., Aktas, E., & Paluvadi, V. S. K. (2024). Guest editorial: Climate neutral supply chains. *The International Journal of Logistics Management, 35*(3), 713–718.

Nagel, R. N., & Dove, R. (1998). *21st century manufacturing enterprise strategy: An industry-led view*. Diane Publishing.

Pettit, T. J., Croxton, K. L., & Fiksel, J. (2013). Ensuring supply chain resilience: Development and implementation of an assessment tool. *Journal of Business Logistics, 34*(1), 46–76.

Richey, R. G., Roath, A. S., Adams, F. G., & Wieland, A. (2022). A responsiveness view of logistics and supply chain management. *Journal of Business Logistics, 43*(1), 62–91.

Roh, J., Swink, M., & Kovach, J. (2022). Linking organization design to supply chain responsiveness: The role of dynamic managerial capabilities. *International Journal of Operations & Production Management, 42*(6), 826–851.

Schoenherr, T., Mena, C., Vakil, B., & Choi, T. Y. (2023). Creating resilient supply chains through a culture of measuring. *Journal of Purchasing and Supply Management, 29*(4), 100824.

Son, B.-G., Chae, S., & Kocabasoglu-Hillmer, C. (2021). Catastrophic supply chain disruptions and supply network changes: A study of the 2011 Japanese earthquake. *International Journal of Operations & Production Management, 41*(6), 781–804.

Zinn, W., & Goldsby, T. J. (2019). Supply chain plasticity: Redesigning supply chains to meet major environmental change. *Journal of Business Logistics, 40*(3), 184–186.

A Model of an Integrated Analytics Decision Support System for Situational Proactive Control of Recovery Processes in Service-Modularized Supply Chain

Dmitry Ivanov, Boris Sokolov, and Valerii Zakharov

Abstract In the supply chain (SC) recovery process, a disruptive event, planning of the recovery control policy and implementation of this policy are distributed in time and subject to SC structural and parametrical dynamics. In other words, the environment, SC structure and its operational parameters may change in the period between the planning of the recovery control policy and its implementation. As such, situational proactive control with combined use of simulation-optimization and analytics is proposed in the paper to improve processes of transition between a disrupted and a restored SC state. Implementation of situational proactive control can reduce investments in robustness and increase resilience by obviating the time traps in transition process control problems. This chapter develops a model of a decision support system for situational proactive control of SC recovery processes based on a combination of optimization and analytics techniques. More specifically, three dynamic models are developed and integrated with each other, i.e. a model of SC material flow control, a model of SC recovery control and a model of SC recovery control adjustment. The given models are developed within a cyber-physical SC framework based on the service modularization approach.

1 Introduction

Supply chain (SC) vulnerability and disruption management have become a prominent research domain over the last two decades (Dolgui et al., 2019; Ivanov et al., 2022; Ivanov & Sokolov, 2019). Literature differentiates preparedness and recovery

D. Ivanov (✉)
Department of Business and Economics, Berlin School of Economics and Law, Berlin, Germany
e-mail: divanov@hwr-berlin.de

B. Sokolov · V. Zakharov
St. Petersburg Federal Research Center of the Russian Academy of Sciences (SPC RAS), St. Petersburg, Russia
e-mail: sokol@iias.spb.su

D. Ivanov et al. (eds.), *Handbook of Ripple Effects in the Supply Chain*, International Series in Operations Research & Management Science 276,
https://doi.org/10.1007/978-3-031-85508-5_9

decisions (Sheffi, 2005; Ivanov, 2017; Dolgui et al., 2018, 2023; Ivanov, 2018; Yoon et al., 2018; Sawik, 2021; Sodhi et al., 2021; Katsaliaki et al., 2022; Ramani et al., 2022; Aldrighetti et al., 2023a). Recovery processes usually follow the disruption and are deployed on the basis of proactive contingency plans and a recovery forecast (Tomlin, 2006; Ho et al., 2015; Pavlov et al., 2018; Macdonald et al., 2018; Sokolov et al., 2020a, b; Sawik, 2022; Sokolov et al., 2022; Gruchmann et al., 2024).

The existing decision support systems utilize the power of optimization and simulation techniques to support decision-makers in protecting the SC using optimal contingency plans such as risk mitigation inventory, capacity reservations, backup sourcing and combinations thereof (Gupta et al., 2015; Behzadi et al., 2017; Ivanov et al., 2017a, b, c; Sawik, 2017; He et al., 2018; Ivanov, 2019, 2023a; Pavlov et al., 2019; Chopra et al., 2021; Echefaj et al., 2024; Schleifenheimer and Ivanov, 2024). In addition, recovery process simulation has recently been introduced in research (Ivanov et al., 2013, 2014a, b, 2016c, 2018b; Schmitt et al., 2017; Scheibe & Blackhurst, 2018; Sokolov et al., 2019; Ivanov, 2023b).

A disruptive event, planning of the recovery control policy and implementation of this policy are distributed in time and subject to SC structural and parametrical dynamics during a SC recovery process (Xia et al., 2004; Shao & Dong, 2012; Sheffi, 2015; Ivanov, 2018; Alikhani et al., 2023; Dolgui & Ivanov, 2023; Ivanov, 2024). In other words, the environment, SC structures and operational parameters may change during the period between the planning of the recovery control policy and its implementation. This fact represents a gap in current research which mostly relies on recovery models based on recovery forecasting. The missing incorporation of real-time data about process execution in the recovery control deployment models reduces their value and efficiency. There are only a few works on recovery process scheduling (Ivanov et al., 2018a; Paul & Chowdhury, 2021; Sokolov et al., 2021a, b; Li et al., 2024).

At the same time, recent studies have provided evidence of the technical applicability and effectiveness of such techniques by using SC event management techniques (Meyer et al., 2014; Sommerfeld et al., 2018). T&T systems and feedback control can be supported by RFID technology (Dolgui & Proth, 2010) and SC event management systems. Both RFID and SC event management systems can be used to effectively communicate disruptions to the other tiers and to help revise initial processes and schedules, respectively (Dolgui & Proth, 2010; Zelbst et al., 2012). A critical issue in this area is detecting disruptions and their scope in real time. Embedding SC visualization and identification technology is crucial for its successful application in practice (Meyer et al., 2014). However, those studies are rather context-specific and did not provide a model of an integrated proactive-reactive decision support system for situational proactive control of SC recovery processes.

As such, situational proactive control with a combined use of simulation-optimization and analytics is proposed in the paper to improve the transition processes between disrupted and restored SC states. Implementation of situational proactive control can reduce investments in robustness and increase resilience by obviating transition process control problems (Sokolov et al., 2024; Zakharov, 2022a, 2022b). This paper develops a model of an integrated proactive-reactive decision support system for situational proactive control of SC recovery processes based on the combination of optimization and analytics techniques. More specifically, three dynamic models of a service-analytics decision support system for supply chain recovery control are developed and integrated with each other, i.e. a model of SC material flow control, a model of SC recovery control and a model of SC recovery control adjustment. The given models are developed within a cyber-physical SC framework based on a service-oriented approach.

The rest of this study is organized as follows. Section 2 presents the problem of situational proactive control. The decision support framework is shown in Sect. 3. Section 4 is devoted to mathematical models of recovery control. The paper is concluded in Sect. 5 with a summary of the findings and a description of limitations and future research avenues.

2 Problem of Situational Proactive Control

The SC recovery process includes some events which start with the disruption and are continued into planning of the recovery control policy (e.g. based on a contingency plan) and the implementation of this policy through its deployment and control (Fig. 1).

It can be observed in Fig. 1 that the aforementioned events are distributed in time. In this time, the SC is subject to structural and parametrical dynamics. In other words, the environment, SC structures and operational parameters may change in the period between the start of the planning of the recovery control policy and its implementation. The recovery process is planned at the point of time "start of recovery planning" subject to the deployment at the point of time "start of recovery implementation" considering as inputs the SC structure at the point of time "start of recovery planning," the internal parameters in this structure (e.g. available capacity and inventory), the external parameters (e.g. demand) and the forecasted states of those inputs at the point of time "start of recovery implementation." More

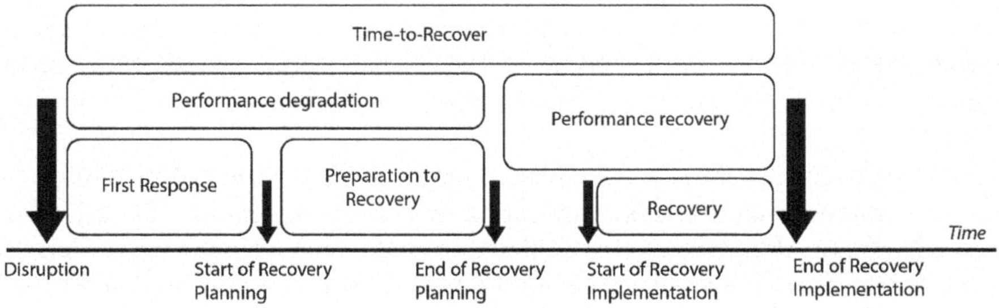

Fig. 1 Supply chain recovery process

specifically, the time gap between the events "start of recovery planning" and "start of recovery implementation" is responsible for the situation in which the actual SC structure and the internal and external parameters may be quite different compared to those forecasted. As such, new methods are needed to obviate the time traps in transition process control problems (Dunke et al., 2018).

As such, the incorporation of real-time data about process execution in recovery control deployment models is necessary to increase their value and efficiency. Such incorporation is needed both for data updating and the adjustment of the recovery control policy during its planning and realization.

3 Cyber-Physical SC Framework and Decision Support System Concept

Success in SC competition has become more and more dependent on analytics algorithms in combination with optimization and simulation modelling. Initially intended for process automation, business analytics techniques now disrupt markets and business models and have had a significant impact on the development of SCM. As such, new disruptive SC business models will arise where SCs are understood not as rigid physical systems with a fixed and static allocation of some processes to some firms (Ivanov et al., 2019; Panetto et al., 2019). Instead, different physical firms will offer services of supply, manufacturing, logistics and sales which will result in the dynamic allocation of processes and dynamic SC structures (Fig. 2).

A new generation of simulation and optimization models extends decision support systems (DSS) towards decision analysis, modelling, control and learning systems (DAMCLS). Recent literature documented the possibility of modelling such integrated service-material flow SCs (Ivanov et al., 2014c; Yang et al., 2017) and service modularization in the SC (Giannakis et al., 2018).

4 Model of Decision Support System

Recall Sect. 2 and the three kinds of models that need to be developed and integrated, i.e. a model of SC material flow control, a model of SC recovery control and a model of SC recovery control adjustment. The given models are developed within a cyber-physical SC framework based on a service-oriented approach, as described in Sect. 3.

The SC is modelled as a networked, controlled system described through a *dynamic* interpretation of the service execution (Ivanov & Sokolov, 2013; Ivanov et al., 2016a, b; Aldrighetti et al., 2023b; Aron et al., 2023). Three joint models of process control are proposed. The control model (M1) is first used to describe the schedule of the material flow control process, and then another control model (M2)

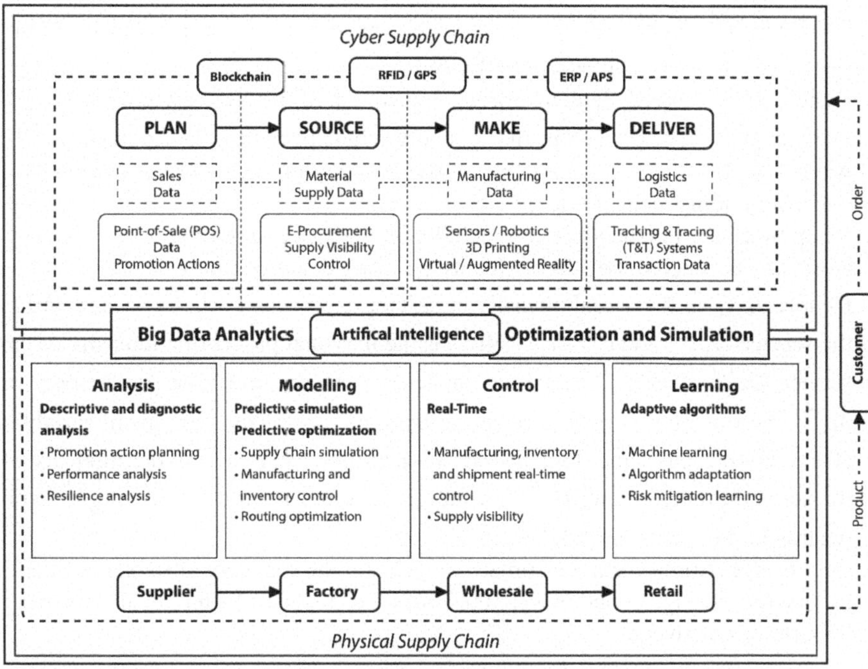

Fig. 2 Cyber-physical SC framework based on a service modularization approach

is used to describe the recovery process. Subsequently, the model (M3) is introduced to describe the adjustment of the recovery process.

General Assumptions and Parameters

- Consider a set of customer orders A_ν $(\nu = 1, \ldots, S)$.
- Denote $D_{\nu\chi}^{(o,j)}$ as services in the SC material flow process (marked as $^{(o)}$), where $\chi = 1, \ldots, \Gamma$ is the running index of the service. Services are logically arranged in the sequence in which they need to be executed to fulfil the customer order.
- Consider a set of SC elements (e.g. factories, warehouses) $M = \{M^{(j)}, j \in N, N = 1, \ldots, n\}$ that can execute services.
- Denote $M_r^{(f,j)}$ as SC elements to be recovered.
- Denote $D_{\nu\chi}^{(f,j)}$ as services in the recovery process (marked as $^{(f)}$), each of which also belongs to order A_ν.
- $a_i^{(o)}$ and $a_i^{(f)}$ are the planned processing volumes of the services.
- Denote $\varepsilon(t)$ as an element of the matrix of time-spatial constraints ($\varepsilon(t) = 1$, if $t_0^k < t \le t_f^k$, $\varepsilon(t) = 0$ otherwise), where k is the number of time windows available for service execution.
- If a service is completed, a channel to the next stage appears. Denote the set of these vendor–buyer relations in the SC as $B = \{B^{(\mu)}, \mu \in \overline{N}\}$, where the subscript η is used for the immediate product acceptor. If a relation exists,

the flows $P^{(i,j)} = \left\{P^{(i,j)}_{<\chi,\rho>}, \chi = 1, \ldots, s_i, \rho = 1, \ldots, p_i\right\}$. appear, where $\rho = 1, \ldots, p_i$ is the enumeration of flows.

- Denote $M_\delta^{(p,j)}$ as SC elements for the adjustment of the recovery policy.
- Denote $D_{rk}^{(p,j)}$ as services in the recovery adjustment process (marked as $^{(p)}$), where $r = 1, \ldots, R_j$ is the number of the recovery process and k is the index of a recovery adjustment service).
- Setup times are independent and included in the processing time.
- Costs are assumed to be a linear function of processed volumes.
- Denote $u_{\nu\chi\mu}^{(f,j)}$ as the actual intensity of service execution at $M_r^{(f,j)}$ with regard to the service $D_{\nu\chi}^{(o,j)}$ and $e_r^{(j)}$, $V_r^{(j)}$, $\Phi_r^{(j)}$ as the maximal processing intensity of the service $D_{\nu\chi}^{(f,j)}$ at $M_r^{(f,j)}$, maximal capacity of M_j and maximal productivity of $M_r^{(f,j)}$ before the recovery, correspondingly; $\overline{e}_r^{(j)}$, $\overline{V}_r^{(j)}$, $\overline{\Phi}_r^{(j)}$ are given variables characterizing the same domains but after the recovery; $b_{r\delta}^{(k)}$ is the intensity of the service processing at $M_\delta^{(p,j)}$, which is required for the service $D_{rk}^{(p,j)}$ with regards to the recovery of $M_j^{(f)}$.
- Let t be the current instant of time, $T = (T_0, T_f]$ the recovery horizon (e.g. time-to-recover, cf. Fig. 1) and T_0 (T_f) the start (end) instant of time for the recovery horizon, respectively.
- $\beta_{\nu\chi}^{(o,j)}(\tau)$ are given time functions for assessing the operated volume of the service $D_{\nu\chi}^{(o,j)}$ or penalties for non-fulfilment.
- $\alpha_{\nu\chi r}^{(f,j)}$ are penalty functions which are assumed to be known and characterize the time points when penalties increase because supply terms are broken.

In order to describe the execution of services, let us introduce the following *state variables*:

$x_{\nu\chi}^{(o,j)}$, $x_{\nu\chi}^{(f,j)}$ which characterize the fulfilment of the services $D_{\nu\chi}^{(o,j)}$,$D_{\nu\chi}^{(f,j)}$ in the order $A_\nu^{(o,j)}$ at M_j

$x_r^{(f,j)}$ which characterizes the total employment time of $M_r^{(f,j)}$

$x_{rk}^{(p,1)}(t)$ which characterizes the current state of $D_{rk}^{(p,j)}$

$x_r^{(p,2)}(t)$ which is an auxiliary variable characterizing the current state of the service processing. Its value is numerically equal to the time interval that has elapsed since the end of the recovery of $M_r^{(f,j)}$;

Decision Variables and Goals

Let us introduce the control variables:

$u_{\nu\chi\mu}^{(o,j)}(t)$ is a control that is equal to 1 if the service $D_{\nu\chi}^{(o,j)}$ is executed at M_j, otherwise $u_{\nu\chi\mu}^{(o,j)}(t) = 0$. $w_{\nu\chi\mu}^{(f,j)}$ is control variable that is equal to 1 if the service $D_{\nu\chi}^{(f,j)}$ is executed at $M_r^{(f,j)}$ and is equal to 0 otherwise.

$\tilde{\nu}_{\nu\chi r}^{(f,j)}$ is an auxiliary control variable that is equal to 1 if the execution of service $D_{\nu\chi}^{(f,j)}$ has been fully completed and is equal to 0 otherwise.

$\nu_{\chi r}^{(p,2)}(t)$ is secondary control variable that is equal to 1 at the instant of time t if a transition (i.e. recovery) from characteristics $(e_r^{(j)}, V_r^{(j)}, \Phi_r^{(j)})$ to new ones $(\overline{e}_r^{(j)}, \overline{V}_r^{(j)}, \overline{\Phi}_r^{(j)})$ occurs; otherwise $\nu_{\chi r}^{(p,2)}(t) = 0$.

$\nu_{r\delta k}^{(p,1)}(t)$ is a control that is equal to 1 if the recovery service $D_{rk}^{(p,j)}$ for $M_r^{(f,j)}$ is executed at $M_\delta^{(p,j)}$; otherwise $\nu_{r\delta k}^{(p,2)}(t) = 0$.

$\nu_r^{(p,2)}(t)$ is an auxiliary control that is equal to 1 if the recovery process of $M_r^{(f,j)}$ is completed; otherwise $\nu_r^{(p,2)}(t) = 0$.

The *model* helps to coordinate three processes during the recovery period, i.e.

- An optimal process recovery
- A SC material flow control under recovery during the recovery time
- The recovery adjustment process

Goals are measured by the order delivery times to customers and the volume of the orders delivered. These goals correspond to practical key performance indicators (KPI) of customer service level and delivery reliability. Customer service level is measured by a function of the actual delivery times. All the orders have to be completed by the time T_f. Otherwise, penalties for delivery breaks or backlog appear, subject to the known penalty functions $\alpha_{\nu\chi r}^{(f,j)}$ and $\beta_{i\eta\rho}(\tau)$, respectively. The reliability of delivery is measured by the volumes $x_{\nu\chi}^{(o,j)}$ of the delivered jobs subject to the planned volumes (i.e. the fullness of the order completing).

4.1 *Mathematical Model of the Material Flow Control Process (Model M1)*

The execution dynamics of services $D_\mu^{(i)}$ can be expressed as (1)

$$\frac{dx_{\nu\chi}^{(o,j)}}{dt} = \sum_{\mu=1}^{m_j} u_{\nu\chi\mu}^{(o,j)}, \tag{1}$$

The economic sense of (1) consists of the representation of the service execution dynamics in which process non-stationary and execution dynamics are reflected. The state variable $x(t)$ accumulates the executed volume of the considered service. The control actions are constrained as follows:

$$\sum_{\nu=1}^{n_j} \sum_{\chi=1}^{s_\nu} u_{\nu\chi\mu}^{(o,j)}(t) \le 1, \forall\mu; \sum_{\mu=1}^{m_j} u_{\nu\chi\mu}^{(o,j)}(t) \le 1, \forall\nu, \forall\chi; \tag{2}$$

$$\sum_{j=1}^{n} u_{\nu\chi\mu}^{(o,j)} \left[\sum_{\alpha \in \Gamma_{i\mu_1}^{-}} \left(a_{i\alpha}^{(o)} - x_{i\alpha}^{(o)} \right) + \prod_{\beta \in \Gamma_{i\mu_2}^{-}} \left(a_{i\beta}^{(o)} - x_{i\beta}^{(o)} \right) \right] = 0 \tag{3}$$

$$\sum_{\mu=1}^{m_j} u_{\nu\chi\mu}^{(o,j)} \left[\left(a_{\nu(\chi-1)}^{(o,j)} - x_{\nu(\chi-1)}^{(o,j)} \right) + \left(a_{\nu\chi}^{(f,j)} - x_{\nu\chi}^{(f,j)} \right) \right] = 0; \tag{4}$$

$$0 \le u_{\nu\chi\mu}^{(o,j)}(t) \le 1, \forall \nu, \forall \chi, \forall \mu; \tag{5}$$

$$u_{\nu\chi\mu}^{(o,j)}(t) \in \{0, 1\}, \tag{6}$$

where $\Gamma_{i\mu_1}^{-}$, $\Gamma_{i\mu_2}^{-}$ are the sets of service numbers which immediately precede a service, subject to completion of all the predecessor services or at least one of these services, respectively. Equation (2) shows that only one service can be executed at a SC element at a time. Constraint (3) brings the natural time logic into the model and determines the precedence relations. Constraint (4) determines the possibility of the delivery of the product ρ-flow to the η-customer. Constraints (5)–(6) reflect the intensity of service processing.

According to Eq. (6), the control contains the values of the *Boolean variables*. In order to assess the results of order execution, we define the following initial and end conditions:

$$x_{\nu\chi}^{(o,j)} \left(T_0^{(j)} \right) \in 0; \tag{7}$$

$$x_{\nu\chi}^{(o,j)} \left(T_f^{(j)} \right) = a_{\nu\chi}^{(o,j)}, \forall \nu, \forall \chi. \tag{8}$$

The constraints (7) reflect that, in the beginning, the volume of executed services is equal to zero (if a certain volume of orders is to be transferred from the previous planning period to the beginning of the current planning period, then this should be reflected in (7)). The conditions (8) reflect the desired end state. The right parts of Eq. (8) are predetermined at the planning stage subject to the volumes (i.e. lot-sizes) of each order.

According to the problem statement, let us introduce the following performance indicators:

$$J_1^{(o,j)} = \sum_{\nu=1}^{n_j} \sum_{\chi=1}^{s_\nu} \sum_{\mu=1}^{m_j} \int_{t_o^{(j)}}^{t_f^{(j)}} \beta_{\nu\chi}^{(o,j)}(\tau)\, u_{\nu\chi\mu}^{(o,j)}(\tau)\, d\tau; \tag{9}$$

$$J_2^{(o,j)} = \frac{1}{2} \sum_{\nu=1}^{n_j} \sum_{\chi=1}^{s_\nu} \left(a_{\nu\chi}^{(o,j)} - x_{\nu\chi}^{(o,j)} \left(t_f^{(j)} \right) \right)^2. \tag{10}$$

The indicator J_1 (9) refers to penalties for breaking delivery terms. The goal indicator J_2 (10) characterizes the accuracy of the end conditions' accomplishment, i.e. the service level.

Corollary 1 Constraints (10) ensure that all orders are fully completed, i.e. the planned service level can be reached.

Proof

Analysis of constraints (3) shows that control $\mathbf{u}(t)$ switches on only when the necessary predecessor services have been executed. $\sum_{\mu=1}^{m_j} u_{\nu\chi\mu}^{(o,j)}\Big[\Big(a_{\nu(\chi-1)}^{(o,j)} - x_{\nu(\chi-1)}^{(o,j)}\Big) = 0$ guarantees the total processing of the predecessor services and $\sum_{\mu=1}^{m_j}\Big(a_{\nu\chi}^{(f,j)} - x_{\nu\chi}^{(f,j)}\Big)\Big] = 0$ of the current services. Herewith, constraints (3) determine the possibility of delivery to the customer $\overline{B}^{(\eta)}$ according to the product flow $P_{<s_i,\rho>}^{(j,\eta)}$ with the use of $M^{(j)}$. The proof is complete.

4.2 Mathematical Model of the Recovery Control Process (Model M2)

Let us describe the execution of services in the recovery process as follows:

$$\frac{dx_{\nu\chi}^{(f,j)}}{dt} = \sum_{r=1}^{R_j} u_{\nu\chi r}^{(f,j)}; \tag{11}$$

$$\frac{dx_{r}^{(f,j)}}{dt} = \sum_{\nu=1}^{n_j}\sum_{\chi=1}^{s_j} w_{\nu\chi r}^{(f,j)}; \tag{12}$$

$$\frac{dx_{r}^{(f,j)}}{dt} = \tilde{\nu}_{\nu\chi r}^{(f,j)}. \tag{13}$$

The control actions are constrained as follows:

$$0 \le u_{\nu\chi r}^{(f,j)}(t) \le \Big[e_{\nu\chi r}\Big(1 - \nu_r^{(p,2)}\Big) + \overline{e}_{\nu\chi r}\nu_r^{(p,2)}\Big] w_{\nu\chi r}^{(f,j)}; \tag{14}$$

$$\sum_{\nu=1}^{n_j}\sum_{\chi=1}^{s_\nu} V_{\nu\chi}^{(j)} w_{\nu\chi r}^{(f,j)} \le V_r^{(j)}\Big(1 - \nu_r^{(p,2)}\Big) + \overline{V}_r^{(j)}\nu_r^{(p,2)}; \tag{15}$$

$$\sum_{\nu=1}^{n_j}\sum_{\chi=1}^{s_\nu} u_{\nu\chi r}^{(f,j)}(t) \le \Phi_r^{(j)}\left(1-\nu_r^{(p,2)}\right)+\overline{\Phi}_r^{(j)}\nu_r^{(p,2)}; \tag{16}$$

$$\tilde{\nu}_{\nu s_\nu}^{(f,j)}\left(a_{\nu s_\nu}^{(f,j)}-x_{\nu s_\nu}^{(f,j)}\right)=0; \tag{17}$$

$$\sum_{r=1}^{R_j} w_{\nu\chi r}^{(f,j)}\left(a_{\nu(\chi-1)}^{(f,j)}-x_{\nu(\chi-1)}^{(f,j)}\right)=0; \tag{18}$$

$$\sum_{r=1}^{R_j} w_{\nu\chi r}^{(f,j)}(t) \le 1, \forall\chi, \forall\nu; \tag{19}$$

$$0 \le w_{\nu\chi r}^{(f,j)}(t) \le 1. \tag{20}$$

Constraints (14)–(16) reflect the possibilities of service processing with regard to $M_r^{(f,j)}$ before and after recovery. Constraint (14) describes the joint functioning of the SC process to be changed and the recovery process. Constraints (17) and (18) determine the processing consequence of the services $D_{\nu\chi}^{(f,j)}$, $D_{\nu(\chi-1)}^{(f,j)}$ subject to the corresponding services $D_{\nu\chi}^{(o,j)}$, $D_{\nu(\chi-1)}^{(o,j)}$ of the material flow control process. Constraints (19) and (20) imply that at the given instant of time the processing of the service $D_{\nu\chi}^{(f,j)}$ can be executed only at $M_r^{(f,j)}$.

Boundary conditions (21)–(24) specify the values of variables $x_{\nu\chi}^{(f,j)}$, $x_r^{(f,j)}$ at the beginning and the end of the planning period $t_0^{(j)}$ и $t_f^{(j)}$ and can be written as follows:

$$x_{\nu\chi}^{(f,j)}\left(t_0^{(j)}\right)=1; \tag{21}$$

$$x_r^{(f,j)}\left(t_0^{(j)}\right)=0; \tag{22}$$

$$x_{\nu\chi}^{(f,j)}\left(t_f^{(j)}\right)=a_{\nu\chi}^{(f,j)}; \tag{23}$$

$$x_r^{(f,j)}\left(t_f^{(j)}\right)\in R^1. \tag{24}$$

Their meaning is similar to those in the model M1. The performance indicators are defined in the form of Eqs. (25)–(28):

$$J_1^{(f,j)}=\sum_{r=1}^{R_{j-1}}\sum_{r_1=r+1}^{R_j}\int_{t_o^{(j)}}^{t_f^{(j)}}\left(x_r^{(f,j)}(\tau)-x_{r_1}^{(f,j)}(\tau)\right)^2 d\tau; \tag{25}$$

$$J_2^{(f,j)} = \sum_{\nu=1}^{n_j} \sum_{\chi=1}^{s_\nu} \sum_{r=1}^{R_j} \int_{t_o^{(j)}}^{t_f^{(j)}} \alpha_{\nu\chi r}^{(f,j)}(\tau)\, w_{\nu\chi r}^{(f,j)}(\tau)\, d\tau; \tag{26}$$

$$J_3^{(f,j)} = \frac{1}{2} \sum_{\nu=1}^{n_j} \sum_{\chi=1}^{s_\nu} \left(a_{\nu\chi}^{(f,j)}(\tau) - x_{\nu\chi}^{(f,j)}\left(t_f^{(j)}\right) \right)^2; \tag{27}$$

$$J_4^{(f,j)} = \sum_{r=1}^{R_j} \left(T^{(j)} - x_r^{(f,j)}\left(t_f^{(j)}\right) \right)^2; \tag{28}$$

Indicator (25) estimates the equal recovery volumes at different SC elements. Indicator (26) characterizes the accuracy of the end conditions' accomplishment, i.e. the service level. Indicator (27) refers to penalties for breaking delivery terms or any other penalties due to disruptions, i.e. the delivery reliability. Indicator (28) estimates the total processing of the orders affected by recovery.

4.3 *Mathematical Model of the Adjustment of Recovery Control Process (M3)*

The model presented in this subsection is similar to the models M1 and M2. This model describes the recovery process adjustment based on feedback information from RFID, sensors and T&T systems concerning changes in the SC and the environment. The mathematical model of recovery control can be presented as (29)–(30):

$$\frac{dx_{rk}^{(p,1)}}{dt} = \sum_{\delta=1}^{\Delta j} b_{r\delta k} v_{r\delta k}^{(p,1)}; \tag{29}$$

$$\frac{dx_r^{(p,2)}}{dt} = v_r^{(p,2)}. \tag{30}$$

The control actions are constrained as follows:

$$\sum_{r=1}^{R_j} v_{r\delta k}^{(p,1)}(t) \le c_{\delta j}^{(p,1)}, \forall \delta, \forall k; \tag{31}$$

$$\sum_{\delta=1}^{\Delta j} v_{r\delta k}^{(p,1)}(t) \le 1, \forall \delta, \forall k, k = 1, \ldots, K_j^{(r)}; \tag{32}$$

$$\sum_{\delta=1}^{\Delta j} \nu_{r\delta k}^{(p,1)} \left(a_{r(k-1)}^{(p,1)} - x_{r(k-1)}^{(p,1)} \right) = 0; \tag{33}$$

$$\nu_r^{(p,2)} \left(a_{rK_j}^{(p,1)} - x_{rK_j}^{(p,1)} \right) = 0; \tag{34}$$

$$0 \le \nu_{r\delta k}^{(p,1)}(t) \le 1; 0 \le \nu_r^{(p,2)}(t) \le 1 \tag{35}$$

Constraints (31) reflect the possibilities of the performing recovery services $D_{rk}^{(p,j)}$ subject to the recovery of $M_r^{(f,j)}$. Constraints (32) set the requirement that the recovery service $D_{rk}^{(p,j)}$ can be performed only by one of $M_\delta^{(p,j)}$ available for the recovery. Constraints (33) determine the prioritizing of the services $D_{rk}^{(p,j)}$ and $D_{r(k-1)}^{(p,j)}$, associated with recovery of $M_r^{(f,j)}$. Constraints (34) define the end time of the recovery process of $M_r^{(f,j)}$. Constraints (35) set the domain of control actions $\nu_{r\delta k}^{(p,1)}(t)$ and $\nu_r^{(p,2)}(t)$ possible values. The end conditions are identical to those in model M2. The performance indicator is defined as (36):

$$\begin{aligned} J_1^{(p,1)} = \sum_{\delta=1}^{\Delta j} \sum_{r=1}^{R_\nu} \sum_{k=1}^{K_j} \Bigg[\left(\int_{t_0^{(j)}}^{t_f^{(j)}} \left(\lambda_1 c_{r\delta k}^{(p,j)}(\tau) + \lambda_2 \beta_{rk}^{(p,j)}(\tau) \right) \nu_{r\delta k}^{(p,1)} d\tau \right) \\ + \lambda_3 \frac{1}{2} \left(a_{rk}^{(p,1)} - x_{rk}^{(p,1)} \right)^2 \Bigg]; \end{aligned} \tag{36}$$

where $c_{r\delta k}^{(p,j)}(\tau)$ are the costs of recovery and λ are the weight coefficients. Indicator (36) estimates the total processing quality of services $D_{\upsilon\chi}^{(f,j)}$. This indicator also enables estimation of total recovery cost, ongoing service of the modernized process, and total penalty for the schedule violation within the process recovery. These values can be calculated using the first component of the indicator (36). The terminal (second) component of indicator (36) estimates the accuracy of the confirmation for the end conditions.

The recovery control process can now be formulated as the following problem of optimal program control (OPC): this is necessary to find an allowable control $\mathbf{u}(t)$, $t \in (T_0, T_f]$ that ensures the vector constraint functions $\mathbf{q}^{(1)}(\mathbf{x},\mathbf{u}) = \mathbf{0}$, $\mathbf{q}^{(2)}(\mathbf{x},\mathbf{u}) \le \mathbf{0}$ (2)–(6), (14)–(20) and (31)–(35) are met for the model governed by (1), (11)–(13) and (29)–(30) and guides the dynamic system (i.e. the SC) $\dot{\mathbf{x}} = \mathbf{f}(\mathbf{x},\mathbf{u},t)$ from the initial state to the specified final state. In terms of OPC, the program control of services execution is simultaneously the optimal recovery process. The formulated model is a linear, non-stationary, finite-dimensional controlled differential system with the convex area of admissible control. Note that this is a standard OPC problem; see (Lee & Markus, 1967). This model is linear in the state and control variables, and the objective is linear. The transfer of non-linearity to the constraint ensures convexity and allows the use of interval constraints. For such kind of models, both sufficient and necessary optimality conditions have been proven (Dolgui et al., 2019; Ivanov et al., 2021).

It can be observed that the elaborated multi-level modelling complex can be applied for the investigation and solution of different problem domains in the analysis and synthesis of SCs with regard to different, interlinked processes. The integration and coordination of the models is ensured by the following basic components:

- M2 influences the M1 model through the constraints (3). A service $D_{\nu\chi}^{(o,j)}$ in the unchanged process can start only after completing the previous service $D_{\nu(\chi-1)}^{(o,j)}$ in this process and the service $D_{\nu\chi}^{(f,j)}$ in the coordinated process.
- M3 influences M2 through Eqs. (14)–(16).
- The M1 model influences the M2 and M3 models through the conjunctive variables by the OPC problem solution with the help of the local cut method (Ivanov & Sokolov, 2013).

The conducted investigations of the proposed models show that the use of the goal indicators (9)–(10), (25)–(28) and (36) makes it possible to model other factors with regard to SC effectiveness and efficiency. For example, with the use of the $c_{r\delta k}^{(p,j)}(\tau)$, the following can be explicitly reflected:

- Changes in the costs of SC recovery regarding SC evolution in dynamic
- Changes in total SC costs regarding the mutual influences of interlinked processes

The elaborated models can also be applied to simulate different scenarios of recovery and the corresponding impact on SC performance including in solving the tasks of proactive planning and scheduling of the functioning and modernization of complex technical objects; development of programs for managing and control of material and information processes; organization of modelling and simulation processes; management of micro-logistics of transportation and logistics systems in modern airports (Sokolov et al., 2024; Gnidenko et al., 2019; Sokolov et al., 2020a, b, 2021a, b; Sokolov and Yusupov, 2022; Zakharov, 2023). The analytical optimality properties of the model presented as well as the efficient computational algorithm have been previously presented and proven in the studies (Ivanov & Sokolov, 2013; Ivanov et al., 2016a, b; Ivanov et al., 2018a). We omit their detailed presentation in this paper.

5 Conclusions

During the SC recovery process, a disruptive event, planning of the recovery control policy and implementation of this policy are distributed in time and subject to SC structural and parametrical dynamics. In other words, the environment, SC structures and operational parameters may change during the period between the planning of the recovery control policy and its implementation.

As such, situational proactive control with combined use of simulation-optimization and analytics has been proposed in the paper to improve the transition processes from a disrupted to a restored SC state. Implementing situational proactive control reduces investments in robustness and increases resilience by obviating the transition process control problems. This paper developed a model for an integrated proactive-reactive decision support system for situational proactive control of SC recovery processes based on a combination of optimization and analytics techniques. More specifically, three dynamic models were developed and integrated with each other, i.e. a model of SC material flow control, a model of SC recovery control and a model of SC recovery control adjustment. The given models are developed within a cyber-physical SC framework based on a service-oriented approach.

The advantages of the proposed approach are evident in two areas. On the one hand, continuous optimization is a convenient engineering tool to model service-oriented, cyber-physical SCs since the services and their accumulation over time can be easily described with the help of continuous state and control variables. On the other hand, dynamic description of the SC recovery process with the help of optimal control enables the incorporation of these results into an axiomatic of feedback control which uses the same methodological principles.

Some limitations of this study and future research can be considered in light of the technical integration of data analytics into the framework and the models developed. The models themselves need to be adapted for individual applications for which specific algorithms need to be developed, analytically analysed, and practically implemented.

Acknowledgements The research of Boris Sokolov and Valerii Zakharov related to this chapter was supported by the Russian Science Foundation grant No 24-19-00823.

References

Aldrighetti, R., Battini, D., & Ivanov, D. (2023a). Efficient resilience portfolio design in the supply chain with consideration of preparedness and recovery investments. *Omega, 117*, 102841.

Aldrighetti, R., Calzavara, M., Martignago, M., Zennaro, I., Battini, D., & Ivanov, D. (2023b). A methodological framework for the design of efficient resilience in supply networks. *International Journal of Production Research, 62*(1–2), 271–290.

Alikhani, R., Ranjbar, A., Jamali, A., Torabi, S. A., & Zobel, C. W. (2023). Towards increasing synergistic effects of resilience strategies in supply chain network design. *Omega, 116*, 102819. https://doi.org/10.1016/j.omega.2022.102819

Aron, C., Sgarbossa, F., Ballot, E., & Ivanov, D. (2023). Cloud material handling systems: A cyber-physical system to enable dynamic resource allocation and digital interoperability. *Journal of Intelligent Manufacturing*. https://doi.org/10.1007/s10845-023-02262-6

Behzadi, G., O'Sullivan, M. J., Olsen, T. L., Scrimgeour, F., & Zhang, A. (2017). Robust and resilient strategies for managing supply disruptions in an agribusiness supply chain. *International Journal of Production Economics, 191*, 207–220.

Chopra, S., Sodhi, M., & Lücker, F. (2021). Achieving supply chain efficiency and resilience by using multi-level commons. *Decision Sciences, 52*(4), 817–832.

Dolgui, A., & Ivanov, D. (2023). Metaverse supply chain and operations management. *International Journal of Production Research, 61*(23), 8179–8191.

Dolgui, A., & Proth, J. M. (2010). *Supply chain engineering: Useful methods and techniques*. Springer.

Dolgui, A., Ivanov, D., & Sokolov, B. (2018). Ripple effect in the supply chain: An analysis and recent literature. *International Journal of Production Research, 56*(1–2), 414–430.

Dolgui, A., Ivanov, D., Sethi, S., & Sokolov, B. (2019). Scheduling in production, supply chain and industry 4.0 systems by optimal control: Fundamentals, state-of-the-art, and applications. *International Journal of Production Research, 57*(2), 411–432.

Dolgui, A., Gusikhin, O., Ivanov, D., Li, X., & Stecke, K. (2023). A network-of-networks adaption for cross-industry manufacturing repurposing. *IISE Transactions, 56*(6), 666–682.

Dunke, F., Heckmann, I., Nickel, S., & Saldanha-da-Gama, F. (2018). Time traps in supply chains: I optimal still good enough? *European Journal of Operational Research, 264*, 813–829.

Echefaj, K., Charkaoui, A., Cherrafi, A., & Ivanov, D. (2024). Design of resilient and viable sourcing strategies in intertwined circular supply networks. *Annals of Operations Research, 337*, 459–498.

Giannakis, M., Doran, D., Mee, D., Papadopoulos, T., & Dubey, R. (2018). The design and delivery of modular legal services: Implications for supply chain strategy. *International Journal of Production Research, 56*(20), 6607–6627.

Gnidenko, A., Sobolevsky, V., Potriasaev, S., & Sokolov, B. (2019). Methodology and integrated modeling technologies for synthesis of cyber-physical production systems modernization programs and plans. *IFAC-PapersOnLine, 52*(13), 642–647.

Gruchmann, T., Stadtfeld, G. M., Thürer, M., & Ivanov, D. (2024). Supply chain resilience as a system quality: Survey-based evidence from multiple industries. *International Journal of Physical Distribution & Logistics Management, 54*(1), 92–117.

Gupta, V., He, B., & Sethi, S. P. (2015). Contingent sourcing under supply disruption and competition. *International Journal of Production Research, 53*(10), 3006–3027.

He, J., Alavifard, F., Ivanov, D., & Jahani, H. (2018). A real-option approach to mitigate disruption risk in the supply chain. *Omega, 88*, 133–149.

Ho, W., Zheng, T., Yildiz, H., & Talluri, S. (2015). Supply chain risk management: A literature review. *International Journal of Production Research, 53*(16), 5031–5069.

Ivanov, D. (2017). Simulation-based ripple effect modelling in the supply chain. *International Journal of Production Research, 55*(7), 2083–2101.

Ivanov, D. (2018). *Structural dynamics and resilience in supply chain risk management*. Springer.

Ivanov, D. (2019). Disruption tails and revival policies: A simulation analysis of supply chain design and production-ordering systems in the recovery and post-disruption periods. *Computers and Industrial Engineering, 127*, 558–570.

Ivanov, D. (2023a). Collaborative emergency adaption for ripple effect mitigation in intertwined supply networks. *Annals of Operations Research*. https://doi.org/10.1007/s10479-023-05408-0

Ivanov, D. (2023b). Intelligent digital twin (iDT) for supply chain stress-testing, resilience, and viability. *International Journal of Production Economics, 263*, 108938. https://doi.org/10.1016/j.ijpe.2023.108938

Ivanov, D. (2024). Exiting the COVID-19 pandemic: After-shock risks and avoidance of disruption tails in supply chains. *Annals of Operations Research, 335*, 1627–1644.

Ivanov, D., & Sokolov, B. (2013). Dynamic coordinated scheduling in the supply chain under a process modernization. *International Journal of Production Research, 51*(9), 2680–2697.

Ivanov, D., & Sokolov, B. (2019). Simultaneous structural–operational control of supply chain dynamics and resilience. *Annals of Operations Research, 283*(1–2). https://doi.org/10.1007/s10479-019-03231-0

Ivanov, D., Sokolov, B., & Pavlov, A. (2013). Dual problem formulation and its application to optimal re-design of an integrated production-distribution network with structure dynamics and ripple effect considerations. *International Journal of Production Research, 51*(18), 5386–5403.

Ivanov, D., Sokolov, B., & Dilou Raguinia, E. A. (2014a). Integrated dynamic scheduling of material flows and distributed information services in collaborative cyber-physical supply networks. *International Journal of Systems Science: Services & Logistics, 1*(1), 18–26.

Ivanov, D., Sokolov, B., & Dolgui, A. (2014b). The ripple effect in supply chains: Trade-off 'efficiency-flexibility-resilience' in disruption management. *International Journal of Production Research, 52*(7), 2154–2172.

Ivanov, D., Sokolov, B., & Pavlov, A. (2014c). Optimal distribution (re)planning in a centralized multi-stage network under conditions of ripple effect and structure dynamics. *European Journal of Operational Research, 237*(2), 758–770.

Ivanov, D., Sokolov, B., Pavlov, A., Dolgui, A., & Pavlov, D. (2016a). Disruption-driven supply chain (re)-planning and performance impact assessment with consideration of pro-active and recovery policies. *Transportation Research: Part E, 90*, 7–24.

Ivanov, D., Sokolov, B., Dolgui, A., Werner, F., & Ivanova, M. (2016b). A dynamic model and an algorithm for short-term supply chain scheduling in the smart factory industry 4.0. *International Journal of Production Research, 54*(2), 386–402.

Ivanov, D., Dolgui, A., & Sokolov, B. (2016c). Robust dynamic schedule coordination control in the supply chain. *Computers and Industrial Engineering, 94*(1), 18–31.

Ivanov, D., Tsipoulanidis, A., & Schönberger, J. (2017a). *Global supply chain and services management* (1st ed.). Springer.

Ivanov, D., Dolgui, A., Sokolov, B., & Ivanova, M. (2017b). Literature review on disruption recovery in the supply chain. *International Journal of Production Research, 55*(20), 6158–6174.

Ivanov, D., Pavlov, A., Pavlov, D., & Sokolov, B. (2017c). Minimization of disruption-related return flows in the supply chain. *International Journal of Production Economics, 183*, 503–513.

Ivanov, D., Dolgui, A., & Sokolov, B. (2018a). Scheduling of recovery actions in the supply chain with resilience analysis considerations. *International Journal of Production Research, 56*(19), 6473–6490.

Ivanov, D., Sethi, S., Dolgui, A., & Sokolov, B. (2018b). A survey on the control theory applications to operational systems, supply chain management and industry 4.0. *Annual Reviews in Control, 46*, 134–147.

Ivanov, D., Dolgui, A., & Sokolov, B. (2019). The impact of digital technology and industry 4.0 on the ripple effect and supply chain risk analytics. *International Journal of Production Research, 57*(3), 829–846.

Ivanov, D., Sokolov, B., Chen, W., Dolgui, A., Werner, F., & Potryasaev, S. (2021). A control approach to scheduling flexibly configurable jobs with dynamic structural-logical constraints. *IISE Transactions, 53*(1), 21–38. https://doi.org/10.1080/24725854.2020.1739787

Ivanov, D., Dolgui, A., Sokolov, B., & Ivanova, M. (2022). Integrated simulation-optimization modeling framework of resilient design and planning of supply chain networks. *IFAC-PapersOnLine, 55*(10), 2713–2718. https://doi.org/10.1016/j.ifacol.2022.10.121

Katsaliaki, K., Galetsi, P., & Kumar, S. (2022). Supply chain disruptions and resilience: A major review and future research agenda. *Annals of Operations Research, 319*, 965–1002.

Lee, E. B., & Markus, L. (1967). *Foundations of optimal control theory*. Wiley.

Li, X., Krivtsov, V., Pan, C., Nassehi, A., Gao, R. X., & Ivanov, D. (2024). End-to-end supply chain resilience management using deep learning, survival analysis, and explainable artificial intelligence. *International Journal of Production Research*, 1–29. https://doi.org/10.1080/00207543.2024.2367685

Macdonald, J. R., Zobel, C. W., Melnyk, S. A., & Griffis, S. E. (2018). Supply chain risk and resilience: Theory building through structured experiments and simulation. *International Journal of Production Research, 56*(12), 4337–4355.

Meyer, G. G., Buijs, P., Szirbik, N. B., & Wortmann, J. C. (2014). Intelligent products for enhancing the utilization of tracking technology in transportation. *International Journal of Services & Production Management, 34*(4), 422–446.

Panetto, H., Iung, B., Ivanov, D., Weichhart, G., & Wang, X. (2019). Challenges for the cyber-physical manufacturing enterprises of the future. *Annual Reviews in Control, 47*, 200–213.

Paul, S. K., & Chowdhury, P. (2021). A production recovery plan in manufacturing supply chains for a high-demand item during COVID-19. *International Journal of Physical Distribution & Logistics Management, 51*(2), 104–125.

Pavlov, A., Ivanov, D., Dolgui, A., & Sokolov, B. (2018). Hybrid fuzzy-probabilistic approach to supply chain resilience assessment. *IEEE Transactions on Engineering Management, 65*(2), 303–315.

Pavlov, A., Ivanov, D., Pavlov, D., & Slinko, A. (2019). Optimization of network redundancy and contingency planning in sustainable and resilient supply chain resource management under conditions of structural dynamics. *Annals of Operations Research*. https://doi.org/10.1007/s10479-019-03182-6

Ramani, V., Ghosh, D., & Sodhi, M. S. (2022). Understanding systematic disruption from the Covid-19-induced semiconductor shortage for the auto industry. *Omega, 113*, 102720.

Sawik, T. (2017). A portfolio approach to supply chain disruption management. *International Journal of Production Research, 55*(7), 1970–1991.

Sawik, T. (2021). On the risk-averse selection of resilient multi-tier supply portfolio. *Omega, 101*, 102267.

Sawik, T. (2022). Stochastic optimization of supply chain resilience under ripple effect: A COVID-19 pandemic related study. *Omega, 109*, 102596.

Scheibe, K. P., & Blackhurst, J. (2018). Supply chain disruption propagation: A systemic risk and normal accident theory perspective. *International Journal of Production Research, 56*(1–2), 43–59.

Schleifenheimer, M., & Ivanov, D. (2024). Pharmaceutical retail supply chain responses to the COVID-19 pandemic. *Annals of Operations Research*. https://doi.org/10.1007/s10479-024-05866-0

Schmitt, T. G., Kumar, S., Stecke, K. E., Glover, F. W., & Ehlen, M. A. (2017). Mitigating disruptions in a multi-echelon supply chain using adaptive ordering. *Omega, 68*, 185–198.

Shao, X. F., & Dong, M. (2012). Supply disruption and reactive strategies in an assemble-to-order supply chain with time-sensitive demand. *IEEE Transactions on Engineering Management, 59*(2), 201–212.

Sheffi, Y. (2005). *The resilient enterprise: Overcoming vulnerability for competitive advantage*. MIT Press.

Sheffi, Y. (2015). Preparing for disruptions through early detection. *MIT Sloan Management Review, 57*, 31.

Sodhi, M. S., Tang, C. S., & Willenson, E. T. (2021). Research opportunities in preparing supply chains of essential goods for future pandemics. *International Journal of Production Research, 61*(8), 2416–2431.

Sokolov, B. V., & Yusupov, R. M. (2022). Scientific Basis of Management and Cybernetics Methodologies Integration. Lecture Notes in Networks and Systems, 442 LNNS. https://doi.org/10.1007/978-3-030-98832-6_5.

Sokolov, B., Trofimova, I., Nazarov, D., & Zakharov, V. (2019). Modification of multiple-model description and planning and update control algorithms of supply chain. *IFAC-PapersOnLine, 52*(13), 1972–1977. https://doi.org/10.1016/j.ifacol.2019.11.492

Sokolov, B. V., Kalinin, V. N., & Zakharov, V. V. (2020a). Integrated planning and scheduling of enterprise information system modernization. *CEUR Workshop Proceedings, 2803*, 3–12.

Sokolov, B., Pavlov, A., Potriasaev, S., & Zakharov, V. (2020b). Methodology and technologies of the complex objects proactive intellectual situational management and control in emergencies. In *Advances in intelligent systems and computing: vol. 1156 AISC*. https://doi.org/10.1007/978-3-030-50097-9_24.

Sokolov, B. V., Potryasaev, S. A., & Yusupov, R. M. (2021a). Proactive Management of Information Processes in the industrial internet. *Journal of Physics: Conference Series, 1864*(1). https://doi.org/10.1088/1742-6596/1864/1/012007

Sokolov, B., Zakharov, V., & Krylov, A. (2021b). Complex scheduling of measurement and calculation systems functioning. *CEUR Workshop Proceedings, 2924*, 58–63.

Sokolov, B. V., Potryasaev, S., Korableva, O. N., & Zakharov, V. V. (2022). Methodology support and algorithms for dynamic synthesis of technologies and programs for proactive management of information processes in the industrial internet under a crisis. *International Journal of Risk Assessment and Management, 25*(3–4). https://doi.org/10.1504/IJRAM.2022.130524

Sokolov, B., Ushakov, V., & Zakharov, V. (2024). Optimal planning and scheduling of information processes during interaction among mobile objects. *International Journal of Production Research, 62*(16). https://doi.org/10.1080/00207543.2024.2302388

Sommerfeld, D., Teucke, M., & Freitag, M. (2018). Effects of sensor-based quality data in automotive supply chains–a simulation study. In M. Freitag, H. Kotzab, & J. Pannek (Eds.), *Dynamics in logistics. LDIC 2018, Bremen 20–22, 2018* (pp. 289–297). Lecture Notes in Logistics). Springer.

Tomlin, B. T. (2006). On the value of mitigation and contingency strategies for man–aging supply chain disruption risks. *Management Science, 52*(5), 639–657.

Xia, Y., Yang, M. H., Golany, B., Gilbert, S. M., & Yu, G. (2004). Real-time disruption management in a two-stage production and inventory system. *IIE Transactions, 36*(2), 111–125.

Yang, Y., et al. (2017). Mitigating supply chain disruptions through interconnected logistics services in the physical internet. *International Journal of Production Research, 55*(14), 3970–3983.

Yoon, J., Talluri, S., Yildiz, H., & Ho, W. (2018). Models for supplier selection and risk mitigation: A holistic approach. *International Journal of Production Research, 56*(10), 3636–3661.

Zakharov, V. (2022a). Combined optimization algorithm of complex technical object functioning and its information system modernization. In Y. S. Vasiliev, N. D. Pankratova, V. N. Volkova, O. D. Shipunova, & N. N. Lyabakh (Eds.), *System analysis in engineering and control* (pp. 487–497). Springer International Publishing.

Zakharov, V. (2022b). Combined optimization algorithm of complex technical object functioning and its information system modernization. In Y. S. Vasiliev, N. D. Pankratova, V. N. Volkova, O. D. Shipunova, & N. N. Lyabakh (Eds.), *Lecture notes in networks and systems: Vol. 442 LNNS* (pp. 487–497). Springer International Publishing. https://doi.org/10.1007/978-3-030-98832-6_43

Zakharov, V. (2023). Method of dynamic multi-criteria quality assessment and optimization for proactive complex objects functioning schedules. In *Lecture notes in networks and systems, 777 LNNS*. https://doi.org/10.1007/978-3-031-43792-2_29

Zelbst, P. J., Green, K. W., Sower, V. E., & Reyes, P. M. (2012). Impact of RFID on manufacturing effectiveness and efficiency. *International Journal of Services & Production Management, 32*(3), 329–350.

Entropy-Based Analysis and Quantification of Supply Chain Recoverability

Dmitry Ivanov

Abstract The problem of designing resilient supply chains at the semantic network level is considered. The entropy method is used to show the interrelations between supply chain design and recoverability. Easy-to-compute quantitative measures are proposed to estimate supply chain recoverability. For the first time, entropy-based supply chain analysis is brought into correspondence with supply chain structural recoverability and flexibility considerations downstream the supply chain. An exact and a heuristic computation algorithm are suggested and illustrated. The developed approach and recoverability measure can be used to select a resilient supply chain design in terms of potential recoverability.

1 Introduction

Designing the resilient supply chains has become a crucial research avenue over the last decade. In essence, the main research thread is to identify supply chain (SC) structures, which are robust and/or recoverable in the event of severe disruptions such as facility breakdowns, strikes, supplier bankruptcies, or financial crises. Significant research advancements have been achieved in this area at the semantic network level, design, planning, and control levels (Martel & Klibi, 2016; Mistree et al., 2017; Ivanov, 2018). While design, planning, and control levels deal with both structural parameters (i.e., the facility locations and connections in between them) and operational parameters (i.e., demand, capacity, and lead time, to name a few), the semantic network level considers the structural domain only. The target of the analysis at the semantic network level is to identify major interdependencies between the network graph forms and the supply chain robustness, flexibility,

D. Ivanov (✉)
Department of Business and Economics, Berlin School of Economics and Law, Berlin, Germany
e-mail: divanov@hwr-berlin.de

D. Ivanov et al. (eds.), *Handbook of Ripple Effects in the Supply Chain*,
International Series in Operations Research & Management Science 276,
https://doi.org/10.1007/978-3-031-85508-5_10

adaptability, and resilience (Zobel, 2014; Ivanov, 2017; Giannoccaro et al., 2018; Ivanov et al., 2017a, 2017b; Dolgui et al., 2018).

Linking supply chain complexity and disruption risk resistance has become more and more important (Nair & Vidal, 2011; Scheibe & Blackhurst, 2018; Levner & Ptuskin, 2018; Birkie et al., 2017; He et al., 2019; Gruchmann et al., 2024; Ivanov, 2024a). Blackhurst et al. (2005), Ho et al. (2015), Jain et al. (2017), and Ivanov et al. (2017a) underline that global sourcing, product individualization, and cross-channel logistics strategies increase supply chain complexity. Scheibe and Blackhurst (2018) identified SC structure as one of the three major drivers of disruption propagations in the SC. Complex networks become more vulnerable to severe disruptions, which change the supply chain structures and are involved with supply chain structural dynamics (Ivanov et al., 2010; Ivanov & Sokolov, 2010; Mistree et al., 2017; Ivanov, 2018; Aldrighetti et al., 2023b; Echefaj et al., 2024; Ruel et al., 2024). Moreover, the ripple effect in the supply chain complicates disruption management (Liberatore et al., 2012; Ivanov et al., 2014a, 2014b; Ivanov, 2017, 2023, 2024b, 2024c; Li et al., 2025).

As an opposite to the well-known bullwhip effect that considers high-frequency-low-impact *operational risks*, the ripple effect studies low-frequency-high-impact *disruptive risks* (Ivanov et al., 2014a; Simchi-Levi et al., 2015; Aqlan & Lam, 2015; Sokolov et al., 2016; Snyder et al., 2016; Han & Shin, 2016; Sawik, 2017; Ramani et al., 2022; Alikhani et al., 2023; Rolf et al., 2023). The ripple effect describes the impact of disruption propagation in the supply chain on structural dynamics and performance. Recent studies extensively considered disruption risks in light of the impact of disruption propagation (Wilson, 2007; Ivanov et al., 2014b; Paul et al., 2014; Ivanov, 2017; Scheibe & Blackhurst, 2018; Levner & Ptuskin, 2018; Ivanov, 2019; Altay et al., 2018; Käki et al., 2015; Lücker & Seifert, 2017; Sawik, 2022, 2023; Ivanov, 2024d). Previous studies also suggested several measures to quantify disruption risks (Zobel, 2011; Basole & Bellamy, 2014; Han & Shin, 2016; Lin et al., 2017; Dolgui et al., 2019; Choi & Shi, 2022; Sardesai & Klingebiel, 2023; Liu et al., 2024). However, single-stage disruption has mostly been considered, and disruption propagation in light of supply chain complexity has been neglected.

In addition, recoverability of the supply chain plays an important role in supply chain resilience and in mitigating the ripple effect. The disruption profile constituted in the work by Sheffi and Rice (2005) includes eight phases: preparation actions, the disruptive event, the first response, the initial impact, the full impact, the recovery preparations, and the recovery and long-term impact. That is why the recoverability is an important part of supply chain resilience. Ivanov et al. (2017b) revealed several major recovery policies in the supply chain. One of them is the backup sourcing and shipment activation in the event of disruption and the resulting structural changes in the supply chain. This study focuses on the said recovery policy and brings it into correspondence to supply chain complexity.

Complexity quantification can be achieved by means of "entropy" which was first introduced by Shannon and Weaver (1963) and applied to supply chain domain in the studies by Harremoës and Topsøe (2001), Isik (2010), Allesina et al. (2010),

Ivanov and Arkhipov (2011), Yu and Xiao (2014), Yuming (2015), Levner and Ptuskin (2015, 2018), and Aldrighetti et al. (2023a).

Levner and Ptuskin (2018) addressed ripple effect analysis using entropy measure from the environmental risk perspective and underlined that it would be "more practicable to study the harmful impact of the environmental risks in more general framework," i.e., in combination with other risk types and classes (Quang & Hara, 2018). This study closes the research gap described above and develops a quantitative entropy-based measure to analyze supply chain recoverability in event of facility disruptions. For the first time, entropy-based supply chain analysis will be brought into correspondence with supply chain structural recoverability and flexibility considerations downstream the supply chain. An exact and a heuristic computation algorithm will be suggested and illustrated.

The rest of this paper is structured as follows. Section 2 is devoted to literature analysis. The method for a recoverability potential assessment is presented in Sect. 3. The exact computation algorithm is presented and illustrated numerically in Sect. 4. Section 5 considers the heuristic method. Section 6 summarizes the outcomes of this study and discusses its limitations and future research avenues (Hosseini & Barker, 2016).

2 State of the Art

2.1 Entropy-Based Studies

Harremoës and Topsøe (2001) developed an entropy-based approach to examining the vulnerability of supply network nodes using maximum entropy (Hosseini & Barker, 2016). They tested the entropic approach on a real-world healthcare supply chain and successfully extracted the most dangerous risks in the supply chain. Isik (2010) applied the entropy concept to measure complexity associated with information and material flows.

Allesina et al. (2010) developed eight indexes based on entropy to measure the level of complexity in the supply chain by mapping the exchanges of goods between the different actors in the network. The impact of possible modifications of the supply chain structure can be evaluated using these tools, providing an evaluation of the different structural dynamics scenarios. Ivanov and Arkhipov (2011) applied the entropy model to the analysis of supply chain adaptation potential. They considered supply chains in virtual enterprises which are formed through the dynamic selection of partners from a pool of available suppliers (the so-called structural–functional reserve) in the supply chain. They also developed some modifications in regard to real-scale supply chain structures.

Yu and Xiao (2014) used the entropy principles for evaluation, analysis, and data processing in problems related to supply chain resilience and risk management. Levner and Ptuskin (2015, 2018) applied the entropy concept to identify vulnerable supply chain nodes and to measure the impact of environmental risks, respectively.

The study by Yuming (2015) explored supply chain flexibility from the dimensions of resources and time when the supply chain is coordinated. Resource allocation with the entropy concept in the event of supply chain coordination was proposed. The entropy-based analysis allows estimation of resource output elasticity and time output elasticity. The paper also proposed a method of measuring supply chain flexibility by integrating resource flexibility and time flexibility. Levner and Ptuskin (2018) addressed the ripple effect analysis using entropy measure from the environmental risk perspective.

2.2 *Related Studies to Quantify the Supply Chain Disruption Risk Resistance*

In a study by Nair and Vidal (2011), 20 supply chains subject to random demand were analyzed using graph-theoretical topology analysis to study the robustness of those supply chains in the event of disruptions. In consideration of varying probabilities of the random failure of nodes during targeted attacks, the authors evaluated the performance of the supply chains. The assessment of severity of the disruptions included the downtime of the nodes affected. The conclusions of the study were based on multi-agent simulation, and the authors detailed the impact of the supply chain structural design on robustness when both demand and disruption were uncertain.

Assuming a sudden onset disruption and linear recovery behavior, Zobel (2011) estimated supply network resilience, while Zobel (2014) extended the study to a general case which accounted for nonlinearity and average loss per unit time. Using the largest connected component and average and minimum path length, Zhao et al. (2011) quantified network connectivity and accessibility. On the other hand, Basole and Bellamy (2014) quantified the level of risk diffusion as the number of functioning nodes at time t relative to the size of the network using the "number of healthy nodes" as a measure. In the study by Simchi-Levi et al. (2015), a risk exposure index was developed for an automotive supply chain. The index was computed using two models, time to recovery and time to survive, and the impact of a disruption on the supply chain's performance was then assessed.

Using a survival model to illustrate the time from when a system failed to the time when it resumed functioning (i.e., recovery), supply chain resilience was analyzed in the study by Raj et al. (2014). The input to the model was the event of failure, while the output was recovery time, which allowed the resilience of the supply chain to be measured in terms of recovery time. Using several indicators from graph theory, Sokolov et al. (2016) developed a model to assess and quantify, using connectivity, reachability, complexity, and centralization as metrics, the performance impact of the ripple effect in a distribution network.

Considering supply chain structural dynamics in a multistage network with cross-docking terminals, the study by Ivanov et al. (2016) presented a method of quantification for the financial impact of disruptions on the supply chain, and then recommended the use of hybrid optimal control, a linear programming model which considers several structural constancy intervals. The supply chain structure remained stable between these intervals, while structural changes occurred during the transition to the next structural constancy interval. The authors modeled the structural dynamics of the supply chain with optimal control and performed optimization between intervals with a linear programming model with two-side constraints.

In random networks, the study by Han and Shin (2016) evaluated the structural robustness of the supply chain and compared this robustness to the random risk of network disruption. Lin et al. (2017) assessed the reliability of a multistage supply chain using the probability that market demand can be met if multiple stations of transit are used with the appropriate time frame, where system reliability, acting as the delivery performance index, is assessed by the number of minimal paths. Pavlov et al. (2018) developed a hybrid fuzzy-probabilistic approach to supply chain resilience estimation with structural dynamics and ripple effect consideration. The genome method was applied with the objective of including the structural properties of the supply chain design into resilience assessment. A supply chain design resilience index was developed that can be used as a method of comparing different supply chain designs regarding the resilience both to disruption propagation and with recovery consideration. Moreover, the developed approach allows the identification of groups of critical suppliers whose failure interrupts the supply chain operation.

3 Method to Quantify Supply Chain Recoverability

We apply the method of network adaptation potential analysis (Ivanov & Arkhipov, 2011). The supply chains are functionally and structurally complex and are multi-stage systems. A recovery policy in the supply chain is formed through dynamic selection of backup sources and paths in the network which results in a higher number of possible alternative recovery trajectories. The realization of such alternative recovery strategies allows a flexible reaction to the disrupted supply chain's structural changes. Thus, even at the planning stage, the supply chain design should be considered "branchy."

The introduction of alternative recovery structures into the supply chain results in different levels of structural complexity. The assessment of this structural characteristic can be intuitively related to the quantity and variety of supply chain recovery trajectories, the quantity of polytypic actions, the variety of logical connections between them, the existence of "branching" or structure decoupling points, and the possibility of choosing between alternative variants of supply chain recoverability. In this paper, the approach to the above-described assessment issue, called supply chain recoverability potential, will be considered.

The recoverability potential concept can be formulated as follows. The recoverability potential is the supply chain's structural property which is characterized by the decoupling or branching degree of the supply chain recovery policy and the possibility to adapt to a real execution environment.

Quantitative estimation of recoverability potential serves for the analysis of the determination of a number of elements in the supply chain design, their variety, and the interrelations between them with regard to the potential ability of a supply chain to survive in the presence of disruption, i.e., to have at least one undisrupted path to the end stage in the supply chain, e.g. customers or markets.

4 Exact Method

4.1 Computational Method

Taking into account the semantic affinity of complexity, uncertainty, and recoverability, the measure of a complex system state's relative variety, i.e. the entropy is suggested for the quantitative estimation of the supply chain's recoverability potential. Assume that the supply chain network structure is designed and can be divided into t-layers, i.e. supplier, factories, and retailers. It is possible to choose between n_t supply chain elements within each t-layer. The probability p of choosing any path in the supply chain network is equal. Recovery paths are designed according to the set system of logic links according to the recovery policies. Then network structure complexity, as a measure of subject possibilities and equivalent to a variety of choices of alternative sets, can be estimated with the indicator known as entropy H of a complex system:

$$H = -\sum_{i=1}^{N} p_i \ln p_i \tag{1}$$

where p_i is the probability of i-state of system or in our case, the selection probability of i-recovery trajectory, $i = 1, 2, \ldots, N$. First, in order to compute the entropy index H, the probabilities of each recovery path in the supply chain should be determined, and, second, the logarithm of this probability should be found. This will be explained in more detail further in the paper using a numerical example (see Fig. 1).

As the entropy assessment of supply chain complexity is performed as a basis for the estimation of further recoverability potential, any of the log bases can be used. We use the natural (normal) *ln* because it is the most convenient way to compute experiments in entropy as shown in Shannon and Weaver (1963).

It is not difficult to compute the index of network entropy when the network is set up and the hypothesis about the choice of equally probable operations at each planning stage is accepted. As a result, an estimation can be received that indirectly characterizes the network recoverability potential. Let us present a simple example (see Fig. 1).

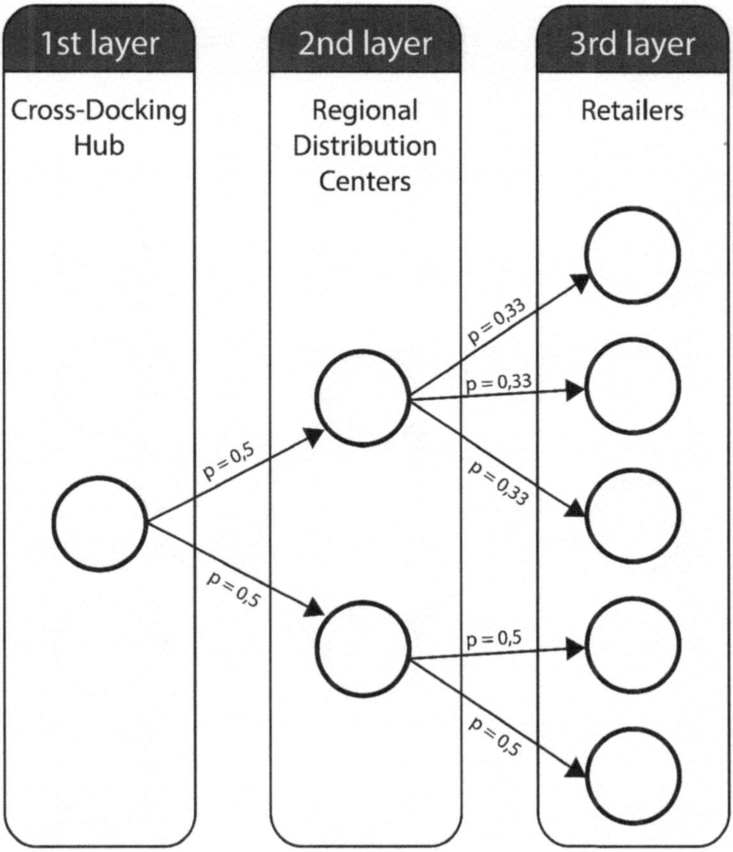

Fig. 1 Example of the supply chain design structure

Let a supply chain design have the structure presented in Fig. 1. The downstream supply chain design includes three layers: a cross-docking hub, two distribution centers, and five retailers. For this type of the network, the entropy index is H = 1.6 according to Eq. (1). There are two options to transit from the first layer (i.e., the hub) to the second layer (i.e., the distribution centers). Therefore, the probability of each path selection in between the first and second layers is 0.5. In considering the upper node of the second layer, there are three alternative links to reach the third layer, i.e. the retailers. Therefore, the probability of each path selection from this node to the retailers at the third layer will be approximately 0.33. Hence, the selection probability of each of the three alternative ways through the upper node of the second layer is $p = 0.5 * 0.33 = 0.165$. Now, the logarithm from this probability can be taken: ln $0.165 = -1.8$. We get $0.165 * (-1.8) = -0.3$. As three alternative links exist, $-0.3 \times 3 = -0.9$. Analogously, in considering the bottom node of the second layer, there are two alternative links to reach the third layer. Therefore, the probability for each will be 0.5. Hence, the selection

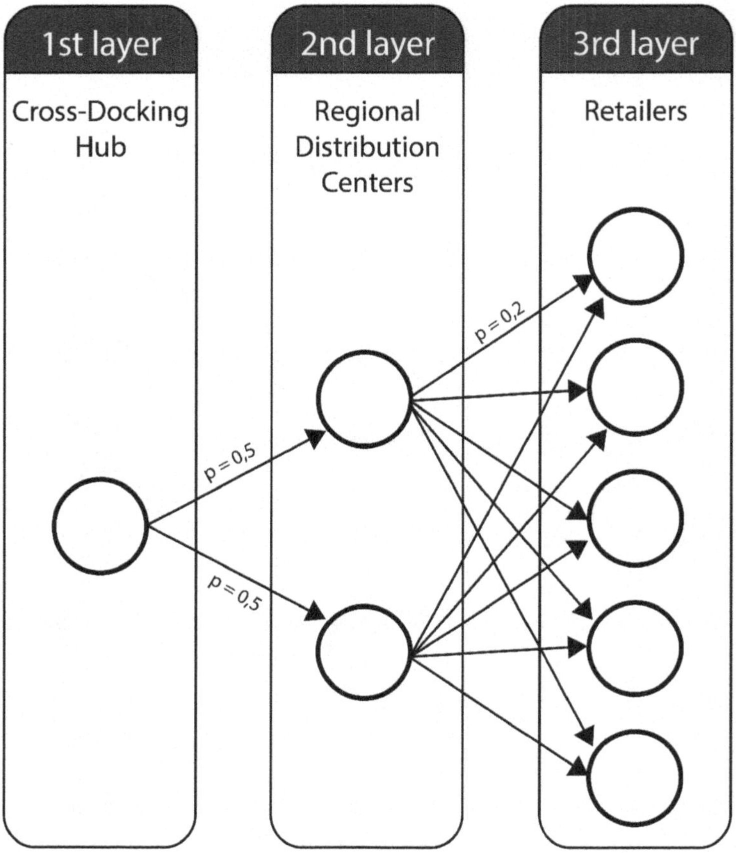

Fig. 2 Supply chain design structure as a complete bipartite graph

probability of each of the two ways through the bottom node of the second layer is $p = 0.5 * 0.5 = 0.25$. Now, the logarithm from this probability can be taken: $\ln 0.25 = -1.39$. We get $0.25 * (-1.39) = -0.35$. As two alternative links exist, we then consider $H = -(-0.35 \times 2) = 0.7$. The sum H of the two nodes in the first stage is, therefore, $0.9 + 0.70 = 1.6$.

The maximum recoverability value for a network with a set number of supply chain elements at each stage will obviously be achieved when all the variants of transitions from i-operation elements to $(i + 1)$-operation elements are admissible and equally probable, i.e., in the case of a complete bipartite graph (see Fig. 2). For this case, we will have $H^{max} = 2.3$.

In analyzing a real supply chain, it is useful to have relative (normal) estimations $H^{(o)} = H/H^{max}$ as well as absolute estimations of a variety level (entropy) H. For the network in Fig. 1, the relative estimation of entropy is $H^{(o)} = 1.6/2.30 = 0.7$. Absolute estimations allow us to compare supply chain design structures that differ in the quantity of the supply chain stages, the quantity of elements at different

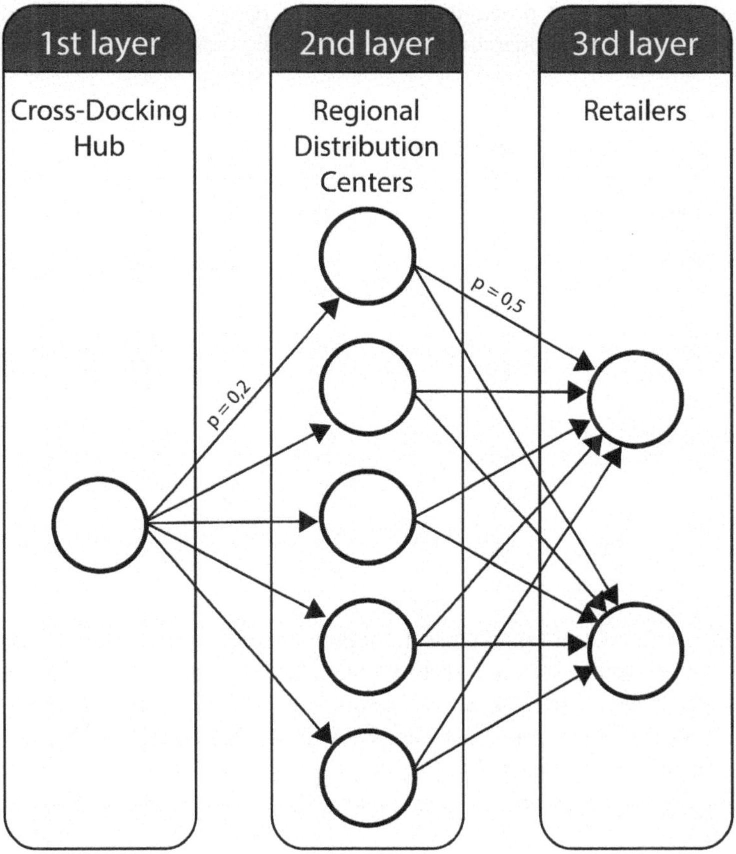

Fig. 3 Modified supply chain design structure

stages, and a variety of links between elements. Relative estimations characterize, in essence, the degree of affinity between the variety of links in a concrete supply chain to the maximum value.

However, index (1) does not reflect one important requirement for the supply chain structural recoverability, i.e., the maintenance of a high service level and the required higher recoverability due to higher flexibility downstream in the supply chain, which is important for supply chain adaptation possibilities on the customer side. Let us introduce another supply chain structure, but with the same number of elements and links (see Fig. 3).

The entropy indexes for both supply chain design structures presented in Figs. 2 and 3 are identical and equal $H = 2.30$. At the same time, these two structures essentially differ, because the recovery path selection variety changes in different ways while moving downstream in the network.

An extended recoverability estimation can be concluded after several transformations of Eq. (1) by presenting the i-trajectory selection probability as the multiplication of its unit probabilities $p_i = p_{il} \times p_{i2} \times \ldots.. \times p_{iT}$. Equation (1) will be transformed as shown in Eq. (2):

$$H = -\sum_{i=1}^{N} p_{i1}xp_{i2}x\ldots xp_{iT}\ln\left(p_{i1}xp_{i2}\ldots xp_{iT}\right) =$$
$$\sum_{i=1}^{N} p_{i1}xp_{i2}x\ldots xp_{iT}\ln\left(p_{i1}xp_{i2}\ldots xp_{iT}\right) =$$
$$\sum_{i=1}^{N}\left(\begin{array}{l} \prod\limits_{\substack{k=0\\k\neq t}}^{T-1} p_{ik}xp_{il}\ln\ p_{il} + \prod\limits_{\substack{k=0\\k\neq t}}^{T-1} p_{ik}xp_{i2}\ln\ p_{i2} + \ldots + \prod\limits_{\substack{k=0\\k\neq t}}^{T-1} p_{ik}xp_{ik}\ln\ p_{ik} \\ N \qquad\qquad T-1 \qquad\qquad T-1 \end{array}\right)$$
$$\sum_{i=1}\sum_{t=0}\left(\prod_{k=0} p_{ik}\right) x\left(p_{it}\ln p_{it}\right) \tag{2}$$

It can be observed in Eq. (2) that the supply chain recovery variety is expressed through the probabilities of a trajectory unit selection. This allows us to introduce weights which reflect the subjective considerations of the selection variety value at each interval of structural constancy. We will designate these weights as w_t and consider them normal, that is, $0 \leq w_t \leq 1$, $\sum w_t = 1$. Let us name the obtained estimation *the supply chain-weighted variety* (the weighted entropy) and denote it as H_w (Eq. 3):

$$H_W = -\sum_{i=1}^{N}\sum_{t=0}^{T-1} w_t \left(\prod_{\substack{k=0\\k\neq t}}^{T-1} p_{ik}\right) x\left(p_{it}\ln p_{it}\right) = -\sum_{i=1}^{N}\prod_{t=0}^{T-1} p_{it}\sum_{t=0}^{T-1} w_t\ln p_{it} \tag{3}$$

The index of weighted variety (Eq. 4) can also be called the supply chain absolute recoverability potential: we denote it as A and $A = H_w$.

For a supply chain design with the maximum value of weighted entropy (i.e., the supply chain network structure with initial functionality and the maximum number of equally probable recovery paths), Eq. (3) becomes simpler (Eq. 4):

$$H_w^{max} = -\sum_{t=0}^{T-1} w_t\ln p_{it} \tag{4}$$

Let us consider the following estimation as the indicator of the supply chain relative recoverability (Eq. 5):

$$A^{(0)} = \frac{H_w}{H_w^{max}} \tag{5}$$

The recoverability potential index is sensitive to changes in the supply chain node structure, the allocation of these knots within intervals of structural constancy, and the variety of recovery path selection.

4.2 Numerical Example

To illustrate the proposed technique, we use the supply chain design structures shown in Figs. 1, 2, and 3. The corresponding values of the necessary estimations of network variety and recoverability potential are shown in Table 1. The weights w_t are considered in direct proportion to the t layer number (i.e., the layers downstream in the supply chain and closer to the customers get higher weights subject to the requirement of a higher service level).

It can be observed that the recoverability potential $A^{(0)}$ of the supply chain design structures shown in Figs. 2 and 3 is equal and is greater compared to that of structure 1a. In addition, the indicator H_w shows a higher responsiveness and agility in the structure as shown in Fig. 2 (see Eqs. 3–5). The higher value of the indicator H_w indicates that supply chain recoverability increases downstream in the supply chain, that is, the variety of recovery paths increases downstream in the supply chain and extends flexibility and responsiveness in the parts of the supply chain near the customers.

The final selection of the supply chain design with recoverability considerations should be performed on the basis of economic efficiency and recoverability potential indicators. In addition, if a supply chain design structure is economically preferable, but lacks the recoverability in event of severe structural disruptions, additional links can be introduced in the supply chain to increase the flexibility. For example,

Table 1 Examples of supply chain variety and adaptation potential estimations for different network structures

Supply chain design structure	Variety estimations, H, (H^{max})	Weighted variety estimation (absolute recoverability potential), $A = H_w$	Maximum estimation of weighted variety, H_w^{max}	Relative recoverability potential estimation, $A^{(o)}$
Figure 1	1.6; (2.30)	0.94	1.31	0.72
Figure 2	2.30; (2.30)	1.31	1.31	1.00
Figure 3	2.30; (2.30)	0.99	0.99	1.00

technology standardization at production plants can be performed so that each of them would be able to produce different (or at least two) product families.

5 Heuristic Method

Equations (1)–(5) are rather simple and can be easily applied in software. However, in the case of complex network models, technical computation difficulties may arise since the number of trajectories in the network can become astronomical. In addition, individual trajectory selection probabilities in complex networks are rather small and, as it is known, arithmetical calculations with small values can lead to serious errors. This is why it is suggested to carry out the recoverability adaptation potential calculations on the basis of network model decomposition.

The analysis of various supply chain design structures shows that the operations within a supply chain layer often does not influence other operational blocks. Such a situation occurs when all the possible links between the elements of adjacent supply chain layers exist. Let us show several examples. We analyze a supply chain design structure as illustrated in Fig. 4

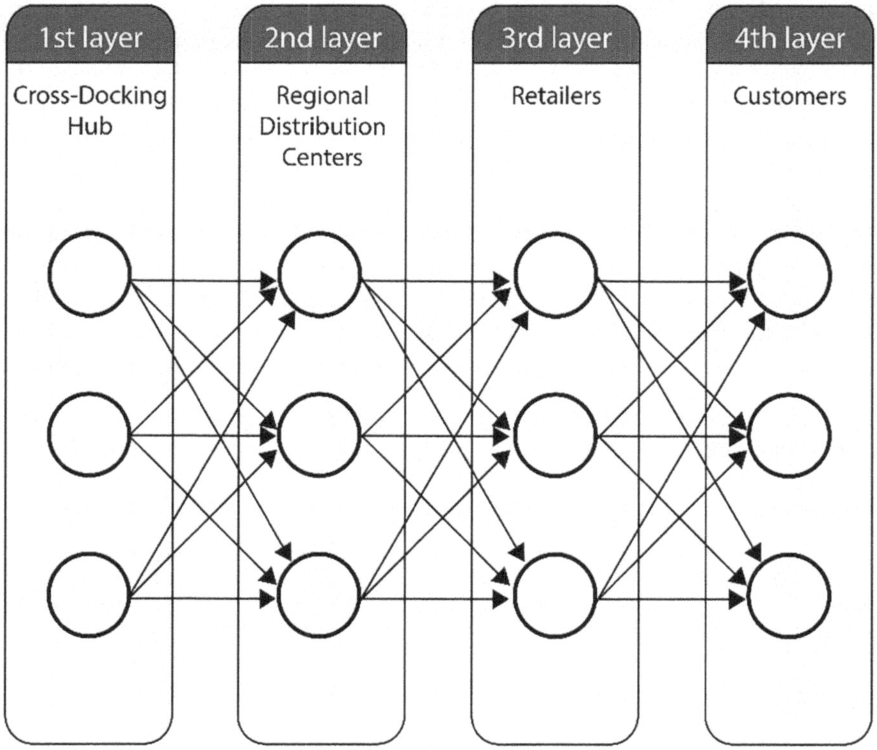

Fig. 4 Exampleofafour-layersupplychain

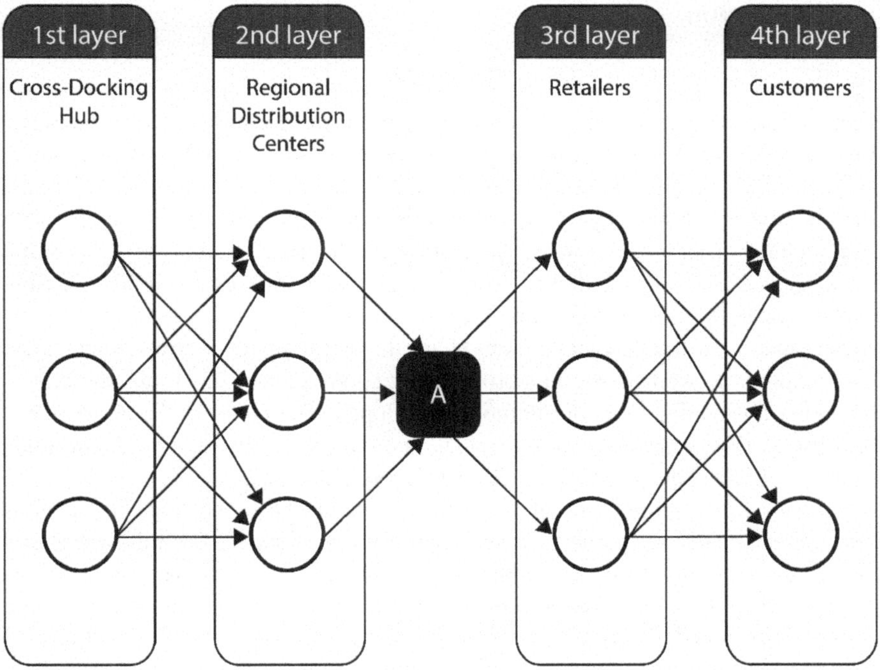

Fig. 5 Example of a four-layer supply chain with auxiliary element *A*

The selection of a node at one supply chain layer into the recovery path does not influence the probability of selecting any node in the subsequent layer into the same recovery path. In this case, the auxiliary "dividing" block can be introduced between the layers, which include one element that divides the supply chain into two fragments (Fig. 5).

The recoverability adaptation potential estimation is then approximately equal to the fragment potential sum. For supply chain designs that have the maximum number of links, this equality is exact, and it is possible to present this formally. The auxiliary action *A* in Fig. 5 divides the supply chain into two fragments, *I*, *II*. The supply chain and its fragment recoverability potentials are as follows: $H_{max} = lnN = 4.394$, $H_I = ln9 = 2.197$, and $H_{II} = ln9 = 2.197$. We can observe that the equation $H_{max} = H_I + H_{II}$ holds true. When the recoverability potentials of the fragments are different, an approximation should take place. The analysis of different supply chain structures showed that computational accuracy with network model decomposition is sufficient to carry out a real data analysis. Hence, it can be concluded that the recoverability adaptation potential estimation of supply chain design structure fragments can be used to simplify practical computations.

6 Conclusion

The problem of designing resilient supply chains at the semantic network level has been considered in this study. The entropy method was used to reveal the interrelations between supply chain design and recoverability. An entropy-based quantitative measure has been suggested to estimate supply chain recoverability. For the first time, the entropy-based supply chain analysis is brought into correspondence with supply chain structural dynamics. An exact and a heuristic computation algorithm are suggested and illustrated. The developed approach and recoverability measure can be used to select a resilient supply chain design.

Advantages of the developed method are the simplicity of recoverability index computations as well as the numerous applications to real-life problems due to the higher abstraction level of the model developed. The nodes in the network can be considered suppliers, logistics, production companies, or even a production line resource allowing almost any level of abstraction in the analysis.

In terms of the limitations of this study, it should be noted that homogenous goods downstream in the supply chain have been considered. The bill-of-material considerations in the upstream supply chain would make it necessary to adapt this concept. Regarding the absolute and relative recoverability potential estimation, it should be noted that the relative estimations with the maximum variety of links are identical and equal to 1, no matter how many stages the supply chain has. This limits the application possibilities to a certain extent, but nevertheless, they still retain an analytical role, supplementing absolute adaptation potential estimations. The proposed supply chain adaptation potential indicators can be used as criteria for selecting supply chain design structures at the configuration stage with consideration of severe disruption risks and supply chain structural dynamics. Future research can include the introduction of bill-of-material considerations and case-study applications of the developed method to reveal specific aspects relevant for different industries and services.

References

Aldrighetti, R., Battini, D., & Ivanov, D. (2023b). Efficient resilience portfolio design in the supply chain with consideration of preparedness and recovery investments. *Omega, 117*, 102841.

Aldrighetti, R., Calzavara, M., Martignago, M., Zennaro, I., Battini, D., & Ivanov, D. (2023a). A methodological framework for the design of efficient resilience in supply networks. *International Journal of Production Research, 62*(1–2), 271–290.

Alikhani, R., Ranjbar, A., Jamali, A., Torabi, S. A., & Zobel, C. W. (2023). Towards increasing synergistic effects of resilience strategies in supply chain network design. *Omega, 116*, 102819.

Allesina, S., Azzi, A., Battini, D., & Regattieri, A. (2010). Performance measurement in supply chains: New network analysis and entropic indexes. *International Journal of Production Research, 48*, 2297–2321.

Altay, N., Gunasekaran, A., Dubey, R., & Childe, S. J. (2018). Agility and resilience as antecedents of supply chain performance under moderating effects of organizational culture within humanitarian setting: A dynamic capability view. *Production Planning and Control, 29*(14), 1158–1174.

Aqlan, F., & Lam, S. S. (2015). Supply chain risk modelling and mitigation. *International Journal of Production Research, 53*(18), 5640–5656.
Basole, R. C., & Bellamy, M. A. (2014). Supply network structure, visibility, and risk diffusion: A computational approach. *Decision Sciences, 45*(4), 1–49.
Birkie, S. E., Trucco, P., & Campos, P. F. (2017). Effectiveness of resilience capabilities in mitigating disruptions: Leveraging on supply chain structural complexity. *Supply Chain Management: An International Journal, 22*(6), 506–521.
Blackhurst, J., Craighead, C. W., Elkins, D., & Handfield, R. (2005). An empirically derived agenda of critical research issues for managing supply-chain disruptions. *International Journal of Production Research, 43*(19), 4067–4081.
Choi, T. M., & Shi, X. (2022). Reducing supply risks by supply guarantee deposit payments in the fashion industry in the "new normal after COVID-19". *Omega, 109*, 102605.
Dolgui, A., Ivanov, D., & Rozhkov, M. (2019). Does the ripple effect influence the bullwhip effect? An integrated analysis of structural and operational dynamics in the supply chain. *International Journal of Production Research, 58*(5), 1285–1301.
Dolgui, A., Ivanov, D., & Sokolov, B. (2018). Ripple effect in the supply chain: An analysis and recent literature. *International Journal of Production Research, 56*(1–2), 414–430.
Echefaj, K., Charkaoui, A., Cherrafi, A., & Ivanov, D. (2024). Design of resilient and viable sourcing strategies in intertwined circular supply networks. *Annals of Operations Research, 337*, 459–498.
Giannoccaro, I., Nair, A., & Choi, T. (2018). The impact of control and complexity on supply network performance: An empirically informed investigation using NK simulation analysis. *Decision Science, 49*(4), 625–659.
Gruchmann, T., Stadtfeld, G. M., Thürer, M., & Ivanov, D. (2024). Supply chain resilience as a system quality: Survey-based evidence from multiple industries. *International Journal of Physical Distribution & Logistics Management, 54*(1), 92–117.
Han, J., & Shin, K. S. (2016). Evaluation mechanism for structural robustness of supply chain considering disruption propagation. *International Journal of Production Research, 54*(1), 135–151.
Harremoës, P., & Topsøe, F. (2001). Maximum entropy fundamentals. *Entropy, 3*(3), 191–226.
He, J., Alavifard, F., Ivanov, D., & Jahani, H. (2019). A real-option approach to mitigate disruption risk in the supply chain. *Omega, 88*, 133–149.
Ho, W., Zheng, T., Yildiz, H., & Talluri, S. (2015). Supply chain risk management: A literature review. *International Journal of Production Research, 53*(16), 5031–5069.
Hosseini, S., & Barker, K. (2016). A Bayesian network for resilience-based supplier selection. *International Journal of Production Economics, 180*, 68–87.
Isik, F. (2010). An entropy-based approach for measuring complexity in supply chains. *International Journal of Production Research, 48*(12), 3681–3696.
Ivanov, D. (2017). Simulation-based ripple effect modelling in the supply chain. *International Journal of Production Research, 55*(7), 2083–2101.
Ivanov, D. (2018). *Structural dynamics and resilience in supply chain risk management*. Springer.
Ivanov, D. (2019). Disruption tails and revival policies: A simulation analysis of supply chain design and production-ordering systems in the recovery and post-disruption periods. *Computers and Industrial Engineering, 127*, 558–570.
Ivanov, D. (2023). Intelligent digital twin (iDT) for supply chain stress-testing, resilience, and viability. *International Journal of Production Economics, 263*, 108938.
Ivanov, D. (2024a). Transformation of supply chain resilience research through the COVID-19 pandemic. *International Journal of Production Research, 62*(23), 8217–8238.
Ivanov, D. (2024b). Two views of supply chain resilience. *International Journal of Production Research, 62*(11), 4031–4045.
Ivanov, D. (2024c). Supply chain resilience: Conceptual and formal models drawing from immune system analogy. *Omega, 127*, 103081.
Ivanov, D. (2024d). Exiting the COVID-19 pandemic: After-shock risks and avoidance of disruption tails in supply chains. *Annals of Operations Research, 335*, 1627–1644.

Ivanov, D., & Arkhipov, A. (2011). Analysis of structure adaptation potential in designing supply chains in an agile supply chain environment. *International Journal of Integrated Supply Management, 6*(2), 165–180.

Ivanov, D., Dolgui, A., Sokolov, B., & Ivanova, M. (2017b). Literature review on disruption recovery in the supply chain. *International Journal of Production Research, 55*(20), 6158–6174.

Ivanov, D., & Sokolov, B. (2010). *Adaptive supply chain management*. Springer.

Ivanov, D., Sokolov, B., & Dolgui, A. (2014a). The ripple effect in supply chains: Trade-off 'efficiency-flexibility-resilience' in disruption management. *International Journal of Production Research, 52*(7), 2154–2172.

Ivanov, D., Sokolov, B., & Kaeschel, J. (2010). A multi-structural framework for adaptive supply chain planning and operations control with structure dynamics considerations. *European Journal of Operational Research, 200*(2), 409–420.

Ivanov, D., Sokolov, B., & Pavlov, A. (2014b). Optimal distribution (re)planning in a centralized multi-stage network under conditions of ripple effect and structure dynamics. *European Journal of Operational Research, 237*(2), 758–770.

Ivanov, D., Sokolov, B., Pavlov, A., Dolgui, A., & Pavlov, D. (2016). Disruption-driven supply chain (re)-planning and performance impact assessment with consideration of pro-active and recovery policies. *Transportation Research Part E, 90*, 7–24.

Ivanov, D., Tsipoulanidis, A., & Schönberger, J. (2017a). *Global supply chain and operations management* (1st ed.). Springer.

Jain, V., Kumar, S., Soni, U., & Chandra, C. (2017). Supply chain resilience: Model development and empirical analysis. *International Journal of Production Research, 55*(22), 6779–6800.

Käki, A., Salo, A., & Talluri, S. (2015). Disruptions in supply networks: A probabilistic risk assessment approach. *Journal of Business Logistics, 36*(3), 273–287.

Levner, E., & Ptuskin, A. (2015). An entropy-based approach to identifying vulnerable components in a supply chain. *International Journal of Production Research, 53*(22), 6888–6902.

Levner, E., & Ptuskin, A. (2018). Entropy-based model for the ripple effect: Managing environmental risks in supply chains. *International Journal of Production Research, 56*(7), 2539–2551.

Li, X., Krivtsov, V., Pan, C., Nassehi, A., Gao, R. X., & Ivanov, D. (2025). End-to-end supply chain resilience management using deep learning, survival analysis, and explainable artificial intelligence. *International Journal of Production Research, 63*(3), 1174–1202.

Liberatore, F., Scaparra, M. P., & Daskin, M. S. (2012). Hedging against disruptions with ripple effects in location analysis. *Omega, 40*(2012), 21–30.

Lin, Y. K., Huang, C. F., Liao, Y.-C., & Yeh, C. T. (2017). System reliability for a multistate intermodal logistics network with time windows. *International Journal of Production Research, 55*(7), 1957–1969.

Liu, M., Ding, Y., Chu, F., Dolgui, A., & Zheng, F. (2024). Robust actions for improving supply chain resilience and viability. *Omega, 123*, 102972.

Lücker, F., & Seifert, R. W. (2017). Building up resilience in a pharmaceutical supply chain through inventory, dual sourcing and agility capacity. *Omega, 73*, 114–124.

Martel, A., & Klibi, W. (2016). *Designing value-creating supply chain networks*. Springer.

Mistree, F., Allen, J., Khosrojerdi, A., & Rasoulifar, G. (2017). *Architecting fail safe supply networks*. CRC Press.

Nair, A., & Vidal, J. M. (2011). Supply network topology and robustness against disruptions: An investigation using multiagent model. *International Journal of Production Research, 49*(5), 1391–1404.

Paul, S. K., Sarker, R., & Essam, D. (2014). Real time disruption management for a two-stage batch production–inventory system with reliability considerations. *European Journal of Operational Research, 237*, 113–128.

Pavlov, A., Ivanov, D., Dolgui, A., & Sokolov, B. (2018). Hybrid fuzzy-probabilistic approach to supply chain resilience assessment. *IEEE Transactions on Engineering Management, 65*(2), 303–315.

Quang, H. T., & Hara, Y. (2018). Risks and performance in supply chain: The push effect. *International Journal of Production Research, 56*(4), 1369–1388.

Raj, R., Wang, J., Nayak, A., Tiwari, W. K., Han, B., Liu, C., & Zhang, W. J. (2014). Measuring the resilience of supply chain systems using a survival model. *IEEE Systems Journal, 9*(2), 377–381.

Ramani, V., Ghosh, D., & Sodhi, M. S. (2022). Understanding systemic disruption from the Covid-19-induced semiconductor shortage for the auto industry. *Omega, 113*, 102720.

Rolf, B., Jackson, I., Müller, M., Lang, S., Reggelin, T., & Ivanov, D. (2023). A review on reinforcement learning algorithms and applications in supply chain management. *International Journal of Production Research, 61*(20), 7151–7179.

Ruel, S., Baz, J. E., Ivanov, D., & Das, A. (2024). Supply chain viability: Conceptualization, measurement, and nomological validation. *Annals of Operations Research, 335*, 1107–1136.

Sardesai, S., & Klingebiel, K. (2023). Maintaining viability by rapid supply chain adaptation by using a process capability index. *Omega, 115*, 102778.

Sawik, T. (2017). A portfolio approach to supply chain disruption management. *International Journal of Production Research, 55*(7), 1970–1991.

Sawik, T. (2022). Stochastic optimization of supply chain resilience under ripple effect: A COVID-19 pandemic related study. *Omega, 109*, 102596.

Sawik, T. (2023). A stochastic optimisation approach to maintain supply chain viability under the ripple effect. *International Journal of Production Research, 61*(8), 2452–2469.

Scheibe, K. P., & Blackhurst, J. (2018). Supply chain disruption propagation: A systemic risk and normal accident theory perspective. *International Journal of Production Research, 56*(1–2), 23–49.

Shannon, C. E., & Weaver, W. (1963). *The mathematical theory of communication*. The University of Illinois Press.

Sheffi, Y., & Rice, J. B. (2005). A supply chain view of the resilient enterprise. *MIT Sloan Management Review*.

Simchi-Levi, D., Schmidt, W., Wei, Y., Zhang, P. Y., Combs, K., Ge, Y., et al. (2015). Identifying risks and mitigating disruptions in the automotive supply chain. *Interfaces, 45*(5), 375–390.

Snyder, L. V., Zümbül, A., Peng, P., Ying, R., Schmitt, A. J., & Sinsoysal, B. (2016). OR/MS models for supply chain disruptions: A review. *IIE Transactions, 48*(2), 89–109.

Sokolov, B., Ivanov, D., Dolgui, A., & Pavlov, A. (2016). Structural quantification of the ripple effect in the supply chain. *International Journal of Production Research, 54*(1), 152–169.

Wilson, M. C. (2007). The impact of transportation disruptions on supply chain performance. *Transportation Research Part E: Logistics and Transportation Review, 43*, 295–320.

Yu, Z., & Xiao, R. (2014). Modelling of cluster supply network with cascading failure spread and its vulnerability analysis. *International Journal of Production Research, 52*(23), 6938–6953.

Yuming, X. (2015). Flexibility measure analysis of supply chain. *International Journal of Production Research, 53*(10), 3161–3174.

Zhao, K., Kumar, A., Harrison, T. P., & Yen, J. (2011). Analyzing the resilience of complex supply network topologies against random and targeted disruptions. *IEEE Systems Journal*.

Zobel, C. W. (2011). Representing perceived tradeoffs in defining disaster resilience. *Decision Support Systems, 50*(2), 394–403.

Zobel, C. W. (2014). Quantitatively representing nonlinear disaster recovery. *Decision Sciences, 45*(6), 1053–1082.

Disruption Tails and Revival Policies in the Supply Chain

Dmitry Ivanov and Maxim Rozhkov

Abstract We study capacity disruption and recovery policy impacts on supply chain (SC) performance. Discrete event simulation methodology is used for analysis with real company data and real disruptions. Two novel findings are shown. First, disruption-driven changes in SC behaviour may result in backlog and delayed orders, the accumulation of which in the post-disruption period we call 'disruption tails'. A transition of these residues into the post-disruption period causes the post-disruption SC instability, resulting in further delivery delays and non-recovery of SC performance. Second, a smooth transition from the contingency policy through a special 'revival policy' to the normal operation mode allows the negative effects of the disruption tails to be partially mitigated. These results suggest three managerial insights. First, contingency policies need to be applied during the disruption period to avoid disruption tails. Second, recovery policies need to be extended towards an integrated consideration of both disruption and the post-disruption periods. Third, revival policies need to be developed for the transition from the contingency to the disruption-free operation mode. A revival policy intends to mitigate the negative impact of the disruption tails and stabilize the SC control policies and performance. The experimental results suggest the revival policy should be included in the SC resilience framework if the performance cannot be recovered fully after the capacity recovery.

D. Ivanov (✉)
Department of Business and Economics, Berlin School of Economics and Law, Berlin, Germany
e-mail: divanov@hwr-berlin.de

M. Rozhkov
HSE University, Moscow, Russia

D. Ivanov et al. (eds.), *Handbook of Ripple Effects in the Supply Chain*, International Series in Operations Research & Management Science 276,
https://doi.org/10.1007/978-3-031-85508-5_11

1 Introduction

Disruptions in supply chain (SC) capacities may happen at both production factories and at logistics facilities, such as distribution centres (DC) (Snyder & Daskin, 2005; Tang, 2006; Klibi et al., 2010; Simangunsong et al., 2012; Simchi-Levi et al., 2015; Gunasekaran et al., 2015; Sokolov et al., 2016; Tukamuhabwa et al., 2015; Spiegler et al., 2016; Kamalahmadi & Mellat-Parast, 2016; Fahimnia et al., 2016; Jain et al., 2017; Rezapour et al., 2017; Ivanov, 2018a; Dolgui et al., 2018; Ivanov et al., 2019a, 2019b; Singh et al., 2021; Ivanov & Rozhkov, 2020; Li et al., 2023). Disruption risks can be caused by natural or man-made catastrophes, political crises, strikes, or legal disputes. These disruptions are rarely localized at the disrupted node and frequently propagate in the SC causing the ripple effect (Liberatore et al., 2012; Ivanov et al., 2014a, 2014b; Han & Shin, 2016; Scheibe & Blackhurst, 2018; Ivanov, 2018a, 2018b, Namdar et al., 2018; Levner & Ptuskin, 2017; Dolgui et al., 2018; Ivanov et al., 2019a, 2019b; Ivanov & Dolgui, 2019; Hosseini et al., 2020; Dolgui & Ivanov, 2021; Ghadge et al., 2022; Sawik, 2022; Pavlov et al., 2022; Kravchenko et al., 2024; Sawik & Sawik, 2024).

In regard to the disruption risks, resilient production backup and contingency inventory control policies became a visible research avenue over the last decade (Snyder & Daskin, 2005; Tomlin, 2006; Song & Zipkin, 2009; Yang et al., 2009; Ivanov et al., 2010; Federgruen & Yang, 2011; Atan & Snyder, 2012; Lim et al., 2012; Kouvelis & Li, 2012; Schmitt & Singh, 2012; Ivanov & Sokolov, 2013; Qi, 2013; Kim & Tomlin, 2013; Hishamuddin et al., 2013; Raj et al., 2015; Choi et al., 2016; Ivanov et al., 2016; Govindan et al., 2016; Sawik, 2017; Mizgier, 2017; Schmitt et al., 2017; Dubey et al., 2017; Namdar et al., 2018; Sawik, 2018; He et al., 2019; Dolgui et al., 2020; Pavlov et al., 2019; Ivanov & Dolgui, 2021; Liu et al., 2021). Despite significant progress in theoretical studies and empirical principles to manage severe disruptions in the SC at proactive and reactive stages, recent literature has usually assumed an immediate transition to a normal operation mode at the time of capacity recovery. Moreover, a full system stabilization after the capacity recovery has typically been assumed without considering any residual effects such as delayed orders and backlogs accumulated over the destabilized system states during the disruption time (Ivanov, 2025a, 2025b).

This study closes the research gap described above with the objective of revealing the dependencies between SC disruptions, contingency policies, and transition to the post-disruption operation mode. First, the analysis is conducted in regard to disruption-driven changes in SC behaviour resulting in delayed orders and backlogs, the accumulation of which can be considered ‘disruption tails’. The influence of these tails during the post-disruption time in the course of transition into the normal operation mode is investigated. A comparison of SC operation with and without contingency policy is performed to compare the impact of disruption tails on SC operational and financial performance. Second, a comparison of SC operational and financial performance between an immediate deactivation of the contingency plans and installation of the normal operation policies after capacity recovery and usage of the revival policy is studied. These experiments aim at providing managerial insights on the application of contingency production and inventory control policies during the disruption period and revival policies during the transition time to normal operation after capacity recovery.

The rest of this study is organized as follows. Section 2 describes recent literature on recovery policies in the SC and simulation studies on SC disruptions. Section 3 is devoted to case study presentation and research methodology description. Experimental settings with production and distribution disruptions are the focus of Sects. 4 and 5, respectively. Section 6 concludes the paper by summarizing the most important insights, limitations of this study, and future research avenues.

2 State of the Art

2.1 Recovery Policies in the Supply Chain

SC resilience has been a prominent research topic for the last 10 years. *Proactive* and *reactive* resiliency policies have been developed with the aim of protecting the SC from disruption before it happens, and ensuring mitigation of the disruption after it happens (Ho et al., 2015; Snyder et al., 2016; Ivanov, 2017; Paul & Chowdhury, 2021; Hosseini & Ivanov, 2022; Pavlov et al., 2022; Ivanov, 2024; Nikookar et al., 2024; Ivanov, 2025c). Given the nature of this study, this literature analysis is focused on contingency inventory control policies and backup sourcing when used during the recovery phase.

Assigning customers to disruption-prone locations to minimize total SC costs, Snyder and Daskin (2005) optimized SC design. To analyse the dynamic effects of inventory buffers when suppliers are unreliable, Federgruen and Yang (2011) created a general periodic review model. To capture the trade-off between inventory policies and disruption risks for an unreliable, dual sourcing supply network, Iakovou et al. (2010) studied a single period stochastic inventory model for both capacitated and incapacitated cases, and evaluated different contingency strategies. In a study by Shao and Dong (2012), an assemble-to-order system with a backup source is analysed. This system offers on-time delivery and a policy in which customers are compensated for waiting in each period of disruption. The results of the study imply that at the start of an SC disruption a backup source strategy is preferable, while a compensation strategy is preferable as time goes on. In the specific example considered, a dynamic mixed strategy with customer choices is better than a backup sourcing strategy. The manufacturer's decision regarding which reactive strategy to choose is determined according to backup costs and customer sensitivity (Mosayebi et al., 2024; Dolgui et al., 2024; Jin et al., 2024).

Considering supplier capacity constraints, Costantino et al. (2012) created an agile reconfiguration approach to resilient SC design. Using diagraph modelling and integer linear programming, this approach optimizes SC design. In the study by Benyoucef et al. (2013), SC design is considered in the case of unreliable suppliers, subject to minimization of fixed location, inventory, and safety stock costs at distribution centres, and ordering and transportation costs throughout the SC. Ivanov et al. (2013) combined linear programming and optimal program control in a model that included SC reconfiguration as a reaction to disruptions.

Hu et al. (2013) studied the incentives, whether *ex ante* (prior to disruption) or *ex post* (after disruption), which drive a supplier's investment in capacity restoration. According to their findings, the *ex ante* commitment is weakly preferred by both the buyer and the supplier when the buyer provides incentives. According to Kim and Tomlin (2013), if recovery capacity is the only option, then firms, in a decentralized setting, overinvest in capacity, which results in higher system availability at a higher cost. Yet, when it is possible to invest in both, then firms tend to underinvest in preventing failure and overinvest in recovery capacity.

In the study by Qi (2013), a continuous review inventory model including random disruptions for the main supplier was developed. Considering transportation disruption, Hishamuddin et al. (2013) showed a recovery model of a two-echelon serial SC, which allowed the determination of optimal ordering and production quantities during the recovery period. This minimized total costs.

Considering structural dynamics, Ivanov et al. (2014b) constructed a multi-period, multi-commodity SC model, the formulation of which distributes static as well as dynamic parameters between the linear programming and control models. In the study by Ivanov et al. (2016), an approach for analysing several proactive SC structures, determining recovery policies to redirect material flows in two disruption scenarios, and measuring the performance impact on service levels and costs was developed. Dupont et al. (2017) applied a mixed-integer linear programming method to develop a model for helping an SC manager to select a supplier portfolio with disruption risk considerations. Their study allows consideration of the risk sensitivity (i.e. risk aversion and loss aversion) of the SC manager in the modelling process.

While the Dupont et al. study (2017) considers deterministic demand, Sawik (2013) developed a stochastic programming model for supplier selection and order allocation in light of disruption risks, and conceptualized a portfolio approach to SC disruption management (Sawik, 2017). Sawik's works also consider an SC manager's risk sensitivity. Khalili et al. (2017) analysed integrated production–distribution planning in the SC considering excessive capacities at the production plants, backup routes for shipments, and pre-positioning of emergency inventory in distribution centres. They also proposed a new indicator for optimizing the SC resilience level based on restoration of lost capacities.

Amiri-Aref et al. (2018) studied a multi-period location-inventory optimization problem in a multi-echelon SC characterized by uncertain demand and a multi-sourcing. The authors integrated inventory planning decisions made under a reorder point order-up-to-level (s, S) policy, with the location-allocation design decisions to cope with demand uncertainty. A two-stage stochastic mathematical model that maximizes the total expected supply chain network profit is proposed. The results show the efficiency of the linear approximation of the (s, S) policy at the strategic level to produce robust design solutions under uncertainty. Further insights from this study underline the sensitivity of the design solution to the demand type and the impact of the inventory holding costs and backorder costs, especially under non-stationary processes.

2.2 *Simulation Studies in Supply Chain Disruptions*

Considering severe SC disruptions for resilience analysis and using discrete event simulation, Carvalho et al. (2012) analysed the behaviour of an SC in four stages considering several recovery strategies and the performance of the SC during disruption. The scenarios the authors studied were different in terms of whether or not there was a disturbance and whether or not a mitigation strategy was in place. The ARENA-based simulation model they developed provided for the determination of lead time rations and total SC costs. In the study by Schmitt and Singh (2012), disruption risk was measured by using 'weeks of recovery' as a proxy for the amplification of disruption. The scenarios developed were both proactive and reactive and satisfied demand by making use of another network location, obtaining material or transport from other sources or routes and retaining inventory reserves along the whole SC. Utilizing the duration and likelihood of closure in a completely observed, exogenous Markov chain model, Lewis et al. (2013) studied risks of disruption at ports of entry. The periodic review inventory control model that the authors developed implies that in the scenarios developed operating margins could either decrease 10% as a result of relatively long port of entry closures or, without contingency plans, be eliminated entirely. In addition, anticipated holding and penalty costs could increase by 20% when port of entry utilization is expected to increase.

In a four-stage SC created in the software anyLogistix, Ivanov et al. (2017a, 2017b, 2017c) used simulation to study the SC's dynamic behaviour and the impact of disruptions on performance. The authors observed the ripple effect as well as studied proactive and reactive strategies. The results of the study imply that disruptions upstream more often lead to the ripple effect when there is a single source policy in place, and safety stock should be increased at facilities downstream from elements of the SC that are risky. In addition, ripple effect propagation towards the customers is decreased when inventory levels are higher in the downstream SC. However, increasing safety stock at disruption-risky facilities must be carefully considered, since when disruption-risky facilities cannot perform outbound operations then this will not effectively decrease the ripple effect. This study also pointed out that the ripple effect impacts service level and order fulfilment more than the duration of the disruption, which indicates that dual sourcing at the bottlenecks of SCs, as well as large inventory holdings downstream from facilities that are disruption risky is of greater importance than quick investments in fast recovery.

A multi-stage SC with suppliers, factory, distribution centres, and customers was studied by Ivanov (2017b). As in previous studies, the findings show a time lag between the start of recovery and the impact of that recovery on the closing of the service-level gap: proactive SC policies must account for the duration of disruptions. Using AnyLogic, Ivanov and Rozhkov (2020) analysed how capacity disruptions impact the performance of policies for ordering and production in a real-life retail SC with considerations of product perishability. The findings of the study imply that SC managers should consider the effects of 'postponed redundancy' when they design resilient SCs. The effect is concerned with how redundant production ordering system behaviour during the period of disruption impacts the production ordering system post-disruption. Redundant behaviour during the disruption period might include redundant production or deliveries which are downstream from the affected part of the SC, or redundant order allocations for upstream facilities which are disrupted. In addition, the authors created and tested a coordinated SC production ordering contingency policy in and after the disruption in order to decrease the negative effects of 'postponed redundancy'.

Schmitt et al. (2017) investigated adjustments in order policies in the framework of a four-stage assembly SC. Simulation experiments revealed that longer lasting impacts occur from disruptions at echelons close to ultimate consumption. Moreover, the results show that expediting inventories in the disrupted mode can trigger unintended bullwhip effects, and hurt rather than help overall performance. As an alternative, dynamic order-up-to policies perform more promisingly as an adaptive mitigation tool. Trucco et al. (2017) analysed an Italian FMCG supply chain and simulated its resilience in AnyLogic. The results suggest that it is important to develop coordinated control strategies in the event of severe supply chain disruptions. These results are in line with the insights provided in the study by Schmitt et al. (2017), Ivanov and Rozhkov (2020), and Ivanov (2019).

3 Case Study Description and Methodology

3.1 Case Studies

Two case studies are considered in this paper in regard to both production and distribution capacity disruptions that occurred in reality and for which the authors observed real company operational policies and performance.

3.1.1 Production Capacity Disruption

An FMCG company, which produces juices and beverages in Russian contracted and proprietary plants, is the first case study (Ivanov & Rozhkov, 2020). The data of the company, covering a period from 2013–2015, was gathered in 2016. The data is primarily concerned with the plants the company owned and the plants which it subcontracted. This includes real and forecasted demand data, collected from internal company report, as well as inventory, production, and shipment control data were obtained by observing the management information system in a joint analysis with factory and DC department managers.

The study concerns part of the FMCG SC. It is a two-stage SC with five DCs and one factory. This factory delivers the product 'juice', which has an average shelf life of 270–360 days, to the DCs. Product perishability is a key factor of the SC. Demand forecasts are the basis for ordering at the DCs, and the inventory policy is (Q, s, r). The inventory level y is forecasted for n periods at each period r. Subject to minimum reorder quantity Q, a new order, size O_r, with a planned delivery period $r + n$ is placed when y is less than reorder point s. The planning is made for $n + m - 1$ periods when production is only possible for each m period. For all customers, shipments follow a first expired first out (FEFO) policy.

Decreases in safety stock and increases in the frequency of transportation are the typical results of the limits of product perishability. Subject to minimum service level, with a targeted service level of 98.5%, the SC planning accounts for the risk of writing off products. The other constraints on shelf life are taken from contract agreements with major food retail companies which operate in Russia (ranging from 62 to 70%). By considering the risk of production capacity disruption, safety stock may be increased. The factory's production capacity is subject to random disruptions. Disruptions of production capacity occurred often in 2015 and 2016, which caused interruptions of deliveries from plants to DCs. These interruptions lasted anywhere from 2 days to 3 weeks.

The primary focus of this study is how disruptions of production capacity impact SC performance, accounting for a two-component demand structure and limited expiration dates. A discrete event simulation model was developed in AnyLogic to show the SC dynamics and ascertain reorder, production, and shipment numbers and times from the DCs to the factory. Each shipment, disruption, and recovery is modelled as an event, while information exchange between DCs and the factory is obtained from state charts and messages. In Appendix 1, the mathematical formulation of this model can be found.

The planning algorithms account for the fact that the stock deteriorates and the batch, which is ready to ship now, might be unusable in a few weeks. A week is the basic time unit. It is assumed that plans are made each week. However, several parameters are measured in days. Heuristics were developed, tried out, and implemented for daily use with the help of the analysts of the company's SC. As is common for the food industry, the demand model is subject to randomness and seasonal factors, which according to the company's data cause demand variations of 50% within the planning horizon. When demand rises or falls by 20%, it is possible to have long term, 4-week duration demand changes. It is possible to describe both factors of demand variation according to uniform or triangular distribution. Along with an aggregation of historic demand from 60 periods, uniform distribution was used in the study experiments. The ordering policy is similar to an economic lot scheduling model's basic planning process (Wagner & Whitin, 1958). Differences include remaining shelf life, based on Nahmia's (1980) mathematical approach, and consideration of the demand structure. In the simulation model, demand non-stationarity is an individual function, and 13 periods constitute the model year. The seasonal demand coefficient k is defined using a basic demand level for each period r. The planning of production was found on a discrete event simulation approach: when a DC's inventory reaches reorder point, then another production order is made. The size of the order is a multiple of the minimum lot size, and cannot be cancelled. The planning of production accounts for lead time from factory to DC. When a batch's computed production period is met, then orders enter the queues in the system. This applies to all products. If an order remains in the queues longer than the DC's planning horizon, then the order goes out of the system, but this does not result in lost order costs. If the waiting time constraint is reached, then the order is moved to the production module. The start times of processing are based on the production week computed. Schedule smoothing, or early production, is not allowed.

The setups of production utilize time and cost containment. Simultaneously, infrequent setups can cause delivery delays and increased variability in lead times. In the model, setups are controlled in planned and disrupted modes. Without capacity disruption, the planned mode uses lot-size-based planning. This means that if five orders for one product, each for 10,000 product units, are in waiting in the queue with a minimum lot size of 40,000 units, then four of the five orders will be batched and produced together as one lot. After this, the setup which was planned will be completed.

When there is a capacity shortage as a result, e.g. of a demand peak or disruption, then flexible setup rules are followed. The queue size, parameter QC, is monitored. The inventory of raw material in the system of production has no limit. First, in the sequence, the allocated order O_r or O_r is sent forward in the queue μ_r. The time for manufacture t_m is calculated on the basis of a function of order quantity O_r and production capacity K.

At the end, the analysis of SC resilience is made according to random disruptions which cause a decrease of 50% in production capacity (cf. Eq. 2). In the model, disruptions are random events. Normal distribution governs the intervals between disruptions t_{dp} and their duration t_{ds}. This hypothetical assumption of normal distribution for the occurrence of disruption and recovery period is in part proved by real data.

3.1.2 Distribution Capacity Disruption

The second case study is based on another company that produces non-perishable products for four regional markets. Without loss of generality, a fragment of the SC considered comprises three production plants and four regional distribution centres (DCs) (Fig. 1).

The DC in region 1 crashed due to construction quality problems. A huge amount of inventory was destroyed. At the day of the DC disruption, the experts estimate that the reconstruction of the DC will take about 4 months.

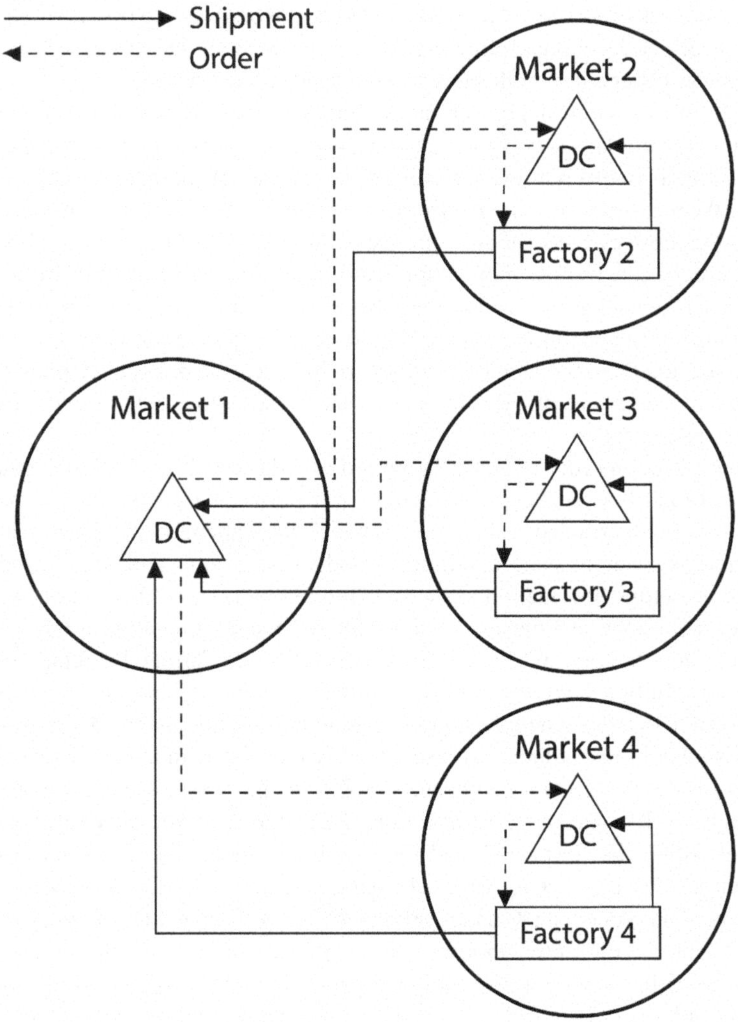

Fig. 1 Supply chain design structure

The following data (but not limited to) has been collected at the company:

- SC design: Locations of SC elements (factories and DCs) and links in between them.
- Demand in the markets and its uncertainty.
- Parameters of SC elements (e.g. production capacities, throughputs, prices, and costs).
- Operating policies of SC elements (e.g. inventory control policy, production control policy, shipment control policy, and sourcing control policy).

This data was observed for a period of 3 years. Moreover, we observed the actual sales and service-level data following the disruption at the DC described. A drastic decrease was observed in both sales and service-level dynamics.

The following control algorithms are implemented in the simulation model. Demand in the markets is considered as an aggregated, normally distributed demand of all customers in this region characterized by a seasonal component subject to four periods. Weekly order placements from the markets to the DCs are considered.

A continuous review system is applied at the DCs. Backordering is allowed so that no orders can be lost. For simplification, an average lead time from DC to the market is considered: it is assumed that all customers within the region can be reached during this lead time. According to demand generation algorithms, orders are placed at the DCs. Subject to inventory-on-hand, safety stock, lead times, reorder point, and the target inventory, shipments to the markets and replenishment from the factories are controlled.

A transportation order aggregation period of 5 days with the less-than-truckload (LTL) control policy has been considered. Transportation time has been considered as a normally distributed value. (cf. Table 2 in Appendix 2). Each factory is considered as a single-stage continuous production system with fixed production time and no setups. Production capacity is limited by the unit production time. For example, a production time of 0.4 days for m^3 means maximum daily capacity of 2.5 m^3 at a factory. Without loss of generality, no further batching rules are considered. Multiple sourcing control with the preference 'closest location' is used. The algorithm decides where to source the demand from the paths 'Markets—DCs' and 'DCs—Factories' subject to closest facility location with available inventory.

A set of key performance indicators (KPI) has been established to analyse the simulation results. The expected lead time (ELT) service level is the ratio of orders delivered within the 'Expected lead time' to total orders. The expected lead time is a parameter that is set up for each market. It measures the time between the placement of an order at a DC and receiving the goods from the DC. Current backlog orders depict the currently unprocessed number of orders, i.e. the orders which were received but have not been shipped yet. It is updated on a daily basis when incoming orders are received, a new shipment is sent, incoming orders are lost, or an incoming shipment is processed (in accordance with processing time set for the facility). Delayed orders show statistics on the quantity of orders, which failed to arrive within the specified, expected lead time. The data is updated each time an order is delayed.

For the problem considered above, a discrete event simulation model and a network optimization model have been developed in anyLogistix (see Appendix 2). anyLogistix is a simulation and optimization tool developed by the AnyLogic Company. The optimization functionality of anyLogistix is implemented on the CPLEX basis in a network optimization module. Simulation functionality in anyLogistix is based on discrete event simulation with agents that can be used either in a standard setting or be customized in AnyLogic. anyLogistix allows a wide range of experiments in regard to facility location planning, multi-stage and multi-period SC design and planning, inventory control, transportation control, and sourcing analysis in both deterministic and stochastic settings. Variation and comparison experiments can be performed. Modelling of the SC disruptions is implemented using events and state change diagrams.

3.2 Research Methodology

Because the problem statements concerning disruption propagation deal with time-dependent settings which include dynamic inventory control, transportation control, sourcing control, and production control policies, the simulation methodology for the given problem domain has earned an important role in academic research (Ivanov, 2017; Boute et al., 2022; Babai et al., 2023). In comparison to analytical closed-form analysis, simulation has the advantage that it can handle complex problem settings with situational behaviour changes in the system over time (Ivanov, 2020). This is inevitable in considering dynamic changes in SC organizational and parametrical structures (Ivanov et al., 2010).

In this study, we use discrete event simulation methods. For validation, the network optimization with and without disruption consideration has been performed in CPLEX using anyLogistix optimization and simulation software. The optimization experiments allowed a determination of the aggregate annual throughputs which are used for validation of the simulation results. The simulations in anyLogistix are run over the optimization results and include additional, time-dependent inventory, production, transportation, and sourcing control policies which are difficult to implement at the network optimization level. In addition, *analytical computations* using standard inventory control models have been done. For *verification*, the following methods have been used: simulation run over network optimization results, output data analysis in the log files, and testing with the help of deterministic demand and lead time data. Moreover, replications and a warming up time with some initial inventory have been applied for the *testing*. The disruptions have been scheduled in the middle of the simulation period in order to avoid the 'noise' of the simulation experiment start. Variation experiments to validate the simulation model have been performed. In particular, mean and standard deviation of demand, safety stock, and production capacity have been varied to confirm model robustness.

The first stage of the experiments contains production disruption analysis with a focus on contingency inventory control policy impacts. The second part of the experiments addresses DC disruptions focusing on analysis of survival policy impacts.

At both stages mentioned above, the analysis is conducted in regard to disruption-driven changes in SC behaviour resulting in delayed orders and backlogs the accumulation of which can be considered 'disruption tails'. The influence of these tails at the time of post-disruption in the course of transition into the normal operation mode is investigated. A comparison of SC operations with and without contingency policy will be made to compare the impact of disruption tails on SC performance. A comparison of SC performance between an immediate deactivation of the contingency plans and installation of the normal operation policies after the capacity recovery and usage of the revival policy will be performed. These experiments aim at providing managerial insights on the application of contingency production and inventory control policies during the disruption period and revival policies at the transition time to normal operation after capacity recovery.

The contingency plan includes the installation of additional links in the SC which lead from the factories directly to the market 1. This implies a longer lead time (cf. Table 2 in Appendix).

The revival policy includes the contingency plan and additional elements such as:

- Backup contractors
- Capacity flexibility (capacities of own plant in region 1) and
- Using capacity of other own plants in neighbourhood countries

that are activated during the disruption period to support the contingency plan actions. The details of revival policy activation will be given in Sect. 5.

4 Production Disruptions Experiments

4.1 No Contingency Inventory Control Policy

In Fig. 2, the simulated dynamics of customer order fulfilment is presented subject to the model and data in Appendix 1.

In period #110, the disruption begins and continues for 26 weeks, and then 30 weeks of a recovery period follow. The DC's inventory is accessible through period #117, because immediately before the disruption period a delivery was completed from the factory to the DC. Because 50% of the capacity remains operable, we know that in periods #120 and #122, two small deliveries from the factory to the DC were

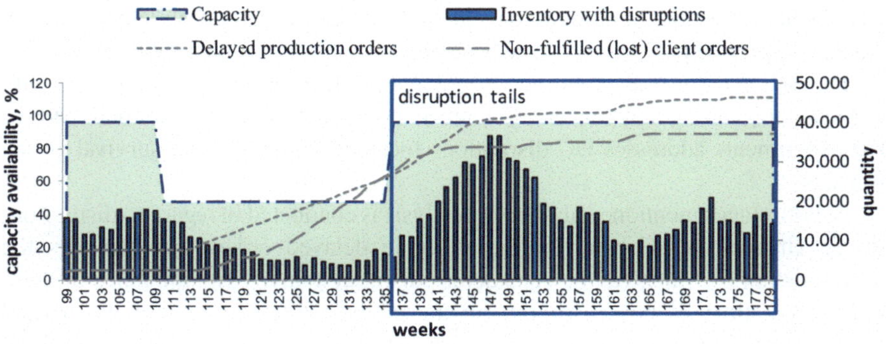

Fig. 2 Dynamics of customer order fulfilment

made. Higher inventory costs result from several delayed production orders being shipped to the DC after the capacity recovery. Following this, the intensity of the order allocations changes again. The result of the high inventory levels is that write risks are increased and the system attempts to allocate fewer order for production. Penalties may occur if deliveries are delayed, e.g. inventory becomes zero in period #165. This indicates lost sales. This in turn shows that *both product shortages and write-off risks are caused by production capacity disruptions.*

Given the higher inventory costs caused by the shipment of the delayed production orders, write-off risks are increased and the system tries to allocate fewer production orders, as stated. These SC dynamics can be called '*postponed redundancy*'. This refers to how the production ordering system behaviour in the after-disruption period is impacted by redundant production ordering system behaviour during the disruption period (Ivanov & Rozhkov, 2020). Examples of redundant SC behaviour during the disruption period might include redundant production or deliveries downstream from the disruption, or redundant order allocations made to facilities upstream from the disruption.

Following production capacity disruption and stabilization, the increases in lost orders rise significantly in several periods after capacity disruption. Nearly directly following production disruption, there is an increase in delayed orders. The period of stabilization, the start of which is closely connected to inventory reaching its maximum level in the SC, lasts longer for lost than delayed orders.

The SC's average inventory does not fall to zero during disruption because the factory remains capable of operating at 50% production capacity. A peak in inventory, related to disruption tails, occurs following the recovery of capacity. As a result, we understand *that when there are increases in delayed orders while there are stabilized service levels, then there will be a significant increase in inventory in the SC soon after.*

4.2 Contingency Inventory Control Policy

Measures can be taken to mitigate increases in inventory while the SC recovers. Particularly, it is recommended to cancel each waiting production order in the period of capacity recovery. Essentially, the orders which are waiting in the queue in the recovery period have a delay of at least one period. However, those orders which are allocated should remain. The results of this hypothesis simulation are shown in Fig. 3.

By making a comparison of Figs. 2 and 3, we understand *that cancelling the waiting orders during the period of capacity recovery means that overstocking and write-off risks can be avoided.* The inventory level is not greater than disruption-free mode levels. As a result, we surmise *that the impact of disruption tails can be mitigated by a contingency production-inventory control policy, i.e. cancelling the waiting orders in the production-inventory system.*

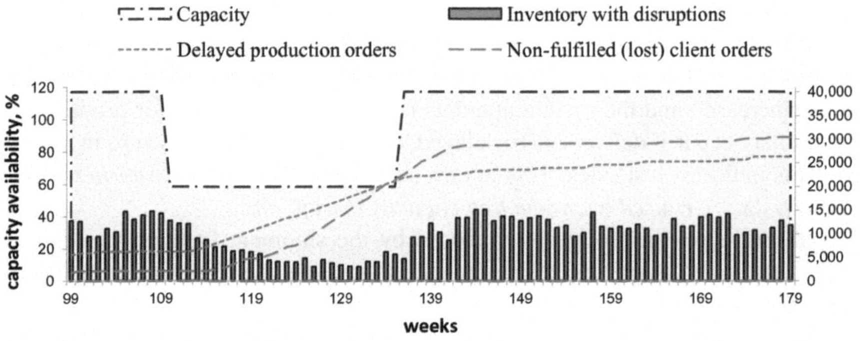

Fig. 3 Impact of waiting order cancellation

5 Distribution Centre Disruption Experiments

Simulation experiments have been conducted subject to the following three settings:

- Without the contingency policy
- With the contingency policy that implies an installation of additional links in the SC from the factories to the market 1 (cf Fig. 1) (These links are activated immediately after the DC1 disruption and function until the DC1 recovery.)
- With the contingency and revival policy that includes such emergency sources as backup contractors, capacity flexibility (capacities of own plant in region 1), and using capacity of other own plants in neighbourhood countries

In line with the study by Ivanov (2017), we assume a time lag between the disruption, activating the contingency policy capacities and the first effects of these contingency policy operations. The emergency sources operate according to the following logic: no initial inventory is available; 2 days after the DC1 disruption, the emergency sources start producing for the market 1; first deliveries to the market 1 arrive in about 18–20 days after the disruption date.

The results are shown in Figs. 4 and 5.

Figures 4 and 5 depict the dynamics of order fulfilment and service level (measured in orders; the diagram represents the average) following the disruption on day 91 and lasting until the DC1 recovery on day 213. The delayed and backlog orders can be observed in the cases without any contingency policy and with the contingency even in the post-disruption period representing the disruption tails. The revival policy helps to improve the service level and reduce the impacts of the disruption tail in terms of delayed and backlog orders in the post-recovery period.

When observing Figs. 4 and 5, an increase in service level, a reduction in the number of delayed orders during the disruption period, and an elimination of delayed orders after the disruption recovery can be observed with a transition from the case without contingency policy, through the introduction of the contingency

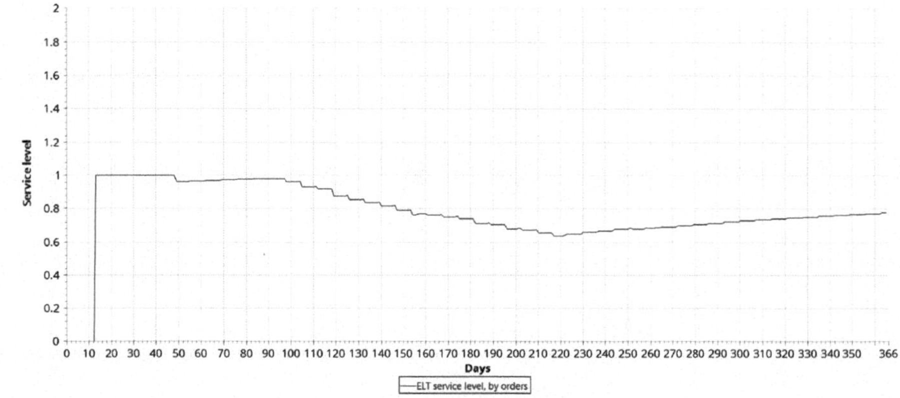

(a) service level dynamics without contingency and revival policy

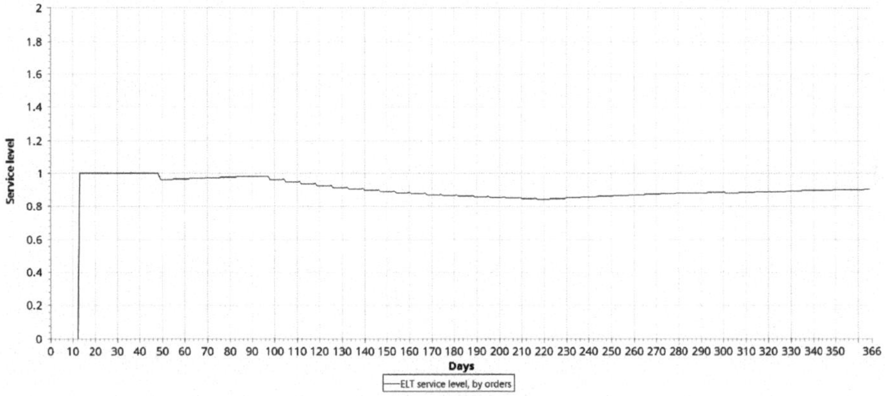

(b) service level dynamics with contingency policy

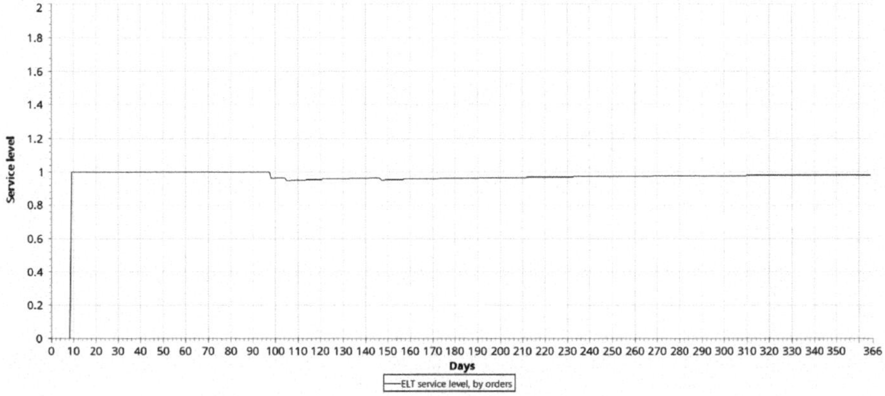

(c) service level dynamics with contingency and revival policy

Fig. 4 Service-level dynamics. **a** Service-level dynamics without contingency and revival policy; **b** Service-level dynamics with contingency policy; **c** Service-level dynamics with contingency and revival policy

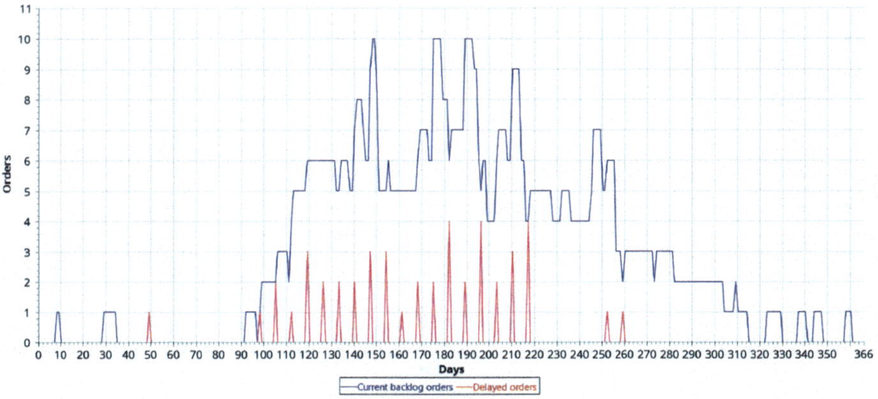

(a) Order fulfillment dynamics without contingency and revival policy

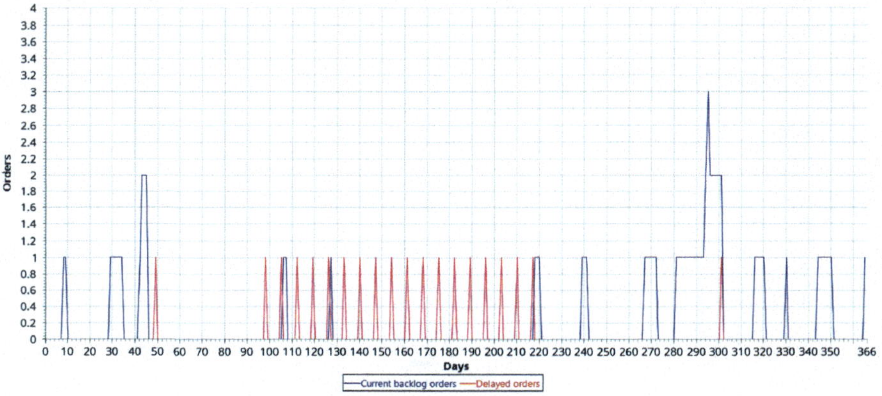

(b) Order fulfillment dynamics with contingency policy

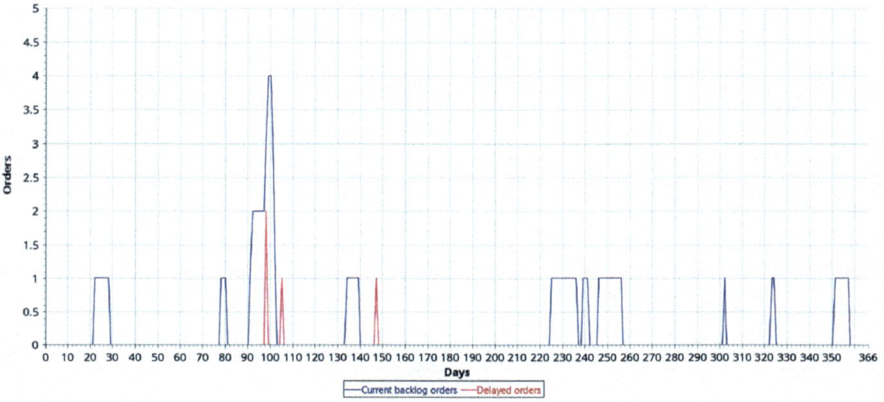

(c) Order fulfillment dynamics with contingency and revival policy

Fig. 5 Order fulfilment dynamics. **a** Order fulfilment dynamics without contingency and revival policy; **b** Order fulfilment dynamics with contingency policy; **c** Order fulfilment dynamics with contingency and revival policy

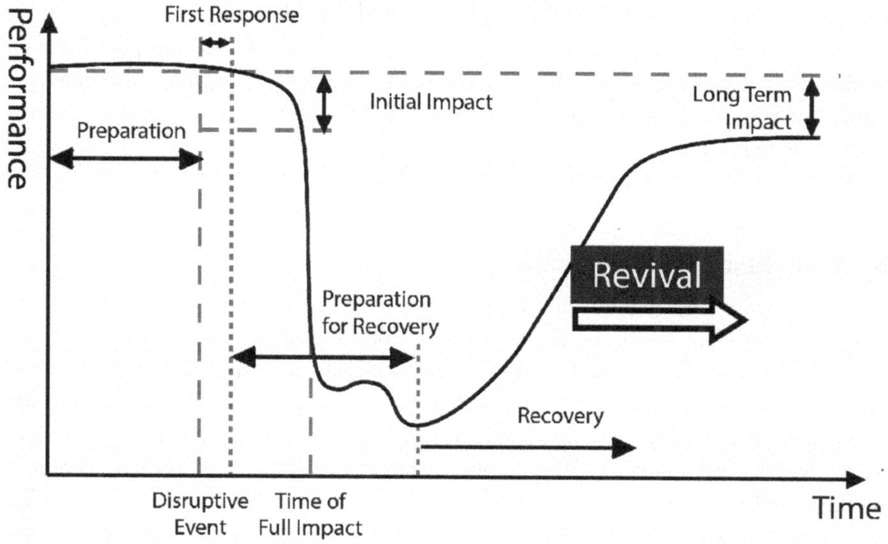

Fig. 6 Extended supply chain resilience framework (extended from Sheffi & Rice, 2005)

policy, to the usage of the revival policy. The revival policy allows stabilization of the order fulfilment dynamics resulting in a positive effect on service-level performance. The *delayed orders accumulated over the disruption period do not influence SC operations and performance since new contracting plants compensate for this with the help of additional production capacity. This allows the service level to recover faster as compared to the usage of recovery policy only* (Figs. 4 and 5). This observation provides the evidence of *disruption tail mitigation with the help of a revival policy based on a production capacity increase in the post-disruption period.* It indicates the necessity of considering not only contingency recovery policies, but also the *revival policies in the SC* which may align normal operation policy and deactivation of the contingency policies. The insights gained recommend inclusion of the revival policy in the SC resilience framework (Fig. 6).

The disruption profile constituted in the work by Sheffi and Rice (2005) includes eight phases: preparation actions, the disruptive event, the first response, the initial impact, the full impact, the recovery preparations and the recovery and long-term impact. Our experimental results suggest including the revival policy into the SC resilience framework if the performance cannot be recovered fully after the capacity recovery. The revival policy extends the SC resilience framework at the stage of transition from recovery to post-disruption period. The rationale for inclusion of a revival policy into the SC resilience framework is the fact that an immediate transition from the contingency plan during the disruption and recovery period to the normal operation mode may be complicated by disruption tails. In addition, of the companies operated with forecasted recovery dates, the inertness of the decisions on activation and deactivation of contingency plans frequently leads to disruption tails. The disruption tails represent residue from the disruption period, such as backlog and delayed orders, which may influence SC operations and performance in the post-disruption mode. The revival policy intends to mitigate the negative impact of these disruption tails and stabilize the SC control policies and long-term performance impact.

6 Conclusion

Despite significant progress in theoretical studies and empirical principles for managing severe disruptions in the SC at proactive and reactive stages, recent literature has mostly assumed an immediate transition to a normal operation mode at the time of capacity recovery. Moreover, full system stabilization after capacity recovery has been assumed without considering any residue, such as delayed orders accumulated over the destabilized system states during the disruption time. The experiments performed allowed revelation of the *disruption tails,* which are the repercussions of the backlogs and delayed orders accumulated over the disruption period.

Post-disruption instability, i.e. the disruption tails in the SC, has been observed in this study as a consequence of the production ordering behaviour during the disruption period. Disruption tails occur in two cases. The *first reason for disruption tails* is non-coordinated inventory and production control policies, i.e. if DCs and customers continue placing new orders even if the production capacity is disrupted. A transition of backlog and delayed order residues into the post-disruption period destabilizes the SC resulting in further delivery delays and non-recovery of the operational and financial SC performance. These orders are accumulated in the waiting lines of production systems and increase the delays and backlogs. After the capacity recovery, the production system needs to produce both the backlog and new incoming orders. If the recovered production capacity is lower than the total of the backlog and new incoming orders, this results in new delays and backlogs. If the recovered production capacity is sufficient to cover both the total of the backlog and new incoming orders, this results into disproportionally high delivery quantities at the DCs. In turn, the DCs stop ordering because of the high inventory, and this results in some periods having new shortages. The *second reason for disruption tails* is the inertness of the decisions on activation and deactivation of contingency plans. An immediate deactivation of the contingency plans and installation of the normal operation policies after capacity recovery in the presence of the delayed orders and backlogs also result in the destabilization of inventory and production systems and in the new delayed orders and backlogs.

Two case studies with different settings, perishable and non-perishable products, and different data sets have been considered and the experiments with both case studies confirmed the disruption tails existence. With the results gained, this study closes the research gap described above with the objective of revealing the dependencies between SC disruptions, contingency policies, and transition to the post-disruption operation mode.

The results obtained suggest two managerial insights. First, contingency production and inventory control policies need to be applied during the disruption period. Moreover, recovery policies should not be limited to the disruption period only, they should also consider the post-disruption period. Second, special actions need to be developed for the transition time from the contingency plan to the disruption-free operation mode. We call these actions 'revival policy'. The experimental results argue in favour of revival policies to mitigate the impact of the disruption tails. Moreover, a revival policy extends the SC resilience framework at the stage of transition from recovery to post-disruption period.

We provide examples of contingency policies for production factory and DC disruptions as well as for the revival SC policy. Disruption tails can be reduced by applying a contingency policy during the disruption time when the production control system cancels excessive DC orders waiting in the production system queues because of disrupted manufacturing capacities. Alternatively, the DCs need to adjust their ordering policy subject to the reduced production capacity to avoid long waiting times and the resulting delayed orders, backlogs, service-level reductions, and the production-inventory control system destabilization after the capacity recovery. As an example of the revival policy, an increase in SC production capacity has been suggested by means of contracting additional plants prior to normal contingency recovery policy deactivation. This allows avoidance of the disruption tails in the post-disruption period.

Concerning the limitations of this study, the contextual insights gained experimentally need to be pointed out. Further research might include analysis of other industries and datasets. Moreover, analytical studies are needed to provide more generalizable theoretical results and practical recommendations, i.e. in the area of recovery policies with the post-disruption period considerations.

Appendix 1. Mathematical Model for Section 4

Indices

α	Priority customer index
β	Non-priority customer index
f	Actual demand index
r	Period index, $r \in [1; T]$
ST	Standard deviation index
l	Trend variation index
i, j	Products 1 and 2, respectively
g	Distribution centre number, $g \in [1; G]$
z	DC order index $n \in [1; N]$, where N is the total number of DC orders
ω	Current forecasting period, $\omega \in [r; r + LT + m - 1]$
w	Index of products with expired date
h	Index of setups at the factory, $h \in [1; Nch]$
ds	Disrupted
ch	Setup

Parameters

T	Number of planning periods in planning horizon
G	Number of DCs
d	Basis demand in a r-period, in units
k	Seasonal demand coefficient in a r-period
δ^{ST}	Demand standard deviation r-period
δ^{l}	Demand trend parameter for the length of r-periods
d_{fr}	Actual demand in a r-period, in units
ν	Priority customer rate
s	Reorder point, in units
Q	Minimum reorder quantity, in units
LT	Lead time, in periods
m	Frequency of batch setups in the factory, in periods
p_{α}	Minimum requirement on the rest product freshness for α-customers
K	Maximum production capacity per period, in units
B	Minimum batch size, in units
QC	Maximum production order queue length, in orders
t_{ch}	Setup time
t_{dp}	Disruption time
t_{ds}	Disruption duration, in periods
ξ	Capacity reduction coefficient, in units
c_h	Unit inventory holding costs per period, in $

(continued)

c_{tr}	Unit transportation costs per delivery, in \$
c_{fix}	Fixed production costs, in \$ per capacity unit
c_{ch}	Setup costs, in \$
p	Unit price, in \$
u	Penalty for non-delivered products, in \$
SL_{min}	Minimum service level, %
η	Shelf life

Variables

O	Order quantity from DC to factory, in units
F	Production date for a z-batch, period
H	Total holding costs
T	Total transportation costs
W	Total write-off costs
U	Total penalty costs
M	Total manufacturing costs
TC	Total costs
μ	Processing queue length, in units
t_m	Production time for a batch, in periods
y	Inventory in a r-period

Objective function

$$\begin{aligned}
&\min TC = H + T + W + U + M = \\
&\mathit{Total\ supply\ chain\ costs} \\
&\quad = H = \sum_{r=1}^{1200} \sum_{g=1}^{5} c_h \cdot y_{ir}^{g} + \sum_{r=1}^{1200} \sum_{g=1}^{5} c_h \cdot y_{jr}^{g} + \\
&\mathit{Total\ inventory\ holding\ costs} \\
&\quad + T = \sum_{n=1}^{N^i} \sum_{g=1}^{5} c_{tr} \cdot L^{g*} O_{in}^{g} + \sum_{n=1}^{N^j} \sum_{g=1}^{5} c_{tr} \cdot LT^{g} \cdot O_{jn}^{g} + \\
&\mathit{Total\ transportation\ costs} \\
&\quad + W = \sum_{r=1}^{1200} \sum_{g=1}^{5} p \cdot y_{wir}^{g} + \sum_{r=1}^{1200} \sum_{g=1}^{5} p \cdot y_{wjr}^{g} + \\
&\mathit{Total\ write-off\ costs} \\
&\quad + U = \sum_{r=1}^{1200} \sum_{g=1}^{5} u_{ir}^{g} + \sum_{r=1}^{1200} \sum_{g=1}^{5} u_{jr}^{g} + \\
&\mathit{Total\ penalty\ costs} \\
&\quad + M = \sum_{h=1}^{N_{ch}^{i}} c_{ch} + \sum_{r=1}^{1200} c_{fix} \cdot K \\
&\mathit{Total\ manufacturing\ costs}
\end{aligned} \tag{1}$$

Constraints

$$K_{ds} = K \cdot \xi \tag{2}$$

$$\forall SL_g \geq SLmin \tag{3}$$

$$\mu_{jr} > QC_1 \tag{4}$$

$$\mu_{jr} - \mu_{ir} > QC_2 \tag{5}$$

$$O_i \geq B_i \tag{6}$$

$$s_{i,j} \in \text{int} \tag{7}$$

$$s_{i,j} \geq 0 \tag{8}$$

Two-stage, multi-period SC planning with multiple constraints on production capacity, setups, shipments, and inventory control is the object under investigation. A two-product system with independent seasonal stochastic demand with high variability is analysed. The planning horizon is 7 weeks. Production planning decisions include inventory dynamics at the DC. We consider both product availability and 'freshness' level requirements in service levels although customers are segmented according to their freshness requirements.

The fixed parameters include minimum order size and inventory level at DCs, minimum production batch size, queue size limits, setup time, production capacity, wastage, inventory holding, production, setup, and transportation costs. Also included are the mean demand and its standard deviation, shelf life and freshness threshold, production order allocation interval, penalties, mean and standard deviation of time duration and interval of capacity breakdown, and the remaining capacity percentage after the disruption.

Production constraints include minimum lot size, maximum capacity, and setup time. Inventory constraints are comprised of minimum inventory levels (i.e. reorder point expressed in days of supply availability) and minimum order size. Outbound deliveries from distribution centres follow the FEFO rule. A continuous review system with fixed order quantity and pull production strategy is considered.

The objective is to minimize total system costs while maintaining the required service level. Total costs are computed as a sum of total holding costs, transportation costs, write-off costs, penalty costs, and manufacturing costs (Eq. 1). Unit inventory holding costs c_h and transportation costs c_{tr} are used to compute total costs. In a case of inventory with expired date y_w, write-off costs increase proportionally to the purchasing prices p. If the customer order size exceeds the inventory at DC, a penalty u is applied. Manufacturing costs depend on the number of setup and fixed costs for capacity units, c_{fix}.

Service level is calculated as a ratio of products shipped divided by products ordered with no backlogging within model period. SC performance is therefore measured with the help of total costs and service level. Total cost metrics are comprised of inventory holding costs at the DCs, write-off costs, transportation costs, production costs, and penalties. Holding costs are computed subject to interest rates. Write-off costs are computed based on the product costs. Transportation costs depend on the distance, order quantity, and shipment tariff. Production costs include fixed equipment-related costs (proportional to the capacity units) and setup costs. Penalties are applied if the order size from the key customer exceeds the available delivery quantity. Service level is computed as a ratio of the delivered and ordered products.

According to Eq. (2), production capacity can be reduced by a disruption coefficient ξ. By default, the following parameters are used: mean interval is 100 periods and mean duration of disruption is 20 periods. Standard deviations are 50 and 10 periods, respectively. At the end of the disruption period, the capacity K returns to normal.

Equation (3) sets the constraint on minimum service level. In the considered practical case, 98.5% has been used as the reference value for minimum service level. Equations (4) and (5) define maximum queue lengths in the production system. According to Eq. (6), production quantity can equal or exceed minimum batch size. Equations (4)–(6) define the rules for production setups. Equations (7) and (8) are binary and non-negativity constraints on the reorder point for products i and j.

Empirical data revealed the average weekly demand of 2500 units. The basic demand in the model is 2541 units multiplied by the seasonal factor. The period demand d_r is therefore defined according to Eq. (9).

$$d_r = k^* d \tag{9}$$

The actual demand d_{fr} may vary in a period with a standard deviation δ_r^{ST} subject to uniform distribution. Additionally, period demand may be corrected by a trend δ_r^l of demand increase or decrease for the length of four periods. Therefore, actual period demand d_{fr} is generated according to Eq. (10):

$$d_{fr} = d_r{}^* \delta_r^{ST*} \delta_r^l \tag{10}$$

Demand is divided into two customer groups, i.e. α customers have higher priority than β customers. Demand share of α customers is defined by parameter ν according to Eq. (11).

$$d_{fr\alpha} = d_{fr}{}^{*}\nu \ ; d_{fr\beta} = d_{fr}{}^{*}(1-\nu) \tag{11}$$

DC operations are modelled using a multi-agent approach. We considered a set Z of production batches that are sorted upwards according to production dates F_z Parameters ρ_α and ρ_β. Then, we defined the minimum requirements for the rest shelf life for both customer groups. Let us consider current forecasting period $\omega\ (\omega \in [r; r+n+m-1])$ in order to define the general outbound delivery planning algorithm for key customers and each period as follows:

$$\begin{aligned}
&for\ z_i \in Z\\
&if\ F_{z_i} > \omega - \rho_\alpha{}^{*}\eta\\
&if\ z_i > d_{\omega\alpha}\\
&z_i = z_i - d_{\omega\alpha}\\
&d_{\omega\alpha} = 0\\
&else\\
&d_{\omega\alpha} = d_{\omega a} - z_i\\
&else\\
&next.
\end{aligned}$$

The algorithm described above allows for consideration of both inventory dynamics and expected shelf life of future deliveries.

The experiments have been performed on the following parameter setting (Table 1).

Table 1 Input data

Parameters	Parameter value
Minimum days of supply	14
Minimum order size in product units	10,000
Rolling planning horizon, in periods	7
Minimum period between production order allocations, in periods	1
Production capacity, product units per period	40,000
Minimum reorder quantity	10,000
Lead time, in periods	2–3
Maximum production capacity per period, in units	40,000
Minimum batch size, in units	20,000
Setup time, periods	0.08
Setup costs per period, in $	7500
Mean of the interval between capacity disruptions, in periods	100
Standard deviation of the interval between capacity disruptions, in periods	50
Mean of disruption duration, in periods	20
Capacity reduction coefficient, in %	50
Minimum service level, %	98.5
Shelf life, in periods	36

Appendix 2. Mathematical Model for Section 5

Indices

f	Actual demand index
α	α-service level
r	Period index, $r \in [1; T]$
ST	Standard deviation index
λ	Market number, $\lambda \in [1; \Lambda]$
i	Production facility number, $i \in [1; H]$
j	Distribution centre number, $j \in [1; G]$)
t	Running time index
T	Length of the planning horizon

Parameters

T	Number of planning periods in planning horizon
G	Number of DCs
H	Number of factories
Λ	Number of markets
D	Mean weekly demand in a r-period, in units
q	Mean basis demand, in units
k	Seasonal demand coefficient in a r-period
δ^{ST}	Weekly demand standard deviation in a r-period
K	Maximum production capacity per day, in units
B	Maximum storage capacity at the DCs per day, in units
L^{in}	Maximum inbound processing capacity at the DCs per day, in units
L^{out}	Maximum outbound processing capacity at the DCs per day, in units
ξ	Capacity reduction coefficient, in units
c_h	Unit inventory holding costs per day, in $
c_{tr}	Unit transportation costs per delivery, in $
c_{fix}	Fixed site costs, in $ per day
c_{man}	Own manufacturing costs, in $ per unit
c_{sub}	Subcontracting manufacturing costs, in $ per unit
c_{in}	Inbound processing costs, in $ per unit
c_{out}	Outbound processing costs, in $ per unit
c_{down}	Penalty for demand non-fulfilment, in $ per unit
p	Unit price, in $

Variables

P	Production quantity at the factory, in units per day
S	Selling quantity in the markets, in units
X^{in}	Processed inbound quantity at the DC, in units per day
X^{out}	Processed outbound quantity at the DC, in units per day
Q	Shipment quantities in between the factory, DC, and the markets, in units per day
H	Total inventory holding costs, in $
T	Total transportation costs, in $
W	Total processing costs, in $
F	Total fixed costs, in $
M	Total manufacturing costs, in $
U	Total penalty for delayed delivery, in $
TC	Total costs, in $
y	Inventory in a r-period, in units
d	Distance, in km (computed based on real routes)

Objective function

$$max Profit = Revenue - TC = (p \cdot S) - (H + T + W + U + M + F), \tag{12}$$

where

$$H = \sum_{t=1}^{T} \sum_{j=1}^{G} c_h \cdot y_{jt}^{g}$$

Total inventory holding costs

$$T = \sum_{j=1}^{G} \sum_{i=1}^{N} c_{tr} \cdot d_{ij} \cdot Q_{ij} + \sum_{i=1}^{N} \sum_{\lambda=1}^{\Lambda} c_{tr} \cdot d_{\lambda i} \cdot Q_{i\lambda}$$

Total transportation costs

$$W = \sum_{j=1}^{G} (c_{in} + c_{out})$$

Total processing costs

$$F = \sum_{j=1}^{G} c_{fix} + \sum_{i=1}^{N} c_{fix}$$

Total fixed costs

$$F = \sum_{i=1}^{N} c_{sub} \cdot P_i + \sum_{i=1}^{N} c_{man} \cdot P_i$$

Total manufacturing costs

$$U = \sum_{\lambda=1}^{\Lambda} c_{down}$$

Total penalty costs

Demand constraints

$$Q_{j\lambda t} \geq d_{t\lambda}$$

$$D_r = k \cdot q_\lambda$$

$$D_{fr} = D_r \cdot \sigma_r^{ST}$$

Shipment constraints

$$Q_{ijt} \leq X_t^{out}$$

$$Q_{\lambda jt} \leq y_{jt}$$

Capacity constraints

$$P_{it} \leq K_{it} \cdot \xi$$

Constraints on inventory holding and processing at the DCs

$$y_j \leq B_j \cdot \xi$$

$$X_{t+1}^{out} \leq L^{out}$$

$$X_{t+1}^{in} \leq L^{in}$$

The experiments have been performed with the following parameters (Table 2).

Table 2 Experimental settings

Parameter	Values
Mean basis weekly demand in the market 1, in m^3	6000
Mean basis weekly demand in the markets 2–4, in m^3	4000
Number of periods	4
Period length, in months	3
Seasonal demand coefficients for four periods	1.00–1.25–0.75–1.0
Standard deviation of weekly demand, in m^3	25% from the mean
Expected lead time in the markets, in days	7
Lead time in between two SC stages within a region, in days	1
Mean lead time in between two SC stages from different regions, in days	5
Standard deviation lead time in between two SC stages from different regions, in days	2
Reorder point at the DC1, the factories, and emergency plants, in m^3	10,000
Target inventory level at the DC1, the factories, and emergency plants, in m^3	20,000
Safety stock at the DC1, in m^3	6000
Reorder point at the DCs 2–4, in m^3	7000
Target inventory level at the DCs 2–4, in m^3	14,000
Initial inventory at the DCs 2–4 and factories, in m^3	10,000
Initial inventory at the DC1, in m^3	20,000
Production time for product unit, in days, in m^3	0.001
Maximum production capacity at own factory, in m^3 per period	90,000
Maximum production capacity at emergency sources, in m^3 per period	10,000
Recovery time after a disruption, in months	4
Time between the disruption and activating the contingency policy such as subcontractor, milk producer capacity, and own factories abroad, in days	20
Mean lead time to the market 1 in the disruption period, in days	8
Standard deviation lead time to the market 1 in the disruption period, in days	2

References

Amiri-Aref, M., Klibi, W., & Babai, M. Z. (2018). The multi-sourcing location inventory problem with stochastic demand. *European Journal of Operational Research, 266*(1), 72–87.

Atan, Z., & Snyder, L. V. (2012). Inventory strategies to manage supply disruptions. In *Supply disruptions: Theory and practice of managing risk* (pp. 115–139). Springer.

Babai, M. Z., Ivanov, D., & Kwon, O. K. (2023). Optimal ordering quantity under stochastic time-dependent price and demand with a supply disruption: A solution based on the change of measure technique. *Omega, 116*, 102817.

Benyoucef, L., Xie, X., & Tanonkou, G. A. (2013). Supply chain network design with unreliable suppliers: A Lagrangian relaxation-based approach. *International Journal of Production Research, 51*(21), 6435–6454.

Boute, R. N., Disney, S. M., Gijsbrechts, J., & Mieghem, J. A. V. (2022). Dual sourcing and smoothing under nonstationary demand time series: Reshoring with speedfactories. *Management Science, 68*(2), 1039–1057.
Carvalho, H., Barroso, A. P., Machado, V. H., Azevedo, S., & Cruz-Machado, V. (2012). Supply chain redesign for resilience using simulation. *Computers & Industrial Engineering, 62*(1), 329–341.
Choi, T. M., Cheng, T. C. E., & Zhao, X. (2016). Multi-methodological research in operations management. *Production and Operations Management, 25*, 379–389.
Costantino, A., Dotoli, M., Falagario, M., Fanti, M. P., & Mangini, A. M. (2012). A model for supply management of agile manufacturing supply chains. *International Journal of Production Economics, 135*, 451–457.
Dolgui, A., & Ivanov, D. (2021). Ripple effect and supply chain disruption management: New trends and research directions. *International Journal of Production Research, 59*(1), 102–109.
Dolgui, A., Ivanov, D., & Rozhkov, M. (2020). Does the ripple effect influence the bullwhip effect? An integrated analysis of structural and operational dynamics in the supply chain. *International Journal of Production Research, 58*(5), 1285–1301.
Dolgui, A., Ivanov, D., & Sokolov, B. (2018). Ripple effect in the supply chain: An analysis and recent literature. *International Journal of Production Research, 56*(1–2), 414–430.
Dolgui, A., Gusikhin, O., Ivanov, D., Li, X., & Stecke, K. (2024). A network-of-networks adaptation for cross-industry manufacturing repurposing. *IISE Transactions, 56*(6), 666–682.
Dubey, R., Gunasekaran, A., Childe, S. J., Papadopoulos, A., Blome, C., & Luo, Z. (2017). Antecedents of resilient supply chains: An empirical study. *IEEE Transactions on Engineering Management, 99*, 1–12.
Dupont, L., Bernard, C., Hamdi, F., & Masmoudi, F. (2017). Supplier selection under risk of delivery failure: A decision-support model considering managers' risk sensitivity. *International Journal of Production Research, 56*(3), 1054–1069.
Fahimnia, B., Tang, C. S., Davarzani, H., & Sarkis, J. (2016). Quantitative models for managing supply chain risks: A review. *European Journal of Operational Research, 247*(1), 1–15.
Federgruen, A., & Yang, N. (2011). Procurement strategies with unreliable suppliers. *Operations Research, 59*(4), 1033–1039.
Ghadge, A., Er, M., Ivanov, D., & Chaudhuri, A. (2022). Visualisation of ripple effect in supply chains under long-term, simultaneous disruptions: A System Dynamics approach. *International Journal of Production Research, 60*(20), 6173–6186.
Govindan, K., Jafarian, A., Azbari, M. E., & Choi, T.-M. (2016). Optimal bi-objective redundancy allocation for systems reliability and risk management. *IEEE Transactions on Cybernetics, 46*(8), 1735–1748.
Gunasekaran, A., Subramanian, H., & Rahman, S. (2015). Supply chain resilience: Role of complexities and strategies. *International Journal of Production Research, 53*(22), 6809–6819.
Han, J., & Shin, K. S. (2016). Evaluation mechanism for structural robustness of supply chain considering disruption propagation. *International Journal of Production Research, 54*(1), 135–151.
He, J., Alavifard, F., Ivanov, D., & Jahani, H. (2019). A real-option approach to mitigate disruption risk in the supply chain. *Omega: The International Journal of Management Science, 88*, 133–149.
Hishamuddin, H., Sarker, R. A., & Essam, D. (2013). A recovery model for a two-echelon serial supply chain with consideration of transportation disruption. *Computers & Industrial Engineering, 64*(2), 552–561.
Ho, W., Zheng, T., Yildiz, H., & Talluri, S. (2015). Supply chain risk management: A literature review. *International Journal of Production Research, 53*(16), 5031–5069.
Hosseini, S., & Ivanov, D. (2022). A multi-layer Bayesian network method for supply chain disruption modelling in the wake of the COVID-19 pandemic. *International Journal of Production Research, 60*(17), 5258–5276.

Hosseini, S., Ivanov, D., & Dolgui, A. (2020). Ripple effect modeling of supplier disruption: Integrated Markov Chain and Dynamic Bayesian Network Approach. *International Journal of Production Research, 58*(11), 3284–3303.

Hu, X., Gurnani, H., & Wang, L. (2013). Managing risk of supply disruptions: Incentives for capacity restoration. *Production and Operations Management, 22*(1), 137–150.

Iakovou, E., Vlachos, D., & Xanthopoulos, A. (2010). A stochastic inventory management model for a dual sourcing supply chain with disruptions. *International Journal of Systems Science, 41*(3), 315–324.

Ivanov, D. (2017). Simulation-based the ripple effect modelling in the supply chain. *International Journal of Production Research, 55*(7), 2083–2101.

Ivanov, D. (2018a). Revealing interfaces of supply chain resilience and sustainability: A simulation study. *International Journal of Production Research, 56*(10), 3507–3523.

Ivanov, D. (2018b). *Structural dynamics and resilience in supply chain risk management*. Springer.

Ivanov, D. (2019). Disruption tails and revival policies: A simulation analysis of supply chain design and production-ordering systems in the recovery and post-disruption periods. *Computers and Industrial Engineering, 127*, 558–570.

Ivanov, D. (2020). Predicting the impacts of epidemic outbreaks on global supply chains: A simulation-based analysis on the coronavirus outbreak (COVID-19/SARS-CoV-2) case. *Transportation Research Part E: Logistics and Transportation Review, 136*, 101922.

Ivanov, D. (2024). Exiting the COVID-19 pandemic: After-shock risks and avoidance of disruption tails in supply chains. *Annals of Operations Research, 335*, 1627–1644.

Ivanov, D. (2025a). When is the supply chain resilient? Customer and operational perspectives. *International Journal of Production Research*, 1–16. https://doi.org/10.1080/00207543.2025.2454331

Ivanov, D. (2025b). Comparative analysis of product and network supply chain resilience. *International Transactions in Operational Research*. https://doi.org/10.1111/itor.13612

Ivanov, D. (2025c). What network and performance indicators can tell us about supply chain and sourcing resilience (and what they cannot). *Journal of Purchasing and Supply Management, 101014*.

Ivanov, D., & Dolgui, A. (2019). Low-Certainty-Need (LCN) Supply Chains: A new perspective in managing disruption risks and resilience. *International Journal of Production Research, 57*(15-16), 5119–5136.

Ivanov, D., & Dolgui, A. (2021). A digital supply chain twin for managing the disruptions risks and resilience in the era of Industry 4.0. *Production Planning and Control, 32*(9), 775–788.

Ivanov, D., Dolgui, A., & Sokolov, B. (2019a). The impact of digital technology and Industry 4.0 on the ripple effect and supply chain risk analytics. *International Journal of Production Research, 57*(3), 829–846.

Ivanov, D., Dolgui, A., Sokolov, B., & Ivanova, M. (2017a). Literature review on disruption recovery in the supply chain. *International Journal of Production Research, 55*(20), 6158–6174.

Ivanov, D., Pavlov, A., Pavlov, D., & Sokolov, B. (2017b). Minimization of disruption-related return flows in the supply chain. *International Journal of Production Economics, 183*, 503–513.

Ivanov, D., & Rozhkov, M. (2020). Coordination of production and ordering policies under capacity disruption and product write-off risk: An analytical study with real-data based simulations of a fast moving consumer goods company. *Annals of Operations Research, 291*(1–2), 387–407.

Ivanov, D., & Sokolov, B. (2013). Control and system-theoretic identification of the supply chain dynamics domain for planning, analysis, and adaptation of performance under uncertainty. *European Journal of Operational Research, 224*(2), 313–323.

Ivanov, D., Sokolov, B., & Dolgui, A. (2014a). The ripple effect in supply chains: Trade-off 'efficiency-flexibility-resilience' in disruption management. *International Journal of Production Research, 52*(7), 2154–2172.

Ivanov, D., Sokolov, B., & Kaeschel, J. (2010). A multi-structural framework for adaptive supply chain planning and operations control with structure dynamics considerations. *European Journal of Operational Research, 200*(2), 409–420.

Ivanov, D., Sokolov, B., & Pavlov, A. (2013). Dual problem formulation and its application to optimal re-design of an integrated production-distribution network with structure dynamics and ripple effect considerations. *International Journal of Production Research, 51*(18), 5386–5403.

Ivanov, D., Sokolov, B., & Pavlov, A. (2014b). Optimal distribution (re)planning in a centralized multi-stage network under conditions of the ripple effect and structure dynamics. *European Journal of Operational Research, 237*(2), 758–770.

Ivanov, D., Sokolov, B., Pavlov, A., Dolgui, A., & Pavlov, D. (2016). Disruption-driven supply chain (re)-planning and performance impact assessment with consideration of pro-active and recovery policies. *Transportation Research Part E, 90*, 7–24.

Ivanov, D., Tsipoulanidis, A., & Schönberger, J. (2017c). *Global supply chain and operations management*. Springer Nature.

Ivanov, D., Tsipoulanidis, A., & Schönberger, J. (2019b). *Global supply chain and operations management* (2nd ed.). Springer Nature.

Jain, V., Kumar, S., Soni, U., & Chandra, C. (2017). Supply chain resilience: Model development and empirical analysis. *International Journal of Production Research, 55*(22), 6779–6800.

Jin, S., He, Y., Ivanov, D., & Chen, W. (2024). To act or wait: The impact of disruption duration estimates on manufacturer's response strategies. *IISE Transactions*. https://doi.org/10.1080/24725854.2024.2417377

Kamalahmadi, M., & Mellat-Parast, M. (2016). Developing a resilient supply chain through supplier flexibility and reliability assessment. *International Journal of Production Research, 54*(1), 302–321.

Khalili, S. M., Jolai, F., & Torabi, S. A. (2017). Integrated production–distribution planning in two-echelon systems: A resilience view. *International Journal of Production Research, 55*(4), 1040–1064.

Kim, S. H., & Tomlin, B. (2013). Guilt by association: Strategic failure prevention and recovery capacity investments. *Management Science, 59*(7), 1631–1649.

Klibi, W., Martel, A., & Guitouni, A. (2010). The design of robust value-creating supply chain networks: A critical review. *European Journal of Operational Research, 203*(2), 283–293.

Kouvelis, P., & Li, P. (2012). Contingency strategies in managing supply systems with uncertain lead-times. *Production and Operations Management, 21*(1), 161–176.

Kravchenko, K., Gruchmann, T., Ivanova, M., & Ivanov, D. (2024). Responding to the ripple effect from systemic disruptions: Empirical evidence from the semiconductor shortage during COVID19. *Modern Supply Chain Research and Applications*, forthcoming.

Levner, E., & Ptuskin, A. (2017). Entropy-based model for the ripple effect: Managing environmental risks in supply chains. *International Journal of Production Research, 56*(7), 2539–2551.

Lewis, B. M., Erera, A. L., Nowak, M. A., & White, C. C., III. (2013). Managing inventory in global supply chains facing port-of-entry disruption risks. *Transportation Science, 47*(2), 162–180.

Li, M. K., Sodhi, M. S., Tang, C. S., & Yu, J. J. (2023). Preparedness with a system integrating inventory, capacity, and capability for future pandemics and other disasters. *Production and Operations Management, 32*(2), 564–583. https://doi.org/10.1111/poms.13887

Liberatore, F., Scaparra, M. P., & Daskin, M. S. (2012). Hedging against disruptions with ripple effects in location analysis. *Omega, 40*, 21–30.

Lim, M. K., Bassamboo, A., Chopra, S., & Daskin, M. S. (2012). Facility location decisions with random disruptions and imperfect estimation. *Manufacturing and Service Operations Management, 15*(2), 239–249.

Liu, M., Liu, Z., Chu, F., Zheng, F., & Chu, C. (2021). A new robust dynamic Bayesian network approach for disruption risk assessment under the supply chain ripple effect. *International Journal of Production Research, 59*(1), 265–285.

Mizgier, K. J. (2017). Global sensitivity analysis and aggregation of risk in multi-product supply chain networks. *International Journal of Production Research, 55*(1), 130–144.

Mosayebi, M., Fathi, M., Hedayati, M. K., & Ivanov, D. (2024). Time-to-Adapt (TTA). *International Journal of Production Economics, 278*, 109432.

Nahmias, S. (1980). Perishable inventory theory: A review. *Operations Research, 30*(4), 680–708.

Namdar, J., Li, X., Sawhney, R., & Pradhan, N. (2018). Supply chain resilience for single and multiple sourcing in the presence of disruption risks. *International Journal of Production Research, 56*(6), 2339–2360.

Nikookar, E., Stevenson, M., & Varsei, M. (2024). Building an antifragile supply chain: A capability blueprint for resilience and post-disruption growth. *Journal of Supply Chain Management, 60*(1), 13–31. https://doi.org/10.1111/jscm.12313

Paul, S. K., & Chowdhury, P. (2021). A production recovery plan in manufacturing supply chains for a high-demand item during COVID-19. *International Journal of Physical Distribution & Logistics Management, 51*(2), 104–125.

Pavlov, A., Ivanov, D., Pavlov, D., & Slinko, A. (2019). Optimization of network redundancy and contingency planning in sustainable and resilient supply chain resource management under conditions of structural dynamics. *Annals of Operations Research*. https://doi.org/10.1007/s10479-019-03182-6

Pavlov, A., Ivanov, D., Werner, F., Dolgui, A., & Sokolov, B. (2022). Integrated detection of disruption scenarios, the ripple effect dispersal and recovery paths in supply chains. *Annals of Operations Research, 319*, 609–631.

Qi, L. (2013). A continuous-review inventory model with random disruptions at the primary supplier. *European Journal of Operational Research, 225*(1), 59–74.

Raj, R., Wang, J. W., Nayak, A., Tiwari, M. K., Han, B., Liu, C. L., et al. (2015). Measuring the resilience of supply chain systems using a survival model. *IEEE Systems Journal, 9*(2), 377–381.

Rezapour, S., Farahani, R., & Pourakbar, M. (2017). Resilient supply chain network design under competition: A case study. *European Journal of Operational Research, 259*(3), 1017–1035.

Sawik, T. (2013). Integrated selection of suppliers and scheduling of customer orders in the presence of supply chain disruption risks. *International Journal of Production Research, 51*(23–24), 7006–7022.

Sawik, T. (2017). A portfolio approach to supply chain disruption management. *International Journal of Production Research, 55*(7), 1970–1991.

Sawik, T. (2018). Two-period vs. multi-period model for supply chain disruption management. *International Journal of Production Research, 57*(14), 4502–4518.

Sawik, T. (2022). Stochastic optimization of supply chain resilience under ripple effect: A COVID-19 pandemic related study. *Omega, 109*, 102596.

Sawik, T., & Sawik, B. (2024). Risk-averse decision-making to maintain supply chain viability under propagated disruptions. *International Journal of Production Research, 62*(8), 2853–2867. https://doi.org/10.1080/00207543.2023.2236726

Scheibe, K. P., & Blackhurst, J. (2018). Supply chain disruption propagation: A systemic risk and normal accident theory perspective. *International Journal of Production Research, 56*(1–2), 43–59.

Schmitt, A. J., & Singh, M. (2012). A quantitative analysis of disruption risk in a multi-echelon supply chain. *International Journal of Production Economics, 139*, 22–32.

Schmitt, T. G., Kumar, S., Stecke, K. E., Glover, F. W., & Ehlen, M. A. (2017). Mitigating disruptions in a multi-echelon supply chain using adaptive ordering. *Omega, 68*, 185–198.

Shao, X. F., & Dong, M. (2012). Supply disruption and reactive strategies in an assemble-to-order supply chain with time-sensitive demand. *IEEE Transactions on Engineering Management, 59*(2), 201–212.

Sheffi, Y., & Rice, J. B. (2005). A supply chain view of the resilient enterprise. *MIT Sloan Management Review*.

Simangunsong, E., Hendry, L. C., & Stevenson, M. (2012). Supply-chain uncertainty: A review and theoretical foundation for future research. *International Journal of Production Research, 50*(16), 4493–4523.

Simchi-Levi, D., Schmidt, W., Wei, Y., Zhang, P. Y., Combs, K., Ge, Y., et al. (2015). Identifying risks and mitigating disruptions in the automotive supply chain. *Interfaces, 45*(5), 375–390.

Singh, S., Kumar, R., Panchal, R., & Tiwari, M. K. (2021). Impact of COVID-19 on logistics systems and disruptions in food supply chain. *International Journal of Production Research, 59*(7), 1993–2008.

Snyder, L. V., Atan, Z., Peng, P., Rong, Y., Schmitt, A. J., & Sinsoysal, B. (2016). OR/MS models for supply chain disruptions: A review. *IIE Transactions, 48*(2), 89–109.

Snyder, L. V., & Daskin, M. S. (2005). Reliability models for facility location: The expected failure cost case. *Transportation Science, 39*, 400–416.

Sokolov, B., Ivanov, D., Dolgui, A., & Pavlov, A. (2016). Structural quantification of the ripple effect in the supply chain. *International Journal of Production Research, 54*(1), 152–169.

Song, J.-S., & Zipkin, P. (2009). Inventories with multiple supply sources and networks of queues with overflow bypasses. *Management Science, 55*(3), 362–372.

Spiegler, V. L. M., Potter, A. T., Naim, M. M., & Towill, D. R. (2016). The value of nonlinear control theory in investigating the underlying dynamics and resilience of a grocery supply chain. *International Journal of Production Research, 54*(1), 265–286.

Tang, C. S. (2006). Perspectives in supply chain risk management. *International Journal of Production Economics, 103*, 451–488.

Tomlin, B. (2006). On the value of mitigation and contingency strategies for managing supply chain disruption risks. *Management Science, 52*(5), 639–657.

Trucco, P., Petrenj, B., & Birkie, S. E. (2017). Assessing supply chain resilience upon critical infrastructure disruptions: A multilevel simulation modelling approach. In Y. Khojasteh (Ed.), *Supply chain risk management* (pp. 311–334). Springer.

Tukamuhabwa, B. R., Stevenson, M., Busby, J., & Zorzini, M. (2015). Supply chain resilience: Definition, review and theoretical foundations for further study. *International Journal of Production Research, 53*(18), 5592–5623.

Wagner, H. M., & Whitin, T. (1958). Dynamic version of the economic lot size model. *Management Science, 5*, 89–96.

Yang, Z., Aydin, G., Babich, V., & Beil, D. (2009). Supply disruptions, asymmetric information, and a backup production option. *Management Science, 55*(2), 192–209.

Epidemiological Models to Predict Infection Epidemic: A Literature Review

Fatemeh Mirsaeedi, Mohammad Sheikhalishahi, Mehrdad Mohammadi, and Dmitry Ivanov

Abstract This chapter provides an in-depth review of the epidemiological models utilized for predicting and managing epidemic infectious diseases, with a particular emphasis on Susceptible-Infected-Recovered (SIR) models and their various facets. The objectives of this review are threefold: (1) the assumptions, equations, and methodologies for estimating critical parameters within these models are evaluated; (2) the challenges associated with implementing disease control interventions, including both medical and non-medical strategies, and their integration into the models are explored; and (3) relation of SIR models and optimization models are investigated. Additionally, a systematic review at the micro-level has identified significant research gaps within the existing literature, leading to recommendations for future research directions. One of the most important findings is that many assumptions need to be reviewed to be better aligned with real conditions, which should hopefully stimulate some useful inquiries into how to model epidemic diseases.

F. Mirsaeedi · M. Sheikhalishahi (✉)
School of Industrial Engineering, College of Engineering, University of Tehran, Tehran, Iran
e-mail: fatemeh.mirsaeedi@ut.ac.ir; m.alishahi@ut.ac.ir

M. Mohammadi
Department of Industrial Engineering and Innovation Sciences, Eindhoven University of Technology, Eindhoven, The Netherlands
e-mail: m.mohammadi1@tue.nl

D. Ivanov
Department of Business and Economics, Berlin School of Economics and Law, Berlin, Germany
e-mail: divanov@hwr-berlin.de

D. Ivanov et al. (eds.), *Handbook of Ripple Effects in the Supply Chain*,
International Series in Operations Research & Management Science 276,
https://doi.org/10.1007/978-3-031-85508-5_12

1 Introduction

Infectious diseases have posed one of the major global challenges to public health for decades. The emergence of diseases like SARS in 2003 (Seto et al., 2003; He et al., 2004), the A/H1N1 influenza epidemic in 2009 (Jones & Salathe, 2009; Khan et al., 2009; Tang et al., 2010), the Ebola virus disease in 2014 (Aylward et al., 2014; Baize et al., 2014), and the recent outbreak of the Coronavirus Disease (COVID-19) in 2019 has profoundly impacted human health, the global economy, and societal behaviors (Huang et al., 2020; Tang et al., 2020a, 2020b; Chintalapudi et al., 2020).

One effective strategy for mitigating the spread of infectious diseases is to reduce the number of infectious contacts. Since most infections are transmitted through interaction with an infected individual, if the infection occurs via airborne particles, then strategies like self-isolation and wearing face masks can considerably reduce the risk of infection (Deng et al., 2021). The prompt and effective implementation of medical interventions, such as vaccinations, along with non-medical measures like wearing face masks and enforcing quarantine, plays a crucial role in eradicating diseases and reducing associated risks. For this purpose, three main programs must be implemented during a disease outbreak: (1) reducing the contact rate between infected and susceptible individuals, (2) vaccinating susceptible populations, and (3) providing timely treatment to infected individuals (Cui et al., 2008a, 2008b; Zhou & Fan, 2012; Tchuenche et al., 2011; Wang & Xiao, 2013; Huang & Li, 2019). Therefore, implementing both medical and non-medical interventions to reduce the possibility of infection—such as social distancing to reduce contact rates and vaccination to build immunity—is particularly important. A powerful tool is needed to monitor disease progression and evaluate the effectiveness of interventions aimed at controlling the outbreak.

Various models have been developed to assess and manage infectious diseases. These models address significant aspects such as emergency planning, policy-making, risk assessment, promoting and improving various health-economic factors, and defining control programs (Al-Sheikh, 2013). Additionally, numerous models are available for forecasting the transmission of infectious diseases within a community. Alongside the widely recognized SIR-type Ordinary Differential Equation (ODE) models (Anderson & May, 1979; Kermack & McKendrick, 1927), there are integral-differential models (Kermack & McKendrick, 1927; Keimer & Pflug, 2020), Bayesian Monte Carlo approaches (Stojanovic et al., 2019), and agent-based models (Kim et al., 2018; Bicher et al., 2020; Ng, 2020). The lack of appropriate decision-making support tools, particularly in the early stages of an epidemic, is emphasized in the literature (McBryde et al., 2020). In the case of the COVID-19 pandemic, SIR models were instrumental in improving the understanding of the virus's spread, surpassing their use in earlier public health crises (Eubank et al., 2020; Holmdahl & Buckee, 2020; Rhodes et al., 2020). These models help forecast the progression of epidemics, assess the impact of medical and non-medical interventions, and predict potential side effects (McBryde et al., 2020). SIR models can be highly effective in predicting disease outbreaks by simulating the implementation of interventions and accounting for social, economic, and political challenges and obstacles. These include social barriers, such as vaccine hesitancy and anti-vaccination groups, which can significantly hinder epidemic management and reduce overall vaccination effectiveness. Economic challenges involve the costs of interventions and maintaining economic stability during an epidemic. Furthermore, political challenges relate to issues of politicization in many countries regarding vaccination efforts.

Since current review articles related to SIR models in infectious diseases tend to focus only on specific aspects, and a comprehensive review is lacking, the present work aims to provide an extensive and detailed review of the use of SIR models in disease control. This article will discuss the modeling process in detail, including assumptions, equations, and parameter estimation for each model by categorizing them. Other key aspects of SIR models, such as the reproduction number, control measures, challenges in implementing interventions, and optimization strategies, will also be covered to provide a holistic overview. Additionally, this study will identify existing research gaps and potential challenges that could help emerging researchers model epidemics accurately and apply these models for trend forecasting and managing future outbreaks.

The paper is structured as follows: Section 2 outlines the search strategy used to gather relevant information. Section 3 assesses epidemiological models based on their assumptions, equations, and parameters. Assumptions are grouped into population, disease, and intervention factors, providing the necessary background on various factors that must be considered when modeling the course of epidemics. Section 4 describes the reproduction number as one of the fundamental parameters in epidemiological models. Sections 5 and 6 discuss the implementation of control interventions and the related challenges, respectively. Section 7 examines the relationship between these models and optimization issues, emphasizing the importance of optimizing interventions and strategies for epidemic control. Section 8 explores the application of methodological models in the supply chain. Finally, Sect. 9 summarizes the key findings, and Sect. 10 highlights areas that require further investigation and provides recommendations for future studies. The main sections are listed in Fig. 1, which provides an overview of the article's structure.

2 Review Methodology

This section discusses the search methodology and the contributions of this study.

2.1 Search Methodology

The relevant literature search was conducted by targeting a combination of terms in the titles, abstracts, and keywords: ("Medical intervention" OR "Non-Medical

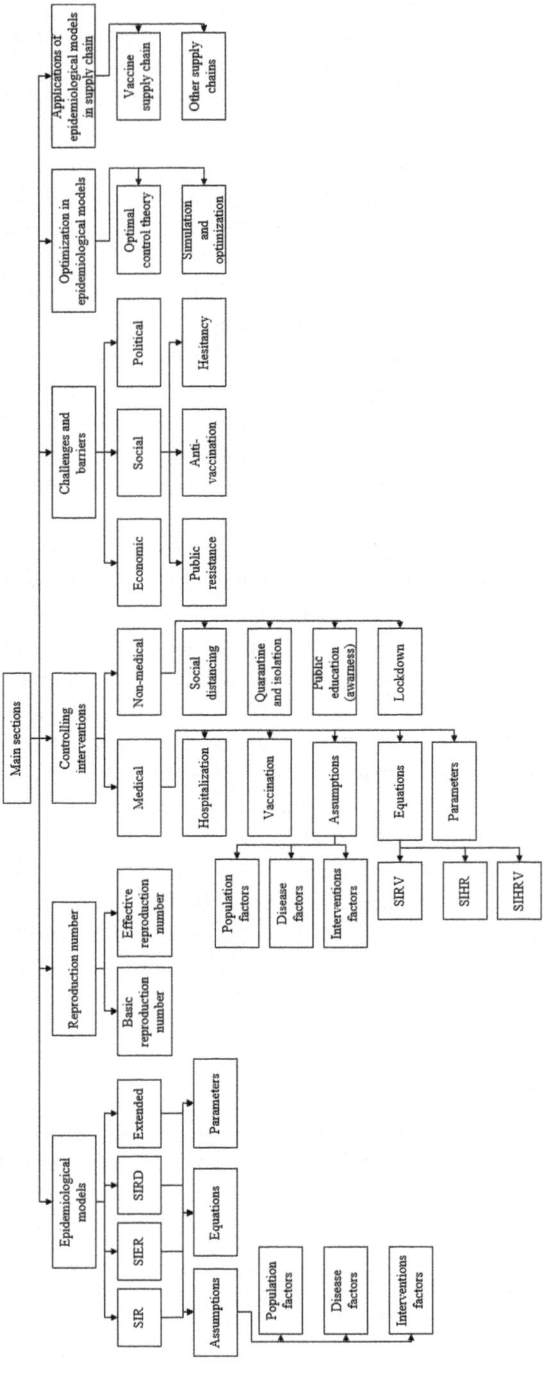

Fig. 1 Overview of the structure of the article

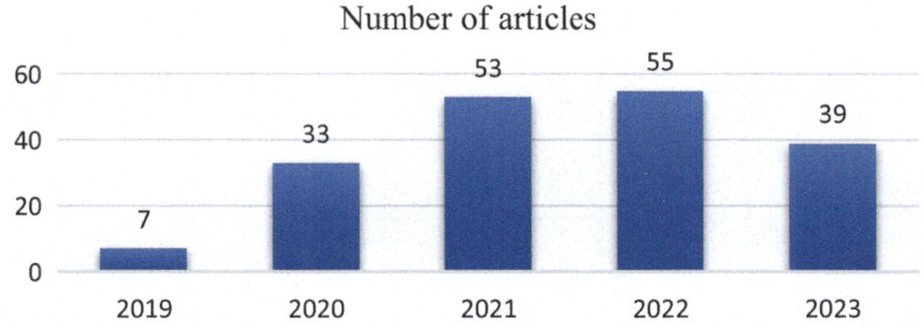

Fig. 2 Number of reviewed studies in this paper

intervention" OR "Vaccine") AND ("Epidemiological Model" OR "SIR Model" OR "SEIR model") for articles published from 2019 through July 2023. This time frame was chosen because the outbreak of COVID-19 in 2019 significantly increased research interest in this subject. The review was performed using the Google Scholar and SCOPUS search engines, along with major databases such as ScienceDirect and PubMed. The search yielded a total of 877 documents that matched the selected keywords within the specified time frame. After removing duplicates, the remaining documents were subjected to full-text screening to filter out irrelevant papers, including those focused on psychological issues, animal infectious diseases, childhood diseases, and sexually transmitted infections, among others. Additionally, studies on SIR models in fractional, spatial, and network forms were excluded to maintain cohesion and relevance among the reviewed documents. During the full-text screening, the reference lists of the selected studies were also reviewed to ensure the relevance of the chosen documents. The final number of documents included for review is 187.

Figure 2 shows the number of articles published each year surveyed in this study. Given that the COVID-19 virus spread at the end of 2019, few articles were published during that year. Due to the spread of the disease and greater attention by researchers to its control in 2021 and 2022, the published articles reached their maximum.

2.2 *Contributions of This Article*

Several review articles have been published in the field of epidemic control. Table 1 compares these articles with the current research.

Mutalik (2017) identified various mathematical models used to predict H1N1 outbreaks and demonstrated their usefulness in estimating the magnitude of these outbreaks. Arefi and Poursadeqiyan (2020) reviewed studies aimed at containing the spread of the coronavirus worldwide. Since no effective vaccine was available at the time of their writing to curb the spread of COVID-19, they discussed the impact of non-medical interventions. Rezapour et al. (2021) provided an extensive review of

Table 1 Comparison of reviewed studies with this research

Reference	Modeling view	Type of view	Year		NMIC	Challenges			OM	RN	Type of models					
			From	To		E	S	P			SIR	SEIR	SIRD	SIRV	SIQR	Extended models
Mutalik (2017)	+	Ma	2004	2015	+	–	–	–	–	±	+	+	–	–	±	±
Arefi and Poursadeqiyan (2020)	–	Ma	2019	2020	+	–	–	–	–	–	–	–	–	–	–	–
Rezapour et al. (2021)	–	Ma	2019	2020	+	+	–	–	–	–	+	+	–	–	–	±
Callegari and Feder (2022)	–	Ma	1970	2021	–	+	–	–	–	–	–	–	–	–	–	–
Current research	+	Mi	2019	2023	+	+	+	+	+	+	+	+	+	+	+	+

+: Satisfied, –: Not satisfied, ±: Partially satisfied, Ma: Macro level, Mi: Micro level, NMIC: Non-Medical Intervention Classification, E: Economic, S: Social, P: Political, OM: Optimization Methods, RN: Reproduction Number economic evaluations related to preventive strategies, programs, and treatments for COVID-19, dividing the studies into three categories: (1) cost-effectiveness studies, (2) cost-utility studies, and (3) cost-benefit studies. Additionally, Callegari and Feder (2022) examined the pandemic's long-term effects on economic development, incorporating financial and historical debates.

Compartmental models, such as SIR models, are commonly used for predicting the spread of diseases and guiding policymakers (Hethcote, 2000; Rock et al., 2014; Barrat et al., 2008; Pastor-Satorra et al., 2015). Therefore, it is essential to understand the various aspects and details of these models. In this context, the present study makes several important contributions to the existing body of literature. First, it reviews articles from a modeling perspective, covering assumptions, equations, and parameters, aspects that were only briefly mentioned by Mutalik (2017), whose study provided a limited examination of models. Other review articles did not specifically focus on compartmental models. Although studies by Tang et al. (2020a, 2020b), Daghiri et al. (2022), and Rahman et al. (2021) addressed these models in distinct ways, they lacked a consistent structure in their reviews, evident in their omission of critical components such as search strategies and article classification.

One key method for controlling epidemics and evaluating interventions is by calculating the reproduction number, a factor mentioned only in Mutalik (2017). Therefore, the second contribution of this study is its detailed consideration of the reproduction number and its calculation. Despite significant advances in global pandemic preparedness, substantial gaps and challenges remain (Madhav et al., 2017). While some review articles have explored the economic challenges of controlling epidemic diseases, this article uniquely examines social dimensions, such as vaccine hesitancy and anti-vaccination movements, as well as the political challenges discussed in the reviewed literature. This exploration represents the third contribution of the study.

Overall, this article distinguishes itself from other reviews by offering a comprehensive and detailed examination of compartmental SIR models from a modeling perspective. It also provides an in-depth discussion of control interventions, related challenges, and additional modeling considerations, such as optimization strategies.

3 Epidemiological Model

Different approaches to predicting epidemic evolution can be applied to various situations, which is why researchers have introduced different categories of epidemiological models. Zhou and Ji (2020) divided epidemiological models into three main groups: curve-fitting models, compartment models, and agent-based models. Similarly, Zhang et al. (2022a, 2022b) classified epidemiological models into three categories: agent-based epidemiological models, Cellular Automata-based (CA) epidemiological models, and compartment models. Iranzo and Pérez González (2021) also highlighted the widespread use of two-compartment models and agent-based models in the study of epidemic diseases. In addition to compartmental models (Arfan et al., 2021), agent-based approaches (Kai et al., 2020), cellular automaton models (White et al., 2007; Bouaine & Rachik, 2018), and curve-fitting models (Adiga et al., 2021), the literature includes a wide array of effective methods for accurately estimating disease propagation. These methods include artificial intelligence techniques such as machine learning (Kushwaha et al., 2020), time series analysis (Betcheva et al., 2021; Qi et al., 2020), metrological and meta-population models (Ma et al., 2021), and deep learning (Yang et al., 2020).

From a technical perspective, aside from compartmental epidemiological models, many other methods rely heavily on data availability, which can make them impractical during the early stages of an epidemic due to a lack of reliable data (Gnanvi et al., 2021). Furthermore, the use of historical data is often limited due to the emergence of novel viruses and the potential for mutations over the course of an epidemic. Additionally, these methodologies frequently overlook the impact of government interventions, such as quarantine measures and public awareness campaigns (Biswas & Alfandari, 2022; Nikolopoulos et al., 2021). Ultimately, these approaches are insufficient for evaluating and optimizing actions related to the implementation of government policies, such as efforts to reduce mortality rates or minimize the economic costs associated with controlling an epidemic's spread (Vahdani et al., 2023).

In contrast, compartmental models are characterized by their minimal data requirements for implementation, which often makes them a preferred choice. This advantage is particularly evident in situations involving sudden outbreaks with high mortality rates, where delays in decision-making can lead to unacceptable increases in fatalities. In such cases, compartmental models prove valuable due to their previous applications in various epidemics and their relatively modest data needs for calibration compared to other modeling approaches. However, obtaining accurate information about parameters at the onset of an unfamiliar disease can be challenging, which affects not only compartmental models but also any predictive model or strategy (Iranzo & Pérez González, 2021). Therefore, compartmental models hold significant importance. Like other modeling methodologies, compartmental models are based on a set of assumptions that either approximate real-world scenarios or simplify the modeling process. Various categories of compartmental models arise from these assumptions. Notable distinctions include Markovian/non-Markovian (Basnarkov et al., 2022), discrete (Zhang et al., 2023)/continuous (Sharov, 2020), and deterministic (Safan, 2019)/stochastic (Kain et al., 2021; Bogdanov, 2023) approaches. This classification allows models to be tailored to the specific conditions of the problem at hand.

Among the many compartmental models, the SIR model stands out as one of the simplest yet most foundational frameworks, from which numerous derivative models have evolved. Initially introduced by Kermack and McKendrick in 1927, the SIR model effectively captures the dynamics of epidemic diseases. The final model depends on the assumptions made during problem formulation (Libotte et al., 2020). According to Keeling and Rohani (2007) and Hethcote (2000), compartmental models can be expanded by:

- Adding new compartments to represent vaccinations or any stage of disease progression;
- Incorporating time-dependent parameters that reflect seasonal effects on the disease;
- Grouping the target population based on factors such as age, health status, and other relevant demographics;
- Considering human demographic factors, as the timescale of human demographics often aligns with disease dynamics. This can include assuming constant rates of immigration, as well as birth and death rates per capita, alongside density-dependent or disease-related mortality rates.

Thus, an epidemiological model can be extended by adding compartments and considering different assumptions. Each category of these models is reviewed in its respective section. In each subsection, the properties of the models—such as assumptions, model structure (components and their relationships), equations, and essential parameters—are analyzed. In the following sections, additional properties of the models are discussed. Table 2 lists the main and detailed features of the reviewed studies and proposed models. Some studies have incorporated varying population characteristics, such as birth and mortality rates. For instance, some models accounted for newborns, dividing them into two categories: susceptible and recovered. If newborns are resistant or vaccinated immediately, they are placed in the recovered category (Chang et al., 2020). Otherwise, they fall into the susceptible compartment (Batistela et al., 2021). The transition rate may vary if rates differ between groups, time periods, or regions (Storlie et al., 2021a; Dias et al., 2022; Hazard-Valdés & Montero, 2023), but it is otherwise assumed to be constant. Therefore, the variable rate of transmission does not solely relate to the implementation of interventions or changes in transmission rates.

The assumption of multiple regions refers to the existence of several areas where people interact with one another. Mobility between these regions is possible and can be influenced by factors such as quarantine measures or travel restrictions. Additionally, the efficacy of interventions—such as lockdowns, quarantine protocols, and mask mandates—varies across regions, primarily due to differences in the administrative capabilities of local authorities and the population's compliance (Qundus et al., 2023). Taking into account regional differences and the assumption of non-uniformity makes the model more reflective of real-world scenarios. Duan et al. (2023) suggested that migration rates significantly impact the scale of infections, which is why the assumption of migration has been incorporated into Table 2. It is also worth noting that greater emphasis should be placed on the incubation period and duration of the disease in practical research (Duan et al., 2023). This article also explores various aspects of epidemic modeling and control, including interventions, associated challenges, and optimization strategies.

Table 2 Main properties of reviewed studies

Property	Type	Details	
Components	Modeling of epidemic evolution	Main model	1) SIR
		Main model + Exposed individual	1) SEIR, 2) SEAIR[a]
		Main model (Main model + Exposed individual) + Died individual	1) SIRD, 2) SEIRD, 3) SEAIRD
	Modeling of epidemic evolution with medical interventions	Main model (Main model + Exposed individual) + Vaccination	1) SIRV, 2) SEIRV, 3) SIRVD, 4) SEIRVD, 5) SEAIRV, 6) SEIARVD
		Main model (Main model + Exposed individual) + Hospitalization	1) SIRH, 2) SEIRH, 3) SIRHD, 4) SEIRHD, 5) SEAIRH, 6) SEIARHD
		3) Combination of Vaccination and Hospitalization components	1) SIRHV, 2) SEIRHV, 3) SIRHVD, 4) SEIRHVD, 5) SEAIRHV, 6) SEIARHVD
	Modeling of epidemic evolution with non-medical interventions (Quarantine component)	1) SIQR, 2) SEIQR, 3) SIQRD, 4) SEIQRD, 5) SEAIQR, 6) SEAIQRD	
	Extended model	Combination of two medical and non-medical interventions	1) SIQRV, 2) SIQRVD, 3) SEIQRV, 4) SEIQRVD, 5) SEAIQRV, 6) SEAIQRVD, 7) SIQRH, 8) SIQRHD, 9) SEIQRH, 10) SEIQRHD, 11) SEAIQRH, 12) SEAIQRHD, etc.
		Additional components	1) SECIHUDR, 2) SEIQPRD, etc.
Assumptions	Population factors	Grouping	1) Multi-group, 2) Single-group
		Composition	1) Homogenous, 2) Heterogeneous
		Births and deaths (due to a cause other than infection)	1) Consider, 2) Ignore
			1) Equal, 2) Unequal
		Number of persons	1) Constant, 2) Variable
		Newborns	1) Recovered, 2) Susceptible
		Migration	1) Consider, 2) Ignore

		Multi-group	1) Grouping based on gender, 2) Grouping based on age, 3) Grouping based on other factors such as socioeconomic status, geographical location, comorbidity	
		Number of regions	1) Single region, 2) multi-region	
	Disease factors	Rate of transition	1) Constant, 2) Variable	
		Secondary infection	1) Consider, 2) Ignore	
		Immunity	1) Permanent, 2) Weak/Loss	
			1) Immediate, 2) Delayed	
		Incubation time	1) Consider, 2) Ignore	
		Number of periods	1) Single-period, 2) Multi-period	
		Number of strains	1) Single strain, 2) Multi-strain	
	Intervention factors	Type of intervention	Medical	1) Vaccination, 2) Hospitalization
			No-medical	1) Social distancing, 2) Quarantine and Isolation, 3) Public education (awareness), 4) Lockdown
		Vaccine	Effectiveness	1) Known, 2) Unknown
				1) Effective, 2) Ineffective
			Number of doses	1) Single dose, 2) Multi-dose
			Types of vaccine	1) Single type, 2) Multi-type

(continued)

Table 2 (continued)

Property	Type	Details
Parameter estimation	Methods	
	Important parameters	
Reproduction number	Basic	
	Effective	
Challenges and barriers	Economic	
	Social	1) Hesitancy, 2) Anti-vaccination, 3) Public resistance
	Political	
Optimization in epidemio-logical models	Optimal control theory	
	Simulation and optimization	
Epidemiological models in supply chains	Vaccine supply chain	
	Other supply chains	

[a] Asymptomatic individuals (A) are a group of infectious people (I), so a separate category is not considered for them

3.1 SIR Model

The SIR model is a basic yet essential framework utilized to study infectious diseases (Erkayman et al., 2023). The model is structured based on three compartments:

- *S*: Denotes the number of susceptible individuals infected through contact with infectious individuals.
- *I*: Characterizes the number of infectious individuals capable of transmitting the disease to those within the susceptible category.
- *R*: Denotes the number of recovered individuals who have either overcome the disease or have transitioned into a recovered state.

Some scholars use the SIR model to predict the prevalence of infections and hospitalizations (Storlie et al., 2021b). This includes analyzing the effect of vaccination on the spread of COVID-19 (Storlie et al., 2021a; Ai et al., 2022; Chang et al., 2020), estimating the impact of community intervention measures (Mehmood et al., 2021), and understanding the influence of both vaccination and social distancing on the epidemic's spread (Batistela et al., 2021). Additionally, the model has been instrumental in managing the number of hospitalized individuals to ensure they remain within a defined threshold (Dias et al., 2022) and in evaluating various factors influencing the rate of COVID-19 transmission (Erkayman et al., 2023). Other applications include studying the impact of gender on influenza dynamics, exploring the indirect and direct effects of mass vaccination (Safan, 2019), and conducting surveys on undetected infectives (Melis & Littera, 2021). Undetected infectives, who play a key role in the spread of infection, can be classified into two categories: (1) pre-symptomatic individuals, who will later show symptoms, and (2) asymptomatic or paucisymptomatic individuals, who do not develop noticeable symptoms during their infection.

3.1.1 Assumptions

As mentioned before, the assumptions of the reviewed models are divided based on factors related to population, disease, and intervention. The assumptions of the SIR model in the relevant literature are organized according to various characteristics that are shown in Table 3.

3.1.1.1 Population Factors

The SIR model has some mathematical constraints. For instance, the sum of all groups must always equal the total population size at every stage (Qundus et al., 2023), and the overall population size can influence the dynamics of disease transmission (Erkayman et al., 2023). Incorporating birth and death rates from

Table 3 Classification of studies considering SIR model based on their assumptions

	Population Factors														Disease Factors									Intervention Factors				
		Number of persons		Newborns		Composition		Multi-Grouping			Migration				Rate of transition		Immunity							Class		Vaccine		
Reference	Birth and death rate	C	V	S	R	Ho	He	MA	MG	MO	BG	BR	Multi-region	Social challenges	C	V	P	W	Im	D	Incubation time	Multi-period	Multi-Strains/wave	M	NM	EFV	MD	MT
Libotte et al. (2020)	*														*									Va				
Iranzo and Pérez (2021)		*				*									*		*								G			
Mehmood et al. (2021)		*														*									G			
Storlie et al. (2021a)														H		*							*	Va		*	*	*
Batistela et al. (2021)	E			*													*							Va	SD			
Dias et al. (2022).		*				*		*								*									SD			
Ai et al. (2022)		*				*																	*	Va	G			
Goodkin-Gold et al. (2022)	E				*	*											*							Va	SD	*		
Zhang et al. (2023)																*									G			
Gros and Gros (2022)															*									Hos	G			
Safan (2019)	*	*		*		*		*	*							*								Va				
Hwang et al. (2020)		*													*									Va		*		
Deka et al. (2020)		*				*		*							*			*					*	Va		*		
Arribas et al. (2022)							*	*						H	R	I		*					*	Va		*		
Nepomuceno et al. (2021)	E	*		*											*		*							Va				
O'Regan et al. (2020)	E	*		*	*																			Va				
Deng et al. (2021)	*	*													*					*				Va				
Loertscher and Muir (2021)		*				*		*								*								Va	L			

Nguyen et al. (2022)		*										*	*		*		*					*			QS			
Sharov (2020)		*														*						*			L			
Kozyreff (2021)		*													*								*	Hos	L			
Chang et al. (2020)		*		*	*	*									*									Va				
Marinov and Marinova (2020)		*				*										*	*								G			
Duan et al. (2023)															*										QS			
Vattiatio et al. (2022)						*		*								*		*				*	*	Va-Hos	QS	*	*	
Huang et al. (2021)		*								*			*		*	*	*						*		G			
Alanazi et al. (2020)		*						*							*										G			
Betti et al. (2021)		*													*		*				*			Va	G			*
Erkayman et al. (2023)		*												H	*					*			*	Va		*	*	
Qundus et al. (2023)		*													*		*							Va		*		
Hussien and Mohammad (2022)	*			*	*										*		*							Va				

E: Equal, C: Constant, V: Variable, S: Susceptible, R: Recovered, Ho: Homogenous, He: Heterogeneous, MG: Grouping based on gender, MA: Grouping based on age, MO: Grouping based on other factors such as socioeconomic status, geographical location, comorbidity, BR: Between regions, BG: Between groups, H: Hesitancy, A: Anti-vaccination, PR: Public resistance, P: Permanent, W: Weak/Loss, Im: Immediate, D: Delayed, M: Medical, NM: Non-medical, Va: Vaccination, Hos: Hospitalization, QS: Quarantine and Isolation, SD: Social distancing, PE: Public education, L: Lockdown, EFV: Effectiveness, MD: Multi-dose, MT: Multi-types causes other than the disease itself can make the model more realistic. However, in certain cases, the model may be simplified, or the outbreak may be assumed to occur over a relatively short time frame, leading to the assumption that birth and death rates from natural or non-disease-related causes are negligible. In some instances, birth and death rates are assumed to be equal, effectively canceling each other out and resulting in a stable population size, both overall and within specific subgroups.

When births are factored into the model, it is important to determine the conditions for newborns and decide which compartment they will be placed in. This decision depends on the characteristics of the disease. For example, in some cases where the virus type is well known, newborns may receive vaccinations immediately after birth and then be assigned to the recovered compartment (Chang et al., 2020). On the other hand, during the COVID-19 pandemic, due to age restrictions on vaccinations, newborns were placed in the susceptible compartment (Batistela et al., 2021).

The population under study can be viewed as a cohesive group or segmented based on demographic factors, geographical characteristics, infection risk, and social conditions. A thoroughly mixed population can be described as a structured population in which, at any given time, each individual can interact with other members of the population (Marinov & Marinova, 2020). This classification is referred to as homogeneity or heterogeneity. Individuals within the population, whether considered as a whole or in subgroups, may be homogeneous or heterogeneous. In the literature reviewed in this section, the inherent simplicity and foundational nature of the SIR model often lead to the assumption of homogeneity. For instance, it is typically assumed that individuals across various age categories spend their time in different locations: children are primarily at school, working-age individuals are mostly found in offices or other workplaces, and retired individuals tend to stay at home (Alanazi et al., 2020). Additionally, the classification of people by region may reflect their homogeneity in terms of the intensity and frequency of epidemics (Huang et al., 2021). When subgroups are determined based on gender, assumptions regarding the gender of newborns should also be considered. For example, in research conducted by Safan (2019), it is assumed that the total number of newborns is equal for both boys and girls, thereby ensuring that the population remains constant within each gender subgroup.

3.1.1.2 Disease Factors

Assumptions regarding the conditions of each compartment and the processes of change within it—such as transmission rate, recovery rate, probability of infection, and type of immunity—can vary across different models. Infection and recovery rates, as essential parameters of epidemiological models, may be classified as known or unknown (Marinov & Marinova, 2020) and can be either constant or variable (Iranzo & Pérez González, 2021). For instance, in Huang et al. (2021), the transmission rate is variable while other rates are held constant. In contrast,

Betti et al. (2021) assume that the population's testing/reporting rate is relatively constant. When the testing/reporting rate is variable, it may be characterized by a statistical distribution (Storlie et al., 2021a) or may fluctuate over time (Zhang et al., 2023; Mehmood et al., 2021). Additionally, these rates can be influenced by compliance with pharmaceutical and non-pharmacological interventions and may vary among individuals based on factors such as age, underlying health conditions, and geographical location (Prem et al., 2020). When infection and recovery rates differ among individuals, society can be divided into groups, with each group assigned a rate that reflects its specific characteristics (Dias et al., 2022). Furthermore, these rates can be calculated based on effective parameters. For example, Dias et al. (2022) calculated the transmission rate based on the number of contacts per unit of time and the probability that such contact with a susceptible individual will lead to transmission.

Another influential assumption in epidemiological models is the possibility of re-infection or secondary infection. Re-infection occurs when an individual contracts a disease for a second time, while a secondary infection develops during or after treatment for another infection, potentially due to the initial treatment or changes in the immune system. Notably, secondary infections are often overlooked in epidemiological models. They are mentioned only by Ai et al. (2022), and in some instances, it is assumed that co-infection or superinfection does not occur (Betti et al., 2021). Regarding re-infection, some studies have asserted that it is not possible for individuals to return from the recovery compartment to the infection compartment (Goodkin-Gold et al., 2022), either due to acquired permanent immunity (Batistela et al., 2021; Iranzo & Pérez González, 2021) or because the immunity lasts longer than the duration of the outbreak (Betti et al., 2021).

Nevertheless, the probability of re-infection is a crucial assumption in modeling, influenced by the emergence of new virus variants and the potential weakening or loss of immunity. New viruses typically arise due to mutations, which can lead to subsequent outbreaks. There have been reports of individuals who have recovered from COVID-19 being re-infected. Marinov and Marinova (2020) included a folded arrow from the recovered class (R) back to the infected class (I). However, they noted that due to the currently low incidence of such cases, it remains challenging to accurately estimate the re-infection rate, leading them to exclude this option from their model. Khalifi and Britton (2023) represented the Susceptible-Infectious-Recovered-Susceptible (SIRS) model, which accommodates the gradual waning of immunity.

The duration of the epidemic and the timing of transitions between states are also significant considerations. In some cases, the entire course of the epidemic may be regarded as short or segmented into smaller intervals to develop a multi-period model. Additionally, the incubation period—an important time interval—may be treated as negligible, suggesting that an individual becomes infectious immediately upon infection (Iranzo & Pérez González, 2021). The level of disease risk, including fatality and non-fatality rates, is another assumption that can impact mortality rates. In some models, this assumption does not influence the calculations. Iranzo and

Pérez González (2021) have assumed that although infectious diseases may be fatal, the recovery rate does not differentiate between individuals who have died and those who successfully recovered. Furthermore, the severity of the disease can be incorporated into the model; for example, Betti et al. (2021) assumed that the probability of a case being mild versus severe is constant, with all severe infections reported while only a fraction of mild infections is documented.

SIR models and their variants exhibit several notable limitations. A primary restriction is their failure to incorporate mobility and displacement, which prevents them from directly considering quarantine strategies and mobility restrictions as measures. Anderson and May (1979) assumed that individuals are uniformly distributed and behave homogeneously. Accurately estimating the number of infected individuals is also challenging due to varying transmission patterns. Symptomatic cases are not treated as unsupervised, and numerous random factors related to disease transmission exist, including the likelihood of infection from contact with contaminated surfaces (Alanazi et al., 2020).

3.1.1.3 Intervention Factors

The foremost assumption regarding interventions is whether medical and non-medical strategies are implemented. At the onset of unknown epidemics, governments typically enforce non-medical interventions due to the lack of vaccines. Alongside these measures, medical interventions such as the use of masks and disinfectants are also employed. Evidence suggests that these interventions can effectively mitigate epidemic peaks (Goodkin-Gold et al., 2022), as their intensity is designed to keep the epidemic's severity below a specified threshold (Dias et al., 2022). In models related to the early stages of an outbreak, it is generally assumed that no vaccine is available (Dias et al., 2022), and non-medical interventions are in effect, including the isolation of detected infected individuals in hospitals or at home (Melis & Littera, 2021). However, in scenarios where a vaccine is available, critical parameters such as the vaccine's effectiveness, safety, required dosage, and target populations for vaccination must be defined. For example, Goodkin-Gold et al. (2022) assumed that vaccinations are administered to newborns, suggesting that if the initial vaccine dose is inadequate, subsequent doses may also be ineffective. Erkayman et al. (2023) outlined several assumptions regarding the effects of the vaccine:

1. Vaccinated individuals are less likely to become infected upon contact with an infectious person.
2. Vaccinated individuals experience a lower mortality rate compared to unvaccinated individuals when infected.
3. Vaccinated individuals transmit the virus to others at a reduced rate compared to transmission between two unvaccinated individuals.

Thus, vaccination affects the model's rates, and this influence can be integrated into the model's assumptions. Furthermore, regarding vaccine risks and efficacy, it

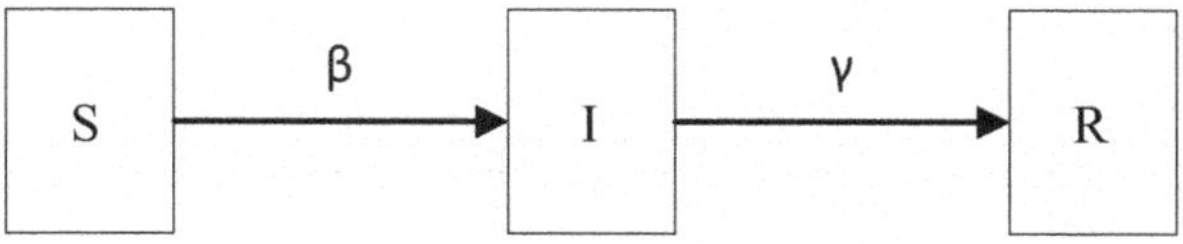

Fig. 3 SIR model

can be posited that public understanding of vaccine-associated risks may diminish exponentially as vaccine coverage increases, while the perceived benefits of vaccination may rise linearly with increasing mortality rates (Deka et al., 2020).

For simplicity, some models assume that the vaccine is entirely effective in preventing infection (Arribas et al., 2022). Additionally, the level of vaccination coverage necessary to impede disease transmission is a critical aspect addressed in epidemiological models (Attanayake, 2022). Various assumptions can be made about other vaccination characteristics, such as the eligible population, vaccination rate, and cost. For instance, it may be assumed that a constant proportion of the population is vaccinated within a specified timeframe (Arribas et al., 2022), and that vaccination is available to both susceptible and recovered individuals. Furthermore, many epidemic models often exclude the cost of vaccination, operating under the assumption that the societal benefits of vaccination outweigh its costs, given the substantial damage inflicted by the disease (Goodkin-Gold et al., 2022).

3.1.2 Equations

The SIR model, which does not account for vital dynamics such as birth and death (also referred to as demographic factors), can be described by the following set of ordinary differential equations (1–3) and illustrated in Fig. 3. In this context, β and γ represent the transmission and recovery rates, respectively.

$$\frac{\mathrm{dS}}{\mathrm{dt}} = -\frac{\beta SI}{N} \tag{1}$$

$$\frac{\mathrm{dI}}{\mathrm{dt}} = \frac{\beta SI}{N} - \gamma I \tag{2}$$

$$\frac{\mathrm{dR}}{\mathrm{dt}} = \gamma I \tag{3}$$

In this subsection, we will analyze the relevant equations, highlighting the most significant ones referenced in various articles and explaining how the effects of interventions can be integrated into these equations. Table 4 summarizes the application of various factors and assumptions in the SIR model. The impact of

Table 4 Equations of modified SIR models

Factor	Reference	Equations		
Birth and death	O'Regan et al. (2020) Nepomuceno et al. (2021)	$\frac{ds'}{dt} = \frac{ds}{dt} + \theta N - \mu S$	$\frac{dI'}{dt} = \frac{dI}{dt} - \mu I$	$\frac{dR'}{dt} = \frac{dR}{dt} - \mu R$
Mobility	Ai et al. (2022)	$\frac{ds'}{dt} = \rho \frac{ds}{dt}$	$\frac{dI'}{dt} = -\rho \frac{ds}{dt} - \frac{dR}{dt}$	
Time periods instead of rates	Kozyreff (2021)	$\frac{dI'}{dt} = -\frac{dS}{dt} - \frac{I}{t_R}$	$\frac{dR'}{dt} = \frac{I}{t_R}$	
Vaccination				
Without known effectiveness	Batistela et al. (2021)	$\frac{ds'}{dt} = -\omega S + \frac{ds}{dt}$	$\frac{dR'}{dt} = \omega S + \frac{dR}{dt}$	
With known effectiveness	Ai et al. (2022) Hwang et al. (2020)	$\frac{ds'}{dt} = -\omega \varepsilon S + \frac{ds}{dt}$	$\frac{dR'}{dt} = \omega \varepsilon S + \frac{dR}{dt}$	
Newborn	O'Regan et al. (2020) Nepomuceno et al. (2021)	$\frac{ds'}{dt} = \theta N (1 - \omega) \frac{ds}{dt}$	$\frac{dR'}{dt} = \frac{dR}{dt} + \omega \theta N$	
The impact of the use of medicine	Qundus et al. (2023)	$\frac{ds'}{dt} = \frac{ds}{dt} - \varepsilon v$	$\frac{dI'}{dt} = \frac{dI}{dt} - CM$	

θ: Birth rate, μ: Death rate, ρ: The strength of personnel mobility, t_R: Duration of recovery, ω: vaccination rate, ε: The effective rate of the vaccine, M: The number of immune individuals, C: The effective rate of the medication

these factors on the equations is compared with Eqs. (1)–(3). When the vaccine's effectiveness is known, it is multiplied by the vaccination rate. For instance, Arribas et al. (2022) developed an age-structured SIR model utilizing age-contact matrices estimated by Mistry et al. (2021), which were adjusted to reflect the population structure of 2019 (Arregui et al., 2018).

Zhang et al. (2023) used a discrete-time SIR model by the equation $I_t - I_{t-1} = (\lambda_t - \mu_t)\Delta t\, I_{t-1}$. This equation indicates that each infected individual can infect $(\lambda_t - \mu_t)$ individuals on day t. The variable ϕ_t is defined as:

$$\phi_t = (\lambda_t - \mu_t) \tag{4}$$

This variable, termed the effective spreading rate on day t, reflects both the infection's intensity and the control policy's effectiveness. A lower value of ϕ_t indicates a stricter control policy.

Table 5 Estimation of transmission rate

Effective factor	Equation	Reference
Original	$a * b$	Erkayman et al. (2023)
Different waves and interventions	$\beta(t) = \begin{cases} \beta_1 t^{-\sigma_1} & t_0 < t \le t_1 \\ \gamma & t_1 < t \le t_2 \\ \beta_2 t^{-\sigma_2} & t_2 < t \end{cases}$	Huang et al. (2021)
Temperature	M [1 − c_1 (1 − expo(−kt))ms] + ξ(t) $m[1 - c_1(1 - \exp(-kt))ms] + \xi(t)$	Ai et al. (2022)

a: Average number of individuals in contact with each person per unit time, b: The possibility of transmitting the disease, $\beta_1(\beta_2)$: Transmission rate for the first (second) outbreaks, σ: Intervention parameter reflecting the force of intervention in the epidemic control, m, c_1, k,and ms: Influence of temperature on the transmission rate, ξ: The Gaussian white noise with the intensity of $\sqrt{2D}$, which can be understood as the difference of transmission rate among different individuals

3.1.3 Parameters

Several parameters are essential in modeling infectious diseases, including the transmission rate (β) and the mean spontaneous recovery time (t_R). However, from a crisis management perspective, predicting the expected number of hospital beds required and the duration of their occupancy is more critical (Zhou et al., 2020; Li et al., 2020; Chowdhury et al., 2020; Chen et al., 2020a, 2020b, 2020c; Wu et al., 2020; Richardson et al., 2020).

In addition to using actual data (Kozyreff, 2021), various approaches for estimating model parameters exist. These include the inverse modeling problem (Marinov & Marinova, 2020; Libotte et al., 2020), the brute force approach (Ahmetolan et al., 2020), the Particle Swarm Optimization (PSO) algorithm (Ai et al., 2022), and the least squares estimation technique (Ji et al., 2021; Husniah et al., 2021; Ai et al., 2022). Solving the inverse problem is crucial for model calibration and parameter control. Inverse problem methodologies can effectively aid in various critical processes, such as tracking the spread of infectious diseases and enabling epidemiologists and public health professionals to forecast epidemics (Bauch et al., 2009). It is also important to note that the transmission rate may vary by age (Dias et al., 2022; Arribas et al., 2022) or gender subgroup (Safan, 2019). Table 5 provides several examples of estimating the transmission rate.

3.2 *SEIR Model*

The SEIR model extends the SIR framework by introducing a fourth compartment known as "exposed." Individuals in this compartment are infected but are not yet capable of transmitting the infection. This aspect is particularly significant for many infections, as there is often a considerable incubation period. In their comparison of the SIR and SEIR models, Roda et al. (2020) demonstrated that the SIR model tends to provide more relevant information for validated data. However, the SEIR model is favored by some researchers due to its incorporation of the incubation period.

3.2.1 Assumptions

Similar to the SIR model, this section categorizes studies on the SEIR model according to their assumptions, as outlined in Table 6. Brett et al. (2020) presented five SEIR models based on various assumptions, including the effects of seasonality, time, and model structure.

3.2.1.1 Population Factors

Due to the similarities between the SEIR and SIR models, mortality rates are typically not incorporated, as noted in studies like Zhang et al. (2022a, 2022b). While Aldila et al. (2020) assume homogeneity, Lai et al. (2021) consider heterogeneity between different countries, reflecting varying intensities and impacts of interventions. They emphasize that the intensity of infection varies across age groups, continents, income levels, and seasons. Therefore, recognizing the influence of socioeconomic disparities and population heterogeneity on epidemic dynamics is crucial.

3.2.1.2 Disease Factors

Earlier studies of the SEIR model (Li & Muldowney, 1995; Biswas et al., 2014) assumed that viruses could infect exposed individuals, but these individuals were not infectious. However, it has been observed that individuals infected with the COVID-19 virus may transmit the infection during the incubation period (Yang et al., 2020; Chen et al., 2020a, 2020b, 2020c). Consequently, various COVID-19 epidemiological models now assume that exposed individuals can potentially infect susceptible ones (Chen et al., 2020a, 2020b, 2020c; Hou et al., 2020). Furthermore, Zhang et al. (2022a, 2022b) suggested that exposed individuals can infect susceptible individuals before exhibiting symptoms. Various factors influence disease rates, such as the impact of temperature on transmission (Liu et al., 2021a, 2021b) and the effect of hospitalization on recovery rates. Individuals in hospitals may recover more quickly than those who do not receive medical treatment, although the additional recovery rate for an infected person is constrained by available hospital resources (Aldila et al., 2020). Demongeot et al. (2022) observed that the transmission rate varies over time while other rates remain constant.

In terms of recovery and mortality rates, studies generally assume that recovered individuals possess long-term immunity after developing antibodies, thereby not becoming susceptible again (Zhang et al., 2022a, 2022b). Lai et al. (2021) did not

Table 6 Classification of studies considering the SEIR model based on their assumptions

	Population factors														Disease factors									Intervention factors				
		Number of persons		Newborns		Composition		Multi-Grouping			Migration		Multi-region	Social challenges	Rate of transition		Immunity				Incubation time	Multi-period	Multi-Strains/wave	Class		Vaccine		
Reference	Birth and death rate	C	V	S	R	Ho	He	MA	MG	MO	BG	BR			C	V	P	W	Im	D				M	NM	EFV	MD	MT
Lai et al. (2021)		*					*		*						*						*				SD-L			
Feng et al. (2020)		*		*	*				*							*					*			Va		*		
Aldila et al. (2020)		*				*									*	*		*			*			Hos-RT	QS			
Zhang et al. (2022a, 2022b)												*	*			*	*				*			H	QS			
Caldwell et al. (2021)																		*			*		*	Va		*	*	
Demongeot et al. (2022)															*	*	*				*			Va		*	*	
Pey et al. (2022)		*											*			*		*		*	*		*	Va	QS	*	*	

E: Equal, C: Constant, V: Variable, S: Susceptible, R: Recovered, Ho: Homogenous, He: Heterogeneous, MG: Grouping based on gender, MA: Grouping based on age, MO: Grouping based on other factors such as socioeconomic status, geographical location, comorbidity, BR: Between regions, BG: Between groups, H: Hesitancy, A: Anti-vaccination, PR: Public resistance,, P: Permanent, W: Weak/Loss, Im: Immediate, D: Delayed, M: Medical, NM: Non-medical, Va: Vaccination, H: Hospitalization, QS: Quarantine and Isolation , SD: Social distancing, PE: Public education, L: Lockdown, EFV: Effectiveness, MD: Multi-dose, MT: Multi-types. RT: Rapid Test account for new susceptibility or the transition from recovered to susceptible. To determine disease severity, symptomatic infectious individuals can be divided into a fraction $0 \leq f \leq 1$ representing severe symptoms and the fraction $(1 - f)$ representing mild symptoms, which are assumed to go undetected. Symptomatic individuals who are reported do not significantly contribute to transmission (Demongeot et al., 2022). Asymptomatic, undetected individuals may socialize more freely than those who are asymptomatic but detected due to the lack of visible symptoms. Additionally, disease-induced mortality is considered to occur only for those in the infected (I) compartment (Aldila et al., 2020).

3.2.1.3 Intervention Factors

Quarantine, as a non-medical intervention, can be enforced both within and between regions. Zhang et al. (2022a, 2022b) assumed that infected individuals do not travel between cities, allowing only susceptible and exposed individuals to move among them. As exposed individuals travel, they can transmit the virus to the destination city, infecting susceptible individuals. Consequently, the Susceptible-Exposed-Infectious-Recovery with Flow and Medical investments(SEIR-FMi) model incorporates intra-city population movement, inter-city population movement, and medical resource investment on the spread of the COVID-19 epidemic. It was observed that reductions in travel and contact had comparable effects on decreasing the risk of COVID-19 exposure across both space and time (Lai et al., 2021). During the COVID-19 epidemic, some areas implemented rapid testing for all individuals except those in the ICU, allowing for the identification of asymptomatic individuals and preventing further disease spread (Aldila et al., 2020). To build an easily adjustable model, a few simple assumptions are necessary. Demongeot et al. (2022) overlooked both the duration of the immunization period (permanent immunity) and the length of the post-vaccination immunity phase, assuming it to be uniform across all vaccinated individuals. They also hypothesized that vaccination efficacy was independent of age, acknowledging that this assumption may be overly restrictive.

3.2.2 Equations

An exponential distribution with parameter α is considered for the latency period (where the average latency period is α^{-1}). Equations of this model are presented by Eqs. (5) to (8), accompanied by Fig. 4.

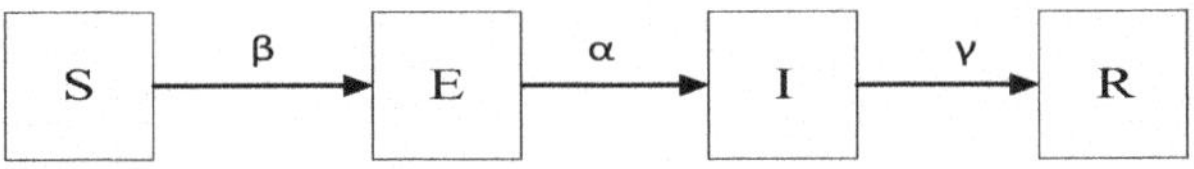

Fig. 4 SEIR model

Table 7 Equations of modified SEIR models

Factor	Reference	Equations		
Infection susceptible by exposed individual	Zhang et al. (2022a, 2022b)	$\frac{ds'}{dt} = \frac{ds}{dt} - \frac{\varepsilon ES}{N}$		$\frac{dE'}{dt} = \frac{dE}{dt} + \frac{\varepsilon ES}{N}$
Asymptomatic detected and undetected individuals with different recoveries	Aldila et al. (2020)	$\frac{ds'}{dt} = \frac{ds}{dt} - \frac{\beta S}{N}(U + A)$ $\frac{dA'}{dt} = \nu U - \gamma_0 A - \eta A$	$\frac{dE'}{dt} = \frac{dE}{dt} + \frac{\beta S}{N}(U + A)$ $\frac{dI'}{dt} = (1 - p)\,\alpha E + \eta A - \left(\gamma_0 + \frac{\gamma_1}{1+bI}\right) I$	$\frac{dU'}{dt} = p\alpha E - \nu U - \gamma_0 U$ $\frac{dR'}{dt} = \gamma_0 (U + A) + \left(\gamma_0 + \frac{\gamma_1}{1+bI}\right) I$
Waning immunity	Aldila et al. (2020)	$\frac{ds'}{dt} = \frac{ds}{dt} + \delta R$		$\frac{dR'}{dt} = \frac{dR}{dt} - \delta R$

ε: Effective transmission rate between the susceptible and the exposed, p: Proportion of exposed individuals who have progressed into asymptomatic individuals, ν: Effort required for early detection of COVID-19 infection, γ_0: Natural recovery rate, γ_1: The enhancement of natural recovery rate owing to treatment in a hospital, η: The rate of hospitalization from Ad to I, δ: Rate of disappearance of temporal immunity

$$\frac{ds}{dt} = -\frac{\beta IS}{N} \tag{5}$$

$$\frac{dE}{dt} = \frac{\beta IS}{N} - \alpha E \tag{6}$$

$$\frac{dI}{dt} = \alpha E - \gamma I \tag{7}$$

$$\frac{dR}{dt} = \gamma I \tag{8}$$

Various factors and assumptions applied in the SEIR model are summarized in Table 7. Interested readers can also refer to Zhang et al. (2022a, 2022b) for disease transmission equations in a multi-regional context.

3.2.3 Parameters

Lai et al. (2021) utilized real data based on reported cases and mobile phone data to calculate the travel and contact rates for each country, rather than focusing solely on transmission rates. Demongeot et al. (2022) determined the transmission rate using various parameters, including the fraction of unreported symptomatic infectious individuals capable of transmitting the pathogen (k), the fraction of reported symptomatic infectious individuals (f), the number of susceptible individuals at time t_0 (S_0), the number of exposed individuals at time t_0 (E0), the number of asymptomatic infectious individuals at time t_0 (I_0), the number of symptomatic infectious individuals who are not reported at a given time t_0 (U_0), and the transition rates between different compartments. According to Aldila et al. (2020), isolation at home or in a hospital reduces the transmission rate. Modelers compute the Herd Immunity Threshold (HIT) to determine the necessary level of population immunity (e.g., through vaccine coverage) required to achieve herd protection. HIT is a valuable tool for guiding vaccination campaigns, but it is unlikely that there is a threshold at which transmission ceases abruptly. The simplest HIT calculation, based on the assumption of homogeneous mixing (where the contact rate is uniform across the population), is given by Eq. (9):

$$\mathrm{HIT} = 1 - \frac{1}{R_0} \tag{9}$$

In practical scenarios, HIT varies depending on context, as contact and transmission rates are not homogenous within populations. HIT can be adjusted by factors like behaviors (e.g., occupation), biological characteristics (e.g., age), mitigation efforts (e.g., physical distancing), connectedness (e.g., contact patterns), vaccine coverage, and natural immunity (Caldwell et al., 2021).

3.3 *SIRD Model*

The model includes Susceptible-Infectious-Recovered-Deceased (SIRD) compartments. It distinguishes between individuals who have recovered (indicating survival and immunity) and those who have deceased (Bailey, 1975).

3.3.1 Assumptions

In this section, assumptions of related studies that employed the SIRD model are considered, as summarized in Table 8.

3.3.1.1 Population Factors

Like the SIR model, the total population size remains approximately constant (Shringi et al., 2021). In this model, the focus is on the number of deaths, considering various factors that impact disease severity and resultant fatalities, including age (Rao & Brandeau, 2021a, 2022) and seasonal effects (Engelbrecht & Scholes, 2021).

Table 8 Classification of studies considering SIRD model based on their assumptions

Ref.	Population factors														Disease factors									Intervention factors				
		Number of persons		Newborns		Composition		Multi-grouping			Migration				Rate of transition		Immunity							Class		Vaccine		
	Birth and death rate	C	V	S	R	Ho	He	MA	MG	MO	BR	BG	Multi-region	Social Challenges	C	V	P	W	Im	D	Incubation time	Multi-period	Multi-Strain	M	NM	EFV	MD	MT
Engelbrecht and Scholes (2021)		*				*									*		*								SD-L			
Rao and Brandeau (2021a)								*								*						*		Va		*		*
Rao and Brandeau (2021b)								*								*						*		Va		*		
Rao and Brandeau (2022).								*								*						*		Va		*		
Shringi et al. (2021)		*														*							*		G			
Šušteršicˇ et al. (2021)		*				*										*					*				QS			

E: Equal, C: Constant, V: Variable, S: Susceptible, R: Recovered, Ho: Homogenous, He: Heterogeneous, MG: Grouping based on gender, MA: Grouping based on age, MO: Grouping based on other factors such as socioeconomic status, geographical location, comorbidity, BR: Between regions, BG: Between groups, H: Hesitancy, A: Anti-vaccination, PR: Public resistance, S(PR): Social (Public resistance), P: Permanent, W: Weak/Loss, Im: Immediate, D: Delayed, M: Medical, NM: Non-medical, Va: Vaccination, H: Hospitalization, QS: Quarantine and Isolation, SD: Social distancing, PE: Public education, L: Lockdown, EFV: Effectiveness, MD: Multi-dose, MT: Multi-types. RT: Rapid Test

3.3.1.2 Disease Factors

Due to the short time frame of the model, it does not account for births, deaths unrelated to infection, or other demographic changes (Rao & Brandeau, 2021a). Individual transmission varies dynamically between compartments as it is time-dependent. Additionally, the frequency of non-medical interventions directly influences this time dependence. Therefore, estimating the simultaneous reduction in the transmission rate can help evaluate the effectiveness of these interventions (Shringi et al., 2021).

3.3.1.3 Intervention Factors

When a limited supply of effective vaccines is available, vaccinating susceptible individuals facilitates their transition to the recovered compartment. Perspectives on the effects of vaccination vary, including the assumption that vaccination does not impact transmission rates between infected and unvaccinated individuals or the recovery rates of infected individuals (Rao & Brandeau, 2021a, 2021b, 2022).

To add complexity to the model, incorporating multiple vaccines with varying efficacies is beneficial (Rao & Brandeau, 2021a), along with considering multi-period vaccination strategies. In the multi-period approach, vaccinations occur at the start of each period (Rao & Brandeau, 2022), whereas in the single-period approach, it is assumed that vaccinations are conducted at time $t = 0$ (Rao & Brandeau, 2021a, 2021b). Furthermore, seasonality and non-pharmaceutical control measures are modeled to influence the reproduction number by modifying transmission rates (Engelbrecht & Scholes, 2021).

3.3.2 Equations

The SIRD model employs the following differential equations (Eqs. 10–13) and is illustrated in Fig. 5.

Fig. 5 SIRD model

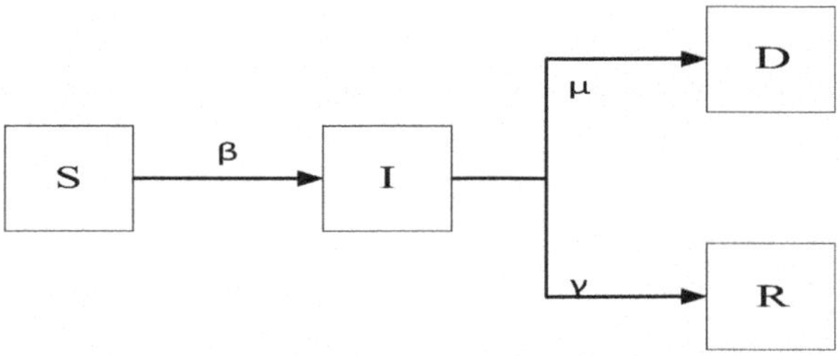

Table 9 Parameter estimations in SIRD models

Effective factor	Estimated parameter	Equation	Reference
Severity of infection	Death rate (μ)	$\mu = \frac{\xi}{d}$ $d = dm + \alpha ds$	Rao and Brandeau (2021a)
	Recovery rate (γ)	$\gamma = {}^{1}/_{d} - \mu$	
Isolation	Transmission rate	$(1 - \psi).\ (\beta_1 + \beta_2 + \beta_3)$	Šušteršicˇ et al. (2021)

ξ: Fraction of infected individuals die, *dm*: Average duration of a mild infection, *ds*: Average duration of a severe infection, α: fraction of infections in group *i* that are severe, Ψ: Effect of isolation, β_1: Transmission rate of mild cases, β_2: Transmission rate of severe cases (hospitalized cases), β_3: Transmission rate of critical cases (ICU cases)

$$\frac{ds}{dt} = -\frac{\beta IS}{N} \tag{10}$$

$$\frac{dI}{dt} = \frac{\beta IS}{N} - \gamma I - \mu I \tag{11}$$

$$\frac{dR}{dt} = \gamma I \tag{12}$$

$$\frac{dD}{dt} = \mu I \tag{13}$$

where β, γ, and μ represent the infection, recovery, and mortality rates, respectively. To enhance accuracy, the model incorporates a multi-group assumption. For example, Rao and Brandeau (2021a) developed an SIR model of people with $n \geq 2$ interacting groups in which an infectious disease spreads. Within each group i, individuals can be categorized as susceptible (S_i), infected (I_i), recovered (R_i), or dead (D_i). Individuals in group i can acquire infection from contacts with individuals in their own population group (at rate β_i) or from individuals in another group j (at rate β_{ij}). Infected individuals in group i can either recover (at rate γ_i) or die (at rate μ_i).

3.3.3 Parameters

Table 9 shows estimations of important parameters based on effective factors.

In addition to the mentioned parameters, deriving an analytical solution for an SIR model involving n interacting populations is inherently challenging, if not unfeasible. To address this complexity, Rao and Brandeau (2021a) approximated the dynamics of the disease at time t using first-order and second-order Taylor series expansions. Similarly, Zaric and Brandeau (2001) utilized a comparable approach for an SI model, assuming insufficient contacts across population groups. Rao and Brandeau (2021a) adapted this approach for the SIR model with n groups, facilitating cross-infection between these groups: $\beta_{ij} \neq 0$ for $i \neq j$. Additionally, Shringi et al. (2021) emphasized the presence of noise when estimating parameters, categorizing it into two types: process noise and measurement noise, both of which should be carefully considered.

3.4 Extended Versions

Several researchers have extended the SIR, SEIR, and SIRD models to enhance the accuracy of epidemic predictions, integrating additional compartments to provide a detailed examination of disease dynamics and yield results that closely align with real-world observations. Notably, they have introduced compartments for individuals who carry the virus but are not yet infectious to others: Carrier (C), asymptomatic and infective (IA), mildly infective (IM), severely infective or hospitalized in wards (IS), and critically infective or admitted to an Intensive Care Unit (ICU) (IC). Kemp et al. (2021) considered two types of quarantine: hospitalization and home recovery. Vakil and Trappe (2022) and Beams et al. (2021) reflected different levels of disease severity and immunity. The SEIQRV model presented by Herrera-Serrano et al. (2022) incorporates two additional assumptions: first, that exposed and symptomatic infected individuals enter quarantine at differing rates, and second, that vaccination occurs alongside the consideration of waning immunity and population migration. Table 10 presents examples of these extended models.

3.4.1 Assumptions

This section outlines the assumptions of extended models, which are summarized in Table 11.

3.4.1.1 Population Factors

In the SIR and SEIR models, it is assumed that the total population remains constant over time, expressed as: $S(t) + E(t) + I(t) + R(t) = N$. However, this assumption holds only over a short duration; as the epidemic extends over a longer period, its validity diminishes. Consequently, adjustments are required to account for transitions that consider births and natural deaths within the respective compartments (Hethcote, 2000). Additionally, to more accurately capture human mobility, migrating individuals could be treated as additional components within the birth and death processes, reflecting their entrance and exit from the system (Bhadauria et al., 2021).

Table 10 Extended models

Reference	Purpose	Model	Description of added components
Kühn et al. (2021)	Assess the effectiveness of the sets of non-pharmaceutical interventions	SECIHUDR	Carrier (*C*): Who carry the virus and are infectious to others but do not yet show symptoms (they may be pre- or asymptomatic); Hospitalized (*H*): Who experience a severe development of the disease; In Intensive Care Unit (*U*)
Parolini et al. (2022)	Investigation of new species and the effect of vaccination on the course of the disease	SUIHTERV	*U*: Number of undetected (both asymptomatic and symptomatic) infected individuals; *I*: Number of infected individuals isolated at home; *H*: Number of infected hospitalized individuals; *T*: Number of infected threatened individuals hosted in ICUs; *E*: Number of extinct individuals; *V*1: Number of individuals partially vaccinated with only one dose; *V*2: Number of individuals fully vaccinated with two doses; *VR*: Number of recovered individuals who have been vaccinated
Bertsimas et al. (2021)	Forecast the progression of COVID-19 and evaluate the impact of various social distancing policies	DELPHI	*UR*: Undetected people who will recover; *UD*: Undetected people who will decease; *DHR*: Detected hospitalized people who will recover; *DHD*: Detected hospitalized people who will decease; *DQR*: Detected quarantined people who will recover; *DQR*: Detected quarantined people who will decease; *DT*: Total Detected Cases; *DD*: Total Detected Deaths

(continued)

Table 10 (continued)

Reference	Purpose	Model	Description of added components
Robinson et al. (2022)	Model more complex disease dynamics and to capture clinically relevant quantities of interest	22-compartment model	Q: Exposed (quarantined); A: Infectious (pre-symptomatic); W: Infectious (pre-symptomatic, isolated); B: Infectious (mild-to-moderate symptoms); N: No access to hospital care; $H1$: Admitted to hospital (pre-ICU); I: Admitted to hospital (ICU); $H2$: Admitted to hospital (post-ICU), etc.
Kemp et al. (2021)	Consider interplay between vaccinations and social measures And estimate vaccination rate to achieve herd immunity	16- compartment model	NI: Recovering at home; H: Hospital; H_1: Longer stay hospital; ICU: Admitted to hospital (ICU); $AICU$: Back to hospital after ICU
O'Dea and Drake (2022)	Forecasts of COVID-19 cases, hospital admissions and deaths	9- Compartmental model	Z_r: Number of cases reported each day; A: New hospital admissions; D_r: The number of newly reported deaths each day;
Pang et al. (2022)	Consider impact of asymptomatic COVID-19 carriers on pandemic policy outcomes	SEOAR	O: Ordinary carriers; A: Asymptomatic carriers

Given the demographic heterogeneity among different regions, as discussed by Robinson et al. (2022), constructing a disease model for a limited area may not yield accurate estimations of the overall disease situation. Government-imposed interventions require that the entire country be regarded as a unified system. Consequently, the multi-regional assumption (Jarumaneeroj et al., 2022) may be a more logical approach. It is crucial to consider the movement of people between regions, noting that the number of incoming travelers from another region depends on their infection status. Specifically, hospitalized individuals cannot travel, and infected persons are less mobile than healthy individuals (Kühn et al., 2021). To address heterogeneity within the population, homogeneous subgroups can be formed by dividing individuals based on various criteria. In addition to age (López et al., 2022), other indicators can also be selected for grouping. Bertsimas et al. (2021) examined the influence of age, gender, and health factors—such as chronic

Table 11 Classification of studies presenting extended model based on their assumptions

Reference	Population factors													Disease factors										Intervention factors				
		Number of persons		Newborns		Composition		Multi-Grouping			Migration				Rate of transition		Immunity							Class		Vaccine		
	Birth and death rate	C	V	S	R	Ho	He	MA	MG	OG	BG	BR	Multi-region	Social challenges	C	V	P	W	Im	D	Incubation time	Multi-period	Multi-strains	M	NM	EFV	MD	MT
Parolini et al. (2022)		*				*		*							*		*				*		*	Va- Hos	G	*	*	
Du et al. (2023)								*										*			*			Va- Hos		*	*	
Dagpunar and Wu (2023)		*													*			*						Va	G	*	*	
Robinson et al. (2022)							*	*								*		*			*			Va-Hos	QS			
Tuite et al. (2020)																*					*			Hos	SD- QS			
Zhan et al. (2023)						*							*		*						*		*	Va	G	*		
Tzamali et al. (2023)		*				*									*	*		*			*			Va	G	*		
Jarumaneeroj et al. (2022)		*						*					*									*		Va	SD-QS	*		*
Vahdani et al. (2023)	*							*									*					*		Va-	QS			*
Bertsimas et al. (2021)								*	*	*			*			*	*					*		Hos	SD			
Booton et al. (2021)								*							*								*	Va-H	G			*
Kühn et al. (2021)							*	*				*	*		*						*	*		Hos	G			
Deo and Grover (2022)		*														*	*								G			
Rodiah et al. (2023)		*		*			*	*								*		*			*	*	*	Va-Hos	G			
Polcz et al. (2023)		*														*		*		*	*		*	Va-Hos		*	*	

E: Equal, C: Constant, V: Variable, S: Susceptible, R: Recovered, Ho: Homogenous, He: Heterogeneous, MG: Grouping based on gender, MA: Grouping based on age, MO: Grouping based on other factors such as socioeconomic status, geographical location, comorbidity, BR: Between regions, BG: Between groups, H: Hesitancy, A: Anti-vaccination, PR: Public resistance, P: Permanent, W: Weak/Loss, Im: Immediate, D: Delayed, M: Medical, NM: Non-medical, Va: Vaccination, H: Hospitalization, QS: Quarantine and Isolation, SD: Social distancing, PE: Public education, L: Lockdown, EFV: Effectiveness, MD: Multi-dose, MT: Multi-types. RT: Rapid Test diseases and blood conditions—on mortality rates due to infection. Furthermore, Kühn et al. (2021) utilized mobility data from the German Federal Employment Agency and Twitter to address spatial heterogeneity in disease transmission.

3.4.1.2 Disease Factors

Time plays a crucial role in epidemic modeling, encompassing various dimensions such as multi-period modeling (Jarumaneeroj et al., 2022), the consideration of the incubation period (Kühn et al., 2021; López et al., 2022), seasonal effects (López et al., 2022), and the emergence of different viral strains over time (Jing et al., 2023). Specific events occurring within designated timeframes can also be influential; for instance, Kühn et al. (2021) assumed that the transition time between different system states is independent of individuals' ages.

3.4.1.3 Intervention Factors

According to Table 11, most studies assumed the implementation of both non-medical and medical interventions. Consequently, extended models aimed to incorporate additional dimensions of epidemic situations, emphasizing the necessity of including assumptions about interventions. Some articles did not explore vaccination due to the timing of the studies, which occurred before an effective vaccine became available.

The effectiveness of vaccines is a widely debated topic, particularly during epidemics and outbreaks. It is important to note that vaccine effectiveness can vary significantly. Additionally, vaccines may not directly reduce a virus's transmissibility or shorten the infection's duration (Jing et al., 2023). To optimize vaccination strategies, increasing the number of doses and recommending booster shots can enhance effectiveness (Dagpunar & Wu, 2023; Du et al., 2023). However, some researchers argue that multiple vaccine doses may yield similar outcomes (Jing et al., 2023). Vaccine effectiveness can also be influenced by factors such as vaccine type and the characteristics of specific subpopulations, including age groups (Jarumaneeroj et al., 2022).

Concerns regarding potential vaccine ineffectiveness and the emergence of new viral variants highlight the risks of re-infection. The possibility of new variants, particularly in COVID-19, has been extensively documented (e.g., Li et al., 2021). Even in studies that do not explicitly address this assumption, it is often recognized as a limitation that merits consideration in future research. These factors underscore the importance of ongoing investigation and vigilance in addressing the dynamic nature of infectious diseases and their implications for vaccine effectiveness. Zhan et al. (2023) presented reasonable assumptions regarding new variants, such as:

- An individual can only be infected by one specific variant at a time, a premise supported by the likelihood that infected individuals are quarantined or hospitalized, thereby reducing their contact with others carrying different variants.
- Individuals with antibodies against a particular variant (designated as variant-*i*) would be protected from re-infection by that variant. However, they may still be susceptible to infections from different variants (designated as variant-*j*), where $i \neq j$.

Jing et al. (2023) assumed that the entire population is eligible for vaccination. However, a significant assumption frequently overlooked in many studies—or inadequately addressed in calculations—is the delay in vaccine rollout, as discussed by Booton et al. (2021).

3.4.2 Equations

This section reviews articles grouped into two categories: (1) those that combine medical and non-medical interventions, and (2) those that include additional components.

3.4.2.1 Combination of Medical and Non-medical Interventions

Jarumaneeroj et al. (2022) utilized the SIQRV model to describe the transmission dynamics of infection in the COVID-19 Vaccine Allocation Problem (CVAP). The model categorizes the population in each region into seven compartments according to their health states: Susceptible (S), Infectious (I), Quarantined (Q), Recovered (R), Dead (D), vaccinated individuals with immunity (V), and vaccinated individuals without immunity ($\overline{V}$). On the other hand, Saha and Samanta (2021) divided the population $N(t)$ at time t into six subclasses: susceptible individuals (S), asymptomatic individuals who have been exposed to the virus (A), quarantined individuals (Q), symptomatically infected individuals (I), hospitalized and isolated individuals (H), and the recovered population (R). Additionally, the SPEIHQRD model proposed by Rajaei et al. (2021) incorporates a compartment P, for individuals who are insusceptible due to public health improvements, such as the cancellation of large public gatherings and adoption of face mask usage. It's worth noting that while insusceptible individuals are acknowledged in Saha and Samanta (2021), they are not assigned separate compartments.

The equations relevant to the interventions detailed in these three articles are summarized in Table 12. In Saha and Samanta (2021), individuals may be hospitalized after being quarantined, whereas in Rajaei et al. (2021) and Jarumaneeroj et al. (2022), recovery or death are the only outcomes following quarantine. Furthermore, Rajaei et al. (2021) and Jarumaneeroj et al. (2022) incorporated dead components, but Saha and Samanta (2021) did not.

Table 12 Equations for interventions

Intervention	Reference	Equations of interventions		
Quarantine and vaccine	Jarumaneeroj et al. (2022)	$dQ/dt = qI - \gamma Q - \delta Q$	$\left(d\overline{V}\right)/dt = (1-\varepsilon)V - \overline{V}.\beta\mu I$	$dV/dt = \varepsilon V$
Quarantine and Hospitalization	Saha and Samanta (2021)	$dQ/dt = qA - \gamma Q - \alpha Q$	$dH/dt = \alpha Q + \omega I - \gamma H - \delta H$	
	Rajaei et al. (2021)	$dQ/dt = qI - \gamma Q - \delta Q$	$dH/dt = \omega I - \gamma H - \delta H$	

q: Quarantine rate, γ: Recovery rate, δ: Death rate, ε: Effectiveness of vaccine, V: Number of doses of vaccine, $\overline{V}$: Number of individuals that have been vaccinated with but with no developed immunity, ω: Admission rate in hospital, α: Transition rate of quarantined individual into the hospitalized

Table 13 Equations for additional components

Added component	Reference	Related equations	
Insusceptible (P)	Rajaei et al. (2021)	$dS'/dt = dS/dt - \alpha S$	$dP/dt = \alpha S$
ICU (T)	Parolini et al. (2022)	$dH/dt = \omega I - \gamma H - \delta H - \xi H + \theta T$	$dT/dt = \xi H - \delta H - \theta T$

α: Effect of non-medical interventions, δ: Death rate, ω: Admission rate in hospital, γ: Recovery rate, ξ: Admission rate in ICU, θ: Rate at which T-individuals improve their health conditions and return to the less critical H compartment

3.4.2.2 Additional Components

The SEUACRD-VP model proposed by Zhan et al. (2023) represents a significant advancement in understanding the multi-directional mutation process of SARS-CoV-2 and the spread of COVID-19, particularly concerning multiple variants. The model evaluates the impact of various public health interventions, including both non-pharmaceutical and pharmaceutical strategies, in the context of these multi-directional mutations (Zhan et al., 2022). Table 13 presents two examples of equations illustrating added components to main models. A fundamental aspect of the mathematical model presented by Parolini et al. (2022) is the transition between hospitalization and ICU care. The model assumes that individuals infected with the virus are first admitted to the hospital and then transferred to the ICU if their condition deteriorates. A percentage of those transferred to the ICU may recover and return to general hospital wards, while others may not survive.

3.4.3 Parameters

The transmission of the virus from one individual to another is primarily influenced by specific interactions between individuals, which can vary based on factors such as

Table 14 Parameter estimations in SIRD models

Effective factor	Estimated parameter	Equation	Reference
Season	Transmission rate	$\beta = sk(t)\beta_0$; $sk(t) := 1 + ksin\left(\pi\left(\frac{t}{182.5} + \frac{1}{2}\right)\right)$	Kühn et al. (2021)
Undetected individuals	Transmission rate	$\beta = \left(I_X/I + \tilde{\beta}\left(I_H + I_R\right)/I\right)$	Kühn et al. (2021)

β_0: Transmission rate without season effect, k: Season effect, $\tilde{\beta}$: Risk of infection from the detected and infected symptomatic patients not yet effectively isolated, I_X: Undetected Infected, I_H: Hospitalized infected, I_R: Infected that recover on their own

age and contact location (Prem et al., 2020). Various methods have been employed to estimate the transmission rate. For instance, Rajendrakumar et al. (2021) utilized back-calculation to derive the transmission rate (β) from the previously computed basic reproduction number, employing the Sequential Bayesian method. Prem et al. (2020) adopted an age-group model to account for variability in social mixing patterns across different settings, including households, workplaces, and schools. Understanding these social structures is crucial for evaluating the effectiveness of physical distancing measures, such as school closures. Models assessing the impact of interventions must accurately capture these social heterogeneities to provide meaningful insights into disease transmission dynamics (Prem et al., 2020; Hilton & Keeling, 2019).

An important consideration in understanding disease transmission is the seasonal shift from outdoor to indoor contact. Various counter-strategies, such as the use of face masks (Viola et al., 2020; Reiner et al., 2020; Cheng et al., 2020), modifications to airflow (Bhagat et al., 2020), or a combination of these approaches (Buonanno et al., 2020), have been studied. However, it is reasonable to assume that these measures may not entirely eliminate the risk of indoor infections. Qian et al. (2020) found that most analyzed infection clusters occurred in enclosed environments, while Nishiura et al. (2020) quantified the risk of indoor infections as being 18.7 times greater than that of outdoor infections. Additionally, Kühn et al. (2021) allowed for slight variations in the transmission rate with changes in seasons, indicating that seasonality is a factor influencing transmission dynamics. The effect of undetected individuals on the transmission rate is further illustrated in Table 14.

Badfar et al. (2022) employed the Long Short-Term Memory (LSTM) algorithm to estimate the unknown parameters of the system. The LSTM algorithm is designed to address the challenge of long-term dependencies in modeling. Due to the inherent uncertainty and measurement noises associated with epidemiological models, their study developed an Extended Kalman Filter (EKF) to estimate the state variables of the proposed model.

4 Reproduction Number

The reproduction number is a pivotal concept in evaluating infection epidemics, which is categorized into two classes: the basic reproduction number and the effective reproduction number.

4.1 Basic Reproduction Number (R_0)

The basic reproduction number (R_0) and the reproduction number (R) are critical parameters in the field of infectious disease epidemiology. R_0 is the expected number of individuals who will become infected by one infectious person in a population where all individuals are initially susceptible. On the other hand, R is the expected number of infected individuals generated by one infectious person in a population that may be partially susceptible (Lim et al., 2020). If $R_0 > 1$, the epidemic is expected to grow exponentially; if $R_0 < 1$, the epidemic is likely to decline exponentially (Iranzo & Pérez González, 2021). Thus, this parameter elucidates the dynamics of the epidemic model and characterizes the number of susceptible individuals that can be infected by an infected person (Batistela et al., 2021). To derive the basic reproduction number for the epidemic model, several researchers have employed the next-generation method (Heffman et al., 2005; Diekmann et al., 1990; Rao & Brandeau, 2021b; Zhang et al., 2022a, 2022b). Additional methods for calculating this parameter are summarized in Table 15.

For more advanced calculations regarding reproduction numbers, refer to Malinzi & Juma (2023); Aldila et al. (2020); Wali et al. (2022); Song et al. (2020); Hussien & Mohammad (2022); Demongeot et al. (2022). The expected number of new infections produced by individuals with such strain k is calculated in Otunuga (2022). Otunuga and Yu (2023) expanded on the model by incorporating the possibility of re-infection among recovered individuals.

4.2 Effective Reproduction Number (R_e)

The effective reproduction number (R_e) is often used to promptly assess the current dynamics of infectious disease transmission. This measure quantifies the effectiveness of real-time disease control measures (Cauchemez et al., 2006). Estimating R_e necessitates using case numbers, the serial interval, and the timing of symptom onset. However, during epidemics, especially for emerging infectious diseases, acquiring these data can be challenging and may introduce biases into the estimation of R_e. Notably, surveillance systems often capture only symptomatic cases, overlooking actual infections. Consequently, R_e may reflect delayed transmis-

Table 15 Basic reproduction number

Equation	Reference
$\frac{\beta}{\gamma}$	Ai et al. (2022); Cherif and Hasanov (2023); Nguyen et al. (2022); Yaladanda et al. (2022); Libotte et al. (2020); Khan and Tanimoto (2023); Chen et al. (2020a, 2020b, 2020c)
$\frac{S_0\beta}{\gamma}$	Iranzo and Pérez-González (2021); Mohd Yunus et al. (2021)
$\frac{S_0\beta+\xi V_0}{\gamma}$	Turkyilmazoglu (2022)
$\frac{\beta k}{(k+\mu)(\gamma+\mu)}$	Choi and Shim (2020)
$\frac{\beta}{\gamma+\mu}$	Goodkin-Gold et al. (2022); Engelbrecht and Scholes (2021); O'Regan et al. (2020); Deo and Grover (2022)
$\frac{\beta}{2}\left(\frac{1}{(\mu+\gamma)}+\frac{g}{(\mu+a\gamma)}\right)$	Safan (2019)
$\frac{\beta}{\gamma}\,\rho\,(\chi M)$	Arribas et al. (2022)
$\left(\frac{\beta N}{\gamma+\theta}\right)\frac{[\alpha+\delta+(1-f)v]h}{[\alpha+\delta+(1-f)v]h+\alpha v}$	Harari and Monteiro (2022)
$\frac{\alpha_e(\gamma+\mu)+\beta\kappa}{(k+\rho)(\gamma+\mu)}$	Nanda et al. (2023)
$\frac{\beta(1-\varepsilon)\kappa}{\mu(\gamma+\mu)}$	Nasution et al. (2021)

β: Transmission rate, γ: Recovery rate, S_0: Number of susceptible individuals at time zero, ξ: Retransmission rate, V_0: Vaccinated individuals, $1/k$: Duration of latent period, μ: Natural death rate, $g = r_1/r_2$ denote relative susceptibility of females with respect to males, a: Relative recovery rate of infected females with respect to infected males, $\rho(\chi M)$: The spectral radius, or largest eigenvalue, of the age-contact matrix, α: The rate of migration from the A-group to the P-group, δ: The rate of migration from the P-group to the A-group, θ: Death rate of I-individuals, f: Fraction of births of P, v: Individuals immunization rate, h: Death rate constant of R-individuals, α_e: Rate of asymptomatically infection, ρ: Recovery rate of exposed population. ε: Vaccination rate, κ: Birth rate

sion dynamics due to time lags, encompassing the incubation period and delayed reporting (Lim et al., 2020).

Vaccination is a standard strategy for reducing the spread of infectious diseases (Saha et al., 2022), as it effectively decreases the number of susceptible individuals, leading to a natural reduction in the reproduction number. This parameter indicates the transmissibility of the virus by infectious individuals and is influenced by factors such as population density, vaccination rates, quarantine measures, social distancing, mask usage, and other preventive actions (Xu et al., 2020; Sy et al., 2021). The reproduction factors can vary over time and across seasons. It is widely acknowledged that the risk of contagion is significantly higher indoors than outdoors (Gros & Gros, 2022), and contracting the Coronavirus outdoors is considered unlikely (Couronne, 2020). For the influenza virus, fluctuations in the reproduction rate correspond to the typical "flu season," which coincides with colder weather (Sajadi et al., 2020). Other seasonal factors have also been suggested for COVID-19, such as fluctuations in UV light (Brunnermeier, 2020; Karapiperis et al., 2020). For an extensive survey on these factors, see Carlson et al. (2020).

Table 16 Effective reproduction number

Effective factor		Equation	Reference
Social isolation		$\frac{\beta(1-\theta)}{\gamma+\mu}$	Batistela et al. (2021)
Social distancing and vaccination		$\frac{\beta k}{(k+\mu)(\gamma+\mu)} \frac{((1-\delta)\psi+(1-\eta)\mu)}{\psi+\mu}$	Choi and Shim (2020)
		$(1-\omega)\frac{\beta(1-\theta)}{\gamma+\mu}$	Batistela et al. (2021)
Vaccination	Perfect	$(1-\omega)\frac{\beta}{\gamma}$	Yaladanda et al. (2022)
	Mass	$\frac{\beta}{2}\left(\frac{\mu+(1-e_2)\omega}{(\mu+\gamma)(\mu+\omega)} + \frac{g\left[\mu+(1-e_1)e_0\omega\right]}{(\mu+a\gamma)(\mu+e_0\omega)}\right)$	Safan (2019)
	Leaky	$(1-\omega)\frac{\beta}{\gamma} + \omega(1-\delta)\frac{\beta}{\gamma}$	Yaladanda et al. (2022)
	AON	$(1-(V*\delta))\frac{\beta}{\gamma}$	Yaladanda et al. (2022)
	Ring	$R_I + R_H + R_F$	Xie (2019)

β: Transmission rate, γ: Recovery rate, μ: Death rate, ω: Vaccination rate, θ: the effect of social isolation, 1/k: duration of latent period, Ψ: vaccination rate, δ: Vaccine efficacy, η: Reduced contact rate of susceptible people due to social distancing. $g = r_1/r_2$ denote relative susceptibility of females with respect to males, a: Rescaling parameter representing the relative recovery rate of infected females with respect to infected males, e_0: Relative response to vaccination campaigns of females with respect to males, V: Vaccinated individuals. $R_I(R_H, R_F)$: Reproduction number due to transmission rate in community (after hospitalization, during traditional burial)

The effective reproduction number, R_e, indicates the average number of secondary cases per infectious case in a population that includes both susceptible and infected individuals. With the availability of a vaccine for a disease, vaccination is an effective method for controlling epidemic spread (Rao & Brandeau, 2021b). Mungkasi (2021) and Rao and Brandeau (2021b) derived the effective reproduction number $R_e$accounting for vaccination. Zhang et al. (2022a, 2022b), Rodiah et al. (2023), Chaturvedi and Chakravarty (2021), and Rajendrakumar et al. (2021) utilized the effective reproduction number $R_e(t)$ to illustrate how time influences R_e. It also includes the impact of interventions and behavioral changes. Social distancing costs are proportional to the reduction in the reproduction parameter (Gros & Gros, 2022).

Table 16 summarizes the various factors affecting the reproduction number, along with related equations. Due to the complexity of calculations discussed in Xie (2019), the desired formula is given as a summary.

5 Controlling Interventions

Compartmental models aim to predict the spread of an epidemic, and as the proportion of recovered individuals increases, fewer susceptible contacts are available to be infected. This results in a decrease in R_0. Various actions can be taken to reduce the value of R_e. Vaccination is an effective solution, as vaccinated individuals are less likely to become infected and spread the disease. Effectiveness can also be enhanced by reducing the duration of the infectious period, decreasing contact frequency, and minimizing transmission probability. Non-medical interventions, such as isolation and hygienic practices like handwashing, are recommended to help mitigate disease spread (Iranzo & Pérez González, 2021). Kain et al. (2021) utilized a time-varying transmission rate that considers the influence of non-pharmaceutical intervention strategies, which may vary over time. Some researchers surveyed the effects of medical and non-medical interventions simultaneously. Zhang et al. (2022a, 2022b) presented the SEASIRD-HV model, which incorporates variations in lockdown timing and duration, vaccination levels, and hospital capacities. In the following subsections, epidemic control methods are surveyed.

5.1 Medical Interventions

In the study by Jadidi et al. (2021), the objective function is defined as maximizing the total number of individuals who do not get infected. Vaccination contributes to increasing this number through two mechanisms: the direct method, where susceptible individuals are vaccinated, and the indirect method, which reduces exposure for unvaccinated individuals due to the presence of vaccinated individuals in the population. Both mechanisms effectively decrease the overall infection rate. Additionally, utilizing medical services in hospitals serves as another medical intervention. Booton et al. (2022) explored the relationship between vaccination rates and hospitalization demand, demonstrating how medical interventions can aid in disease control.

5.1.1 Hospitalization

To evaluate this intervention, fluctuations in the number of hospital beds in each city over time are employed to measure the availability of medical resources. This information is then utilized to establish the recovery rate as a function of the number of beds and current infection levels (Zhang et al., 2022a, 2022b). Primary models incorporate a hospital compartment (H), also referred to as quarantine, which minimizes contact between infected individuals and susceptible members of the community. However, it is important to note that medical personnel remain at risk of infection. Some models also distinguish between hospitalized individuals who recover and those who do not (Kain et al., 2021). Additionally, they consider recovery rates among hospitalized individuals, the need for intensive care, and the expected recovery rates for patients in intensive care (Debrabant et al., 2021).

5.1.2 Vaccination

Recent research has introduced new variants of the SIR model, including Improved SIR (ISIR) (Zhu & Shen, 2021), SIRS (Liu & Cui, 2021; Rao et al., 2019; Gumus et al., 2022; Nill, 2023), Neuro-SIR (Lymperopoulos, 2021; Mu et al., 2019), Generalized SIR (GSIR) (Singh & Gupta, 2021), a stochastic SIR model based on deep learning (Liu et al., 2021a, 2021b), and the asymptomatic SIR model (ASIR) (Gaeta, 2020), among others.

Successful epidemic models that incorporate vaccination (SIRV) traditionally assume complete, long-lasting immunity following vaccination. However, recent findings from COVID-19 vaccination programs indicate that existing vaccines may not provide full protection, leaving some vaccinated individuals susceptible to infection. This insight necessitates revisions to epidemic models to account for potential waning immunity and transitions between vaccinated and infected compartments. Eichner et al. (2017) previously introduced a similar modification to quantify both the direct and indirect impacts of vaccination on seasonal influenza. This led to the development of a new model, SIRVI, which considers weak or lost immunity due to interactions between vaccinated and infected individuals (Turkyilmazoglu, 2022). Studies that treat vaccination as a separate compartment have explored various issues, including vaccination programs (Kabir & Tanimoto, 2020), waning immunity against multi-strain diseases (Ahumada et al., 2023), scenarios of vaccine efficacy (Yaladanda et al., 2022), full immunization (Angeli et al., 2022), and the timing of the second dose (Geoffroy et al., 2022).

5.1.3 Assumptions

This section reviews the assumptions made in articles concerning medical interventions, summarized in Table 17.

5.1.3.1 Population Factors

Like other models, some researchers assumed a constant population throughout the study, neglecting deaths (excluding those caused by the pandemic) and births (Yaladanda et al., 2022). While these numbers are anticipated to be relatively small compared to the total population (Nanda et al., 2023), certain models do account for a constant recruitment rate (through birth or immigration) within the susceptible class (Saha et al., 2022). Various factors can influence transmission rates among compartments, including time-dependent changes in hospitalization and mortality rates, as well as transmission dynamics related to individuals in different health states, such as asymptomatic, mild, severe, and critical infections (Albani et al., 2021a, 2021b). Additionally, social challenges significantly impact vaccination modeling, a consideration acknowledged in several studies (Gollier, 2020; Harari & Monteiro, 2022; Mesa et al., 2022).

Table 17 Classification of studies incorporating SIRV model based on their assumptions

Ref.	Population factors														Disease factors									Intervention factors				
		Number of persons		Newborns		Composition		Multi-grouping			Migration				Rate of transition		Immunity							Class		Vaccine		
	Birth and death rate	C	V	S	R	Ho	He	MA	MG	MO	BR	BG	Multi-region	Social Challenges	C	V	P	W	Im	D	Incubation time	Multi-period	Multi-Strain	M	NM	EFV	MD	MT
Saad-Roy et al. (2021)	*	*					*									*		*						Va	MW- L- SD	*	*	
Harari and Monteiro (2022)	*	*		*	*	*						*		A	*			*						Va	G	*		
Gollier (2021)		*					*	*						A						*		*		Va-H	L-SD	*		
Ruiz et al. (2023)	*							*										*			*			Va	L	*		
Nanda et al. (2023)	*	*													*			*			*			Va	G	*		
Fudolig (2023)		*													*			*					*	Va		*		
Fudolig and Howard (2020)		*													*			*					*	Va		*		
Tamasiga et al. (2023)						*											*				*			Va		*		
Neofotistos et al. (2022)		*				*												*						Va		*		
Angeli et al. (2022)		*												H-A			*						*	Va		*		
Geoffroy et al. (2022)															*				*		*			Va			*	
Mesa et al. (2022)						*	*	*						H				*		*				Va-H	G	*	*	*
Wali et al. (2022)	*														*			*			*			Va		*		
Hawkes and Good (2022)	*							*				*												Va-H	G			
Chen and Stanciu (2021)																		*						Va		*		
Malinzi and Juma (2023)	*																	*			*		*	Va	G	*		
Zhang et al. (2022a, 2022b)											*		*			*		*			*			H	QS			
Ahumada et al. (2023)	*	*				*									*			*		*			*	Va	QS	*		
Turkyilmazoglu (2022)																		*						Va		*		

(continued)

Table 17 (continued)

Ref.	Population factors														Disease factors									Intervention factors				
		Number of persons		Newborns		Composition		Multi-grouping			Migration				Rate of transition		Immunity							Class		Vaccine		
	Birth and death rate	C	V	S	R	Ho	He	MA	MG	MO	BR	BG	Multi-region	Social Challenges	C	V	P	W	Im	D	Incubation time	Multi-period	Multi-Strain	M	NM	EFV	MD	MT
Choi and Shim (2020)	*	*																						Va	SD			
Kain et al. (2021)																*								H	G			
Kabir and Tanimoto (2020)															*			*					*	Va		*		
Caulkins et al. (2023)	E	*					*	*								*								Va- H	L			
Yaladanda et al. (2022)		*				*									*			*					*	Va-				
Gorji et al. (2023)								*										*					*	Va- H		*		
Filho et al. (2021)																		*						Va				
Albani et al. (2021a, 2021b)								*								*								Va- H	L	*		
Song et al. (2020)	*	*		*	*										*			*					*	Va		*		
Barua and Dénes (2023)	*	*		*	*											*		*			*			Va		*		
Markovic et al. (2021)							*				*			*	*	*		*			*			Va-H	G			
León et al. (2022)		*				*												*			*			Va		*	*	

E: Equal, C: Constant, V: Variable, S: Susceptible, R: Recovered, Ho: Homogenous, He: Heterogeneous, MG: Grouping based on gender, MA: Grouping based on age, MO: Grouping based on other factors such as socioeconomic status, geographical location, comorbidity, BR: Between regions, BG: Between groups, S (H): Social (Hesitancy), S (A):Social (Anti-vaccination), S(PR): Social (Public resistance), P: Permanent, W: Weak/Loss, Im: Immediate, D: Delayed, M: Medical, NM: Non-medical, Va: Vaccination, H: Hospitalization, QS: Quarantine and Isolation , SD: Social distancing, PE: Public education, L: Lockdown, EFV: Effectiveness, MD: Multi-dose, MT: Multi-types. RT: Rapid Test

5.1.3.2 Disease Factors

In most epidemics, multiple strains typically coexist. Ahumada et al. (2023) considered two strain classes: (1) Strain Class 1 (SC1), characterized by moderate contagiousness and relatively low lethality, effectively controlled by high-efficiency vaccination, and (2) Strain Class 2 (SC2), which consists of strains capable of evading vaccine-induced immunity, displaying higher contagiousness and lethality. The model incorporates the probability of mutation from SC1 to SC2. Ahumada et al. (2023) did not account for simultaneous infection by both strain classes, as COVID-19 patients usually isolate after becoming ill, thus reducing the likelihood of co-infection during their isolation period. Conversely, Fudolig (2023) and Fudolig and Howard (2020) assumed that initial infections arise from existing strains, with available vaccines providing complete immunity. They posited that emergent strains would be resistant to the vaccine, allowing re-infection of individuals who have recovered from the initial strain. These models also assume no direct superinfection or co-infection by the emergent strain, with complete immunity conferred after recovery from the emergent strain. Additionally, rare assumptions, such as disease transmission from deceased individuals, were mentioned by Barua and Dénes (2023) and Xie (2019), including models that account for burial rates of corpses.

5.1.3.3 Intervention Factors

For simplicity, Tamasiga et al. (2023) excluded the effects of social distancing, lockdowns, and other restrictive protocols implemented during the COVID-19 pandemic. Caulkins et al. (2023) assumed that sufficient population willingness to be vaccinated will eliminate the need for broad lockdowns. Meanwhile, Choi and Shim (2020) proposed a model where individuals either get vaccinated or practice social distancing, leading susceptible individuals to reduce their contact rates through social distancing measures.

In the model presented by Mesa et al. (2022), vaccination has two protective stages: (1) reducing the likelihood of infection upon exposure (efficacy against infection), and (2) decreasing the likelihood of hospitalization or death after contracting the disease (efficacy against severe outcomes). Fully vaccinated individuals receive two doses of the vaccine, and their transition to the immunized or infected stage is dependent on the vaccine effectiveness (η_k). This term represents the relative risk reduction for an individual becoming infected (Sharma et al., 2021). In simpler terms, if an individual is vaccinated with a vaccine that has $\eta_k = X\%$, their risk of infection is reduced by X% compared to non-vaccinated individual (Siqueira et al., 2022). For a more comprehensive explanation of the concepts and methods used to assess vaccine effectiveness, please refer to the WHO (2020).

5.1.4 Equations

The equations are analyzed in separate subsections according to the type of medical intervention, i.e., vaccination, hospitalization, or a combination of both.

5.1.4.1 Vaccination (SIRV)

The equations from models discussed in related articles are considered according to the immunity states listed in Table 18. This table presents only the modified equations, expressed in a simplified form for clarity. For example, in Caulkins et al. (2023), the compartment of vaccinated individuals is indicated by R_2, but here, V is used for more straightforward comparison across models. Although birth and natural death rates are included in the original model, they have been omitted here to concentrate on vaccination. In addition, an unmarked enclosure in Angeli et al. (2022) is excluded for the sake of model simplicity.

(a) Permanent Immunity

In the state of permanent immunity, it is assumed that immunity lasts for the individual's lifetime or does not diminish during the simulation period. Choi and Shim (2020) extended the SIR-type model by categorizing individuals into various epidemiological statuses, including Susceptible (S), Vaccinated (V), Exposed (E), Infected symptomatic (I), Asymptomatic (A), and Recovered (R). Disease-free individuals are continuously added to the susceptible class through birth or immigration at a constant rate, and their natural death rate is represented by the symbol μ.

(b) Weak/Loss Immunity

Recovery from COVID-19 does not guarantee immunity to other variants, and there have been reports of waning immunity to the virus (Pérez-Alós et al., 2022). Understanding the effects of population immunity on the dynamics of the COVID-19 pandemic has become crucial, particularly given the combined challenges of waning vaccine protection and the emergence of new variants of concern (Gorji et al., 2023).

Waning immunity from previous infection increases individuals' susceptibility to re-infection with SARS-CoV-2 over time. Vattiatio et al. (2022) examined a model that accounted for waning immunity arising from both vaccination and prior infections, considering varying levels of behavioral changes compared to the initial Omicron wave. The durability of immunity from previous infection remains a significant uncertainty in epidemiological models. In Yaladanda et al. (2022), the equations are similar to Turkyilmazoglu (2022), but they used $(1 - V_{eff})\,\beta$ instead of θ, where β and V_{eff} are transmission rate and vaccine effectiveness, respectively.

Table 18 SIRV equations

Immunity state	Effective factor	Equations			Reference
Permanent	Same rate of vaccination on susceptible people and those who have been infected and recovered	$\frac{dS'}{dt} = \frac{dS}{dt} - \frac{\omega S}{S+R}$	$\frac{dR'}{dt} = \frac{dR}{dt} - \frac{\omega R}{S+R}$	$\frac{dV}{dt} = \omega$	Caulkins et al. (2023)
Loss/Weak	Certain rate of re-infection	$\frac{dS'}{dt} = \frac{dS}{dt} - \omega S$	$\frac{dI'}{dt} = \frac{dI}{dt} + \theta VI$	$\frac{dV'}{dt} = \omega S - \theta VI$	Turkyilmazoglu (2022)
Delayed	Due to the time delay, infection of vaccinated people before reaching complete immunity	$\frac{dS'}{dt} = \frac{dS}{dt} - \omega S/N + (1 - \lambda) \epsilon V$	$\frac{dI'}{dt} = \frac{dI}{dt} + \xi IV/N$ $dR'/dt = dR/dt + \lambda \epsilon V$	$\frac{dV'}{dt} = \omega S/N - \xi IV/N - \epsilon V$	Angeli et al. (2022)

ω: Vaccination rate, θ: Parameter of re-infection disease transmits via weak or loss immune host in contact with the infected population, λ: Vaccine efficacy, ϵ^{-1} is the mean amount of time an individual spends in the vaccinated compartment before reaching immunity and moving to the removed compartment, ξ: The rate at which a vaccinated (but still not immune) individual enters in contact with an infected

5.1.4.2 Hospitalization (SIHR)

In various models, the role of hospitalization varies significantly, being incorporated as a fundamental element in some models while merely being an assumption in others, such as in the works of Mesa et al. (2022) and Caulkins et al. (2023). Gorji et al. (2023) utilized compartmental models to analyze epidemic dynamics, integrating age-stratified social mixing alongside considerations for vaccination status, type, and waning efficacy. Their model is structured as an SIR type, including compartments for susceptible (*S*), infected (*I*), hospitalized in a general ward (*H*), hospitalized in ICU (*IU*), and removed (*R*). Their study showcased the model's capability to capture critical aspects of epidemic dynamics and the resulting healthcare needs. The simplified equations of *H* and *IU* are presented by Eqs. (14) and (15):

$$dH/dt = \xi I + \varepsilon IU - \eta H - \gamma H \tag{14}$$

$$dIU/dt = \eta H - \varepsilon IU - \eta H \tag{15}$$

where ξ is the hospitalization rate of infected individuals, ε is the transition rate from compartment *IU* to *H*, and η is vice versa. Γ is the removal rate. There is no removal rate from the ICU (IU) compartment; individuals are only removed from the hospitalized (H) compartment.

5.1.4.3 Combinations of H and V Components (SIHVR)

Filho et al. (2021) introduced the SEIAHRV model (Susceptible, Exposed, Infected, Asymptomatic, Hospitalized, Recovered, and Vaccinated), building upon the model utilized by Moret et al. (2021). This enhanced model introduces compartments for individuals vaccinated with one or two doses of various vaccines, each defined by known efficacy for one or two doses and including a time delay between doses. Each vaccine type is denoted by a superscript, and the model does not account for primary vaccination failure. To include cases of primary vaccination failure, an additional compartment, U, is introduced.

Albani et al. (2021a, 2021b) proposed an epidemiological model that considers the distribution of the population into *n* age ranges. Within each age range, individuals are categorized into nine compartments: susceptible (*S*), vaccinated (*V*), exposed (*E*), asymptomatic and infective (*IA*), mildly infective (*IM*), severely infective or hospitalized in wards (*IS*), critically infective or admitted to the ICU (*IC*), recovered (*R*), and deceased (*D*). Symptomatic infective individuals who do not require hospitalization are classified as mildly infective, while those needing hospitalization are placed in regular hospital beds, and those in the ICU are labeled as critically infective. In this model, only individuals classified as susceptible are considered for vaccination; should someone receive the vaccine after exposure, they will transition to the asymptomatic or mildly infective compartments.

In the work by Saha et al. (2022), the entire population (*N*) is divided into several subpopulations: the susceptible population (*S*), the asymptomatically or pre-symptomatically infected population (I_1), the symptomatically infected and quarantined population (I_2), the hospitalized population (*H*), the recovered population (*R*), and the vaccinated population (*V*). In this model, individuals in the susceptible class contract the infection and transition to the asymptomatic state upon contact with individuals in both the asymptomatically infected class (I_1) and the symptomatically infected class (I_2). The equations governing the vaccinated and hospitalized populations are represented as follows:

$$dH/dt = \phi I_2 - \psi H \tag{16}$$

$$dV/dt = q_1 S + q_2 I_1 \tag{17}$$

In the Saha et al. (2022) model, individuals from the symptomatically infected class transition to the hospital at a rate denoted by φ. Once hospitalized, individuals then move to the recovered compartment at a rate denoted by ψ following appropriate medical treatment. Additionally, it is assumed that a proportion q_1 of the susceptible population and q_2 of the asymptomatically infected population transition to the vaccinated compartment upon completing an entire two-dose vaccination procedure.

5.1.5 Parameters

Angeli et al. (2022) employed a semi-supervised machine learning approach to approximate the parameters and initial conditions for the SAIVR model, which were subsequently updated using a stochastic gradient descent algorithm. Albani et al. (2021a, 2021b) introduced an innovative methodology to estimate stable rates of hospitalization and death associated with COVID-19, utilizing publicly available reports from diverse communities. These rates were then used to estimate underreported infections in the respective areas by analyzing reported daily hospitalizations and deaths. The study also evaluated the implications of underreporting on vaccination strategies across various disease transmission scenarios using a SEIRVD model. Musa et al. (2022) utilized the SEIHRD model, both with and without vaccination, incorporating a time-varying flexible transmission rate to estimate the Infection Fatality Rate (IFR), Infection Attack Rate (IAR), and the reproduction number (R_0) of COVID-19. Table 19 presents an estimation of the transmission rate.

Table 19 Estimation of transmission rate in SIRV model

Effective factor	Equation	Reference
Age and lockdown	$\alpha_{ij}\left(\beta q\left(1-\kappa\right)+\left(\beta b_{jt}+\overline{\beta}\left(1-b_{jt}\right)\right)\kappa\right)\left(1-b_{jt}\right)$	Gollier (2021)
Different efficacies against different strains	$\beta_i(1-\eta_i)$	Ahumada et al. (2023)

κ: Proportion of infected people is asymptomatic, b_{jt}: Intensity of the lockdown imposed to age class j, α_{ij}: Interactions between age-classes i and j, β: Contagion index, β : Daily contagion rate of confined persons, $\overline{\beta}$: Daily contagion rate of working persons, β_i:transmission coefficient for the ith strain class, η_i: Efficiency of the vaccine against strains from the ith class

5.2 *Non-medical Interventions*

The transmission of SARS-CoV-2 is influenced by a multitude of factors, some intrinsic to the virus and others shaped by social contact patterns and non-pharmaceutical interventions (NPIs) implemented by authorities (Kühn et al., 2021). To mitigate potentially risky interactions, two primary approaches can be adopted: complete avoidance of contact and the use of protective measures, such as mask-wearing, maintaining physical distance, and ensuring proper ventilation in enclosed spaces (Kühn et al., 2021). Studies on non-medical interventions pursued different goals. Some focused on investigating the effectiveness of NPIs (Ng, 2020; Pang et al., 2022), while others aimed to reduce the size of the susceptible population (Mehmood et al., 2021) and estimate the cumulative effects of NPIs (Sen et al., 2023). Additionally, factors such as community isolation, population mobility control, and medical resource aid have also been considered (Zhang et al., 2022a, 2022b). This section analyzes various types of non-medical interventions separately.

5.2.1 Social Distancing

Social distancing, alongside school closures, encompasses measures such as reducing class sizes, implementing fixed seating arrangements, ensuring regular ventilation, and conducting pooled testing in school settings (Sunjaya & Sunjaya, 2020). Contact reductions are also feasible in various environments, including bars, restaurants, supermarkets, and public transport (Kühn et al., 2021). The positive outcomes associated with social distancing measures have been documented, indicating that these interventions can mitigate the peak incidence of COVID-19 outbreaks, resulting in fewer overall cases (Prem et al., 2020). Additionally, social distancing can help contain the spread of contagious diseases (Köhler et al., 2020; Dias et al., 2022) and reduce disease fatality rates (Saha & Samanta, 2021). It is essential to recognize that social distancing policies play a critical role in limiting the spread of the virus. Neglecting these policies may lead to an overestimation of the pandemic's spread. Bertsimas et al. (2021) modeled this effect, demonstrating a decline in the transmission rate over time, as outlined in Eq. (18).

$$\frac{dS}{dt} = -\alpha\gamma(t)S(t)I(t), \tag{18}$$

where α represents a constant baseline rate, and $\gamma(t)$ is a time-dependent function that characterizes the policies of each state, formulated as follows:

$$\gamma(t) = \frac{2}{\pi}\tan^{-1}\left(\frac{-\left(t-(t_0)\right)}{k}\right) + 1 \tag{19}$$

The relationship between the implementation of social distancing measures and government responses is captured using the inverse tangent function, delineating three distinct phases. During Phase I, most activities continue as individuals adapt their behavior. As Phase II begins, the transmission rate experiences a significant decrease due to the implementation of policies. In Phase III, the decline in the transmission rate reaches a saturation point. The parameters t_0 and k act as the median date and the intensity of the response, respectively. Alvarez et al. (2021) also described the effects of social distancing on infection transmission and the transmission rate as follows, where σ is the effectiveness of social distancing.

$$\beta' = (1-\sigma)\beta \tag{20}$$

To calculate the attack rate for clinical trials of COVID-19 vaccine candidates, Berry et al. (2020) utilized the Susceptible-Infected-Resolving-Dead-ReCovered with social distancing (SIRDC-SD) model developed by Fernandez-Villaverde and Jones (2020).

5.2.2 Quarantine and Isolation

The standard model of virus transmission inadequately captures the unique dynamics of COVID-19, particularly the need to categorize individuals based on test results. It is essential to classify individuals into two groups: those requiring quarantine and those who do not (Anand et al., 2020). Quarantine and isolation are pivotal measures for controlling the COVID-19 outbreak, involving the removal of exposed or infectious individuals from the population to prevent further transmission (Dias et al., 2022). Understanding the distinction between isolation and quarantine is important. Isolation separates individuals with an infectious illness from those who are not infected, while quarantine protects individuals who have been exposed to a contagious disease (Schwaid, 2017). The effects of isolation have been explored by Cherif and Hasanov (2023) and Duan et al. (2023).

Quarantine represents a more stringent form of social distancing. During quarantine, individuals remain at home, whereas social distancing allows for limited movement while maintaining distance. The movement of individuals through central locations becomes crucial, serving as networks of interconnected areas (Cherif & Hasanov, 2023). Quarantine can be divided into two categories: (1) intra-regional quarantine, which reflects regional contact patterns (Kühn et al., 2021; Nguyen et al., 2022), and (2) general and inter-regional quarantine, which restricts or prohibits movement between different regions (Kühn et al., 2021; Anwarud et al., 2020; Akinlar et al., 2020). Añonuevo et al. (2023) presented a modified SEIR model incorporating intervention protocols during community quarantine. Additionally, a quarantine compartment is sometimes added to the SIR model; for instance, Cherif and Hasanov (2023) and Amiri Mehra et al. (2020) used the SIQR and SQAIR models, respectively.

5.2.3 Public Education (Awareness)

Raising public awareness is vital for effectively controlling the virus's spread, highlighting recommended measures such as wearing face masks, practicing social distancing, adhering to non-pharmaceutical interventions, and ensuring hand hygiene. The role of print and electronic media in disseminating such awareness is well-recognized (Nanda et al., 2023).

The media plays an essential role in mitigating the spread of infectious diseases by providing up-to-date information on disease outbreaks through various channels. This information can enhance individuals' self-protection awareness and motivate them to take action to reduce infection risks. Consequently, media coverage can significantly influence disease transmission dynamics. Several studies have analyzed the impact of media coverage through mathematical models (Cui et al., 2008a, 2008b; Tchuenche et al., 2011; Xiao et al., 2013; Wang & Xiao, 2014; Cui et al., 2017; Chen et al., 2018). Different studies have formulated various terms to describe contact rate reduction, such as $\beta_1 - \beta_2 I(m_1 + I)$ in Tchuenche et al. (2011) or the saturation incidence rate $\beta SI/(1 + \alpha I)$ in Cui et al. (2017). Additionally, several exponential decay functions have been proposed to describe the transmission coefficient, including $\beta e^{-a_1 E - a_2 I - a_3 H}$ (Liu et al., 2007), μe^{-mI} (Cui et al., 2008a, 2008b), and $\beta e^{-\alpha I}$(Wang & Xiao, 2014). These contact terms are all associated with the count of infected populations, with transmission intensity diminishing as this number increases.

Deng et al. (2021) introduced a novel incidence rate model with an exponential decreasing factor $\beta e^{-P_1 I - P_2 \frac{dI}{dt}}$, which depends on I and the changing rate dI/dt. This term was initially derived from studying the media's impact on non-smooth systems (Xiao et al., 2013), suggesting that enhanced media influence leads to lower transmission rates during outbreaks (Deng et al., 2021). Moreover, the effects of awareness diffusion on epidemic spreading have been explored by researchers such as Purushwani and Sinha (2019), Wang et al. (2019), Kabir and Tanimoto (2019), Shanta and Biswas (2020), and Kabir et al. (2019). Zuo et al. (2021) presented a model with six primary states: US (unawareness susceptibility); AS (awareness susceptibility); UI (unawareness infection); AI (awareness infected); UR (unawareness recovered); and AR (awareness recovered).

5.2.4 Lockdown

Lockdowns and social distancing measures, mandated by governments, primarily impact the services sector, including restaurants, bars, and shops. The reduction in economic activity resulting from these measures is directly linked to decreased opportunities for contagion, thereby affecting the effective reproduction rate (Gros & Gros, 2022). For related approaches, see Gros et al. (2020), Strong and Welburn (2020), Krueger et al. (2020), and Eichenbaum et al. (2020). The literature examining the simultaneous effects of lockdowns and vaccinations as control measures is limited, with relevant studies conducted by Buratto et al. (2022), Fu et al. (2022), Garriga et al. (2022), Caulkins et al. (2023), and Saxena et al. (2021). Much of the existing research has focused on either vaccination or lockdown. In the early stages of the COVID-19 pandemic, researchers concentrated on optimizing the timing and duration of lockdowns, assuming a constant intensity (Bosi et al., 2021; Caulkins et al., 2020).

During the COVID-19 pandemic, lockdown measures were widely implemented to mitigate virus transmission before vaccines became available. These measures had substantial economic repercussions and led to societal disruptions. While early re-opening was desirable, it raised concerns about a potential resurgence of infections. Although the World Health Organization (WHO) has established guidelines for re-opening, decisions primarily rely on epidemiological criteria. To address this, Qian et al. (2021) developed the SEIHR(Q) model to assess the impact of different re-opening strategies on COVID-19 transmission. While lockdowns have proven effective in curbing the virus's spread, they have also resulted in severe economic consequences, particularly in developing countries. Ghaffar et al. (2021) proposed an intelligent smart lockdown approach, implementing dynamic, multi-stage lockdowns combined with social distancing and Standard Operating Procedures (SOPs) for businesses, by calculating the R_0 value after a designated period using the SIR model.

6 Challenges and Barriers

The global health agenda increasingly includes the eradication and elimination of epidemics. However, the implementation of control measures often encounters various challenges (Klepac et al., 2015). Despite the rapid development of highly effective COVID-19 vaccines, which represents a remarkable scientific feat (Hu & Zhang, 2013; Graham, 2020; Lurie et al., 2020), the global vaccination effort faces numerous hurdles, encompassing production, distribution, deployment, and societal acceptance (Vahdani et al., 2023). The challenges associated with implementing control interventions are categorized into economic, social, and political dimensions.

6.1 Economic Challenges

While some measures, such as mask-wearing and telecommuting, are cost-effective, government-imposed lockdowns have resulted in significant economic losses and travel restrictions (Caulkins et al., 2023). The financial implications of non-pharmaceutical interventions during outbreaks should be carefully considered, as epidemic prevention and control incur substantial costs that necessitate economic evaluation (Forsythe et al., 2020). Vaccines are relatively inexpensive compared to the health risks of COVID-19 and the economic consequences of quarantine. In affluent nations, the monetary costs are minimal, while lower-income countries often receive vaccines either for free or at subsidized rates (Caulkins et al., 2023). Cost-effectiveness analyses can aid decision-makers in ensuring optimal use of health resources (Ruiz et al., 2023). Such analyses have explored various COVID-19 vaccination scenarios (Ruiz et al., 2023), assessed the impact of influenza vaccines (Crépey et al., 2020; Shin et al., 2022), evaluated mass vaccination with fractional doses (Du et al., 2023), and calculated the Incremental Cost-Effectiveness Ratio (ICER) for significant COVID-19 policies. Major costs include lost economic productivity, reduced future earnings due to educational disruptions, healthcare expenditures, and costs associated with vaccine development. Benefits include reductions in COVID-19-related mortality, although some studies indicate potential increases in cancer deaths due to delayed care (Atherly & Broek Altenburg, 2023).

The global allocation of COVID-19 vaccines has been marked by severe inequities, with high- and upper-middle-income countries achieving significantly higher vaccination rates compared to lower-middle- and low-income countries (LMICs). Furthermore, the rollout in wealthier countries commenced earlier and affected their economies less severely. In this context, Gozzi et al. (2023) investigated the effects of COVID-19 vaccine inequities across twenty LMICs, quantifying the potential impacts of improved or accelerated vaccine availability. The COVID-19 pandemic has profoundly influenced the economy, leading to unemployment, bankruptcies, and disruptions in production and demand. This situation necessitated shifts in economic policies and the development of new budget financing mechanisms. Additionally, the pandemic exacerbated income inequality, particularly affecting lower-educated workers. Studies recommend government support targeting specific economic sectors, localized supply chain development, advanced technologies, and the use of big data analytics. Recovery in the tourism sector is anticipated to be prolonged, requiring innovative risk management strategies and enhanced coordination between private and public policies for industry resilience during crises (Akaev et al., 2022).

6.2 Social Challenges

While news reports recognize the challenges of vaccination, vaccine promotion strategies that reduce vaccine anxiety have evolved slowly (Wood & Schulman, 2021). This subsection addresses key social challenges related to vaccination.

6.2.1 Public Resistance

The emergence of SARS-CoV-2 mutations has led to an unprecedented surge in COVID-19 cases worldwide. Substantial resistance to vaccination remains prevalent in many countries, heightening the risk of increasing COVID-19 cases and creating conditions conducive to the emergence of mutant strains capable of evading vaccine-induced protection (Jing et al., 2023). Social dynamics play a crucial role in shaping individual compliance with non-medical interventions and vaccination efforts. In this context, Chang et al. (2020) introduced a coefficient called "imitation" to quantify the societal impact on individual behavior and adherence to these measures.

6.2.2 Anti-vaccination

Harari and Monteiro (2022) categorized the population into four groups: pro-vaccine susceptible individuals (P-group), anti-vaccine susceptible individuals (A-group), infected individuals (I-group), and recovered individuals (R-group). Khan and Tanimoto (2023) employed a behavioral model to analyze how individuals choose between anti-vaccine and pro-vaccine stances based on the number of infected individuals within each group. Their findings showed that, at Nash equilibrium, both groups exhibit similar behavior. While increasing vaccination rates can help resolve the social dilemma, higher rates of waning immunity tend to exacerbate the situation. Gollier (2021) examined the welfare impact of having 30% of the population as anti-vaccine individuals, concluding that while it does not worsen the economic crisis, it does increase the death toll by 60,000, predominantly affecting anti-vaccine individuals. Moreover, this dynamic generates negative externalities for senior pro-vaccine individuals, with projections suggesting that approximately 5000 elderly people may die from additional infections in the first three months of the vaccination campaign before becoming immunized.

6.2.3 Hesitancy

Vaccine hesitancy—the reluctance or refusal to immunize despite the availability of vaccines—poses a significant threat to the success of global vaccination campaigns (Mesa et al., 2022). Surveys have revealed that 50% of American adults are hesitant to receive a COVID-19 vaccine, with a lack of understanding of vaccination benefits being a key obstacle to acceptance (Guidry et al., 2021; Szilagyi et al., 2020). Given the challenges posed by vaccine hesitancy, it is essential to understand the effects of varying vaccination rates on pandemic control (Storlie et al., 2021a). Vaccine hesitancy not only weakens individual immunity but also threatens community health (Erkayman et al., 2023).

Researchers have examined vaccine hesitancy from various perspectives, focusing on its impact on pandemic control measures and the relaxation of non-medical interventions (Mesa et al., 2022). Additional studies have explored the connections between vaccine hesitancy and factors such as virus mutations (Choi & Shim, 2020), mortality rates (Arribas et al., 2022), and the number of susceptible individuals (Ortiz-Robinson & Foster-Bey, 2021).

6.3 Political Challenges

Balmford et al. (2020) revealed the implicit costs national governments are willing to incur to protect their citizens, reflected in the economic activity sacrificed to save lives. These estimates of the "price of LIF" show significant variation across countries. Differences in policy interventions have outweighed socioeconomic factors in explaining the wide range of death rates observed. It's essential to recognize that the politicization of vaccination in many countries, along with debates surrounding vaccine safety and effectiveness, has been compounded by the efforts of anti-vaccination groups, which have undermined awareness and uptake, creating widespread uncertainty (Ruiz et al., 2023). This political climate has led to delays in vaccination campaigns, negatively affecting estimates of virus-related mortality (Albani et al., 2021a, 2021b).

Vaccine nationalism, particularly in the Western world, prioritizes low-risk individuals in vaccine-rich countries over high-risk individuals in vaccine-poor nations, exacerbating the global welfare consequences of the pandemic. While vaccine-rich countries benefit from their stockpiles, this allocation is considered undesirable. In a two-country model, Gollier (2021) demonstrated that extreme vaccine nationalism, where countries prioritize their populations before global distribution, could result in up to a 20% increase in the global death toll.

7 Optimization in Epidemiological Models

In practice, the probability of infection fluctuates dynamically due to disease prevalence and age-related differences in susceptibility. For instance, the most effective vaccination rates for all age groups are seen at the onset of a seasonal influenza outbreak, although optimal vaccination coverage varies among different age groups (Shim, 2013). Furthermore, as the severity of a disease increases and government interventions—such as school closures and travel restrictions—are implemented, individuals may adjust their behavior by increasing social distancing. To conduct a more comprehensive analysis of optimal personal strategies, both vaccination rates and social distancing measures should be treated as state variables (Choi & Shim, 2020). Consequently, determining the best approach to managing an epidemic through medical and non-medical interventions has become a key area of research.

Game theory offers one method to address this issue. It includes cost-efficiency analyses of voluntary vaccination (Kabir & Tanimoto, 2020), determining optimal control strategies like vaccination and social distancing (Choi & Shim, 2020), and evaluating the combination of vaccination with non-medical interventions in the context of public opposition to vaccines (Jing et al., 2023). Advanced techniques like evolutionary algorithms and machine learning are also applied. For instance, Alkhamis and Hosny (2022) employed a Controlled Elitism Non-Dominated Sorting Genetic Algorithm to assess pulsive vaccination strategies, while Haq et al. (2022) compared the outcomes of the SIR model with machine learning techniques to forecast pandemic trends and determine appropriate intervention levels. Wang et al. (2022) introduced T-SIRGAN, which integrates epidemiological theories with deep learning to capture complex epidemic dynamics and illustrate non-linear relationships, resulting in more accurate predictions of COVID-19 spread.

The theory of optimal control aims to improve policy implementation by applying control variables within epidemiological models. Additionally, the combination of simulation and optimization represents another approach, where outputs from epidemiological models are fed into optimization frameworks to enhance decision-making.

7.1 Optimal Control Theory

Accurate parameter estimation is crucial for the optimal control technique known as model predictive control. However, developing a state estimator for non-linear systems presents considerable challenges. By integrating compartmental epidemiological models with the optimal control problem (OCP), researchers have adopted a robust dynamic optimization approach to tackle these difficulties (Castilho, 2006). This method provides an endemic equilibrium point for assessing the impact of COVID-19 contagion (Gaff & Schaefer, 2009). Despite its potential, the complexity involved in implementing these methods has limited their widespread adoption among researchers (Vahdani et al., 2023). Table 20 outlines several studies that have utilized this approach.

7.2 Simulation and Optimization

Previous studies have combined simulation and optimization for various purposes. Loertscher and Muir (2021) expanded their analysis to incorporate adaptive lockdown policies, focusing on identifying the optimal dynamic strategy while accounting for capacity constraints. López et al. (2022) applied a non-linear programming optimization method for model fitting, aiming to estimate the minimum of a constrained non-linear function.

Medical resource shortages present significant challenges in underdeveloped regions, particularly in remote rural areas of developing countries that lack adequate facilities. Efficient allocation of these limited resources is critical for epidemic control (Deng et al., 2021). In some studies, vaccine allocation begins with an assessment of the community's disease status—considering the numbers of sick, susceptible, and recovered individuals—based on an epidemiological model. Vahdani et al. (2023) introduced methods for optimal vaccine allocation and distribution, proposing a novel multi-vaccine, multi-depot location-inventory-routing problem model to evaluate vaccine demand. Their approach dynamically determined vaccine demand to achieve local equilibrium within the optimal control problem. Similar studies, such as those by Jarumaneeroj et al. (2022) and Hazard-Valdés and Montero (2023), referenced components and outputs from the epidemiological model as constraints in their mathematical frameworks, where estimated vaccine requirements were derived from these models.

8 Epidemiological Models in the Supply Chain

The global health crisis resulting from pandemics has significantly disrupted supply chains, with hoarding and panic buying amplifying downstream effects. Travel restrictions and lockdowns imposed by many countries have further disturbed the supply-demand balance, emphasizing the urgent need for short-term, real-time forecasts to manage the pandemic's impact on supply chains effectively (Nikolopoulos et al., 2021). The COVID-19 pandemic has challenged traditional views of supply chain resilience, causing a sharp decline in economic indicators such as productivity and global GDP (Harris, 2020). Studies report a 16.24% decrease in labor demand due to travel and trade restrictions and workplace closures (Castro et al., 2020; Baveja et al., 2020; Nikolopoulos et al., 2021).

Araz et al. (2020) highlighted COVID-19 as the most severe disruption to the global supply chain in recent memory. Several studies have addressed this issue,

Table 20 Equations of optimal control

Objective	Control variable	Equations	Reference
Minimize the number of infected people while minimizing vaccine immunization costs	Vaccination rate	$dS'/dt = dS/dt - \varepsilon uS$ $dR'/dt = dR/dt + \varepsilon uS$	Hwang et al. (2020)
Minimize the number of infected individuals as well as the number of vaccines used.		$dS'/dt = dS/dt - uS$ $dV'/dt = uS$	Libotte et al. (2020)
Minimize the cost incurred by infected and quarantined individuals and the immunization program in order to maximize the number of recovered individuals		$dS'/dt = dS/dt - uS$ $dR'/dt = dR/dt + uS$	Vahdani et al. (2023)
Minimize cost of lockdown	Lockdown intensity adjustment	$dS'/dt = \beta(u)SI/N \ldots$ $dI'/dt = -\beta(u)SI/N \ldots$	Caulkins et al. (2023)
Minimize the total number of infected, the total use of interventions (vaccines, treatment, social distancing) and the final number of susceptible	Effects of vaccination (u_1), treatment (u_2), and social distancing (u_3)	$dS'/dt = -(1 - u\_3)\beta SI - \varepsilon u\_1S$ $dI'/dt = (1 - u\_3)\beta SI - dR/dt - u\_2I$ $dR'/dt = dR/dt + \varepsilon u\_1S + u\_2I$ $dW/dt = u\_1S$	Ortiz-Robinson and Foster-Bey (2021)

u: Control variable, ε: Vaccination efficacy, W: The number of susceptible willing to vaccinate including simulations predicting its impact on supply chains (Ivanov, 2020), assessments of stress on hospitals and ICUs (Team & Murray, 2020a), and implications for food supply chains (Hobbs, 2020). Proposed solutions to mitigate supply chain disruptions include increasing the supply of critical products and services (Team & Murray, 2020b) and optimizing healthcare supply and demand using Fuzzy Inference Systems (Govindan et al., 2020).

Epidemiological models can also be used to identify and mitigate supply chain disruptions. The SIR model, typically applied to analyze infection transmission, is adapted to dynamically assess and predict supply chain risk status over time (Liang et al., 2022; Yu et al., 2022a, 2022b; Chen & Yin, 2023). These models focus on the transmission of risk rather than disease. Since many epidemic control models rely on differential equations, optimal control offers a practical approach to represent both epidemic and supply chain dynamics within a unified theory of continuous optimization. For example, Brusset et al. (2022a) illustrated how production managers can optimize efforts in implementing prophylactic and social distancing measures to balance costs and demand in a plant or facility workforce. Brusset et al. (2022b) analyzed the ripple effects of epidemics on supplier production capacity, and Brusset et al. (2023b) outlined models detailing necessary prophylactic measures over time.

8.1 Vaccine Supply Chain

Azam et al. (2021) developed an integrated modeling framework that combines a vaccine supply chain model with an SEIR model. This framework captures the preparation time required to initiate and conclude vaccination campaigns, evaluates vaccination coverage, and clarifies the simultaneous dynamics of disease outbreaks. Vahdani et al. (2023) introduced a new model that addresses the multi-vaccine, multi-depot location-inventory-routing problem, leveraging an epidemiological model to forecast vaccine demand. Enayati and Ozaltin (2020) employed a compartmental model to study influenza transmission, developing a mathematical program designed to minimize vaccine dose distribution during the initial phases of an outbreak. Gomes et al. (2023) proposed a two-stage approach for an agile and lean supply chain, optimizing the vaccine distribution network while using the SEIR model to assess the impact of vaccination interventions.

8.2 Other Supply Chains

Guttieres et al. (2021) introduced a framework for developing an end-to-end antibody strategy that addresses clinical needs during COVID-19 and future pandemics. This framework integrates product design, optimization, epidemiology, demand, and supply as interconnected elements. Nikolopoulos et al. (2021) provided predictive analytics tools for forecasting and planning during pandemics, addressing

COVID-19 growth rates and excess demand for products and services through statistical, epidemiological, machine- and deep-learning models, alongside a new hybrid forecasting method based on nearest neighbors and clustering. Sharif et al. (2022) extended the SEIR model with four sub-models to simulate the effects of non-pharmaceutical interventions on outbreak patterns and healthcare system responses to surges in demand. Brusset et al. (2023a) proposed a dynamic approach, featuring an optimal control model for supply chain reconfiguration, which is integrated with epidemic dynamics to analyze ripple effects.

9 Discussion

The substantial volume of literature produced in response to the COVID-19 pandemic highlights the critical importance of research aimed at managing infectious disease outbreaks. The pandemic has exposed the extensive damage that infectious diseases can inflict on society (Bloom et al., 2021). During the outbreak, all levels of society experienced significant disruptions, with the resultant mortality exacerbated by systemic weaknesses in healthcare, further increasing the death toll from unrelated diseases (Roser et al., 2020). The pandemic has also led to long-term socioeconomic consequences, including heightened burdens on healthcare systems, interruptions to routine immunizations, the closure of educational institutions and workplaces, disruptions in international supply chains, and the worsening of social inequalities (Curiel & Ramirez, 2021; Ali & Albahri, 2020; Lima et al., 2020; Bloom et al., 2021). Effective epidemic control requires a thorough understanding of societal conditions and disease dynamics, enabling policymakers to determine the necessary intensity and type of interventions. Thus, a comprehensive grasp of epidemic disease modeling is critically important, as discussed in this review.

Through comparative analysis, the SIR model emerges as the most frequently utilized model, likely due to its clarity and simplicity in calculations. Additionally, the SIR model has demonstrated superior performance over the SEIR model in accurately reflecting data from confirmed cases (Nakamura et al., 2021). According to Table 10, extended models are used to investigate the effects of interventions, considering more details, such as the number of vaccinated and quarantined individuals in the ICU. The predominant assumptions across the analyzed models include homogeneous, fixed population and age-based group classifications. Other scenarios that could broaden the model are discussed in the conclusion section. Medical and non-medical interventions, encompassing vaccination, hospitalization, and quarantine, can serve as foundational assumptions or compartments within the models. To assess the efficacy of multiple vaccine doses and types, models such as SIRV and their extensions are employed, differentiating among individuals vaccinated with various dosages.

The basic reproduction number, a pivotal metric in epidemic evaluation, frequently uses a simplified formula (β/γ) in research. However, some studies have incorporated additional coefficients based on specific problem contexts, as outlined in Table 15. The effective reproduction factor analyzes the impact of interventions, showing how vaccination rates, vaccine effectiveness, and social distancing can reduce reproduction numbers, as presented in Table 16. Epidemiological models may be enhanced through optimal control theory, integrating decision variables as control variables into ordinary differential equations (ODEs). Table 21 shows the objectives, control variables, and related equations of this method, often focused on minimizing affected populations and intervention costs. Few studies reviewed have combined mathematical and epidemiological approaches to inform vaccine allocation decisions. ODEs can be incorporated as constraints within mathematical models or used as parameters, such as vaccine demand or susceptible individual counts.

Disease outbreaks have the potential to significantly disrupt supply chains. One strategy to manage these disruptions during pandemics is to leverage epidemiological models to forecast demand impacts for goods and services, thereby integrating these insights into the supply chain framework. While further research is needed, this study briefly explores this topic.

10 Conclusion

This paper provides a comprehensive review of epidemiological models from a modeling perspective, focusing specifically on various SIR models detailed in the literature, including research published in prominent journals from 2019 to July 2023. Summarizing and examining similar review articles from a modeling standpoint is invaluable for researchers in this domain, particularly as existing reviews have not extensively covered all SIR models. Given the COVID-19 pandemic and the subsequent increase in research activity, there is a pressing need to synthesize and summarize these studies. Furthermore, there are evident research gaps in the assumptions tables that can be addressed. Here are some recommendations for future research in this field:

- Many articles assume homogeneity among individuals, positing that contact patterns and disease susceptibility are uniform across the population. However, this assumption is unrealistic. Achieving homogeneity typically involves categorizing individuals into subgroups, commonly based on age, though various other factors may also influence this categorization. Critical system parameters such as immunity, susceptibility, and recovery rates vary significantly across the population. For instance, individuals' occupations can affect their contact patterns, with medical staff having the highest exposure to infected individuals. Additionally, individuals with underlying health conditions are at a higher risk for infection and mortality. Thus, relying solely on age to define homogeneous groups is insufficient; a broader range of influential factors must be considered.
- Many models designed for short-term analysis often overlook birth and death rates. However, this approach is impractical for long-term diseases like COVID-

Table 21 Integration of simulation and optimization approaches

	Epidemic model		Mathematical model			
Reference	Components	Outputs	Objectives	Constraints	Decision variables	Solution approach
Rao and Brandeau (2021a)	SIRD	Susceptible individuals	Minimize new infections, deaths, life years lost, or quality-adjusted life years (QALYs) lost due to death	Limited supply of vaccine- Proportion of eligible people - Equity	Proportion of the entire population that is vaccinated	Exhaustive search
Jarumaneeroj et al. (2022)	SIQRV	ODEs	Minimize total weighted strain on the whole healthcare systems	Flow balancing- vaccine administration capacity- Equity- Allowed number of vaccinations- ODEs	The number of doses of vaccine - the remaining doses of vaccine- number of vaccinated, susceptible, infection, recovered, and died individuals number	Proposed iterative approach by Lsqcurvefit and IntGen

(continued)

Table 21 (continued)

Reference	Epidemic model		Mathematical model			
	Components	Outputs	Objectives	Constraints	Decision variables	Solution approach
Vahdani et al. (2023)	SIQUR	Demand of vaccine	Minimizes the total costs of the system	Distribution-Inventory-Capacity-Fairness-Routing	Inventory level of vaccine-Amount of vaccine shipped-Amount of fulfilled demand of population-Amount of demand of population-Variables related to location and routing	Bender's decomposition algorithm
Hazard-Valdés and Montero (2023)	SEAIR	ODEs	Minimize the number of infectious people in all the periods and minimize of the number of infectious people at the same time	Contact rate-vaccination constraints- ODEs	Number of persons from subgroup i who will be vaccinated in period p.	Tabu search

19, and a more comprehensive investigation should include these demographic factors. Some studies have explored the epidemic's impact on birth rates (Pomar et al., 2022; Rose et al., 2022), which can fluctuate across different waves depending on the nature of control interventions. Additionally, mortality from non-infection-related causes has also been affected, as individuals may avoid seeking medical care due to fears of COVID-19 infection (Woolf et al., 2020).

- Given the implementation of government interventions and the heterogeneity of populations across regions, incorporating the assumption of multiple regions and the potential for population movement between them can offer decision-makers a more comprehensive understanding of intervention outcomes. This approach is particularly crucial when evaluating quarantine measures and travel restrictions on both regional and local scales.
- Different diseases manifest in specific periods; thus, rather than assuming multiple strains of a virus, models can be structured to account for multiple time periods. This approach allows for the inclusion of disease parameters, vaccine effectiveness against various strains, and the intensity of interventions at different times.
- Investigating social, economic, and political challenges can significantly influence the implementation of interventions and disease control. Addressing these challenges, particularly social ones, is vital for effectively managing epidemics. Susceptible individuals can be categorized based on their reactions to interventions, allowing for an assessment of the effects of vaccine hesitancy and opposition on epidemic dynamics and determining the necessary vaccine quantities.
- The movement of individuals between different groups and regions is often overlooked in research; however, accurately estimating intervention effects requires consideration of population movement. Long-term planning could involve incorporating the movement of people across different age groups. Additionally, transitions among social challenge groups (prone, anti-vaccine, and hesitant) could be influenced by vaccination effectiveness. By considering migration between regions, the accurate estimation of the required number of vaccines for each region, as well as the effects of travel restrictions on the spread of disease, can be further analyzed. Therefore, both epidemic and mathematical models should integrate the complexity of multiple regions. The decision variables of the problem include the implementation or non-implementation of inter-regional quarantines and the specific number of vaccines needed within each region.
- The integration of epidemiological and mathematical models remains underutilized, despite its significant potential to produce valid and practical solutions. Mathematical models can facilitate critical decisions regarding vaccine allocation and distribution, as well as the implementation of non-medical interventions. By leveraging the results of the epidemiological model over multiple periods, disease conditions and required vaccine quantities can be determined. Policymakers can then make optimal decisions based on the outcomes of the mathematical model. This iterative framework allows for continual refinement of decisions, where conditions are updated for the subsequent period and the epidemiological

model is reevaluated. This cycle of simulation and optimization continues until a predetermined stopping condition is fulfilled, which could involve reaching the mathematical model's objectives or completing a set number of iterations.

- Certain interventions, such as tracing and screening, have received insufficient attention in existing research. Rusu et al. (2021) considered tracing status in addition to infection state, while Rabil et al. (2022) highlighted the advantages of integrated screening and vaccination for infection control. Incorporating insights from these studies can enhance epidemiological models by including these interventions, thereby enabling a comprehensive investigation into their impacts on disease management and control.
- This research highlights the use of epidemiological models in supply chain management. By carefully reviewing relevant literature, researchers can identify how these models are integrated into the supply chain structure and pinpoint research gaps for further development. Additionally, recognizing the vital role of the vaccine supply chain during disease outbreaks, employing epidemiological models to analyze disruptions and their ripple effects can significantly enhance disruption management and control.

References

Adiga, A., Wang, L., Hurt, B., Peddireddy, A. S., Porebski, P., Venkatramanan, S., et al. (2021). All models are useful: Bayesian ensembling for robust high resolution COVID-19 Forecasting. In *Proceedings of the 27th ACM SIGKDD Conference on Knowledge Discovery and Data Mining* (pp. 2505–2513).

Ahmetolan, S., Bilge, A. H., Demirci, A., Peker-Dobie, A., & Ergonul, O. (2020). What can we estimate from fatality and infectious case data using the Susceptible-Infected-Removed (SIR) Model? A case study of Covid-19 pandemic. *Frontiers in Medicine, 7*, 556366.

Ahumada, M., Ledesma-Araujo, A., Gordillo, L., & Marín, J. F. (2023). Mutation and SARS-CoV-2 strain competition under vaccination in a modified SIR model. *Chaos, Solitons and Fractals, 166*, 112964.

Ai, H., Wang, Q., & Liu, W. (2022). A mathematical prediction model of infectious diseases considering vaccine and temperature, and its prediction in Hong Kong. *Heliyon, 8*, e12469.

Akaev, A., Zvyagintsev, A. I., Sarygulov, A., Devezas, T., Tick, A., & Ichkitidze, Y. (2022). Growth recovery and COVID-19 pandemic model: Comparative analysis for selected emerging economies. *Mathematics, 10*, 3654.

Akinlar, M. A., Inc, M., Gómez-Aguilar, J. F., & Boutarfa, B. (2020). Solutions of a disease model with fractional white noise. *Chaos, Solitons, and Fractals, 137*(Article ID 109840).

Alanazi, S. A., Kamruzzaman, M. M., et al. (2020). Measuring and preventing COVID-19 using the SIR model and machine learning in smart health care. *Journal of Healthcare Engineering*. https://doi.org/10.1155/2020/8857346

Albani, V., Loria, J., Massad, E., & Zubelli, J. (2021b). COVID 19 underreporting and its impact on vaccination strategies. *BMC Infectious Diseases, 21*, 1111.

Albani, V. V., Loria, J., Massad, E., & Zubelli, J. P. (2021a). The impact of COVID-19 vaccination delay: A data-driven modeling analysis for Chicago and New York City. *Vaccine, 39*, 6088–6094.

Aldila, D., Khoshnaw, H. A., et al. (2020). A mathematical study on the spread of COVID-19 considering social distancing and rapid assessment: The case of Jakarta, Indonesia. *Chaos, Solitons and Fractals, 139*, 110042.

Ali, I., & Albahri, O. M. (2020). COVID-19: Disease, management, treatment, and social impact. *Science of the Total Environment, 728*, 138861.

Alkhamis, A. K., & Hosny, M. (2022). A synthesis of pulse influenza vaccination policies using an efficient Controlled Elitism Non-Dominated Sorting Genetic Algorithm (CENSGA). *Electronics, 11*, 3711.

Al-Sheikh, S. A. (2013). Modeling and analysis of an SEIR epidemic model with a lim- ited resource for treatment resource for treatment. *Global Journal of Science Frontier Research Mathematics and Decision Sciences, 12*, 56–66.

Alvarez, M. M., Bravo-González, S., & Trujillo-de Santiago, G. (2021). Modeling vaccination strategies in an Excel spreadsheet: Increasing the rate of vaccination is more effective than increasing the vaccination coverage for containing COVID-19. *PLoS ONE, 16*(7), e0254430.

Amiri Mehra, A. H., Shafieirad, M., Abbasi, Z., & Zamani, I. (2020). Parameter estimation and prediction of COVID-19 epidemic turning point and ending time of a case study on SIR/SQAIR epidemic models. *Computational and Mathematical Methods in Medicine, 1465923.*

Anand, N., Sabarinath, A., Geetha, S., & Somanath, S. (2020). Predicting the spread of COVID-19 using SIR model augmented to incorporate quarantine and testing. *Transactions of the Indian National Academy of Engineering, 5*(2), 141–148.

Anderson, R. M., & May, R. M. (1979). Population biology of infectious diseases: Part I. *Nature, 280*, 361–367.

Angeli, M., Neofotistos, G., Matthhekis, M., & Kaxiras, E. (2022). Modeling the effect of the vaccination campaign on the COVID-19 pandemic. *Chaos, Solitons and Fractals, 111621*, 154.

Añonuevo, L. E., Lachica, Z. P., Amistas, D. A., Lato, J. E., Bontilao, H. L., Catalan, J. M., et al. (2023). Transmission dynamics and baseline epidemiological arameter estimates of Coronavirus disease 2019 pre-vaccination: Davao City, Philippines. *PLoS ONE, 18*(4), e0283068.

Anwarud, D., Khan, A., & Baleanu, D. (2020). Stationary distribution and extinction of stochastic coronavirus (COVID-19) epidemic model. *Chaos, Solitons & Fractals, 139*, Article ID 110036.

Araz, O. M., Choi, T. M., Olson, D., & Salman, F. (2020). Data Analytics for Operational Risk Management. *Decision Sciences, 51*. https://doi.org/10.1111/deci.12443

Arefi, M. F., & Poursadeqiyan, M. (2020). A review of studies on the COVID-19 epidemic crisis with a preventive approach. *Work, 66*, 717–729.

Arfan, M., Shah, K., Abdeljawad, T., Mlaiki, N., & Ullah, A. (2021). A Caputo power law model predicting the spread of the COVID-19 outbreak in Pakistan. *Alexandria Engineering Journal, 60*(1), 447–456.

Arregui, S., Aleta, A., Sanz, J., & Moreno, Y. (2018). Projecting social contact matrices to different demographic structures. *PLoS Computational Biology, 14*(12), e1006638.

Arribas, A. M., Aleta, A., & Moreno, Y. (2022). Impact of vaccine hesitancy on secondary COVID 19 outbreaks in the US: An age structured SIR model. *BMC Infectious Diseases, 22*, 511.

Atherly, A. J., & Broek Altenburg, E. M. (2023). The effect of medical innovation on the cost effectiveness of Covid 19 related policies in the United States using a SIR model. *BMC Health Services Research, 23*(1), 372.

Attanayake, A. (2022). Vaccination coverage for COVID-19 in Sri Lanka: With and without age stratification on susceptible-infectious-recovered simulation. *Journal of Occupational Health and Epidemiology, 11*(2), 91–98.

Aylward, B., Barboza, P., et al. (2014). Ebola virus disease in West Africa – The first 9 months of the epidemic and forward projections. *The New England Journal of Medicine, 371*, 1481–1495.

Azam, J. M., Saitta, B., Bonner, K., Ferrari, M. J., & Pulliam, J. R. (2021). Modelling the relative benefits of using the measles vaccine outside cold chain for outbreak response. *Vaccine, 39*, 5845–5853.

Badfar, E., Zaferani, E. J., & Nikoofard, A. (2022). Design a robust sliding mode controller based on the state and parameter estimation for the nonlinear epidemiological model of Covid-19. *Nonlinear Dynamics, 109*, 5–18.

Bailey, N. T. l. (1975). *The mathematical theory of infectious diseases and its applications*. Charles Griffin & Company Ltd.

Baize, S., Pannetier, D., et al. (2014). Emergence of Zaire Ebola virus disease in Guinea. *The New England Journal of Medicine, 371*, 1418–1425.

Balmford, B., Annan, J. D., Hargreaves, J. C., Altoè, M., & Bateman, I. J. (2020). Cross country comparisons of Covid 19: Policy, politics and the price of life. *Environmental and Resource Economics, 76*, 525–551.

Barrat, A., Barthelem, M., & Vespignan, A. (2008). *Dynamical processes on complex networks*. Cambridge University Press.

Barua, S., & Dénes, A. (2023). Global dynamics of a compartmental model to assess the effect of transmission from deceased. *Mathematical Biosciences, 364*, 109059.

Basnarkov, L., Tomovski, I., Sandev, T., & Kocarev, L. (2022). Non-Markovian SIR epidemic spreading model of COVID-19. *Chaos, Solitons and Fractals, 160*, 112286.

Batistela, C. M., Romas, M. M., Cabrera, M. A., Dieguea, G. M., & Piqueira, J. R. (2021). Vaccination and social distance to prevent COVID-19. *IFAC-PapersOnLine, 54*(15), 151–156.

Bauch, C., Szusz, E., & Garrison, L. (2009). Scheduling of measles vaccination in low- income countries: Projections of a dynamic model. *Vaccine, 27*(31), 4090–4098.

Baveja, A., Kapoor, A., & Melamed, B. (2020). Stopping COVID-19: A pandemic- management service value chain approach (SSRN Scholarly Paper No. ID 3555280). https://doi.org/10.2139/ssrn.3555280

Beams, A. B., Bateman, R., & Alder, F. R. (2021). Will SARS-CoV-2 become just another seasonal coronavirus? *Viruses, 13*, 854.

Berry, D. A., Berry, S., Hale, P., Isakov, L., Lo, A. W., Siah, K. W., & Wong, C. H. (2020). A cost/benefit analysis of clinical trial designs for COVID-19 vaccine candidates. *PLoS ONE, 15*(12), e0244418.

Bertsimas, D., et al. (2021). From predictions to prescriptions: A data-driven response to COVID-19. *Health Care Management Science, 24*, 253–272.

Betcheva, L., Erhun, F., Feylessoufi, A., Fryers, P., Goncalves, P., Jiang, H., et al. (2021). An adaptive research approach to COVID-19 forecasting for regional health systems in England. Available at SSRN3695258.

Betti, M., et al. (2021). Integrated vaccination and nonpharmaceutical interventions based strategies in Ontario, Canada, as a case study: A mathematical modelling study. *Journal of the Royal Society Interface, 18*, 20210009.

Bhadauria, A., Pathak, R., & Chaudhary, M. (2021). A SIQ mathematical model on COVID-19 investigating the lockdown effect. *Infectious Disease Modelling, 6*, 244–257.

Bhagat, R. K., Wykes, M. S., Dalziel, S. B., & Linden, P. F. (2020). Effects of ventilation on the indoor spread of COVID-19. *Journal of Fluid Mechanics, 903*.

Bicher, M. R., Rippinger, C., Urach, C., Brunmeir, D., Siebert, U., & Popper, N. (2020). *Agent-based simulation for evaluation of contact-tracing policies against the spread of SARS-CoV-2*. MedRxiv. https://doi.org/10.1101/2020.05.12.20098970

Biswas, D., & Alfandari, L. (2022). Designing an optimal sequence of non-pharmaceutical interventions for controlling COVID-19. *European Journal of Operational Research, 303*(3), 1372–1391.

Biswas, M. H., Paiva, L. T., & De Pinho, M. (2014). A SEIR model for control of infectious diseases with constraints. *Mathematical Biosciences and Engineering, 11*(4), 761–784.

Bloom, D. E., Cadarette, D., & Ferranna, M. (2021). The societal value of vaccination in the age of COVID-19. *American Journal of Public Health, 111*(6), 1049–1054.

Bogdanov, O. (2023). Variants of the stochastic SIR models and vaccination strategies. *Cybernetics and Systems Analysis, 59*(2), 325–330.

Booton, R. D., MacGregor, L., Vass, L., et al. (2021). Estimating the COVID-19 epidemic trajectory and hospital capacity requirements in South West England: A mathematical modelling framework. *BMJ Open, 11*, e041536.

Booton, R. D., Powell, A. L., Turner, K. M., & Wood, R. M. (2022). Modelling the effect of COVID-19 mass vaccination on acute hospital admissions. *International Journal for Quality in Health Care, 34*(2), 1–7.

Bosi, C., Camacho, C., & Desmarchelier, D. (2021). Optimal lockdown in altruistic economies. *Journal of Mathematical Economics, 93*, 102488.

Bouaine, A., & Rachik, M. (2018). Modeling the impact of immigration and climatic conditions on the epidemic spreading based on cellular automata approach. *Ecological Informatics, 46*, 36–44.

Brett, T., Ajelli, M., Liu, Q. H., Krauland, M. G., et al. (2020). Detecting critical slowing down in high-dimensional epidemiological systems. *PLoS Computational Biology, 16*(3), e1007679.

Brunnermeier, M. (2020). *Bengt Holmstrom: The seasonality of Covid- 19*. Princeton University Bendheim Center for Finance.

Brusset, X., Davari, M., Kinra, A., & Torre, D. L. (2022a). Modelling ripple effect propagation and global supply chain workforce productivity impacts in pandemic disruptions. *International Journal of Production Research, 61*(8), 2493–2512.

Brusset, X., Ivanov, D., Jebali, A., Torre, D. L., & Repetto, M. (2023a). A dynamic approach to supply chain reconfiguration and ripple effect analysis in an epidemic. *International Journal of Production Economics, 263*, 108935.

Brusset, X., Jebali, A., & Torre, D. L. (2022b). Production optimization in a pandemic context. *International Journal of Production Research, 61*(5), 1642–1663.

Brusset, X., Jebali, A., Torre, D. L., & Liuzzi, D. (2023b). Production optimization in the time of pandemic: An SIS-based optimal control model with protection effort and cost minimization. *Annals of Operations Research*, 1–24. https://doi.org/10.1007/s10479-023-05206-8

Buonanno, G., Stabile, L., & Morawska, L. (2020). Estimation of airborne viral emission: Quanta emission rate of SARS-CoV-2 for infection risk assessment. *Environment International, 141*, 205794.

Buratto, A., Muttoni, M., Wrzaczek, S., & Freiberger, M. (2022). Should the COVID-19 lockdown be relaxed or intensified in case a vaccine becomes available? *PLoS ONE, 15*(2), E0243413.

Caldwell, J. M., Le, X., McIntosh, L., Meehan, M. T., Ogunlade, S., Ragonnet, R., et al. (2021). Vaccines and variants: Modelling insights into emerging issues in COVID-19 epidemiology. *Paediatric Respiratory Reviews, 39*, 32–39.

Callegari, B., & Feder, C. (2022). A literature review of pandemics and development: The long-term perspective. *Economics of Disasters and Climate Change, 6*, 183–212.

Carlson, C., Gomez, A. C., Bansal, S., & Ryan, S. J. (2020). Misconceptions about weather and seasonality must not misguide covid-19 response. *Nature Communications, 11*(1), 1–4.

Castilho, C. (2006) Optimal control of an epidemic through educational campaigns. *Electronic Journal of Differential Equations, 125*, 1–11.

Castro, M. F., Duarte, J. B., & Brinca, P. (2020). Measuring sectoral supply and demand shocks during COVID-19. https://doi.org/10.20955/wp.2020.011

Cauchemez, S., Boëlle, P. Y., Thomas, G., & Valleron, A. J. (2006). Estimating in real time the efficacy of measures to control emerging communicable diseases. *American Journal of Epidemiology, 164*(6), 591–597.

Caulkins, J. P., Grass, D., Feichtinger, G., Hartl, R., et al. (2020). How long should the COVID-19 lockdown continue? *PLoS ONE, 15*(2), E0243413.

Caulkins, J. P., Grass, D., Feichtinger, G., Hartl, R. F., Kort, P. M., Kuhn, M., et al. (2023). The hammer and the jab: Are COVID-19 lockdowns and vaccinations complements or substitutes? *European Journal of Operational Research, 311*, 233–250.

Chang, S. L., Piraveenan, M., & Prokopenko, M. (2020). Impact of network assortativity on epidemic and vaccination behaviour. *Chaos, Solitons & Fractals, 140*.

Chaturvedi, D., & Chakravarty, U. (2021). Predictive analysis of COVID-19 eradication with vaccination in India, Brazil, and U.S.A. *Infection, Genetics and Evolution, 92*, 104834.

Chen, C., Chong, N. S., & Smith, R. (2018). A Filippov model describing the effects of media coverage and quarantine on the spread of human influenza. *Mathematical Biosciences, 296*, 98–112.
Chen, C. M., & Stanciu, A. C. (2021). Simulating influenza epidemics with waning vaccine immunity. *Medicine, 100*(36), e27169.
Chen, J., Qi, T., Liu, L., Ling, Y., Qian, Z., Li, T., et al. (2020a). Clinical progression of patients with COVID-19 in Shanghai. *China. Journal of Infection, 80*(5), e1–e6.
Chen, J., Xiong, Y., Tong, J., & Wang, Y. (2020b). Analysis and prediction of COVID-19 in the US based on the time-varying parameters SIR model. *Journal of Physics: Conference Series, 1678*, 012082.
Chen, J., & Yin, T. (2023). Transmission mechanism of Post-COVID-19 emergency supply chain based on complex network: An improved SIR model. *Sustainability, 15*, 3059.
Chen, M., Li, M., Hao, Y., Liu, Z., Hu, L., & Wang, L. (2020c). The introduction of population migration to SEIAR for COVID-19 epidemic modeling with an efficient intervention strategy. *Information Fusion, 64*, 252–258.
Cheng, V. C., Wong, S. C., Chuang, V. M., et al. (2020). The role of community-wide wearing of face mask for control of coronavirus disease 2019 (COVID-19) epidemic due to SARS-CoV-2. *Journal of Infection, 81*(1), 107–114.
Cherif, R., & Hasanov, F. (2023). Testing our way out of pandemics. *Health Policy and Technology, 12*, 100714.
Chintalapudi, N., Battineni, G., & Amenta, F. (2020). COVID-19 virus outbreak forecasting of registered and recovered cases after sixty day lockdown in Italy: A data driven model approach. *Journal of Microbiology, Immunology and Infection, 53*, 396–403.
Choi, W., & Shim, E. (2020). Optimal strategies for vaccination and social distancing in a game-theoretic epidemiologic model. *Journal of Theoretical Biology, 505*, 110422.
Chowdhury, R., Heng, K., Shawon, M., et al. (2020). Dynamic interventions to control COVID-19 pandemic: A multivariate prediction modelling study comparing 16 worldwide countries. *European Journal of Epidemiology, 35*(5), 389–399.
Couronne, I. (2020). *Catching coronavirus outside is rare but not impossible*. MedicalXpress.
Crépey, P., et al. (2020). Impact of quadrivalent influenza vaccines in Brazil: A cost-effectiveness analysis using an influenza transmission model. *BMC Public Health, 20*, 1374.
Cui, J., Mu, X., & Wang, H. (2008a). Saturation recovery leads to multiple endemic equilibria and backward bifurcation. *Journal of Theoretical Biology, 254*, 275–283.
Cui, J., Sun, Y., & Zhu, H. (2008b). The impact of media on the control of infectious diseases. *Journal of Dynamics and Differential Equations, 20*, 31–53.
Cui, Q., Qiu, Z., Liu, W., & Hu, Z. (2017). Complex dynamics of an SIR epidemic model with nonlinear saturate incidence and recovery rate. *Entropy, 19*.
Curiel, R. P., & Ramirez, H. G. (2021). Vaccination strategies against COVID-19 and the diffusion of anti-vaccination views. *Scientific Reports, 11*(6629), 1–13.
Daghiri, T., Proctor, M., & Matthews, S. (2022). Evolution of select epidemiological modeling and the rise of population sentiment analysis: A literature review and COVID-19 sentiment illustration. *International Journal of Environmental Research and Public Health, 19*, 3230.
Dagpunar, J., & Wu, C. (2023). Sensitivity of endemic behaviour of COVID-19 under a multi-dose vaccination regime, to various biological parameters and control variables. *Royal Society Open Science, 10*, 221277.
Debrabant, K., Grønbæk, L., & Kronborg, C. (2021). The cost effectiveness of a COVID 19 vaccine in a Danish context. *Clinical Drug Investigation, 41*, 975–988.
Deka, A., Pantha, B., & Bhattacharyya, S. (2020). Optimal management of public perceptions during a Flu outbreak: A game-theoretic perspective. *Bulletin of Mathematical Biology, 82*, 139.
Demongeot, J., Griette, Q., Magal, P., & Webb, G. (2022). Modeling vaccine efficacy for COVID-19 outbreak in New York City. *Biology, 11*, 345.

Deng, J., Tang, S., & Shu, H. (2021). Joint impacts of media, vaccination and treatment on an epidemic Filippov model with application to COVID-19. *Journal of Theoretical Biology, 523*, 110698.

Deo, V., & Grover, G. (2022). Method based on state-space epidemiological model for cost-effectiveness analysis of non-medical interventions- A study on COVID-19 in California and Florida. In *24th Annual Conference* (pp. 69–82). Consulté le February 23–27, 2022

Dias, S., Queiroz, K., & Araujo, A. (2022). Controlling epidemic diseases based only on social distancing level: General case. *ISA Transactions, 124*, 21–30.

Diekmann, O., Heesterbeek, J. A., & Metz, J. A. (1990). On the definition and the computation of the basic reproduction ratio *R*0 in models for infectious diseases in heterogeneous populations. *Journal of Mathematical Biology, 28*, 365–382.

Du, Z., Wang, L., et al. (2023). Cost effectiveness of fractional doses of COVID-19 vaccine boosters in India. *Med, 4*, 182–190.

Duan, A., Li, J., Yang, Z., & He, Y. (2023). The defense of Shangri-La: Protecting isolated communities by periodic infection screening in the worst future pandemic. *Infection, Genetics and Evolution, 113*, 105479.

Eichenbaum, M., Rebelo, S., & Trabandt, M. (2020). *The macroeconomics of epidemics*. NBER Working Paper, 26882.

Eichner, M., Schwehm, M., Eichner, L., & Gerlier, L. (2017). Direct and indirect effects of influenza vaccination. *BMC Infectious Diseases, 17*(308), 2–8.

Enayati, S., & Ozaltin, O. Y. (2020). Optimal influenza vaccine distribution with equity. *European Journal of Operational Research, 283*, 714–725.

Engelbrecht, F. A., & Scholes, R. J. (2021). Test for Covid-19 seasonality and the risk of second waves. *One Health, 12*, 100202.

Erkayman, B., Ak, F., & Codur, S. (2023). A simulation approach for COVID-19 pandemic assessment based on vaccine logistics, SARS-CoV-2 variants, and spread rate. *Simulation: Transactions of the Society for Modeling and Simulation International, 99*(2), 127–135.

Eubank, S., Eckstrand, I., Lewis, B., Venkatramanan, S., Marathe, M., & Barrett, C. L. (2020). Commentary on Ferguson, & et al., Impact of non-pharmaceutical interventions (NPIs) to reduce COVID-19 mortality and healthcare demand. *Bulletin of Mathematical Biology, 82*(4), 52.

Feng, Z., Feng, Y., & Glasser, J. W. (2020). Influence of demographically-realistic mortality schedules on vaccination strategies in age-structured models. *Theoretical Population Biology, 132*, 24–32.

Fernandez-Villaverde, J., & Jones, C. I. (2020). *Estimating and simulating a SIRD model of COVID-19 for many countries, states, and cities*. National Bureau of Economic Research.

Filho, T. R., Moret, M. A., Chow, C. C., Phillips, J. C., Cordeiro, J. J., Scorza, F. A., et al. (2021). A data-driven model for COVID-19 pandemic –Evolution of the attack rate and prognosis for Brazil. *Chaos, Solitons and Fractals, 152*, 111359.

Forsythe, S., Cohen, J., Neumann, P., Bertozzi, S. M., & Kinghorn, A. (2020). The economic and public health imperatives around making potential coronavirus disease–2019 treatments available and affordable. *Value Health, 23*(11), 1427–1431.

Fu, Y., Jin, H., Xiang, H., & Wang, N. (2022). Optimal lockdown policy for vaccination during COVID-19 pandemic. *Finance Research Letters, 45*, 102123.

Fudolig, M. (2023). Effect of transmission and vaccination on time to dominance of emerging viral strains: A simulation-based study. *Microorganisms, 11*, 860.

Fudolig, M., & Howard, R. (2020). The local stability of a modified multi-strain SIR model for emerging viral strains. *PLoS ONE, 15*(12), e0243408.

Gaff, H., & Schaefer, E. (2009) Optimal control applied to vaccination and treatment strategies for various epidemiological models. *Mathematical Biosciences and Engineering, 6*, 469–492.

Gaeta, G. (2020). A simple SIR model with a large set of asymptomatic infectives. *Mathematics in Engineering, 3*, 1–39.

Garriga, C., Manuelli, R., & Sanghi, S. (2022). Optimal management of an epidemic: Lockdown, vaccine and value of life. *Journal of Economic Dynamics and Control, 140*, 104351.

Geoffroy, F., Traulsen, A., & Uecker, H. (2022). Vaccination strategies when vaccines are scarce: On conflicts between reducing the burden and avoiding the evolution of escape mutants. *Journal of the Royal Society Interface, 19*, 20220045.

Ghaffar, A., Alanazi, S., Alruwaili, M., Sattar, M. U., Ali, W., Humayun, M., et al. (2021). Multi-stage intelligent smart lockdown using SIR model to control COVID 19. *Intelligent Automation & Soft Computing, 28*(2), 429–445.

Gnanvi, J. E., Salako, K. V., Kotanmi, G. B., & Kakaï, R. G. (2021). On the reliability of predictions on COVID-19 dynamics: A systematic and critical review of modelling techniques. *Infectious Disease Modelling, 6*, 258–272.

Gollier, C. (2020). Pandemic economics: Optimal dynamic confinement under uncertainty and learning. *The Geneva Risk and Insurance Review, 45*(2), 80–93.

Gollier, C. (2021). The welfare cost of vaccine misallocation, delays and nationalism. *Journal of Benefit-Cost Analysis, 12*(2), 199–226.

Gomes, K. R., Perera, H. N., Thibbotuwawa, A., & Sunil-Chandra, N. P. (2023). Comparative analysis of lean and agile supply chain strategies for effective vaccine distribution in pandemics: A case study of COVID-19 in a densely populated developing region. *Supply Chain Analytics, 3*, 100022.

Goodkin-Gold, M., Kremer, M., Snyder, C. M., & Williams, H. (2022). Optimal vaccine subsidies for endemic diseases. *International Journal of Industrial Organization, 84*, 102840.

Gorji, H., Stauffer, N., Lunati, I., Caduff, A., Bühler, M., Engel, D., et al. (2023). Projection of healthcare demand in Germany and Switzerland urged by Omicron wave (January–March 2022). *Epidemics, 43*, 100680.

Govindan, K., Mina, H., & Alavi, B. (2020). A decision support system for demand management in healthcare supply chains considering the epidemic outbreaks: A case study of coronavirus disease 2019 (COVID-19). *Transportation Research Part E: Logistics and Transportation Review*, Article 101967.

Gozzi, N., Chinazzi, M., Dean, N. E., Longini, I. J., Halloran, M. E., Perra, N., & Vespignani, A. (2023). Estimating the impact of COVID-19 vaccine inequities: A modeling study. *Nature Communications, 14*, 3272.

Graham, B. S. (2020). Rapid COVID-19 vaccine development. *Science, 368*(6494), 945–946.

Gros, C., & Gros, D. (2022). The economics of stop-and-go epidemic control. *Socio-Economic Planning Sciences, 81*, 101196.

Gros, C., Valenti, R., Valenti, K., & Gros, D. (2020). Strategies for controlling the medical and socio-economic costs of the corona pandemic. Récupéré sur https://clausen.berkeley.edu/wp-content/uploads/2020/04/Corona.pdf

Guidry, J. D., Laestadius, L. I., Vraga, E. K., et al. (2021). Willingness to get the COVID-19 vaccine with and without emergency use authorization. *American Journal of Infection Control, 49*(2), 137–142.

Gumus, O. A., Salvem, A., & G, & Rajendar, J. (2022). Dynamics of the mathematical model related to COVID-19 pandemic with treatment. *Thai Journal of Mathematics, 10*(2), 957–970.

Guttieres, D., Sinskey, A. J., & Springs, S. L. (2021). Models to inform neutralizing antibody therapy strategies during pandemics: The case of SARS-CoV-2. *Antibody Therapeutics, 4*(1), 60–71.

Haq, I., Hossain, M. I., Saleheen, A. A., Nayan, M. I., & Mila, M. S. (2022). Prediction of COVID-19 pandemic in Bangladesh: Dual application of Susceptible-Infective-Recovered (SIR) and Machine Learning Approach. *Interdisciplinary Perspectives on Infectious Diseases, 8570089.*

Harari, G. S., & Monteiro, H. A. (2022). An epidemic model with pro and anti-vaccine groups. *Acta Biotheoretica, 70*(3), 20.

Harris, R. (2020). *Covid-19 and productivity in the UK*. Durham University Business School. Récupéré sur https://www.dur.ac.uk/research/news/item/?itemno=41707

Hawkes, M. T., & Good, M. (2022). Vaccinating children against COVID-19: Commentary and mathematical modeling. *mBio, 13*(1), e0378921.

Hazard-Valdés, C., & Montero, E. (2023). A heuristic approach for determining efficient vaccination plans under a SARS-CoV-2 epidemic model. *Mathematics, 11*, 834.

He, J., Peng, G., & al., e. (2004). Molecular evolution of the SARS coronavirus, during the course of the SARS epidemic in China. *Science, 303*, 1666–1669.

Heffman, J. M., Smith, R. J., & Wahl, L. M. (2005). Perspectives on the basic reproductive ratio. *Journal of the Royal Society Interface, 2*(4), 281–293.

Herrera-Serrano, J. E., Macías-Díaz, J. E., Medina-Ramírez, I. E., & Guerrero, J. A. (2022). An efficient nonstandard computer method to solve a compartmental epidemiological model for COVID-19 with vaccination and population migration. *Computer Methods and Programs in Biomedicine, 221*, 106920.

Hethcote, H. W. (2000). The mathematics of infectious diseases. *SIAM Review, 42*(4), 599–653.

Hilton, J., & Keeling, M. J. (2019). Incorporating household structure and demography into models of endemic disease. *Journal of the Royal Society Interface, 16*(157), 20190317.

Hobbs, J. E. (2020). Food supply chains during the COVID-19 pandemic. *Canadian Journal of Agricultural Economics/Revue Canadienned'agroeconomie*.

Holmdahl, I., & Buckee, C. (2020). Wrong but useful—What Covid-19 epidemiologic models can and cannot tell us. *New England Journal of Medicine, 383*(4), 303–305.

Hou, C., et al. (2020). The effectiveness of quarantine of Wuhan city against the Corona Virus Disease 2019 (COVID-19): A well-mixed SEIR model analysis. *Journal of Medical Virology, 92*(7), 841–848.

Hu, X. M., & Zhang, J. (2013). *Optimizing vaccine distribution for different age groups of population using DE algorithm*. Computational Intelligence and Security (CIS).

Huang, C., Wang, Y., et al. (2020). Clinical features of patients infected with 2019 novel coronavirus in Wuhan, China. *Lancet, 395*, 497–506.

Huang, D., Tao, H., Wuq, Q., Huang, S. Y., & Xiao, Y. (2021). Modeling of the long-term epidemic dynamics of COVID-19 in the United States. *International Journal of Environmental Research and Public Health, 18*, 7594.

Huang, Y., & Li, C. (2019). Backward bifurcation and stability analysis of a network based SIS epidemic model with saturated treatment function. *Physica A, 527*.

Husniah, H., Ruhanda, & Supriatna, A. K. (2021). SIR mathematical model of convalescent plasma transfusion applied to the COVID-19 pandemic data in Indonesia to control the spread of the disease. *Journal of Physics: Conference Series, 2084*, 012022.

Hussien, A. M., & Mohammad, A. A. (2022). New SIR Model and Vaccine rate with application. *Neuro Quantology, 20*(7), 3050–3059.

Hwang, Y., Kwon, H., & Lee, J. (2020). Feedback control problem of an SIR epidemic model based on the Hamilton-Jacobi-Bellman equation. *Mathematical Biosciences and Engineering, 17*(3), 2284–2301.

Iranzo, V., & Pérez González, S. (2021). Epidemiological models and COVID 19: A comparative view. *History and Philosophy of the Life Sciences, 43*(104).

Ivanov, D. (2020). Predicting the impacts of epidemic outbreaks on global supply chains: A simulation-based analysis on the coronavirus outbreak (COVID-19/SARS-CoV-2) case. *Transportation Research Part E, 136*, 101922.

Jadidi, M. M., Jamshidiha, S., Masroori, I., Moslemi, P., & Mohammadi, A. P. (2021). A two-step vaccination technique to limit COVID-19 spread using mobile data. *Sustainable Cities and Society, 70*, 102886.

Jarumaneeroj, P., Dusadeerungsikul, P. O., Chotivanich, T., Nopsopon, T., & Pongpirul, K. (2022). An epidemiology-based model for the operational allocation of COVID-19 vaccines: A case study of Thailand. *Computers & Industrial Engineering, 167*, 108031.

Ji, Q., Zhao, X., Ma, H., Liu, Q., Liu, Y., & Guan, Q. (2021). Estimation of COVID-19 transmission and advice on public health interventions. *Mathematics, 9*, 2849.

Jing, S., Milne, R., Wang, H., & Xuo, L. (2023). Vaccine hesitancy promotes emergence of new SARS-CoV-2 variants. *Journal of Theoretical Biology, 570*, 111522.

Jones, J. H., & Salathe, M. (2009). Early assessment of anxiety and behavioral response to novel Swine-Origin influenza A(H1N1). *PLoS One, 4*.

Kabir, K. A., Kuga, K., & Tanimoto, J. (2019). Effect of information spreading to suppress the disease contagion on the epidemic vaccination game. *Chaos, Solitons and Fractals, 119*, 180–187.

Kabir, K. M., & Tanimoto, J. (2019). Analysis of epidemic outbreaks in two-layer networks with different structures for information spreading and disease diffusion. *Communications in Nonlinear Science and Numerical Simulation, 72*, 565–574.

Kabir, K. M., & Tanimoto, J. (2020). Cost-efficiency analysis of voluntary vaccination against n-serovar diseases using antibody-dependent enhancement: A game approach. *Journal of Theoretical Biology, 503*, 110379.

Kai, D., Goldstein, G. P., Morgunov, A., Nangalia, V., & Rotkirch, A. (2020). Universal masking is urgent in the COVID-19 pandemic: SEIR and agent based models, empirical validation, policy recommendations. *arXiv preprint arXiv*, 13553.

Kain, M. P., Childs, M. L., Becker, A. D., & Mordecai, E. A. (2021). Chopping the tail: How preventing superspreading can help to maintain COVID-19 control. *Epidemics, 34*, 100430.

Karapiperis, C., Kouklis, P., Papastratos, S., Chasapi, A., & Ouzounis, C. A. (2020). Assessment for the seasonality of covid-19 should focus on ultraviolet radiation and not 'warmer days.

Keeling, M. J., & Rohani, P. (2007). *Modeling infectious diseases in humans and animals*. Princeton University Press.

Keimer, A., & Pflug, L. (2020). *Modeling infectious diseases using integro-differential equations: Optimal control strategies for policy decisions and applications in COVID-19*. Friedrich-Alexander-Universität Erlangen-Nürnberg.

Kemp, F., Proverbio, D., Aalto, A., Mombaerts, L., et al. (2021). Modelling COVID-19 dynamics and potential for herd immunity by vaccination in Austria, Luxembourg and Sweden. *Journal of Theoretical Biology, 530*, 110874.

Kermack, W. O., & McKendrick, A. G. (1927). A contribution to the mathematical theory of epidemics. *Proceedings of the Royal Society A, 115*, 700–721.

Khalifi, M. E., & Britton, T. (2023). Extending susceptible-infectious recovered-susceptible epidemics to allow for gradual waning of immunity. *Journal of the Royal Society Interface, 20*, 20230042.

Khan, K., Arino, J., et al. (2009). Spread of a novel influenza A (H1N1) virus via global airline transportation. *The New England Journal of Medicine, 361*, 212–214.

Khan, M. M., & Tanimoto, J. (2023). Investigating the social dilemma of an epidemic model with provaccination and antivaccination groups: An evolutionary approach. *Alexandria Engineering Journal, 75*, 341–349.

Kim, Y., Ryu, H., & Lee, S. (2018). Agent-based modeling for super-spreading events: A case study of MERS-CoV transmission dynamics in the Republic of Korea. *International Journal of Environmental Research and Public Health, 15*(11), 2369.

Klepac, P., Funk, S., Hollingsworth, T. D., Metcalf, C. J., & Hampson, K. (2015). Six challenges in the eradication of infectious diseases. *Epidemics*. PMC7612385.

Köhler, J., Schwenkel, L., Koch, A., Berberich, J., Pauli, P., & Allgöwer, F. (2020). Robust and optimal predictive control of the COVID-19 outbreak. *Annual Reviews in Control*.

Kozyreff, G. (2021). Hospitalization dynamics during the first COVID-19 pandemic wave: SIR modelling compared to Belgium, France, Italy, Switzerland and New York City data. *Infectious Disease Modelling, 6*, 398–404.

Krueger, D., Uhlig, H., & Xie, T. (2020). Macroeconomic dynamics and reallocation in an epidemic. *Covid Economics*, 21–55.

Kühn, M. J., Abele, D., Mitra, T., Koslow, W., Abedi, M., Rack, K., et al. (2021). Assessment of effective mitigation and prediction of the spread of SARS-CoV-2 in Germany using demographic information and spatial resolution. *Mathematical Biosciences, 339*, 108648.

Kushwaha, S., Bahl, S., Bagha, A. K., Parmar, K. S., Javaid, M., Haleem, A., & Singh, R. P. (2020). Significant applications of machine learning for COVID-19 pandemic. *Journal of Industrial Integration and Management, 5*(04), 453–479.

Lai, S., Ruktanonchai, N. W., et al. (2021). Assessing the effect of global travel and contact restrictions on mitigating the COVID-19 pandemic. *Engineering, 7*, 914–923.

León, U. P. D., Avila-Vales, E., & Huang, K. (2022). Modeling the transmission of the SARS-CoV-2 delta variant in a partially vaccinated population. *Viruses, 14*, 158.

Li, M. Y., & Muldowney, J. S. (1995). Global stability for the SEIR model in epidemiology. *Mathematical Biosciences, 125*(2), 155–164.

Li, Q., Guan, X., Wu, P., Wang, X., Zhou, L., Tong, Y., et al. (2020). Early transmission dynamics in Wuhan, China, of novel coronavirus infected pneumonia. *New England Journal of Medicine, 382*(13), 1199–1207.

Li, R., Li, Y., Zou, Z., Liu, Y., Li, X., Zhuang, G., et al. (2021). Evaluating the impact of SARS-CoV-2 variants on the COVID-19 epidemic and social restoration in the United States: A mathematical modelling study. *Frontiers in Public Health, 9*, 801763.

Liang, D., Bhamra, R., Liu, Z., & Pan, Y. (2022). Risk propagation and supply chain health control based on the SIR epidemic model. *Mathematics, 10*, 3008.

Libotte, G. B., Lobato, F. S., Platt, G. M., & Silva Neto, A. J. (2020). Determination of an optimal control strategy for vaccine administration in COVID-19 pandemic treatment. *Computer Methods and Programs in Biomedicine, 196*, 105664.

Lim, J. S., Cho, S. I., Ryu, S., & Pak, S. (2020). Interpretation of the basic and effective reproduction number. *Journal of Preventive Medicine & Public Health, 53*, 405–408.

Lima, C. K., Carvalho, P. M., Lima, I. A., Nunes, J. V., Saraiva, J. S., Souza, R. I., et al. (2020). The emotional impact of Coronavirus 2019-nCoV (new Coronavirus disease). *Psychiatry Research, 287*, 112915.

Liu, C., & Cui, R. (2021). Qualitative analysis on an SIRS reaction–diffusion epidemic model with saturation infection mechanism. *Nonlinear Analysis: Real World Applications, 62*, 103364.

Liu, R., Wu, J., & Zhu, H. (2007). Media/psychological impact on multiple outbreaks of emerging infectious diseases. *Computational and Mathematical Methods, 8*, 153–164.

Liu, T., Bai, Y., Du, M., Gao, Y., & Liu, Y. (2021a). Susceptible-infected-removed mathematical model under deep learning in hospital infection control of novel coronavirus pneumonia. *Journal of Healthcare Engineering, 1535046*.

Liu, X., Huang, J., Li, C., Zhao, Y., Wang, D., Huang, Z., & Yang, K. (2021b). The role of seasonality in the spread of covid-19 pandemic. *Environmental Research, 195*, 110874.

Loertscher, S., & Muir, E. V. (2021). Road to recovery: Managing an epidemic. *Journal of Mathematical Economics, 93*, 102482.

López, L., Paul, R. E., Cao-Lormeau, V., & Rodó, X. (2022). Considering waning immunity to better explain dengue dynamics. *Epidemics, 41*, 100630.

Lurie, N., Saville, M., Hatchett, R., & Halton, J. (2020). Developing COVID-19 vaccines at pandemic speed. *New England Journal of Medicine, 382*(21), 1969–1973.

Lymperopoulos, I. N. (2021). Stayhome to contain Covid-19: Neuro-SIR – Neurodynamical epidemic modeling of infection patterns in social networks. *Expert Systems with Applications, 165*, 113970.

Ma, Y., Pei, S., Shaman, J., Dubrow, R., & Chen, K. (2021). Role of meteorological factors in the transmission of SARS-CoV-2 in the United States. *Nature Communications, 12*(1), 1–9.

Madhav, N., Oppenheim, B., Gallivan, M., Mulembakani, P., Rubin, E., & Wolfe, N. (2017). Chapter 17 Pandemics: Risks, impacts, and mitigation. In D. D. T. Jamison, H. Gelband, S. Horton, et al. (Eds.), *Disease control priorities: improving health and reducing poverty* (3rd ed.). The International Bank for Reconstruction and Development / The World Bank.

Malinzi, J., & Juma, V. O. (2023). COVID-19 transmission dynamics and the impact of vaccination: Modelling, analysis and simulations. *Royal Society Open Science, 10*, 221656.

Marinov, T. T., & Marinova, R. S. (2020, Sep 20). Adaptive SIR model with vaccination: Simultaneous identification of rates and functions illustrated with COVID 19. *Scientific Reports, 12*(1).

Markovic, R., Sterk, M., Marhl, M., Perc, M., & Gosak, M. (2021). Socio-demographic and health factors drive the epidemic progression and should guide vaccination strategies for best COVID-19 containment. *Results in Physics, 26*, 104433.

McBryde, E. S., Meehan, M. T., Adegboye, O. A., Adekunle, A. I., Caldwell, J. M., Pak, A., et al. (2020). Role of modelling in COVID-19 policy development. *Paediatric Respiratory Reviews, 35*, 57–60.

Mehmood, K., Bao, Y., Petropoulos, G. P., Abbas, R., et al. (2021). Investigating connections between COVID-19 pandemic, air pollution and community interventions for Pakistan employing geoinformation technologies. *Chemosphere, 272*, 129809.

Melis, M., & Littera, R. (2021). Undetected infectives in the Covid-19 pandemic. *International Journal of Infectious Diseases, 104*, 262–268.

Mesa, D. O., et al. (2022). Modelling the impact of vaccine hesitancy in prolonging the need for Non-Pharmaceutical Interventions to control the COVID-19 pandemic. *Communications Medicine, 2*, 14.

Mistry, D., Litvinova, M., Piontti, A. P., et al. (2021). Inferring high-resolution human mixing patterns for disease modelling. *Nature Communications, 12*(1), 1–12.

Mohd Yunus, A., Mohd Yunus, A., & A., Ibrahim, M. S., & Ismail, S. (2021). Future of mathematical modelling: A review of COVID-19 infected cases using S-I-R model. *Baghdad Science Journal, 18*(1), 824–829.

Moret, M., et al. (2021). WHO vaccination protocol can be improved to save more lives. *Research square*. Consulté le Preprint. Available at https://www.researchsquare.com/article/rs-148826/v1

Mu, X., Zhang, Q., & Rong, L. (2019). Optimal vaccination strategy for an SIRS model with imprecise parameters and Lévy noise. *Journal of the Franklin Institute, 356*(18), 11385–11413.

Mungkasi, S. (2021). Variational iteration and successive approximation methods for a SIR epidemic model with constant vaccination strategy. *Applied Mathematical Modelling, 90*, 1–10.

Musa, S. S., Tariq, A., Yuan, L., Haozhen, W., & He, D. (2022). Infection fatality rate and infection attack rate of COVID 19 in South American countries. *Infectious Diseases of Poverty, 11*, 40.

Mutalik, A. V. (2017). Models to predict H1N1 outbreaks: A literature review. *International Journal of Community Medicine and Public Health, 4*(9), 3068–3075.

Nakamura, G., Grammaticos, B., & Badoual, M. (2021). Vaccination strategies for a seasonal epidemic: A simple SIR model. *Open Communications in Nonlinear Mathematical Physics, 1*, 20–40.

Nanda, S. K., Kumar, G., Bhatia, V., & Singh, A. K. (2023). Kalman-based compartmental estimation for covid-19 pandemic using advanced epidemic model. *Biomedical Signal Processing and Control, 84*, 104727.

Nasution, H., Sitompul, P., & Sinaga, L. P. (2021). Effect of the vaccine on the dynamics of spread of tuberculosis SIR Models. *Journal of Physics: Conference Series, 1819*, 012062.

Neofotistos, G., et al. (2022). Susceptibility to resurgent COVID-19 outbreaks following vaccine rollouts: A modeling study. *Viruses, 14*, 2237.

Nepomuceno, E. G., Peixoto, M. L., et al. (2021). Application of optimal control of infectious diseases in a model free scenario. *SN Computer Science, 2*, 405.

Ng, W. L. (2020). To lockdown? When to peak? Will there be an end? A macroeconomic analysis on COVID-19 epidemic in the United States. *Journal of Macroeconomics, 65*, 103230.

Nguyen, T. K., Hoang, N. H., Currie, G., & Vu, H. L. (2022). Enhancing Covid-19 virus spread modeling using an activity travel model. *Transportation Research Part A, 161*, 186–199.

Nikolopoulos, K., Punia, S., Schäfers, A., Tsinopoulos, C., & Vasilakis, C. (2021). Forecasting and planning during a pandemic: COVID-19 growth rates, supply chain disruptions, and governmental decisions. *European Journal of Operational Research, 290*(1), 99–115.

Nill, F. (2023). Endemic oscillationsforSARS-CoV-2Omicron—ASIRS model analysis. *Chaos, Solitons and Fractals, 173*, 113678.

Nishiura, H., Oshitani, H., Kobayashi, T., Saito, T., Sunagawa,, T., Matsui, T., et al. (2020). Closed environments facilitate secondary transmission of coronavirus disease 2019 (COVID-19). MedRxiv. https://doi.org/10.1101/2020.02.28.20029272,

O'Dea, E. B., & Drake, J. M. (2022). A semi-parametric, state-space compartmental model with time-dependent parameters for forecasting COVID-19 cases, hospitalizations and deaths. *Journal of the Royal Society Interface, 19*(187), 20210702.

O'Regan, S. M., et al. (2020). Transient indicators of tipping points in infectious diseases. *Journal of the Royal Society Interface, 17*, 20200094.

Ortiz-Robinson, N., & Foster-Bey, C. (2021). Control strategies to contain SARS-CoV-2 in a data driven SIR model for the State of Michigan, USA. *Letters in Biomathematics, 8*(1), 179–189.

Otunuga, O. M. (2022). Analysis of multi-strain infection of vaccinated and recovered population through epidemic model: Application to COVID-19. *PLoS ONE, 17*(7), e0271446.

Otunuga, O. M., & Yu, A. (2023). Vaccine breakthrough and rebound infections modeling: Analysis for the United States and the ten U.S. HHS regions. *Infectious Disease Modelling, 8*, 717–741.

Pang, W., Chehaitli, H., & Hurd, T. R. (2022). Impact of asymptomatic COVID-19 carriers on pandemic policy outcomes. *Infectious Disease Modelling, 7*, 16–29.

Parolini, N., Dede', L., Ardenghi, G., & Quarteroni, A. (2022). Modelling the COVID-19 epidemic and the vaccination campaign in Italy by the SUIHTER model. *Infectious Disease Modelling, 7*, 45–63.

Pastor-Satorra, R., Castellan, C., Van Mieghe, P., & Vespignan, A. (2015). Epidemic processes in complex network. *Reviews of Modern Physics, 87*(3), 925–979.

Pérez-Alós, L., et al. (2022). Modeling of waning immunity after SARS-CoV-2 vaccination and influencing factors. *Nature Communications, 13*, 1614.

Pey, Y., Li, J., Xu, S., & Xu, Y. (2022). Adaptive multi-factor quantitative analysis and prediction models: Vaccination, virus mutation and social isolation on COVID-19. *Frontier in Medicine, 9*, 828691.

Polcz, P., Tornai, K., Juhász, J., Cserey, G., et al. (2023). Wastewater-based modeling, reconstruction, and prediction for COVID-19 outbreaks in Hungary caused by highly immune evasive variants. *Water Research, 241*, 120098.

Pomar, L., Favre, G., et al. (2022). Impact of the first wave of the COVID-19 pandemic on birth rates in Europe: A time series analysis in 24 countries. *Human Reproduction, 37*(12), 2921–2931.

Prem, K., Liu, Y., Russell, T. W., Kucharski, A. J., Eggo, R. M., Davies, N., et al. (2020). The effect of control strategies to reduce social mixing on outcomes of the COVID-19 epidemic in Wuhan, China: A modelling study. *Lancet Public Health*. https://doi.org/10.1016/S2468-2667(20)30073-6

Purushwani, H., & Sinha, P. (2019). Mathematical modeling on successive awareness policies for swine flu. *International Journal of Scientific & Technology Research, 8*(8), 310–320.

Qi, H., Xiao, S., Shi, R., Ward, M. P., Chen, Y., Tu, W., & Zhang, Z. (2020). Covid-19 transmission in mainland China is associated with temperature and humidity: A time-series analysis. *Science of the Total Environment, 728*, 138778.

Qian, H., Miao, T., Liu, L., Zheng, X., Luo, D., & Li, Y. (2020). Indoor transmission of SARS-CoV-2. *Indoor Air, 31*(3), 639–645.

Qian, Y., Xie, W., Zhao, J., Xue, M., Liu, S., Wang, L., et al. (2021). Investigating the effectiveness of reopening policies before vaccination during a pandemic: SD modelling research based on COVID-19 in Wuhan. *BMC Public Health, 21*, 1638.

Qundus, J. A., Gupta, S., Abusaimeh, H., Peikert, S., & Paschke, A. (2023). Prescriptive analytics-based SIRM model for predicting Covid-19 outbreak. *Global Journal of Flexible Systems Management, 24*(2), 235–246.

Rabil, M. J., Tunc, S., Bish, D. R., & Bish, E. K. (2022). Benefits of integrated screening and vaccination for infection control. *PLoS ONE, 17*(4), e0267388.

Rahman, A., Kuddus, M. A., Ip, R. H., & Bewong, M. (2021). A review of COVID-19 modelling strategies in three countries to develop a research framework for regional areas. *Viruses, 2021*, 2185.

Rajaei, A., Raeiszadeh, M., Azimi, V., & Sharifi, M. (2021). State estimation-based control of COVID-19 epidemic before and after vaccine development. *Journal of Process Control, 102*, 1–14.
Rajendrakumar, A. L., Nair, A. T., Nangia, C., Chourasia, P. K., Chourasia, M. K., Syed, M. G., et al. (2021). Landscape and forecasting of SARS-CoV-2 in India. *Journal of Epidemiology and Global Health, 11*(1), 55–59.
Rao, F., Mandal, P. S., & Kang, Y. (2019). Complicated endemics of an SIRS model with a generalized incidence under preventive vaccination and treatment controls. *Applied Mathematical Modelling, 67*, 38–61.
Rao, I. J., & Brandeau, M. L. (2021a). Optimal allocation of limited vaccine to control an infectious disease: Simple analytical conditions. *Mathematical Biosciences, 337*, 108621.
Rao, I. J., & Brandeau, M. L. (2021b). Optimal allocation of limited vaccine to minimize the effective reproduction number. *Mathematical Biosciences, 339*, 108654.
Rao, I. J., & Brandeau, M. L. (2022). Sequential allocation of vaccine to control an infectious disease. *Mathematical Biosciences, 351*, 108879.
Reiner, R. C., Barber, R. M., Collins, J. K., et al. (2020). IHME COVID-19 forecasting team, modeling COVID-19 scenarios for the United States. *Nature Medicine*, 1–12.
Rezapour, A., Souresrafil, A., Peighambari, M. M., Heidarali, M., & Tashakori-Miyanroudi, M. (2021). Economic evaluation of programs against COVID-19: A systematic review. *International Journal of Surgery, 85*, 10–18.
Rhodes, T., Lancaster, K., Lees, S., & Parker, M. (2020). Modelling the pandemic: Attuning models to their contexts. *BMJ Global Health, 5*(6), e002914.
Richardson, S., Hirsch, J. S., Narasimhan, M., Crawford, J. M., McGinn, T., & al., e. (2020). Presenting characteristics, comorbidities, and outcomes among 5700 patients hospitalized with COVID-19 in the New York city area. *Journal of the American Medical Association, 323*(20), 2052.
Robinson, B., et al. (2022). Comprehensive compartmental model and calibration algorithm for the study of clinical implications of the population-level spread of COVID-19: A study protocol. *BMJ Open, 12*, e052681.
Rock, K., Brand, S., Moir, J., & Keeling, M. J. (2014). Dynamics of infectious diseases. *Reports on Progress in Physic, 77*(2), 026602.
Roda, W. C., Varughese, M. B., Han, D., & Li, M. Y. (2020). Why is it difficult to accurately predict the COVID-19 epidemic? *Infectious Disease Modelling, 5*, 271–281.
Rodiah, I., Vanella, P., Kuhlmann, A., Jaeger, V. K., Harries, M., Krause, G., et al. (2023). Age specific contribution of contacts to transmission of SARS CoV 2 in Germany. *European Journal of Epidemiology, 38*, 39–58.
Rose, A. F., Mantica, G., et al. (2022). COVID-19 impact on birth rates: First data from Metropolitan City of Genoa, Northern Italy. *International Journal of Impotence Research, 34*, 111–112.
Roser, M., Ritchie, H., Ortiz-Ospina, E., & Hasell, J. (2020). Coronavirus pandemic (COVID-19). *Our World in Data*. https://ourworldindata.org/coronavirus
Ruiz, F. J., Torres-Rueda, S., et al. (2023). What, how and who: Cost-effectiveness analyses of COVID-19 vaccination to inform key policies in Nigeria. *PLoS Global Public Health, 3*(3), e0001693.
Rusu, A. C., Emonet, R., & Farrahi, K. (2021). Modelling digital and manual contact tracing for COVID-19. Are low uptakes and missed contacts deal-breakers? *PLoS ONE, 16*(11), e0259969.
Saad-Roy, C. M., Morris, S. E., & Metcalf, C. J. (2021). Epidemiological and evolutionary considerations of SARS-CoV-2 vaccine dosing regimes. *Science, 372*.
Safan, M. (2019). Mathematical analysis of an SIR respiratory infection model with sex and gender disparity: Special reference to influenza A. *Mathematical Biosciences and Engineering, 16*(4), 2613–2649.
Saha, S., Samanta, G., & Nieto, J. J. (2022). Impact of optimal vaccination and social distancing on COVID-19 pandemic. *Mathematics and Computers in Simulation, 200*, 285–314.

Saha, S., & Samanta, G. P. (2021). Modelling the role of optimal social distancing on disease prevalence of COVID-19 epidemic. *International Journal of Dynamics and Control, 9*, 1053–1077.

Sajadi, M. M., Habibzadeh, P., Vintzileos, A., Shokouhi, S., Miralles-Wilhelm, F., & Amoroso, A. (2020). Temperature and latitude analysis to predict potential spread and seasonality for covid-19. Available at: SSRN 3550308.

Saxena, R., Jadeja, M., & Bhateja, V. (2021). Propagation analysis of COVID-19: An SIR model-based investigation of the pandemic. *Arabian Journal for Science and Engineering*. https://doi.org/10.1007/s13369-021-05904-0

Schwaid, G. (2017). Chapter Three - Epidemiology and Biostatistics. In *Dans Board Review in Preventive Medicine and Public Health* (pp. 79–185). Academic Press.

Sen, A., Baker, J. D., Zhang, Q., Agarwal, R. R., & Lam, J. P. (2023). Do more stringent policies reduce daily COVID-19 case counts? Evidence from Canadian provinces. *Economic Analysis and Policy, 78*, 225–242.

Seto, W., Tsang, D., Yung, R., Ching, T., Ng, T., Ho, M., et al. (2003). Effectiveness of precautions against droplets and contact in prevention of nosocomial transmission of severe acute respiratory syndrome (SARS). *Lancet, 361*, 1519–1520.

Shanta, S. S., & Biswas, M. H. (2020). The impact of media awareness in controlling the spread of infectious diseases in terms of SIR model. *Mathematical Modelling of Engineering Problems, 7*(3), 368–376.

Sharif, S. V., Moshfegh, P. H., Morshedi, M. A., & Kashani, H. (2022). Modeling the impact of mitigation policies in a pandemic: A system dynamics approach. *International Journal of Disaster Risk Reduction, 82*, 103327.

Sharma, K., Koirala, A., Nicolopoulos, K., Chiu, C., Wood, N., & Britton, P. N. (2021). Vaccines for COVID-19: Where do we stand in 2021? *Paediatric Respiratory Reviews, 39*, 22–31.

Sharov, K. S. (2020). Creating and applying SIR modified compartmental model for calculation of COVID-19 lockdown efficiency. *Chaos, Solitons and Fractals, 141*, 110295.

Shim, E. (2013). Optimal strategies of social distancing and vaccination against seasonal influenza. *Mathematical Biosciences and Engineering, 10*, 1615–1634.

Shin, G., Kang, D., Cheong, H. J., & Choi, S. E. (2022). Cost-effectiveness of extending the National Influenza Vaccination Program in South Korea: Does vaccination of older adults provide health benefits to the entire population? *Vaccines, 10*, 932.

Shringi, S., Sharma, H., Rathie, P. N., Bansal, J. C., & Nagar, A. (2021). Modified SIRD model for COVID-19 spread prediction for northern and southern states of India. *Chaos, Solitons and Fractals, 148*, 111039.

Singh, P., & Gupta, A. (2021). Generalized SIR (GSIR) epidemic model: An improved framework for the predictive monitoring of COVID-19 pandemic. *ISA Transactions, 124*, 31–40.

Siqueira, P. G., Duarte, H. O., & Moura, M. C. (2022). Risk-based cost-benefit analysis of alternative vaccines against COVID-19 in Brazil: Coronavac vs. Astrazeneca vs. Pfizer. *Vaccine, 40*, 3851–3860.

Song, X., Liu, M., Song, H., & Ren, J. (2020). Dynamical behavior of an SVIR epidemiological model with two stage characteristics of vaccine effectiveness and numerical simulation. *Advances in Intelligent Systems and Interactive Applications, 1084*, 238–242.

Stojanovic, O., Leugering, J., Pipa, G., Ghozzi, S., & Ullrich, A. (2019). A Bayesian Monte Carlo approach for predicting the spread of infectious diseases. *PLoS One, 14*(12), e225838.

Storlie, C. B., Pollock, B. D., Rojas, R. L., et al. (2021a). Quantifying the importance of COVID-19 vaccination to our future outlook. *Mayo Clinic Proceedings, 96*(7), 1890–1895.

Storlie, C. B., Rojas, R. L., Demuth, G. O., et al. (2021b). A hierarchical Bayesian model for stochastic spatiotemporal SIR modeling and prediction of COVID-19 cases and hospitalizations. *arXiv*.

Strong, A., & Welburn, J. (2020). *An estimation of the economic costs of social-distancing policies*. RAND.

Sunjaya, A. F., & Sunjaya, A. P. (2020). Pooled testing for expanding COVID-19 mass surveillance. *Disaster Medicine and Public Health Preparedness, 14*(3), e42–e43.

Šušteršicˇ, T., Blagojević, A., et al. (2021). Epidemiological predictive modeling of COVID-19 infection: Development, testing, and implementation on the population of the Benelux Union. *Frontiers in Public Health, 9*, 727274.

Sy, K. T., White, L., & Nichols, B. (2021). Population density and basic reproductive number of COVID-19 across United States counties. *PLoS ONE, 16*, e024927.

Szilagyi, P. G., Thomas, K., Shah, M. D., et al. (2020). National trends in the US public's likelihood of getting a COVID-19 vaccine. *JAMA, 325*(4), 396–398.

Tamasiga, P., Onyeaka, H., Umenweke, G. C., & Uwishema, O. (2023). An extended SEIRDV compartmental model: Case studies of the spread of COVID-19 and vaccination in Tunisia and South Africa. *Annals of Medicine & Surgery, 85*(6), 2721–2730.

Tang, B., Bragazzi, N. L., Li, Q., Tang, S., Xiao, Y., & Wu, J. (2020a). An updated estimation of the risk of transmission of the novel coronavirus (2019-nCov). *Infectious Diseases, 5*, 248–255.

Tang, L., Zhou, Y., Wang, L., et al. (2020b). A review of multi-compartment infectious disease models. *International Statistical Review, 88*(2), 462–513.

Tang, S., Xiao, Y., Yang, Y., Zhou, Y., Wu, J., & Ma, Z. (2010). Community-based measures for mitigating the 2009 H1N1 pandemic in China. *PLoS One, 5*.

Tchuenche, J. M., Dube, N., Bhunu, C. P., Smith, R. J., & Bauch, C. T. (2011). The impact of media coverage on the transmission dynamics of human influenza. *BMC Public Health, 11*.

Team, I. C.-L., & Murray, C. J. (2020a). *Forecasting COVID-19 impact on hospital bed-days, ICU-days, ventilator-days and deaths by US state in the next 4 months*. MedRxiv. https://doi.org/10.1101/2020.03.27.20043752

Team, I. C.-L., & Murray, C. J. (2020b). *Forecasting the impact of the first wave of the COVID-19 pandemic on hospital demand and deaths for the USA and European economic area countries*. MedRxiv. https://doi.org/10.1101/2020.04.21.20074732

Tuite, A. R., Fisman, D. N., & Greer, A. L. (2020). Mathematical modelling of COVID-19 transmission and mitigation strategies in the population of Ontario, Canada. *CMAJ, 192*(19), e497–e505.

Turkyilmazoglu, M. (2022). An extended epidemic model with vaccination: Weak-immune SIRVI. *Physica A, 598*, 127429.

Tzamali, E., et al. (2023). Mathematical modeling evaluates how vaccinations affected the course of COVID-19 disease progression. *Vaccines, 11*, 722.

Vahdani, B., Mohammadi, M., Thevenin, S., Gendreau, M., Dolgui, A., & Meyer, P. (2023). Fair-split distribution of multi-dose vaccines with prioritized age groups and dynamic demand: The case study of COVID-19. *European Journal of Operational Research, 310*, 1249–1272.

Vakil, V., & Trappe, W. (2022). Projecting the pandemic trajectory through modeling the transmission dynamics of COVID-19. *International Journal of Environmental Research and Public Health, 19*, 4541.

Vattiatio, G., Lustig, A., Maclaren, O. J., & Plank, M. J. (2022). Modelling the dynamics of infection, waning of immunity and re-infection with the Omicron variant of SARS-CoV-2 in Aotearoa New Zealand. *Epidemics, 41*, 100657.

Viola, I. M., Peterson, B., Pisetta, G., Pavar, G., Akhtar, H., Menoloascina, F., et al. (2020). Ace coverings, aerosol dispersion and mitigation of virus transmission risk. arXiv:2005.10720 [physics.med-ph].

Wali, M., Arshad, S., & Huang, J. (2022). Stability analysis of an extended SEIR COVID-19 fractional model with vaccination efficiency. *Computational and Mathematical Methods in Medicine, 3754051*.

Wang, A., & Xiao, Y. (2013). Sliding bifurcation and global dynamics of a Filippov epidemic model with vaccination. *International Journal of Bifurcation and Chaos, 23*.

Wang, A., & Xiao, Y. (2014). A Filippov system describing media effects on the spread of infectious diseases. *Nonlinear Analysis: Hybrid Systems, 11*, 84–97.

Wang, H., Tao, G., Ma, J., Jia, S., Chi, L., Yang, H., et al. (2022). Predicting the epidemics trend of COVID-19 using epidemiological-based generative adversarial networks. *IEEE Journal of Selected Topics in Signal Processing, 16*(2), 276–288.

Wang, W., Liu, Q. H., Liang, J., Hu, Y., & Zhou, T. (2019). Coevolution spreading in complex networks. *Physics Reports, 820*, 1–51.

White, S. H., del Rey, A. M., & Sánchez, G. R. (2007). Modeling epidemics using cellular automata. *Applied Mathematics and Computation, 186*(1), 193–202.

WHO. (2020). Corona virus disease (COVID-19): Herd immunity, lockdowns and COVID-19. Récupéré sur https://www.who.int/news-room/q-a-detail/herd-immunity-lockdowns-and-covid-19 (accessed 10.9.21).

Wood, S., & Schulman, K. (2021). When vaccine apathy, not hesitancy, drives vaccine disinterest. *JAMA, 325*(24), 2435–2436.

Woolf, S. H., Chapman, D. A., Sabo, R. T., Weinberger, D. M., & Hill, L. (2020). Excess deaths from COVID-19 and other causes. *JAMA, 324*(5), 510–513.

Wu, J. T., Leung, K., Bushman, M., Kishore, N., Niehus, R., et al. (2020). Estimating clinical severity of COVID-19 from the transmission dynamics in Wuhan, China. *Nature Medicine, 26*(4), 506–510.

Xiao, Y., Zhao, T., & Tang, S. (2013). Dynamics of an infectious diseases with media/psychology induced non-smooth incidence. *Mathematical Biosciences and Engineering, 10*, 445–461.

Xie, Z. (2019). Data fitting and scenario analysis of vaccination in the 2014 Ebola Outbreak in Liberia. *Osong Public Health and Research Perspectives, 10*(3), 187–201.

Xu, Z., Wu, B., & Topcu, U. (2020). Control strategies for COVID-19 epidemic with vaccination, shield immunity and quarantine: A metric temporal logic approach. *PLoS ONE, 16*, e0247660.

Yaladanda, N., Mopuri, R., Vavilala, H. P., & Mutheneni, S. R. (2022). Modelling the impact of perfect and imperfect vaccination strategy against SARS CoV-2 by assuming varied vaccine efficacy over India. *Clinical Epidemiology and Global Health, 15*, 101052.

Yang, Z., Zeng, Z., Wang, K., Wong, S. S., et al. (2020). Modified SEIR and AI prediction of the epidemics trend of COVID-19 in China under public health interventions. *Journal of Thoracic Disease, 12*(3), 165.

Yu, P., Wang, P., Wang, Z., & Wang, J. (2022b). Supply chain risk diffusion model considering multi-factor influences under hypernetwork vision. *Sustainability, 14*, 8420.

Yu, P., Wang, Z., Sun, Y., & Wang, P. (2022a). Risk diffusion and control under uncertain information based. *Mathematics, 10*, 4344.

Zaric, G. S., & Brandeau, M. L. (2001). Resource allocation for epidemic control over short time horizons. *Mathematical Biosciences, 171*(1), 33–58.

Zhan, C., Zheng, Y., Shao, L., Chen, G., & Zhang, H. (2023). Modeling the spread dynamics of multiple-variant coronavirus disease under public health interventions: A general framework. *Information Sciences, 628*, 469–487.

Zhan, C., et al. (2022). Estimating unconfirmed COVID-19 infection cases and multiple waves of pandemic progression with consideration of testing capacity and non-pharmaceutical interventions: A dynamic spreading model. *Information Sciences, 607*, 418–439.

Zhang, W., Hugginst, Zheng, W., Liu, S., Du, Z., Zhu, H., et al. (2022a). Assessing the dynamic outcomes of containment strategies against COVID-19 under different public health governance structures: A comparison between Pakistan and Bangladesh. *International Journal of Environmental Research and Public Health, 19*, 9239.

Zhang, W., Xie, R., Dong, X., Li, J., Peng, P., & Gonzalez, E. S. (2022b). SEIR-FMi: A coronavirus disease epidemiological model based on intra-city movement, inter-city movement and medical resource investment. *Computers in Biology and Medicine, 149*, 106046.

Zhang, Z., Fu, D., & Wang, J. (2023). How containment policy and medical service impact COVID-19 transmission: A cross-national comparison among China, the USA, and Sweden. *International Journal of Disaster Risk Reduction, 91*, 103685.

Zhou, F., Yu, T., Du, R., Fan, G., Liu, Y., Liu, Z., et al. (2020). Clinical course and risk factors for mortality of adult inpatients with COVID-19 in Wuhan, China: A retrospective cohort study. *The Lancet, 395*(10229), 1054–1062.

Zhou, L., & Fan, M. (2012). Dynamics of an SIR epidemic model with limited medical resources revisited. *Nonlinear Analysis: Real World Applications, 13*, 312–324.

Zhou, T., & Ji, Y. (2020). Semiparametric Bayesian inference for the transmission dynamics of covid-19 with a state-space model. *Contemporary Clinical Trials, 97*, 106146.

Zhu, W. J., & Shen, S. F. (2021). An improved SIR model describing the epidemic dynamics of the COVID-19 in China. *Results in Physics, 25*, 104289.

Zuo, C., Wang, A., Zhu, F., Meng, Z., & Zhao, X. (2021). A new coupled awareness-epidemic spreading model with neighbor behavior on multiplex networks. *Complexity*.

Managing Disruptions and the Ripple Effect in Digital Supply Chains: Empirical Case Studies

Ajay Das, Simone Gottlieb, and Dmitry Ivanov

Abstract This chapter studies the impact of accelerating digitalization on supply chain risk management. The interrelationships between digital technologies and supply chain disruption risk are analyzed using multiple case studies from various industries. The empirical analysis guided a conceptual framework based on extant theory and specific hypotheses. The chapter concludes with a discussion of research opportunities for future study. In particular, the discussion involves perspectives and future transformations that can be expected in the transition toward cyber-physical supply chains.

1 Introduction

To keep pace with current trends, companies and supply chains develop products and processes that meet new requirements in terms of productivity, sustainability, competitiveness, and risk management. Digitalization strategies and technologies are being increasingly identified, evaluated, tested, and applied to meet such requirements (Li et al., 2025). Disruptive innovations such as Blockchain, Industry 4.0, and additive manufacturing catalyze the development of new paradigms, principles, and models in supply chain management (SCM) (Cai et al., 2021; Choi, Dolgui, et al., 2022; Choi, Kumar, et al., 2022; Dolgui et al., 2020; Fragapane et al., 2022; Ivanov, 2017; Ivanov et al., 2013; Ivanov et al., 2022; Ivanov & Dolgui, 2021; Li et al., 2022; Meier et al., 2023; Rai et al., 2021; Sauer et al., 2022; Tiwari et al.,

A. Das
Narendra Paul Loomba Department of Management, Zicklin School of Business, CUNY-Baruch, New York, NY, USA
e-mail: ajay.das@baruch.cuny.edu

S. Gottlieb · D. Ivanov (✉)
Department of Business and Economics, Berlin School of Economics and Law, Berlin, Germany
e-mail: divanov@hwr-berlin.de

D. Ivanov et al. (eds.), *Handbook of Ripple Effects in the Supply Chain*,
International Series in Operations Research & Management Science 276,
https://doi.org/10.1007/978-3-031-85508-5_13

2023; Xu & Choi, 2021). The Internet of Things (IoT), cyber-physical systems, and smart, connected products, facilitate the development of digital supply chains (SC) and smart operations (Dolgui et al., 2019; Dolgui et al., 2020; Dolgui & Ivanov, 2022; Fazili et al., 2017; Ivanov, 2024; Ivanov et al., 2016; Ivanov et al., 2018; Ivanov & Dolgui, 2019; Liao et al., 2017; Panetto et al., 2019; Qu et al., 2017; Rossit et al., 2018; Saberi et al., 2018; Strozzi et al., 2017; Tran-Dang et al., 2017; Yang et al., 2017). Recent surveys by Addo-Tenkorang and Helo (2016), Gunasekaran et al. (2016, 2017, 2018), Nguyen et al. (2018), Moghaddam and Nof (2018), Choi et al. (2018), and Ben-Daya et al. (2017) proposed classifications of different digital technologies and discussed potential impacts on SCM. Such digital technologies include big data analytics, advanced manufacturing technologies with sensors, decentralized agent-driven control, advanced robotics, augmented reality, advanced tracking and tracing technologies, and additive manufacturing.

While the individual technology contributing elements (e.g., robots, sensors, RFID—radio frequency identification, agents, modular factories, etc.) are not really new, they are becoming more practical, and companies are becoming more receptive to their adoption (Dolgui & Ivanov, 2021, 2025; Xue et al., 2025; Zheng et al., 2025). In addition, attempts to interconnect and integrate such local solutions, using current advances in data processing technologies, can be observed in practice. Digitalization demands change in existing operations. As such, it calls for new principles and models to support SC risk management (SCRM) in the future (Ivanov, 2018a, 2018b; Ivanov et al., 2017).

The investigation of the interrelations between digital technology and SC risks is still at a preliminary stage, and requires new conceptual frameworks and taxonomies (Brintrup et al., 2023; Dolgui & Ivanov, 2022; Ivanov et al., 2019; Kumar & Sharma, 2023; Schlüter et al., 2017; Zheng et al., 2023). Papadopoulos et al. (2017) pointed out that data analytics can help in improving SC risk management and disaster resistance. Choi and Lambert (2017) and Choi et al. (2017) provided evidence of how data analytics can be used to improve the resilience of SC operations by utilizing firms' databases and large volumes of data to predict risks, assess vulnerability, and enhance their SCs. Simchi-Levi et al. (2015) presented a data-driven system to analyze supplier exposure in the automotive sector. Ivanov et al. (2019) showed that data analytics can be used at the planning stage to identify supplier risk exposure and can help at the reactive stage to monitor and identify disruptions. They proposed a framework of integrated cyber-physical SC simulation and optimization and related this framework to system-cybernetics principles. Their results echo those in the study by Choi (2018) that presented a new practical perspective on how big data related technologies can be used for global SCs with a system of systems (SoS) mind-set. Baryannis et al. (2018) summarized recent AI applications to SC risk management and showed some future research directions in risk identification, assessment, and response. Jackson, Ivanov, et al. (2024); Jackson, Saenz, and Ivanov (2024) provided the analysis of Generative AI in SC management. Priore et al. (2018) applied machine learning to the dynamic selection of replenishment policies according to SC environmental dynamics. Dolgui and Ivanov (2025) examined the impacts of Internet-of-behaviors on SC operations.

This chapter seeks to move the discussion forward and develop a framework for a detailed analysis of SC digital technology and disruption risk effects (Dolgui et al., 2018; Ivanov et al., 2014, b; Pavlov et al., 2018; Sokolov et al., 2016) using a multiple case study methodology (Blackhurst et al., 2011).

2 Empirical Analysis

This section develops hypotheses regarding the mutual interrelations of digital SC and the ripple effect and analyzes these empirically. The first set of hypotheses (H1–H3) focuses on disruption risks in SCM assuming that supplier disruptions are more likely to occur than other risks. The ripple and bullwhip effects are typical manifestations to be analyzed within this hypothesis set. Causes of disruption, disruption management experience and risk mitigation methods, and disruption recovery are therefore examined in hypotheses H1, H2, and H3.

Hypotheses H4 and H5 are developed to explore the application of digital technologies in practice. The opportunities and benefits of digital technologies are often emphasized in literature. However, such are often surmised or anecdotal. Applications of digital technology may be different in reality and might be characterized by many challenges and obstacles.

The third set of hypotheses (H6–H9) deals with the obvious gap in knowledge about the influence of digital technologies on SCRM. The contribution of digital technologies in creating a resilient SC in the pre- and post-disruption stages, their impact on SC efficiency, and the role of digital technologies in ripple effect control in SCRM are important research questions. The following hypothesis examines these questions:

- **Hypothesis H1**: Supplier disruptions have a higher likelihood of appearance, relative to other types of disruptions.
- **Hypothesis H2**: Disruptions have a serious impact on a large part of the SC.
- **Hypothesis H3**: Flexible SC networks are associated with successful SCRM.
- **Hypothesis H4**: SCRM opportunities and potential benefits become visible through the use of digital technologies.
- **Hypothesis H5**: Obstacles to implementation and data security concerns are the most likely impediments to the use of digital technologies.
- **Hypothesis H6**: Digital technologies contribute to SC resilience by improving risk mitigation capabilities at the pre-disruption stage.
- **Hypothesis H7**: Digital technologies contribute to SC resilience by improving disruption recovery capabilities at the post-disruption stage.
- **Hypothesis H8**: Digital technologies in SCRM are associated with SC efficiency.
- **Hypothesis H9**: Digital technologies contribute to ripple effect control in SCRM.

The above hypotheses emerge in a conceptual framework that broadly examines the impact of digital technologies on SCRM. Figure 1 organizes the SCM factors, disruption risks, and digital technologies into a suggested framework. The generic framework refined and validated during the course of this study intends to provide managerial guidance for the creation of successful SCRM through the use of digital technologies.

3 Case Studies

3.1 Sample Selection

The persons interviewed in this research were key decision-makers in their firms' SCRM and had extensive experience in dealing with disruptions. The data were collected through semi-structured interviews in order to achieve a certain degree of comparability and, at the same time, ensure an unimpeded flow of narration. This method combines the flexibility of an open interview with the focus of a structured survey. Open-ended questions were used to collect qualitative text data, offering participants an opportunity to elaborate and contextualize their experiences with risk management. Structured questions provided ranking options. Interview questions were formulated based on identified gaps in the literature and resultant hypotheses. The collected information was analyzed in the context of the hypotheses and the conceptual framework (Fig. 1).

Experts in SCRM were approached via email and asked to share their experiences through a developed questionnaire provided in English (cf. Appendix). A total of nine completed questionnaires were collected within a period of 4 months. Data confidentiality and respondent anonymity were assured and implemented. The first

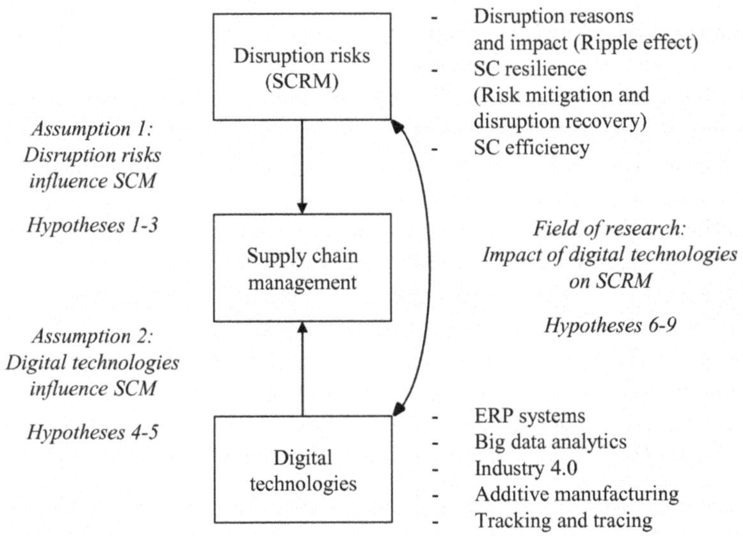

Fig. 1 Conceptual framework

phase of data collection therefore obtained preliminary but detailed knowledge through filled out questionnaires. The second phase of data collection obtained secondary data from public databases to triangulate the interview data. Follow-up phone interviews were then conducted to gain more in-depth knowledge.

Construct validity was pursued using a variety of approaches. The research and interview questions were elaborated in exchange with academics. The result of the first interviews was discussed in order to adjust the interview questions where necessary. Support for internal validity was provided by the triangulation process together with comparisons with existing literature. External validity was examined using a multiple case study approach. Support for both internal and external validities materialized from the comparative case studies. Each case had been carefully selected to meet either similarity or variance goals and subsequently investigated with within-case and cross-case analyses. In addition, a tested and refined interview protocol (questionnaire) provided added reliability. To further ensure a high degree of reliability and traceability of the data, additional questions were asked by telephone in the second phase of the interview, duly transcribed into text format.

The data analysis was conducted by analyzing the individual cases and comparing them with one another. As part of the within-case analysis, each case or respective company and the related data were analyzed individually. In the cross-case analysis, the questionnaire responses were reviewed across the entire sample. Cases were compared to identify similarities and differences. The aim was to find recurring patterns and make qualitative statements.

3.2 Cases

Nine case studies from medium-sized to large companies from various industries were created based on the in-depth interviews. A majority of the companies are from the automotive and semiconductor industry, as these industries, in particular, are expected to have an SC network also in high-risk areas such as Japan, which is frequently affected by extreme nature events. The other case studies are from climate systems, food retail, rail vehicles drive systems, and the consulting industry (Tables 1, 2, and 3).

3.2.1 Case A (Food Industry)

The food retail SC consists of up to three echelons. The hypermarket branch has one federal distribution center, which is controlled by the hypermarket. Each of the 93 hypermarkets has a bakery and salad shop. In the second SC stage there are two federal distribution centers, which are shared with and controlled by the supermarkets. The regional distribution network consists of six distribution centers, also controlled by the supermarkets. The federal distribution centers specialize in

Table 1 Overview of investigated companies

Case	Industry	Main product	Country	Company's size (employees)	Company sales (2017)
A	Food retail	Food	Russia	>500	16–18 Billion Euro as a retail group 1.6–1.8 Billion Euro as a hypermarket business branch
B	Semiconductor	Automotive, Power Management and Multimarket, Industrial Power Control, Chip Card and Security	Germany	>500	7 Billion Euro
C	Rail vehicles drive systems	Mechanical gears	Germany	50–500	50–100 Million Euro
D	Automotive supplier	Automotive electronics	Germany	>500	40, 5 Billion Euro
E	Aerospace, automotive, railway and engineering	Supply chain collaboration Sourcing, quality solutions and transport management Solutions	Germany	50–500	50 Million Euro
F	Semiconductor	Chips, wafer	Netherlands	>500	8 Billion Euro
G	Automotive supplier	Automotive electronics	Germany	>500	120 Million Euro
H	Automotive	Vehicles	Germany	>500	98 Billion Euro
I	Climate systems	Air conditioners, heat pumps, chillers	Netherlands	50–500	175–250 Million Euro

dry, fresh, and frozen products as well as fruit and vegetable products. Regional distribution centers, on the other hand, have the entire product range. Three types of warehouse processing are applied: cross-docking (the goods are pre-packed by the vendor, no longer stored, but processed directly and delivered to the customer), pick-by-line, and picking.

3.2.2 Case B (Semiconductor Industry)

The business processes in case B are based on the SCOR model: planning, sourcing, making, delivery, and return. The company strives for a self-optimized SC with a perfect mix of automated processes and human decision-making processes according to the roadmap being descriptive, diagnostic, predictive, and prescriptive. Operations vision and mission are communicated top-down throughout the opera-

Table 2 Overview of interviewed experts

Case	informant (professional position)	Department	Years of experience	Specialization
A	Supply chain optimization manager	Department of Supply Chain Optimization and Supplier Relationship	1, 5	Supply chain optimization, Analytics, supply chain network optimization
B	Head of supply chain innovation	Supply Chain Innovation	>20	Principal logistics systems
C	Business process manager	Large Traction Drives Technologies	>7	Process optimization
D	Risk and process manager	Risk and Allocation Management	7	Corporate supply chain Management and risk & allocation management
E	Manager consulting	Consulting	23	Automotive industry, IT consulting, supplier portal collaboration
F	Vice president and business unit supply chain and operations manager	Supply Chain Management	30	Global operations management automotive
G	Processes and tools manager	Supply Chain Management	15	Processes and Tools Mexico
H	Risk manager	Risk Management	20	Strategy purchasing and supplier network, risk management, crisis and insolvency management
I	Supply chain manager	Supply Chain Management	15	Supply chain optimization, supply and demand planning

tions community. Operations strategy is documented in various ways to reach all levels. High-level workshops every year and monthly board meetings ensure that all activities, from lighthouse projects to projects and tasks, are aligned with the strategy. The SC has evolved from rigid patterns to a global and highly flexible supply network as the smart factory approach has improved the manufacturing flexibility realized through superior planning processes.

3.2.3 Case C (Rail Vehicles Drive Systems Industry)

The general European SC scope of Company C is component manufacturers, proprietary value creation, bogie manufacturers, and vehicle manufacturers. Approximately ten main components are predominantly procured from Europe with one supplier per component. The main component "gearing" is usually manufactured at Company C's own site (other business segments). Components are then delivered

Table 3 Reasons for experienced disruptions

Disruption type	Causes for disruption
Supplier disruptions	Single sourcing
Production capacity disruptions	Production system inflexibility Low data visibility Missing buffers in capacity utilization Force majeure Single sourcing
Logistics disruptions	Force majeure Single sourcing
Demand disruptions	Missing buffers in capacity utilization Low data visibility No real-time monitoring Customers (first-tier) not knowing their demand because their customers (original equipment manufacturer) do not know their demand
Product/technology disruptions	Complex mix of product specifications and changing customer requirements

to the bogie manufacturer. Some components are provided by the customer. The proprietary value creation of Company C consists of assembly, testing, and completion. Bogie manufacturers and vehicle manufacturers may be identical or separate in terms of logistics or business.

3.2.4 Case D (Automotive Supplier Industry)

At the location surveyed, the group operates the world's largest production plant for automotive electronics. Products for eight different business units are manufactured there, from sensor systems and transmission controls to engine components. A total of around 360,000 units leave the plant every day, which supplies major German and international automobile manufacturers. The leading German automotive supplier group has 427 locations in 56 countries, with a total of more than 3000 suppliers.

3.2.5 E (Aerospace, Automotive, Railway, Engineering Consulting Industry)

Case E is an SC collaboration platform that connects companies with their business partners and assists their ERP systems. The dynamic corporate network connects 65,000 companies worldwide. Of these, 50,000 are suppliers who work with major customers. The platform enables a quick reaction to market changes and fluctuations and thus ensures the long-term success of the SC. The focus of Company E is on the manufacturing industry and covers industry-specific requirements of the aerospace, automotive, railway, and engineering industries with consulting solutions.

3.2.6 Case F (Semiconductor Industry)

Company F produces chips and wafers as automotive products for advanced driver assistance systems, vehicle networking, or safety applications, for example. The respondent of the semiconductor company limited the SC overview to the automotive business. There are nine internal factories for wafers and chips, about ten wafer foundry suppliers, and ten chip suppliers. The number of material suppliers is estimated at over 100.

3.2.7 Case G (Automotive Supplier Industry)

The Group's site has one factory and one logistics center where electronics for the automotive industry are produced. The German plant from which the expert was interviewed has around 370 customers worldwide.

3.2.8 Case H (Automotive Industry)

The Group has over 20 automotive plants worldwide. Most of the goods are procured via single and global sourcing from 5000 suppliers. In addition, just-in-time and just-in-sequence delivery concepts are applied. The interviewed expert does not focus the SC overview on the own location, since the respondent is an SC risk manager for the entire German automotive group.

3.2.9 Case I (Climate System Industry)

Company I, headquartered in the Netherlands, has three suppliers from Asia. Products are air conditioning systems, heat pumps, and chillers for the air conditioning solutions market. There are five distribution centers in Europe, three of which serve several countries. High-running models are delivered directly from Asia to all five distribution centers and low-running models to the three main warehouses supplying all of Europe.

4 Evaluation

4.1 Disruption Causes, Concerns About Risks, and Experiences with Disruptions

In order to examine the situation of the individual cases, concerns about disruption risks in the SC are first surveyed. The two most commonly cited concerns were external and supplier risks. External risks, such as fire, floods at the site, severe weather affecting the power supply, political instability, the risk of terrorism, or earthquakes at the supplier, were mentioned by the experts. In addition, there is a seasonal risk for Company I, which serves the climate solutions market. When summer is not hot, a lot of stock goes unsold and ends up as overstock, incurring costs. On the other hand, when the summer is unexpectedly warm, the company can face a shortage. Five out of nine companies fear supplier disruptions due to changes in product quality or supplier insolvency. Besides external and supplier risks, underestimation of demand, logistics disruptions that occur en route to the site or to an external warehouse, and time risks, such as the failure of bottleneck machines, production capacity disruptions, and reliance on ocean freight, were identified in the questionnaire responses. The companies' concerns are consistent with the categories identified in literature, though information disruption and the ripple effect were not explicitly mentioned.

All respondents had experienced supplier disruptions, production capacity disruptions, logistics disruptions, and demand disruptions. In every case, there were supplier and production capacity disruptions with high to very high impact. Further disruptions were referenced in terms of force majeure, tax, and regulation. Single sourcing, missing buffers in capacity utilization, and low data visibility are the most frequent causes of disruptions. Respondent B added product and technology disruptions caused by a complex mix of product specifications and changing needs in customer product specifications.

Single sourcing, missing buffers in capacity utilization, and low data visibility are the most frequent causes of disruptions. Respondent B added product and technology disruptions caused by a complex mix of product specifications and changing customer product specification needs.

4.2 Disruption Effects on Supply Chain Processes

In assessing the impact of one type of disruption on another, almost all respondents stated that an interruption in production capacity, and thus an interruption in delivery to the customer, was caused by disruptions to suppliers and demand. To gain more insight in empirical studies into measures to prevent the ripple and bullwhip effects, the participants were asked their opinion on the integration of suppliers into their risk management or on the establishment of a risk management system with their suppliers. The risk manager from Company H explains: "We only look at the risks that affect our direct suppliers (first-tier), because the next supplier level (second-tier) is usually not known. It is not revealed by the first-tier supplier or it is disclosed, but not recorded in our system. In principle, we cannot afford to monitor all value-added levels by using our own resources. Our purchasing conditions also stipulate that first-tier suppliers must establish their own risk management. In addition, if we handled the risk management of our suppliers, then they would no longer be responsible. Suppliers would be taken out of their responsibility if we handled their risk management and suppliers were then no longer responsible for disruptions caused into our production."

The risk manager also emphasized that any disruption in the supply of components impedes the construction of a car, but not every component is equally critical. Risk management must therefore be oriented particularly toward supply-relevant components.

4.3 Risk Mitigation and Disruption Recovery Methods

In addition to the opinion on supplier integration in the company's own risk management, the methods used for risk mitigation and disruption recovery are identified. The participants were asked to describe a recent major SC disruption and how this disruptive event was managed. Table 4 presents the experiences according to the type of disruption.

The presentation of disruptions experienced supports hypothesis H1, as four out of nine companies were affected by supplier disruptions. In most cases, it seemed essential in terms of disruption recovery to have second-source suppliers or to be able to replace missing parts with similar parts.

As an SCRM strategy, Company C aims to reduce its risk exposure and avoid the repetition of previously experienced risks. However, general market risks can only be influenced to a limited extent. The SCRM process of Company H is carried out by risk management and crisis management departments: "Our classic risk management comprises the evaluation of the probability of risk occurrence and possible damage potential. In addition, suppliers are evaluated both at the time of tender submission and on an ongoing basis. The crisis management department reacts actively to disruptions. Financial losses, such as insolvencies, are dealt with, or in the event of fire or flooding at the supplier, a task force team is sent to check fire protection conditions or earthquake safety. In the event of supplier interruptions, the replenishment times of the actual supplier and of other suppliers are assessed first. The crisis team then ranks the damages at the supplier and implements measures for the most severe damages. It is important to find the right balance between risk management and crisis management. You can invest a lot of money in risk management so that nothing happens, but at some point the marginal benefit will be small. Then it is better to have a good crisis management team, because something can always happen. But the right balance is gut feeling."

Despite multiple sourcing and checking material requirements on a daily basis, Company D feels hardly able to protect itself against disruptions and a possible ripple effect, as other suppliers can barely help with the shortage of special materials. The disruption recovery measures of Company D in the event of a material shortage begin with an allocation process of limited resources from all other locations of the disturbed supplier and suppliers from the multiple sourcing strategies. The worst-case scenario is to be dependent on a broker as a last resort.

Table 4 Disruption experiences and recovery measures

Disruption type	Case	examples	Disruption recovery measures
Demand disruption	A	*A federal distribution center for fresh and frozen product categories became stymied during the high season mid-December. The results included 100% utilized warehouse space, queues of trucks, and delayed orders. Demand was much higher than planned. This led to a loss of service level and service level time KPIs and increased costs*	*First, we assigned suppliers to ship to regional distribution centers if it was applicable* *Further recovery measures included temporal decentralization of suppliers (direct deliveries to hypermarkets instead of to the distribution center), the use of a discounter branch's distribution center for shipments of frozen products, and a review of the assortment matrix* *Supply Chain Operations Director spent all of December in the distribution center until the situation stabilized, enabling faster management decisions and providing additional expertise*
	I	*Last year Portugal suffered an extreme heat wave and sales spiked heavily*	*We supplied products from all over Europe to Portugal to remedy the shortage*
External disruption	B	*In the third week of February 2015, three of the four production lines at the manufacturing plant of our supplier in Malaysia caught fire. Products for our plant were affected. This incident led to direct influence on the relationships between us, our supplier, and our customer. 30 Million Euro of our company's turnover was affected, together with around 50 products and 100 customers*	*We found a second-source supplier, used products from both suppliers and got the disruption under control*
External disruption	F	*A year and half ago we had an earthquake at one of our wafer foundry suppliers*	*After being notified of the disruption, we inventoried the affected products and evaluated the effects on the supply chain (enough buffers yes/no, alternative supply/dual sourcing yes/no). After the conclusion, we worked with the supplier on the most critical cases without affecting the supply to the end customer*
Supplier disruption	C	*Bankruptcy of a second-source supplier*	*We qualified further potential suppliers as a recovery measure and expanded our second-source strategies*
	D	*Force majeure at supplier*	*We set up a task force team (purchasing, SCM, Business Units) as an interface to the plants and had meetings to solve the problem (tools for identifying the affected raw and customer part number on site). Return on capital was calculated within 12 h*
	G	*One supplier had problems in production; the second-source supplier could not deliver the material in time*	*We bought the same material from a distributor at a higher price*
	H	*Three weeks of production stoppage due to supply interruption at supplier*	*Missing parts were substituted with similar parts*

The responses about risk mitigation methods and disruption recovery measures show that the focus is once again on dealing with supplier disruptions (H1) and that an established crisis management team is necessary.

In order to further investigate the SCRM measures of the cases surveyed, the participants were asked to describe what could have been done better in the event of a disruption. Company B's SC innovation manager, who experienced a fire at the supplier's production site, realized that a systematic SC incident management routine was missing, and this led to nontransparent information exchange and miscommunication. The respondent learned that a company's roles and the human behavior of both the company and their suppliers in disruption risk management have a significant impact on final performance. The company has therefore implemented a formula for considering the performance of human behavior in its discrete event simulation model.

Further findings from the disruptions experienced are that (1) data accuracy in the systems used is absolutely essential for enabling quick reactions on the customer side (Case D), (2) coordination between different departments in large companies is difficult (Case G), and (3) the management of a higher safety stock level in a central warehouse is necessary to ensure a Europe-wide supply in case of material shortage (Case I). The empirical data confirm how important it is to communicate clearly within a company and to work closely with suppliers so that it is possible to react quickly and flexibly depending on the type of disruption and minimize its effects on overall SC performance. These findings reinforce hypothesis H3.

4.4 *Chances of Digital Technologies in Supply Chain Management*

To explore digital technologies as resilience and efficiency drivers in SCM, the participants were asked about the use of digital technologies in their company operations and support in case of disturbances. Digital technologies were evaluated in terms of utility on a scale from 1 (highest degree of utility) to 6 (lowest degree of utility), and in terms of cost of implementation, using a scale from 1 (very high costs) to 6 (very low costs). Table 5 provides an overview of digital technologies used by the participants. Empty fields in the table represent cases where respondents did not have knowledge of costs or did not apply the technology.

BDA and ERP systems were used by all nine companies surveyed while trace and tracking systems were used by eight companies. These three technologies have high utility for the participants but are also associated with medium to high costs. Industry 4.0 applications and additive manufacturing technology are only used by five companies. Apart from Company D, which sees a very high utility in both applications, the other four companies evaluate these only with a medium to low degree of usefulness and generally see medium to high costs in the application of Industry 4.0 applications and additive manufacturing technology. The

Table 5 Digital technology application to SC disruption risk management

Case digital technology	A	B	C	D	E	F	G	H	I	Sum
Big data analytics	++	++	++	+++	++	++	++	+++	+++	9
Industry 4.0 applications		+		+++		++	++	+		5
Additive manufacturing technology		+		+++		+	+	++		5
Advanced T&T technologies	+++	++	+++	+++	++	+++	+++	++		8
ERP system	+++	++	+++	+++	+++	+++	+++	+++	+++	9

Ranking 1–2: (+++) high degree of utility; ranking 3–4: (++) medium degree of utility; ranking 5–6: (+) low degree of utility

Case digital technology	A	B	C	D	E	F	G	H	I	Sum
Big data analytics	++	++	++	+++	++	++	++	+++	+++	9
Additive manufacturing technology		+				+++	+++	++		4
Advanced T&T technologies	++	++	+++		++	+++	+	+		7
ERP system	+++	++	+++		+++	+	++	+++	+++	8

Ranking 1–2: (+++) high costs; ranking 3–4: (++) medium costs; ranking 5–6: (+) low costs

interviewee from the semiconductor industry reported that an advanced planning system, advanced global detailed visibility, advanced detailed reporting, deep learning, simulation, and advanced communication tools are also used in Company B. Apart from deep learning, the respondent evaluates the applications with high to very high utility, but also with medium to high implementation and usage costs. The manager in consulting, Company E, highlighted the collaboration platform connecting customers with suppliers, noting significant benefits for cooperating partners and medium implementation and usage costs. Cited advantages include the optimization of production planning, support for operational demand processes, and transparency in the placing of transport orders—enabling timely intervention that led to the avoidance of risks and impending bottlenecks.

ERP systems, earlier mentioned as a common industry application, have also been identified by seven companies as the most supportive technology for disruption management. According to the respondents, ERP systems make it possible to redirect the material flow of suppliers, warn of bottlenecks, and gain insight into stock levels. ERP is also considered useful for SC analysis and mid-term demand planning.

The business process manager of Company C stated: "*ERP and trace and tracking systems support operational disruption management. Big data analytics are mainly used for strategic risk reduction by evaluating supplier performance (especially quality performance). The main purpose of digitalization is to optimize operational performance.*"

The widespread use and evaluation of the very high utility of digital technologies certainly bring opportunities to implement them in SCRM. Analogous to findings in

Table 6 Influence of digital technology on SC resilience

Do these digital technologies help to increase SC resilience (risk mitigation at the pre-disruption stage) by improving the following?	BDA	Industry 4.0	Additive manufacturing	T&T systems	ERP systems
Real-time monitoring	5	5	0	8	7
Visibility, predictability	7	5	0	5	7
Material supply reliability	8	5	0	5	6
Production capacity reliability	6	7	0	3	6
Do these digital technologies help to increase SC resilience (disruption recovery at the post-disruption stage) by improving the following:	BDA	Industry 4.0	Additive manufacturing	T&T systems	ERP
Real-time monitoring	3	5	0	5	6
Visibility, predictability	5	5	1	5	6
Material supply reliability	5	4	1	3	6
Production capacity reliability	5	5	2	2	6

Ranking from 0 (low) to 10 (high)

the literature, empirical data confirm hypothesis H4, the opportunities, and potential benefits of digital technologies.

4.5 Impact of Digital Technologies on Supply Chain Resilience

To answer the research question regarding digitalization's influence on SC's resilience, respondents reported on how digital technology supports, or could support, the process of risk mitigation and disruption recovery. Table 6 displays the summary of the individual results (ranked from 0: low to 10: high).

In particular, T&T and ERP systems, but also BDA and Industry 4.0 applications, are seen to help increase SC resilience in the pre- and post-disruption phases, through real-time monitoring. Visibility and predictability in the SC are achieved primarily through BDA and ERP systems in both phases. ERP and also T&T systems provide Participant E with a clear overview of the current system status and the effects of the management measures implemented. The ability to mitigate and resist risk is supported by BDA, which improves the reliability of material supply, and by Industry 4.0 applications, which improve the reliability of production systems. The ability to recover from disruption is mainly driven by BDA, Industry 4.0 applications, and ERP systems. Rapid prototyping is also assessed as a factor in increasing SC resilience in the post-disruption stage. The importance and popularity as well as the inflexibility of ERP systems in SCRM are once again emphasized by the following statement of Respondent F: "*Our ERP system (and the linked Advanced Planning Tool) is the most important tool. All necessary information can be found here. It would help if we could do a better scenario analysis (what-if analysis), but ERP systems are usually not built for it."*

Besides the technologies used so far, Respondent D would benefit from a risk prewarning system in SCRM so that they can react quickly to earthquakes, hurricanes, and capacity bottlenecks and form emergency teams.

Although individual answers do not provide fully conclusive results, it is notable that additive manufacturing was almost never mentioned. Interviewee H said that they do not know the technology well enough but consider 3D printing to be inconceivable in risk management and unhelpful in automotive production. Possible advantages are the production of small batches, customer-specific components, or rare spare parts, but it might not be suitable for mass production, according to the interviewee. The insights offered by the respondents show that hypotheses H6 and H7 are supported. However, some digital technologies are still inflexible and not yet mature, or companies have not yet had contact with additive manufacturing in a way that best supports their company's ability to minimize risks and recover from disruptions.

The insights offered by the respondents generally support hypotheses H6 and H7. However, some digital technologies are still immature and inflexible, or like additive manufacturing, not tried out or well understood.

4.6 Contribution of Digital Technologies to Ripple Effect Control in Supply Chain Disruption Risk Management

Since resilience contributes to ripple effect control, this section summarizes these findings and supplements them with an explicit questioning of the participants about their ripple control measures. BDA supports ripple effect control in both the pre- and post-interruption phases. Participants emphasized the visibility and predictability of SCs through data availability. Data enable better demand forecasting and reduce the risk of demand interruptions. Real-time coordination allows faster decision-making, which is critical for responding to disruptive events. According to the findings from the literature, respondents find advanced T&T systems in combination with ERP systems helpful for collecting real-time data. The possibility of real-time monitoring reduces time risks caused by delays in the SC processes and enables better emergency planning in the event of a malfunction.

The risk manager, Company H, explains that BDA's data helped him to react more quickly when a plant caught fire. BDA informed him about the event one day earlier than the media and gave him a decisive time advantage over competitors in buying components from distributors before all products were sold out. The storm surge in Fukushima, 2011, was also mentioned by interviewee H. As a rule, other stages of the value chain are not known, but semiconductors that are not purchased directly are an exception. This allowed the risk manager to react to the storm surge at an early stage, to supply their own suppliers with purchased products and thus to ensure their own automobile production. A possible ripple effect could be prevented in this way.

Participants were concerned about data security not only in relation to BDA. At this point, the consulting manager (Company E) emphasizes the advantages of the SC collaboration platform. The data exchange is outsourced on the platform and no access to the ERP systems of the customers and their suppliers is necessary. The data provided visualizes the current future demand and the current stock situation of the customer and also simulates future shipments. According to the experiences of the respondent, the increased SC transparency reduces the risks of demand and supply interruptions as well as ripple effect risks and improves SC coordination and SC performance.

The ripple effect control and SC collaboration of Company C are supported by ERP systems. Data availability enables preventive SCM and improved supplier selection. Suppliers are integrated into the ERP system through an interface that receives reports on the quality of suppliers. If the supplier is involved and identifies possible quality management deficiencies at an early stage, it is possible to reschedule production. In addition, the main suppliers are audited regularly. Risks and possible measures are jointly defined within the framework of process failure mode and effects analysis.

3D printing can assist companies in controlling the ripple effect through faster response to demand fluctuations and other disturbances, as well as by reducing adverse impacts through greater production flexibility and shorter lead times. However, additive manufacturing is not perceived as suitable for mass production in the empirical study. This perception may be due to the focus on large companies in the study that primarily produce in large batches. Empirical studies demonstrate the contribution of digital technologies to ripple effect control. Nevertheless, many participants were still concerned about data security.

5 Empirically Derived Contribution of Digital Technologies in Supply Chain Disruption Risk Management

The hypotheses tested by questionnaires and supplemented by theory allow the following qualitative statements to be derived. Empirical data from multiple case studies prove the literature-based hypotheses in the segment of disruption risks in SCM (Hypotheses H1-3) to be true. There is a clear focus on supplier disruptions in the experiences of the participants and their risk management measures. Effective collaboration with suppliers, transparent exchange of information, and consideration of the human behavior of all actors appear to be key factors for successful risk management.

The second segment of hypotheses (H4–H5) studied the influence of digital technologies on SCM. BDA, Industry 4.0 applications, additive manufacturing technology, T&T systems, and ERP systems were identified in the literature as the main groups of digital applications and could also be confirmed by the experiences of the participants. Especially, BDA, ERP, and T&T systems appeared to be most useful. In literature, however, the utilization and popularity of ERP systems have been underestimated. In the empirical analysis, ERP systems have been rated as the most common and supportive application for dealing with disruptions. But it has also been discovered that ERP systems are not flexible enough when it comes to rapid data changes. The inflexibility of ERP systems could be a driving force behind the implementation of new technologies. In Industry 4.0 applications and additive manufacturing technology, though, it is primarily the high costs of investment and use that are seen as impediments. The cost factor and inestimable benefits discourage companies from acquiring new digital technologies, even when the potential for considerable gain exists. Another finding from empirical data is the companies' need for trust, which is necessary for data exchange with external SC collaborators.

The impact of digital technologies on SCRM was analyzed in the third set of hypotheses (H6–H9). Digital technologies help to create resilient SCs by increasing the ability to mitigate risks in the pre-disruption phase and recover from disruptions in the post-disruptions phase, and enhance SC efficiency and ripple effect control. Exploratory research confirmed the benefits of BDA, Industry 4.0 applications, T&T systems, and ERP systems. In particular, digital technologies enable better forecasting of disturbances, faster reaction to disruptive events, and improved SC collaboration. The uses and, more importantly, the advantages of additive manufacturing technology, which are highlighted in the literature in terms of simplifying production processes, attaining a competitive advantage or increasing production flexibility and speed, were hardly mentioned. Related reasons could be that companies have not yet had any contact with additive manufacturing technology, the technology does not fit their business, or established systems fear business and SC redesign.

6 Conclusions

The study's empirical analysis developing case studies from a series of interviews with experts from various industries formed a basis for developing a conceptual framework that integrates the various issues in digitalization in SCs. The examination suggests that while most companies would have had experiences with disruptions, supplier disruptions constitute the most severe and significant form of disruption. The focus on supplier disruptions was also reflected in the risk management measures and the digital technologies used for this purpose. BDA, advanced T&T technologies, and ERP systems are frequently used in the pre- and post-disruption phases.

Exploratory research has shown better forecasting of disturbances, faster response to disruptive events, and improved SC collaboration through digital technologies. ERP systems, BDA, T&T technologies, and Industry 4.0 applications, in particular, enable the collection of real-time data and thus guarantee a resilient and efficient SC as well as control of the ripple effect. However, the high costs for the implementation and use of new digital technologies support the popularity of widely used ERP systems despite their inflexibility. The high implementation costs, unsuitability for mass production, and the immense task of SC redesign discourage companies from 3D printing, despite published benefits. Other challenges identified in the multiple case study approaches are the need for standardized interfaces of digital technologies, the need to take into account the human behavior of all actors, and the need for trust between SC actors to enable efficient cooperation between SCs and transparent exchange of information. These aspects seem to be key factors for successful risk management and should be actively considered by SC managers.

Appendix Hypotheses and Corresponding Interview Questions

Key element studied	Hypotheses and corresponding interview questions
Disruption risks in SCM	
Disruptions (Causes, concerns, experiences)	**H1**: Supplier disruptions have a higher likelihood of appearance
	What are the most important supply chain disruption risks that your company is concerned about? *Which disruption risks did your company experience in the past?* *What are the reasons for the disruptions experienced at your company?*
Disruption impacts (Ripple effect)	**H2**: Disruptions have a serious impact on a large part of the SC
	Identify/estimate the impact of one type of disruption on another, e.g., supplier disruptions led to production capacity disruptions, which led to .... *Which processes of your supply chain (e.g., outbound logistics) are mainly affected by disruptions and what KPIs (e.g., on-time delivery) do you use to measure the deviations caused by disruptions?*
Risk mitigation and disruption recovery	**H3**: Flexible SC networks are required for successful SCRM
	Please describe how a recent major supply chain disruption was managed. *Is there a disruption recovery process in your company? How it looks like?* *Have you experienced a supply chain disruption that could have been better managed? Please describe the situation and what you would do differently next time.*

(continued)

Key element studied	Hypotheses and corresponding interview questions
Application of digital technologies in SCM	
Chances of digital technologies	**H4**: Chances become visible through the use of digital technologies to date
	Which of the following digital technologies do you use in your company's supply chain operations? *Which digital technology (if any) best supported your managing a disruptive event, and how?*
Challenges of digital technologies	**H5**: Challenges in the use of digital technologies are more likely to be associated with obstacles to implementation and data security concerns
	Which functionality in the digital technology was missing when you applied it to disruption mitigation and recovery? *Has digital technology hindered you from making a better decision in case of a disruptive event? Can you describe the situation and the obstacle of digital technology?*
Impact of digital technologies on SCRM	
Resilience by risk mitigation	**H6**: Digital technologies contribute to creating resilient SCs by improving risk mitigation capabilities at the pre-disruption stage
	How digital technology does/could support your risk mitigation process? *Do digital technologies help to increase SC resilience at the pre-disruption stage?*
Resilience by disruption recovery	**H7**: Digital technologies contribute to creating resilient SCs by improving disruption recovery capabilities at the post-disruption stage
	How digital technology does/could support your disruption recovery process? *Do digital technologies help to increase SC resilience at the post-disruption stage?*
Supply chain efficiency	**H8**: Applying digital technologies in SCRM increases SC efficiency
	Do digital technologies help to increase SC efficiency?
Ripple effect control	**H9**: Digital technologies contribute to ripple effect control in SCRM
	Via *additional questions by telephone*

References

Addo-Tenkorang, R., & Helo, P. T. (2016). Big data applications in operations/supply-chain management: A literature review. *Computers & Industrial Engineering, 101*, 528–543.

Baryannis, G., Validi, S., Dani, S., & Antoniou, G. (2018). Supply chain risk management and artificial intelligence: State of the art and future research directions. *International Journal of Production Research, 57*(7), 2179–2202.

Ben-Daya, M., Hassini, E., & Bahroun, Z. (2017). Internet of things and supply chain management: A literature review. *International Journal of Production Research., 57*(15–16), 4719–4742. https://doi.org/10.1080/00207543.2017.1402140

Blackhurst, J., Dunn, K., & Craighead, C. W. (2011). An empirically derived framework of global supply resiliency. *Journal of Business Logistics, 32*, 374–391.

Brintrup, A., Kosasih, E., Schaffer, P., Zheng, G., Demirel, G., & MacCarthy, B. L. (2023). Digital supply chain surveillance using artificial intelligence: Definitions, opportunities and risks. *International Journal of Production Research, 62*(13), 4674–4695.

Cai, Y. J., Choi, T. M., & Zhang, J. (2021). Platform supported supply chain operations in the blockchain era: Supply contracting and moral hazards. *Decision Sciences, 52*(4), 866–892.

Choi, T.-M. (2018). A system of systems approach for global supply chain management in the big data era. *IEEE Engineering Management Review, 46*(1), 91–97.

Choi, T. M., Chan, H. K., & Yue, X. (2017). Recent development in big data analytics for business operations and risk management. *IEEE Transactions on Cybernetics, 47*(1), 81–92.

Choi, T. M., Dolgui, A., Ivanov, D., & Pesch, E. (2022). OR and analytics for digital, resilient, and sustainable manufacturing 4.0. *Annals of Operations Research, 310*(1), 1–6.

Choi, T.-M., Kumar, S., Yue, X., & Chan, H.-L. (2022). Disruptive technologies and operations management in the industry 4.0 era and beyond. *Production and Operations Management, 31*, 9–31. https://doi.org/10.1111/poms.13622

Choi, T. M., & Lambert, J. H. (2017). Advances in risk analysis with big data. *Risk Analysis, 37*(8), 1435–1442.

Choi, T. M., Wallace, S. W., & Wang, Y. (2018). Big data analytics in operations management. *Production and Operations Management, 27*(10), 1868–1883.

Dolgui, A., & Ivanov, D. (2021). 5G in digital supply chain and operations management: Fostering flexibility, end-to-end connectivity and real-time visibility through internet-of-everything. *International Journal of Production Research, 60*(2), 442–451.

Dolgui, A., & Ivanov, D. (2022). 5G in digital supply chain and operations management: Fostering flexibility, end-to-end connectivity and real-time visibility through internet-of-everything. *International Journal of Production Research, 60*(2), 442–451.

Dolgui, A., & Ivanov, D. (2025). Internet of behaviors: Conceptual model, practical and theoretical implications for supply chain and operations management. *International Journal of Production Research, 63*(1), 1–8.

Dolgui, A., Ivanov, D., Potryasaev, S., Sokolov, B., Ivanova, M., & Werner, F. (2020). Blockchain-oriented dynamic modelling of smart contract design and execution control in the supply chain. *International Journal of Production Research, 58*(7), 2184–2199.

Dolgui, A., Ivanov, D., Sethi, S., & Sokolov, B. (2019). Scheduling in production, supply chain and Industry 4.0 systems by optimal control: Fundamentals, state-of-the-art, and applications. *International Journal of Production Research, 57*(2), 411–432.

Dolgui, A., Ivanov, D., & Sokolov, B. (2018). Ripple effect in the supply chain: An analysis and recent literature. *International Journal of Production Research, 56*(1–2), 414–430.

Fazili, M., Venkatadri, U., Cyrus, P., & Tajbakhsh, M. (2017). Physical Internet, conventional and hybrid logistic systems: A routing optimisation-based comparison using the Eastern Canada road network case study. *International Journal of Production Research, 55*(9), 2703–2730.

Fragapane, G., Ivanov, D., Peron, M., Sgarbossa, F., & Strandhagen, J. O. (2022). Increasing flexibility and productivity in Industry 4.0 production networks with autonomous mobile robots and smart intralogistics. *Annals of Operations Research, 308*, 125–143.

Gunasekaran, A., Papadopoulos, T., Dubey, R., Wamba, S. F., Childe, S. J., Hazen, B., et al. (2017). Big data and predictive analytics for supply chain and organizational performance. *Journal of Business Research, 70*, 308–317.

Gunasekaran, A., Tiwari, M. K., Dubey, R., & Wamba, S. F. (2016). Big data and predictive analytics applications in supply chain management. *Computers & Industrial Engineering, 101*, 525–527.

Gunasekaran, A., Yusuf, Y. Y., Adeleye, E. O., & Papadopoulos, T. (2018). Agile manufacturing practices: The role of big data and business analytics with multiple case studies. *International Journal of Production Research, 56*(1–2), 385–397.

Ivanov, D. (2017). Simulation-based single vs dual sourcing analysis in the supply chain with consideration of capacity disruptions, Big Data and demand patterns. *International Journal of Integrated Supply Management, 11*(1), 24–43.

Ivanov, D. (2018a). *Structural dynamics and resilience in supply chain risk management*. Springer.

Ivanov, D. (2018b). Revealing interfaces of supply chain resilience and sustainability: A simulation study. *International Journal of Production Research, 56*(10), 3507–3523.

Ivanov, D. (2024). Digital supply chain management and technology to enhance resilience by building and using end-to-end visibility during the COVID-19 pandemic. *IEEE Transactions on Engineering Management, 71*, 10485–10495.

Ivanov, D., & Dolgui, A. (2019). Low-certainty-need (LCN) supply chains: A new perspective in managing disruption risks and resilience. *International Journal of Production Research, 57*(15–16), 5119–5136.

Ivanov, D., & Dolgui, A. (2021). A digital supply chain twin for managing the disruptions risks and resilience in the era of Industry 4.0. *Production Planning and Control, 32*(9), 775–788.

Ivanov, D., Dolgui, A., & Sokolov, B. (2013). Multi-disciplinary analysis of interfaces "Supply Chain Event Management—RFID—Control Theory". *International Journal of Integrated Supply Management, 8*, 52–66.

Ivanov, D., Dolgui, A., & Sokolov, B. (2019). The impact of digital technology and Industry 4.0 on the ripple effect and supply chain risk analytics. *International Journal of Production Research, 57*(3), 829–846.

Ivanov, D., Dolgui, A., & Sokolov, B. (2022). Cloud supply chain: Integrating industry 4.0 and digital platforms in the "Supply Chain-as-a-Service". *Transportation Research – Part E Logistics and Transportation Review, 160*, 102676.

Ivanov, D., Dolgui, A., Sokolov, B., & Ivanova, M. (2017). Literature review on disruption recovery in the supply chain. *International Journal of Production Research, 55*(20), 6158–6174.

Ivanov, D., Sethi, S., Dolgui, A., & Sokolov, B. (2018). A survey on the control theory applications to operational systems, supply chain management and Industry 4.0. *Annual Reviews in Control, 46*, 134–147.

Ivanov, D., Sokolov, B., & Dolgui, A. (2014). The Ripple effect in supply chains: Trade-off 'efficiency-flexibility-resilience' in disruption management. *International Journal of Production Research, 52*(7), 2154–2172.

Ivanov, D., Sokolov, B., Dolgui, A., Werner, F., & Ivanova, M. (2016). A dynamic model and an algorithm for short-term supply chain scheduling in the smart factory Industry 4.0. *International Journal of Production Research, 54*(2), 386–402.

Ivanov, D., Sokolov, B., & Pavlov, A. (2014). Optimal distribution (re)planning in a centralized multi-stage network under conditions of ripple effect and structure dynamics. *European Journal of Operational Research, 237*(2), 758–770.

Jackson, I., Ivanov, D., Dolgui, A., & Namdar, J. (2024). Generative artificial intelligence in supply chain and operations management: a capability-based framework for analysis and implementation. *International Journal of Production Research, 62*(17), 6120–6145.

Jackson, I., Saenz, M., & Ivanov, D. (2024). From natural language to simulations: Applying AI to automate simulation modelling of logistics systems. *International Journal of Production Research, 62*(4), 1434–1457.

Kumar, S., & Sharma, S. C. (2023). Integrated model for predicting supply chain risk through machine learning algorithms. *International Journal of Mathematical, Engineering and Management Sciences, 8*(3), 353–373.

Li, G., Xue, J., Li, N., & Ivanov, D. (2022). Blockchain-supported business model design, supply chain resilience, and firm performance. *Transportation Research – Part E Logistics and Transportation Review, 163*, 102773.

Li, X., Krivtsov, V., Pan, C., Nassehi, A., Gao, R. X., & Ivanov, D. (2025). End-to-end supply chain resilience management using deep learning, survival analysis, and explainable artificial intelligence. *International Journal of Production Research, 63*(3), 1174–1202.

Liao, Y., Deschamps, Y., de Freitas, E., Loures, R., & Ramos, L. F. P. (2017). Past, present and future of Industry 4.0—a systematic literature review and research agenda proposal. *International Journal of Production Research, 55*(12), 3609–3629.

Meier, O., Gruchmann, T., & Ivanov, D. (2023). Circular supply chain management with blockchain technology: A dynamic capabilities view. *Transportation Research: Part E, 176*, 103177.

Moghaddam, M., & Nof, S. Y. (2018). Collaborative service-component integration in cloud manufacturing. *International Journal of Production Research, 56*(1–2), 677–691.

Nguyen, T., Zhou, L., Spiegler, V., Ieromonachou, P., & Lin, Y. (2018). Big data analytics in supply chain management: A state-of-the-art literature review. *Computers & Operations Research, 98*, 254–264.

Panetto, H., Iung, B., Ivanov, D., Weichhart, G., & Wang, X. (2019). Challenges for the cyber-physical manufacturing enterprises of the future. *Annual Reviews in Control, 47*, 200–213. https://doi.org/10.1016/j.arcontrol.2019.02.002

Papadopoulos, T., Gunasekaran, A., Dubey, R., Altay, N., Childe, S. J., & Wamba, S. F. (2017). The role of Big Data in explaining disaster resilience in supply chains for sustainability. *Journal of Cleaner Production, 142*(2), 1108–1118.

Pavlov, A., Ivanov, D., Dolgui, A., & Sokolov, B. (2018). Hybrid fuzzy-probabilistic approach to supply chain resilience assessment. *IEEE Transactions on Engineering Management, 65*(2), 303–315.

Priore, P., Ponte, B., Rosillo, R., & de la Fuente, D. (2018). Applying machine learning to the dynamic selection of replenishment policies in fast-changing supply chain environments. *International Journal of Production Research, 57*(11), 3663–3677.

Qu, T., Thürer, M., Wang, J., Wang, Z., Fu, H., Li, C., et al. (2017). System dynamics analysis for an Internet-of-Things-enabled production logistics system. *International Journal of Production Research, 55*(9), 2622–2649.

Rai, R., Tiwari, M. K., Ivanov, D., & Dolgui, A. (2021). Machine learning in manufacturing and Industry 4.0 applications. *International Journal of Production Research, 59*(16), 4773–4778.

Rossit, D. A., Tohmé, F., & Frutos, M. (2018). Industry 4.0: Smart scheduling. *International Journal of Production Research, 57*(12), 3802–3813.

Saberi, S., Kouhizadeh, M., Sarkis, J., & Shen, L. (2018). Blockchain technology and supply chain management. *International Journal of Production Research, 57*(7), 2117–2135.

Sauer, P. C., Orzes, G., & Culot, G. (2022). Blockchain in supply chain management: A multiple case study analysis on setups, contingent factors, and evolutionary patterns. *Production Planning & Control, 35*(9), 901–916.

Schlüter, F., Hetterscheid, E., & Henke, M. (2017). A simulation-based evaluation approach for digitalization scenarios in smart supply chain risk management. *Journal of Industrial Engineering and Management Science, 1*, 179–206.

Simchi-Levi, D., Schmidt, W., Wei, Y., Zhang, P. Y., Combs, K., Ge, Y., et al. (2015). Identifying risks and mitigating disruptions in the automotive supply chain. *Interfaces, 45*(5), 375–390.

Sokolov, B., Ivanov, D., Dolgui, A., & Pavlov, A. (2016). Structural quantification of the ripple effect in the supply chain. *International Journal of Production Research, 54*(1), 152–169.

Strozzi, F., Colicchia, C., Creazza, A., & Noè, C. (2017). Literature review on the 'Smart Factory' concept using bibliometric tools. *International Journal of Production Research, 55*(22), 6572–6591.

Tiwari, S., Sharma, P., Choi, T. M., & Lim, A. (2023). Blockchain and third-party logistics for global supply chain operations: Stakeholders' perspectives and decision roadmap. *Transportation Research Part E: Logistics and Transportation Review, 170*, 103012.

Tran-Dang, H., Krommenacker, N., & Charpentier, P. (2017). Containers monitoring through the physical internet: A spatial 3D model based on wireless sensor networks. *International Journal of Production Research, 55*(9), 2650–2663.

Xu, X., & Choi, T. M. (2021). Supply chain operations with online platforms under the cap-and-trade regulation: Impacts of using blockchain technology. *Transportation Research Part E: Logistics and Transportation Review, 155*, 102491.

Xue, J., Li, G., & Ivanov, D. (2025). Digital transformation in the blockchain era: Balancing efficiency and resilience in operations management. *International Journal of Production Economics, 282*, 109525.

Yang, Y., Pan, S., & Ballot, E. (2017). Innovative vendor-managed inventory strategy exploiting interconnected logistics services in the physical internet. *International Journal of Production Research, 55*(9), 2685–2702.

Zheng, G., Kong, L., & Brintrup, A. (2023). Federated machine learning for privacy preserving, collective supply chain risk prediction. *International Journal of Production Research, 61*(23), 8115–8132.

Zheng, G., Ivanov, D., & Brintrup, A. (2025). An adaptive federated learning system for information sharing in supply chains. *International Journal of Production Research*. https://doi.org/10.1080/00207543.2024.2432469

Contingent Sourcing Ripples Considering Hoarding Behaviour Under Supply Disruptions in a Three-Tier Supply Chain

Shanshan Li and Yong He

Abstract This chapter considers a make-to-order system where production gets disrupted due to a supply failure. To avoid a potential stockout risk and a responding price increase during disruption, customers might decide to stockpile extra units for future consumption, which may trigger procurement ripples. We investigate the optimal sourcing decisions for both the customers and the manufacturer to cope with random disruptions. To this end, both the once-off stockpiling and the continuous stockpiling (two-stage) are considered. The optimal stockpiling behaviour in these two modes is analytically provided. Next, a main model is formulated for the manufacturer to optimize the contingent sourcing quantities under different patterns of customer hoarding behaviour. Finally, by conducting numerical analysis, we generate further insights into the role of relative factors and provide specific managerial suggestions on how to adapt contingent sourcing to alleviate different random disruptions under different patterns of customers' hoarding behaviours.

1 Introduction

With trends like global purchases and lean production, the risk of supply chain disruption increases. Unexpected events (such as natural disasters) appearing at any nodes could lead to partial or full breakdowns of the system (Ivanov, 2020; Ivanov, 2024). For example, due to the COVID-19 pandemic, 94% of the Fortune 1000 companies have undergone coronavirus-driven SC disruptions by March 2020 (Fortune, 2020). In order to alleviate the negative impact caused by supply chain failures, a large number of strategies have been proposed, for instance, inventory

S. Li
School of Finance, Nanjing Audit University, Nanjing, China

Y. He (✉)
School of Economics and Management, Southeast University, Nanjing, China
e-mail: hy@seu.edu.cn

D. Ivanov et al. (eds.), *Handbook of Ripple Effects in the Supply Chain*, International Series in Operations Research & Management Science 276,
https://doi.org/10.1007/978-3-031-85508-5_14

policies, contingent sourcing, customer compensation (Li et al., 2021; Chen et al., 2015), supplier diversification (Golmohammadi & Hassini, 2020), dynamic scheduling, recovery, etc. (Dolgui et al., 2020, 2024; Rozhkov et al., 2022).

In designing an effective supply risk mitigation strategy, it is critical to understand the dynamics of the post-disruption demand. However, to the best of our knowledge, the existing literature has not paid enough attention to this direction. In the context of supply chain disruption management, customers' post-disruption behaviour is mainly characterized as backorders or lost sales and quantified as demand that follows a given distribution or function (Li et al., 2024). In fact, in the presence of a disruption, managers could face fast-changing demands. In particular, after the occurrence of supply disruptions, consumers often stockpile products through alternative channels to avoid future shortages or price rises. For instance, in 2011, after the radioactive leak at Japan's Fukushima nuclear plant, consumers went on panic buying of various groceries to prevent radiation poisoning. More recently, during the COVID-19 pandemic, various patterns of hoarding behaviours have been observed. For instance, at the beginning of the pandemic, residents hoard a lot of groceries like canned food. In the semiconductor industry, chipmakers stockpile essential raw materials and semi-products (Kim et al., 2020).

Accordingly, consumer stockpiling has been gaining growing attention in marketing research, mainly focusing on price changes of suppliers under such a consideration. However, in the context of supply chain disruption management, the design of dynamic mitigation strategies is still in its infancy, taking into account the consumer post-disruption stockpiling and the ripples triggered by this behaviour. We thus attempt to fill this gap in this study.

To this end, we consider an MTO production system with one manufacturer who sources from a single supplier. After the occurrence of a supply disruption, production pauses immediately if no countermeasure is taken. In the meantime, the selling price of alternative products provided by competitors in the market rises. As a reaction, customers switch to available competitive manufacturers and might stockpile by purchasing more alternative products than their real-time needs. To hedge against the negative impacts, the manufacturer employs supply-side contingent sourcing. Contingent sourcing is a widely utilized emergency countermeasure where an enterprise turns to a backup supplier at a high price in the event of a failure at its main supplier (Li et al., 2017; Ivanov & Dolgui 2019; Gupta & Ivanov, 2020; Echefaj et al., 2024; Dolgui et al., 2024). As pointed out by a survey, 63% of their surveyed companies employed this strategy in 2008 (Li et al., 2021).

Therefore, several research questions arise: Firstly, how will consumers purchase when there is a stockpiling willingness? What mode of hoarding should be adopted? To stock up only once or continuously? How to optimise the quantity to stock up? Secondly, how should the manufacturer make emergency purchases to resume production under different customer hoarding modes?

In answering these two questions, this study provides both academics and practitioners with two unique contributions. First, in view of disruption durations and price rises, two patterns of customer hoarding behaviour are analysed, i.e., once-off and two-stage. The analysis provides analytic evidence that the post-disruption dynamics cannot be comprehensively described as a single distribution or function, as assumed in many existing studies. We analytically present two commonly seen patterns of post-disruption stockpiling behaviour, indicating the conditions and dynamics under which customers will choose to stockpile, which pattern to choose, and in what quantities. Second, we propose contingent sourcing strategies accordingly for the manufacturer who is exposed to fluctuations in supply and demand, as well as price volatility. Through numerical analysis, we present further managerial insights on how to adjust contingent sourcing in accordance with relative factors, especially based on customers' behaviour.

The remainder of the paper is organized as follows. In Sect. 2, the related work is briefly reviewed. Section 3 describes the problem. Customer stockpiling behaviour is discussed in Sect. 4. Based on the dynamics of customer stockpiling, contingent sourcing from the perspective of the manufacturer is proposed in Sect. 5. Finally, conclusions are drawn and future research is suggested in Sect. 6.

2 Related Literature

The relevant literature mainly falls into two streams: contingent sourcing strategies to cope with supply chain disruptions and customer stockpiling behaviour under supply chain disruption.

Following Tomlin (2006), the contingent sourcing strategy comes with no underlying contracts in this study: the manufacturer will place an emergency order from a secondary supplier when the primary supplier fails. Extensive work investigates the optimal time and quantity of the second source. For example, considering a supply chain with two competing manufacturers, Gupta et al. (2015) study several firms' buying decisions and reveal that supply disruption and procurement times jointly impact the optimal order quantities. Some works further investigate the benefits of contingent sourcing in comparison with other tactics. Chen and Yang (2014) compare emergency backup sourcing decisions under various scenarios of relative channel power and reveal that the decentralized operation is rather dependent on backup sourcing products. By constructing an optimal control model that describes the dynamic response of customers to the joint implementation of contingent sourcing, customer compensation and inventory buffers, Li et al. (2021) analyse the demand and inventory state at which a manufacturing system should adopt these three strategies.

Recent research on contingent sourcing strategies concerns the dynamic aspect. Kouvelis and Li (2012) investigate the ex-post dynamic emergency response that identifies the timing and size of an emergency order that needs to be placed. Saghafian and Van Oyen (2016) investigate the optimal design of flexibility in a backup system under dynamic disruption risks. On the basis of a post-disruption customer behaviour forecast, He et al. (2020) propose dynamic contingent sourcing strategies optimizing the sourcing time. Considering that product price and demand evolve randomly over time, Babai et al. (2023) present the optimal ordering quantity under supply disruption and dual sourcing.

Regarding customer stockpiling behaviour, a stream of literature examines optimal consumer stockpiling behaviour under changing sales prices (Guo & Villas-Boas, 2007). For example, by considering the storage cost of storable products as a step function of inventory, Ching and Osborne (2020) discuss how forward-looking customers react to price promotions. However, only a few papers are found considering consumer behaviour in the context of supply disruption management. Considering that a wholesaler purchasing two brands of a product from two suppliers sells to multiple retailers, Tsao et al.(2019) discuss the optimal ordering policy for the wholesaler when he/she hoards inventory for possible supply disruptions. The literature most relevant to our paper is the one by Yoon et al. (2018) and Zheng et al. (2021). They also study the sourcing strategy under consumers' stockpiling behaviour. Different from their works that focus on two periods, we pay attention to a supply disruption that has already occurred and will last a random length of periods. Billore and Anisimova (2021) present a systematic review of panic buying research.

More recently, inspired by the emerging phenomena during the Covid-19 pandemic, a growing body of research focuses on hoarding behaviour induced by potential future shortages during disaster-type events (Putri et al., 2021). Capturing post-disruption price dynamics in four disruption-related periods: responding, rising, recovering, and recovered, Li et al. (2024) provide optimal hoarding policies for customers. As can be seen from the brief review above, despite that various patterns of customer hoarding and various ripple effects have been commonly observed after the occurrence of diversified disruptions (Brusset et al., 2023), little research has been done to examine the ripples of differentiated behaviours brought by customer hoarding patterns in supply chain systems. We will address this important research gap.

3 Problem Description

A firm produces and sells a single product to customers and sources from a regular supplier who is unreliable. There is an emergency supplier who is reliable and expensive and a competitive manufacturer selling alternative products. The normal demand rate of finished products is deterministic and normalized to be "1." The firm practices make-to-order manufacturing where no inventory of finished products is kept on hand. Before the appearance of a supply disruption, the production is realized at the demand rate "1." We assume that a supply disruption occurs at time "0" and will last T periods. Here T is a random variable with a mean value $E(T)$. To simplify the analysis, we assume that the disruption length T follows a uniform distribution on the interval $[0, 2E(T)]$. That is, the probability density function of the random variable T is determined as $f_T = 1/2E(T)$. Due to the event of supply disruption, the selling price of product increases over time, denoted as

Table 1 Notation

	Notations	Description
Decisions	Q_i	Manufacturer's contingent sourcing quantity under the i pattern of customers' hoarding
	T_m	The time when customers' stockpiled products are fully consumed
Parameters	T	Length of disruption duration,
	$E(T)$	Mean value of the random variable T
	T_P	Time when customers stop stockpiling under continuous hoarding mode
	$p(t)$	Selling price per unit of product (including alternative product)
	v	Consumers' valuation of one unit of product
	v_0	Consumers' valuation of one unit of an alternative product, $v_0 < v$.
	$q(t)$	Customers' purchase quantity at time t after the occurrence of disruption
	c_s	Unit extra sourcing cost from the emergency supplier
	c_h	Unit inventory holding cost per unit of time for customers
	c_H	Unit inventory holding cost per unit of time for the manufacturer
	$I_c(t), I(t)$	Inventory held by customers and the manufacturer

$$p(t) = p_0 + \theta t, \tag{1}$$

where $0 \leq t \leq T$. p_0 represents the selling price before the supply disruption occurs, and θ stands for the rate of price increase.

If no countermeasures are adopted, the production stops immediately, leading to a stockout for customers. Facing stockout, customers purchase alternative products from the competitive manufacturer. To avoid a loss of utility caused by the later increase in price, the consumers may stockpile by purchasing more than needed during the early stages.

In this study, we examine two questions: How the relative factors such as expected disruption length and price rise affect consumers' post-disruption purchasing behaviour, and what is the optimal sourcing time and quantity for the manufacturer to cope with disruption, taking customers' stockpiling dynamics into consideration. The notations are defined in Table 1.

Next, we start with exploring the customer's purchase behaviour during disruption.

4 Customers' Stockpiling Behaviour During Disruption

We discuss two patterns of stockpiling in the following.

4.1 Customers' Once-Off Stockpiling

In order to avoid price rise, customers stockpile T_m units of alternative products at the beginning of disruption for future consumption during time interval $(0, T_m)$. In other words, customers purchase zero product until the stock is depleted at time T_m. If the disruption continues, customers do not consider another once-off stockpiling plan but will purchase alternative products from manufacturers in real time, i.e., $q(t) = 1$.

$$q(t) = \begin{cases} T_m, if\ t = 0 \\ 0, if\ 0 < t < T_m \\ 1, if\ t \geq T_m \end{cases}. \tag{2}$$

Compared to non-stockpiling (real-time buying), customers receive a utility difference U^o during the time period $(0, T_m)$.

$$U^o = \begin{cases} U_1^o = U_0 - \int_0^{T_m} [v_0 - p(t)]\,dt, \text{if } T > T_m \\ U_2^o = U_0 - \int_0^T [v_0 - p(t)]\,dt - \int_T^{T_m} (v - p_0)\,dt, \text{if } T < T_m \end{cases}. \tag{3}$$

Where $U_0 = (v_0 - p_0)\,T_m - c_h \int_0^{T_m} I_c(t)dt$ represents the utility that customers gain from stockpiling T_m units of alternative products at price p_0. $I_c(t)$ represents customers' inventory level, consumed according to their real-time needs, i.e., $I_c(t) = T_m - \int_0^t 1d\tau$. It is worth noting that, without stockpiling, customers receive two patterns of utility, $\int_0^{T_m} [v_0 - p(t)]\,dt$ if $T > T_m$ and $\int_0^T [v_0 - p(t)]\,dt + \int_T^{T_m} (v - p_0)\,dt$ if $T < T_m$. The reason is that, if the disruption lasts shorter than T_m, the customers who choose non-stockpiling can return to the original manufacturer immediately at the end of disruption, whereas the stockpiling customers have to keep consuming their stockpiled alternative products.

The customers' optimal once-off stockpiling decision is formulated as the following problem.

Model 1 (Optimizing customers' once-off stockpiling)

$$T_m{}^* = \arg\ \max E\left(U^o\right), \tag{4}$$

$$\text{subject to } E\left(U^o\right) \geq 0. \tag{5}$$

The objective function (4) maximizes the customers' expected utility difference, i.e., $E\left(U^o\right) = \int_{T_m}^{2E(T)} U_1 f_T dT + \int_0^{T_m} U_2 f_T dT$. (5) guarantees that stockpiling is superior to non-stockpiling.

Solving Model 1, customers' optimal once-off stockpiling behaviour is provided in Proposition 1.

Proposition 1 In the pattern of once-off stockpiling, it is preferable for customers to stockpile ${T_m}^*$ units of alternative products if $2E(T) > \frac{1+\frac{v-v_0}{\theta}}{1-\frac{c_h}{\theta}}$. No stockpiling should be utilized otherwise.

$$ {T_m}^* = 2E(T) - 2E(T)\frac{c_h}{\theta} - \frac{v - v_0}{\theta}. \tag{6} $$

As indicated in Proposition 1, the customers' stockpiling behaviour is critically linked to the following factors: the rate θ of price increase, the inventory holding cost c_h, the disruption length represented by the mean value $E(T)$, and customers' preference for the original product indicated by the value difference $v - v_0$ between the original and substitute products. The results reveal several interesting findings. First, under the pattern of once-off stockpiling, it is superior for the customers to instantly stockpile at the beginning of the disruption, if the disruption is expected to last longer than the threshold. Further, the threshold increases with respect to $v - v_0$, but decreases with respect to θ. That is, the customers' preference for the original product will reduce their willingness to stock up in the event of a short interruption. However, an upward trend in price will drive them to stock up even in the case of short interruptions. The result is consistent with common sense. Third, similar trends can be further seen in the quantity of hoarding.

4.2 *Customers' Two-Stage Stockpiling*

4.2.1 Stockpiling Plan

In the case of two-stage stockpiling, customers decide to purchase products at a constant rate b during the first phase $[0, T_P]$ so that there is no need for them to purchase any products at higher prices at the late period $[T_P, T_m]$. In other words, customers' stockpiling terminates at time T_P, and their stockpiled goods are consumed at the rate of "1" until they are fully used up at time T_m. T_m is defined by $I_c(T_P) = (b - 1)T_P = T_m - T_P$, i.e., $T_m = bT_P$. Where, $b > 1$. Summing up, customers' purchase quantity can be described as

$$ q^T(t) = \begin{cases} b, if\ 0 < t < T_P \\ 0, if\ T_P < t < T_m \\ 1, if\ t \geq T_m \end{cases}, \text{if } T > T_P. \tag{7} $$

It is worth noting that, unlike once-off stockpiling, the two-stage stockpiling plan can only be fully realized if the random interruption exceeds T_P. In view of the random disruption length, customers' utility difference U^T between stockpiling and non-stockpiling falls into two scenarios, as follows.

$$U^T = U_1^T = b\int_0^{T_P} [v_0 - p(t)]\,dt - c_h\int_0^{T_m} I_c(t)dt - \int_0^{T_m} [v_0 - p(t)]\,dt, \text{if } T > T_m;$$

$$\begin{aligned} U^T = U_2^T = b\int_0^{T_P} [v_0 - p(t)]\,dt - c_h\int_0^{T_m} I_c(t)dt \\ - \int_0^{T} [v_0 - p(t)]\,dt - (v - p_0)\,(T_m - T)\,, \text{if } T_P < T < T_m. \end{aligned} \tag{8}$$

Where $\int_0^{T_m} I_c(t)dt = \frac{1}{2}T_m I_c\,(T_P) = \frac{1}{2}\,(b-1)\,bT_P{}^2$. If $T > T_m$, the random production disruption continues when customers' stockpiled goods are used up. If $T_P < T < T_m$, the random production disruption continues when customers terminate their hoarding behavior but ends before customers' inventory is used up. Encountering such a shorter disruption, stockpiling customers continue to consume the stockpiled goods at the rate of "1" even when the original manufacturer is available. On the other hand, like (3), those non-stockpiling customers can buy products from the original manufacturer from time T, leading to a customer utility $(v - p_0)(T_m - T)$ during the time period (T, T_m).

4.2.2 Deviation of Stockpiling Plan

If $T < T_P$, the random production disruption terminates during the customer's stockpiling period. There is no need for them to adhere to the stockpiling plan, but to deviate from it, thereby choosing to prematurely end the hoarding when the original manufacturer resumes production at time T. Accordingly, the stockpiled goods of quantity $I_c(T)$ are consumed at the maximum rate until they are fully used up at time T_n. Where, $I_c(T) = (b - 1)T = T_n - T$, i.e., $T_n = bT$. Clearly, $T_n < T_m$.

$$q^T(t) = \begin{cases} b, if\ 0 < t < T \\ 0, if\ T < t < T_n \\ 1, if\ t \geq T_n \end{cases}, \text{if } T < T_P. \tag{9}$$

Due to the plan deviation, the utility difference between stockpiling and non-stockpiling is incurred during the time period $(0, T_n)$.

$$\begin{aligned} U^T = U_3^T = b\int_0^{T} [v_0 - p(t)]\,dt - c_h\int_0^{T_n} I_c(t)dt \\ - \int_0^{T} [v_0 - p(t)]\,dt - (v - p_0)\,(T_n - T)\,, \text{if } T < T_P. \end{aligned} \tag{10}$$

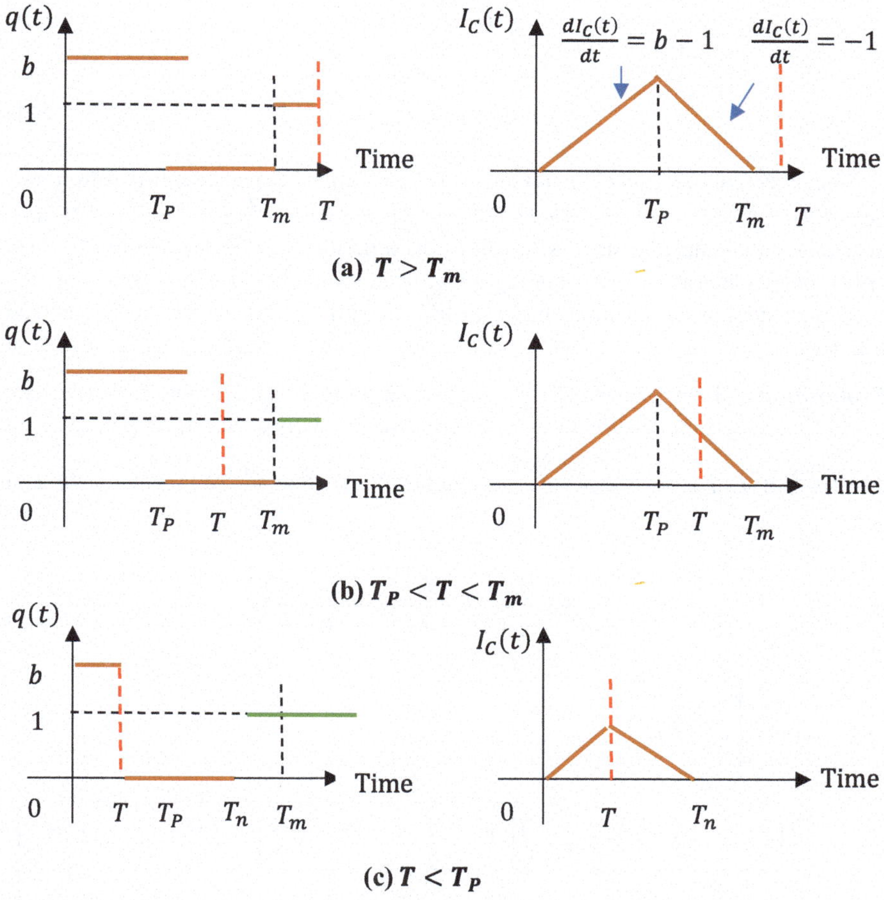

Fig. 1 Customers' purchase quantity and inventory under the two-stage stockpiling

Where $\int_0^{T_n} I_c(t)dt = \frac{1}{2}T_n I_c(T) = \frac{1}{2}(b-1)\,bT^2$. The third and fourth terms give customers' the utility gained from non-stockpiling. Figure 1 summarizes customers' purchase quantity and their inventory dynamics under the two-stage stockpiling.

The customers' optimal two-stage stockpiling decision is formulated as the following problem.

Model 2 (Optimizing customers' two-stage stockpiling)

$$\left\{b^*, T_m{}^*\right\} = \arg\max E\left(U^T\right), \tag{11}$$

$$\text{subject to } E\left(U^T\right) \geq 0. \tag{12}$$

$$1 \le T_m \le 2E(T), \tag{13}$$

$$1 \le b \le T_m. \tag{14}$$

The objective function (11) maximizes the customers' expected utility difference, i.e., $E\left(U^T\right) = \int_{T_m}^{2E(T)} U_1^T f_T dT + \int_{T_P}^{T_m} U_2^T f_T dT + \int_0^{T_P} U_3^T f_T dT$. (12) also guarantees that the two-stage stockpiling is superior over non-stockpiling. (13)–(14) provides the range of stockpiling quantity and time. In particular, if $b = T_m$, the two-stage stockpiling becomes the once-off stockpiling. No stockpiling is realized if $b = 0$.

Proposition 2 In the pattern of two-stage stockpiling, it is preferable for customers to continuously purchase $b*$ units of alternative products at any time point during $\left(0, \frac{T_m{}^*}{b^*}\right)$ if $E(T) \le \frac{\left(\sqrt{3}-1\right)^2 \frac{\Delta v}{\theta}(2-k)}{(1-k)^2}$ and $T_m{}^* \le \frac{3E(T)-1}{\frac{3-k}{1-k}}$. Where $k = \frac{c_h}{\theta}$ and $\Delta v = v - v_0$.

$$\frac{1}{b*} = \max\left\{\frac{1}{T_m{}^*}, \frac{3E(T)}{T_m{}^*} - \frac{3-k}{1-k}\right\},$$

$$T_m{}^* = \max\left\{2E(T), \frac{(1-k)E(T)+2\frac{\Delta v}{\theta}\frac{2-k}{1-k} - \sqrt{\left[(1-k)E(T)+2\frac{\Delta v}{\theta}\frac{2-k}{1-k}\right]^2 - 12\frac{\Delta v}{\theta}(2-k)E(T)}}{2(2-k)}\right\} \tag{15}$$

As indicated in Proposition 2, the hoarding behaviour shows some interesting and non-intuitive findings in the pattern of two-stage stockpiling. First, inconsistent with the once-off stockpiling, the results suggest that customers only use stockpiling strategies to cope with the interruptions lasting shorter than the threshold. Further, the threshold increases with respect to $v - v_0$. In other words, opposite to common sense, customers' preference for the original product increases their willingness to utilize the two-stage stockpiling strategy in the event of longer interruptions. The possible reason for this interesting result is as follows. By consistently stocking up in small quantities, customers hold lower inventories, which in turn allows them to return to the original manufacturer earlier when the random disruption ends prematurely. However, if using once-off stockpiling, the customer would need to hold a large amount of inventory, thus losing some flexibility if the random disruption terminates early. Third, the threshold decreases with respect to θ. That is, if the price increases at a faster rate, the two-stage stockpiling strategy is only advisable for shorter disruptions. The result is reasonable. Compared with the once-off stockpiling strategy which is performed before price increases, the continuous stockpiling strategy is not advantageous in reducing the negative impact of future price increases. This disadvantage becomes more pronounced when prices continue to rise rapidly during the disruption.

5 Manufacturer's Contingent Sourcing Ripple

In this section, based on customers' stockpiling behaviour, we propose contingent sourcing policies for the manufacturer, identifying the optimal sourcing quantity. The ripple effect of purchasing behaviour is then observed by comparing the changes in manufacturers' purchasing strategies under different customer hoarding patterns.

Suppose the manufacturer reroutes to a secondary source at time $t = 0$ and purchases Q_i units of raw material to alleviate the disruption when customers show stockpiling pattern "i." Here $i = N, O, T$, standing for non-stockpiling, once-off stockpiling, and two-stage stockpiling.

With contingent sourcing, the manufacturer resumes production immediately. As a result, demand can be partially or fully satisfied, depending on the quantity Q_i and customers' buying patterns. Note that, compared with the alternative products provided by competitors in the market, customers evaluate a larger utility v on the original product provided by the manufacturer. Thus, we assume that the customers will choose to order from the original manufacturer in priority, if the manufacturer decides to contingently resume production via secondary sources.

The inventory held by the manufacturer exhibits complicated patterns, depending on and the quantity of contingent sourcing and the three main stockpiling patterns of customers. The manufacturer's profits gained from contingent sourcing are described accordingly. Accordingly, the model for determining the optimal contingent sourcing quantity Q_i can be formulated as follows.

Model 3 (Optimizing manufacture's sourcing under customer stockpiling)

$$Q_i^* = \arg\max E\left(\Delta P_i\right), \tag{16}$$

$$\text{Subject to } E\left(\Delta P_i\right) \geq 0 \text{ and } Q_i^* \geq 0. \tag{17}$$

where (17) guarantees the positive profit gained from resuming production via contingent sourcing.

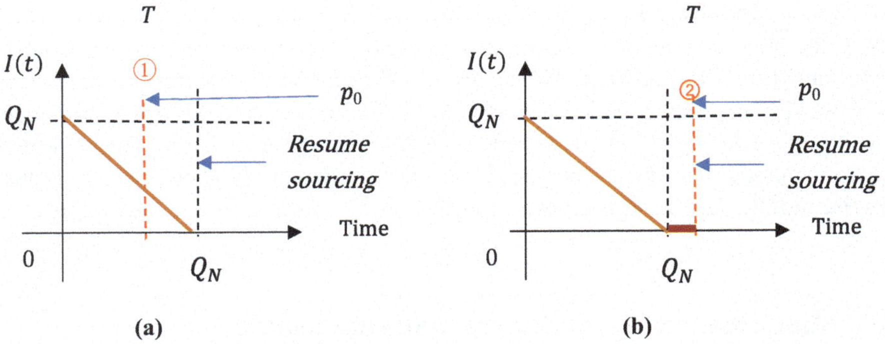

Fig. 2 Manufacturer's inventory $I(t)$ under contingent sourcing Q_N and customers' non-stockpiling

5.1 *Manufacturer's Sourcing Under Customers' Non-Stockpiling*

Without stockpiling, demand maintains at $d(t) = 1$.Therefore, manufacturer's inventory $I(t)$ drops at the constant rate " 1 " until it reaches zero at time Q_N. Then, the manufacturer purchases from the main supplier once she/he is available. On the other hand, the selling price stops increasing and returns back to p_0 immediately at the end of supply disruption.

As depicted in Fig. 2, compared with the decision of doing nothing but passive waiting for supply restoration, the manufacturer achieves extra sales revenue R_1 in the period $[0, \min\{T, Q_N\}]$ by selling products at an increasing selling price $p(t) = p_0 + \theta t$. However, to resume production, two costs are incurred: the contingent sourcing cost $c_s Q_N$ generated from the mark-up price of contingent sources and an inventory holding cost $\frac{c_H Q_N^2}{2}$. Summing up, the manufacturer's expected profit is derived.

$$R_{\mathrm{N}} = \begin{cases} R_N^a = \int_0^T [p(t) - c_1]\,dt \; if \; 0 \le T \le Q_N \\ R_N^b = \int_0^Q [p(t) - c_1]\,dt \; if \; T > Q_N \end{cases},$$

$$E(\Delta P_N) = \int_0^{Q_N} f_T R_{\mathrm{N}}^a dT + \int_{Q_N}^{2E(T)} f_T R_N^b dT - c_s Q_N - \frac{c_H Q_N^2}{2} \tag{18}$$

Solving model 3 in this case, Proposition 3 is provided.

Proposition 3 Without customers' stockpiling, it is preferable for the manufacturer to purchase $Q_N{}^* = \frac{2E(T)(\theta - c_H) - (p_0 - c_1) + \sqrt{[2E(T)(\theta - c_H) - (p_0 - c_1)]^2 + 4(p_0 - c_1 - c_s)\theta}}{2\theta}$ units of contingent sources for production resumption during the disruption of the main supplier.

As stated in Proposition 3, consistent with common sense, customers' stockpiling, the manufacture's contingent sourcing decision mainly depends on his/her inventory holding cost, sourcing cost, and the length of the disruption. The decision here provides a baseline for subsequent analyses of how the manufacturer performs contingency purchases when there are customers stocking up in the market.

5.2 *Manufacturer's Sourcing Under Customers' Once-Off Stockpiling*

In the case of customers' once-off stockpiling, if the manufacturer waits passively, he/she can only gain profit after customers run out of stock and his/her main supplier restores supply, i.e., from time point $max\{T_m, T\}$. By utilizing contingent sourcing, the manufacturer gains an extra profit ΔP_O before this time point, discussed in the following. In view of the sourcing quantity Q_o, manufacturer's inventory $I(t)$ falls into two patterns of dynamics.

$$I(t) = 0, \text{ if } Q_o \leq T_m;$$

$$I(t) = \begin{cases} Q_o - T_m, \text{ if } t \in (0, T_m] \\ Q_o - T_m - t, \text{ if } t \in [T_m, Q_o] \end{cases}, \text{ if } Q_o > T_m. \tag{19}$$

First, if $Q_o \leq T_m$, the manufacturer decides not to satisfy all customers' once-off hoarding demand, thus holding no inventory and receiving an extra profit $Q_o(p_O - c_1) - c_s Q_o$.

Second, if $Q_o > T_m$, the manufacturer decides to implement massive emergency sourcing. By doing so, he/she expects to gain revenue R_O in two phases, that is, by satisfying customers' stockpiling needs at the beginning of the disruption and providing them with high-priced goods in the second cycle $(T_m, \max\{T_m, T\}]$ after they use up their stockpiles. However, the fulfilment of the manufacturer's expectations depends on the duration of the interruption. If $T < T_m$, he/she can only gain the first revenue, i.e., $(p_O - c_1)T_m$. The manufacturer holds inventory $Q_o - T_m$ during time interval $(0, T_m]$ and gradually decreases the inventory thereafter until time Q_o. An inventory holding cost $c_H\left[\left[\frac{1}{2}(Q_o - T_m)^2 + (Q_o - T_m)\, T_m\right]\right.$ is required. Summing up,

$$\Delta P_O = \begin{cases} Q_o\,(p_O - c_1 - c_s)\,, \text{ if } Q_o \leq T_m \\ R_O - c_s Q_o - c_H\left[\left[\frac{1}{2}(Q_o - T_m)^2 + (Q_o - T_m)\, T_m\right]\right., \text{ if } Q_o > T_m \end{cases};$$

$$R_O = \begin{cases} (p_0 - c_1)\,T_m, \text{if } T < T_m \\ (p_0 - c_1)\,T_m + \int_{T_m}^{T} [p(t) - c_1]\,dt, \text{if } T_m < T < Q_o \\ (p_0 - c_1)\,T_m + \int_{T_m}^{Q_o} [p(t) - c_1]\,dt, \text{if } T > Q_o \end{cases}. \tag{20}$$

5.3 Manufacturer's Profit Under Customers' Two-Stage Stockpiling

As depicted in Fig. 2, customers stockpile products during time interval $(0, \min(T, T_P))$ and will return to the normal purchase from time T_m or T_n.

Manufacturer's inventory $I(t)$ falls into the following cases, depending on the manufacturer's sourcing plan, the customers' stockpiling plan, and the deviation situation triggered by the state of the random disruption.

5.3.1 In Case 1 where $Q_T \leq bT_P$

The manufacturer purchases a relatively small amount of contingent sources. As a result, the inventory drops at the demand rate b and reaches zero at time $t_Q^1 = Q_T/b$ before time T_P (see Fig. 3(a-c)). In other words, the manufacturer's contingent sources are fully used to meet customers' hoarding demand, and his/her inventory is depleted when there are some stockpiling demands in the market. Nonetheless, similar to the discussion in Sect. 4, the manufacturer inventory dynamic might deviate from the above pattern.

If $T < t_Q^1$, the random disruption terminates before t_Q^1. Customers stop hoarding products before the manufacturer's inventory is used up. Accordingly, the manufacturer has to hold the remaining inventory until customers resume buying at time T_n. In the time period $\left(T_n, t_Q^2\right)$, manufacturer's inventory drops at the rate of "1" and reaches zero at time t_Q^2. Note $t_Q^2 - T_n = I(T) = (Q_T - bT)$. Compared with doing nothing, the manufacturer receives extra revenue by meeting stockpiling demands during $(0, T)$ at an increasing selling price. Nonetheless, an inventory holding cost during $\left(0, t_Q^2\right)$ and the contingent sourcing cost are required, leading to profit

$$\Delta P_T^1 = b\int_0^T [p(t) - c_1]\,dt - c_s Q_T - c_H\left[\int_0^T (Q_T - bt)\,dt + I(T)\,(T_n - T) + \frac{1}{2}I^2(T)\right]. \tag{21}$$

If $T > t_Q^1$, the manufacturer's contingent sources are fully used to meet customers' hoarding demand. That is, compared with doing nothing, the manufacturer receives a profit:

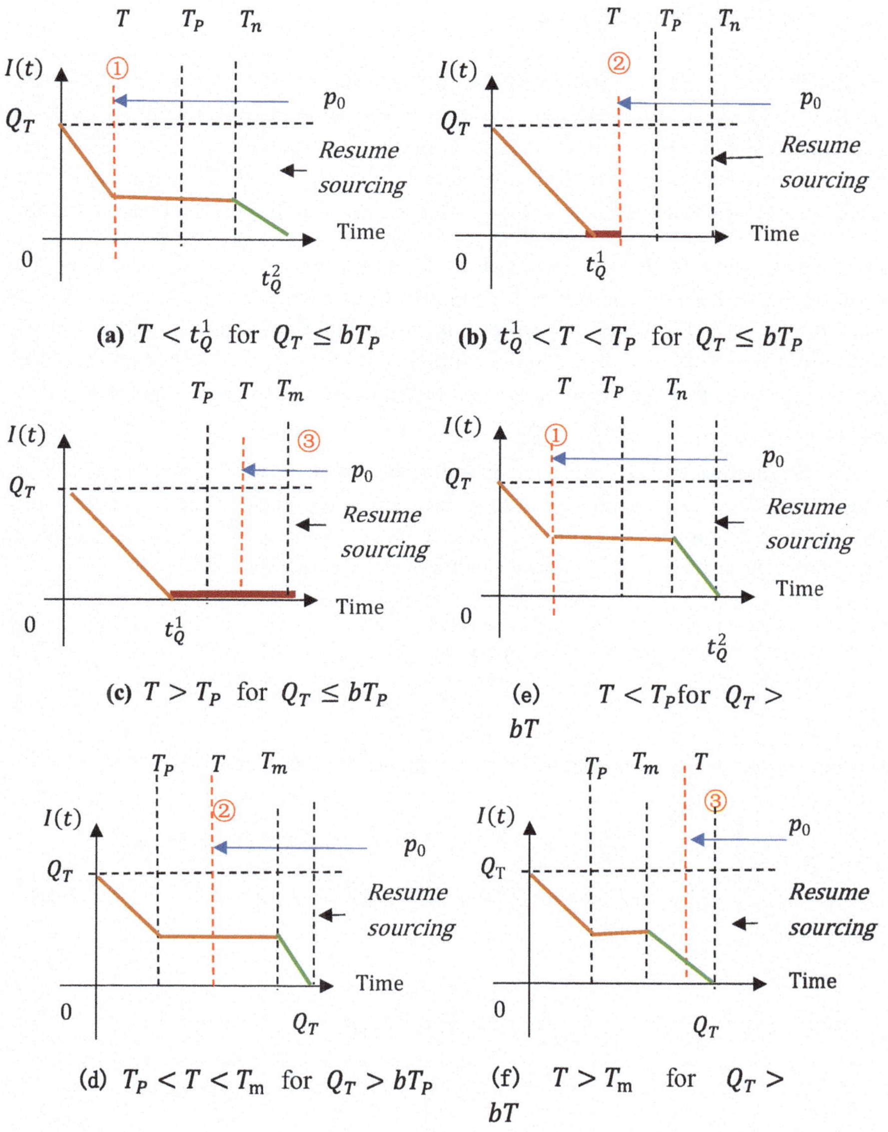

(a) $T < t_Q^1$ for $Q_T \leq bT_P$

(b) $t_Q^1 < T < T_P$ for $Q_T \leq bT_P$

(c) $T > T_P$ for $Q_T \leq bT_P$

(e) $T < T_P$ for $Q_T > bT$

(d) $T_P < T < T_{\mathrm{m}}$ for $Q_T > bT_P$

(f) $T > T_{\mathrm{m}}$ for $Q_T > bT$

Fig. 3 Manufacturer's inventory $I(t)$ under contingent sourcing Q_T and customers' two-stage stockpiling

$$\Delta P_T^2 = b \int_0^{t_Q^1} [p(t) - c_1]\, dt - c_s Q_T - \frac{1}{2} c_H Q_T t_Q^1. \tag{22}$$

5.3.2 In Case 2 Where $Q_T > bT_P$

As discussed in Sect. 5.2, if the manufacturer intends to satisfy both the customers' need to stock up in the first stage and their real-time demand after their stockpiles run out, he/she chooses to make a large contingency purchase, i.e., $Q_T > bT_P$. As a result, the manufacturer still holds inventory $I(T_P) = Q_T - bT_P$ at the moment when the customer stops stocking up at time T_P. (see Fig. 3(d-f)). However, the remaining inventory is only likely to provide an additional benefit $\int_{T_m}^{\min\left\{T, t_Q^3\right\}} [p(t) - c_1]\, dt$ (comparing with doing nothing) to the manufacturer when the disruption lasts longer than T_m, i.e., $T > T_m$. Where t_Q^3 represents the critical time when the manufacturer' remaining inventory is depleted in the following situation: customers' hoarding does not end early so that they resume normal buying patterns from time point T_m. $t_Q^3 - T_m = Q_T - bT_P$, that is, $t_Q^3 = Q_T$.

On the contrary, if $T < T_P$, the manufacturer can only receive profit ΔP_T^1 by meeting customer's hoarding for a short time period $(0, T)$ (see Fig. 3). As the length of the interruption becomes longer, i.e., if $T_P < T < T_m$, by satisfying all of the customer's hoarding needs, the manufacturer receives profit ΔP_T^3.

$$\Delta P_T^3 = b\int_0^{T_p} [p(t) - c_1]\, dt - c_s Q_T - c_H\left[\int_0^{T_p} (Q_T - bt)\, dt + I(T_P)(T_m - T_p) + \frac{1}{2} I^2(T_P)\right]. \tag{23}$$

Summing up, the manufacturer's expected profit is derived as $E(\Delta P_T)$.

$$E(\Delta P_T) = \begin{cases} \int_0^{t_Q^1} f_T \Delta P_T^1 dT + \int_{t_Q^1}^{2E(T)} f_T \Delta P_T^2 dT, if\ Q_T \le bT_P \\ \int_0^{T_P} f_T \Delta P_T^1 dT + \int_{T_P}^{T_m} f_T \Delta P_T^3 dT + \int_{T_m}^{2E(T)} f_T \Delta P_T^4 dT, if\ Q_T > bT_P \end{cases}. \tag{24}$$

Where $\Delta P_T^4 = \Delta P_T^3 + \int_{T_m}^{\min\left\{T, t_Q^3\right\}} [p(t) - c_1]\, dt$.

5.4 *The Ripple Effect of Stockpiling*

Due to the complex expressions of $E(\Delta P_i)$ under the two patterns of customers' stockpiling, the contingent sourcing quantity cannot be analytically examined for the manufacturer. Therefore, we will conduct numerical analyses for further discussion.

To this end, we establish a basic set of parameter values as: $p_0 = 10$, $c_1 = 2$, $c_s = 1$, $c_h = 1$, $c_H = 1$, $\theta = 1$. Then, solving model 4 under the given setting, the optimal sourcing policies that indicate sourcing quantities Q_{O^*} and Q_{T^*} can be visually provided for the manufacturer to cope with disruptions with random lengths. We let the relevant parameters vary and observe how the interaction between the customers' optimal stockpiling decision and the manufacturer's responsive contingent sourcing will change. Note that we focus on the variation trends of relative factors and have run abundant analyses on other basic values of these factors. Our main findings generated from the given setting would not change.

Figure 4 shows the optimal purchasing decisions for customers under the two commonly used patterns of stockpiling, i.e., once-off (not continuous) and two-stage (continuous). The purpose is to cope with disruptions of different random lengths and future price rise. Figure 4(a) presents the different scopes of application of these two types of hoarding strategies. The x-axis represents the important factor, i.e., the real-time of price increases, and the y-axis gives the mean length of the disruption. Intuitively, the two-stage stockpiling is more advisable than the once-off strategy if the price rise is relatively slow and the disruption is short. It is worth noting that there is an area in the figure where both strategies are applicable.

Next, in order to investigate how the customers' stockpiling behaviour affects the manufacturer's sourcing decision, i.e. the souring ripples, Fig. 5 depicts the manufacturers' optimal sourcing quantities under different custoemrs' stockpling. Where $Q_N{}^*$ without customers' stockpiling, $Q_O{}^*$ with once-off stockpiling, and $Q_T{}^*$ with two-stage stockpilng. The extra profit gained from contingent sourcing is also presented.

Several interesting findings are indicated. First, if customers adopt a two-stage stocking mode, i.e., stocking at a constant rate during the first stage and consuming the inventory during the second stage, there is no need for the manufacturer to make a large stockpile/contingent sourcing. The reason is as follows. Under the two-stage stocking, the customer will not stock a large amount of inventory, so there is no need for the manufacturer to stock a large amount of responsive inventory. But it is interesting that, in this process of resuming production with a small amount of emergency purchases, the manufacturer instead earns a higher amount of profit than under other models. The economic reason behind this is as follows. Due to the continuous price increase, the small amount of goods stocked by customers during the price increasing phase will also bring a high profit to the manufacturer. Further, as the price increase becomes quicker, the manufacturer should increase their contingent sourcing amount.

Second, if customers adopt a once-off stocking mode, the results suggest that the manufacturer contingently buys a large amount of secondary sources in the following case: the price of a product increases slowly, i.e., small θ. The possible reasons are stated as follows. As can be seen from Fig. 4 (b), customers choose to hoard a small amount of inventory in this case. As a result, there is a possibility that the disruption continues at the critical time when customers' stockpile is fully consumed. In the meantime, the price increase could reach a significant level due to the long-lasting shortage of the products. Thereby, by selling products to customers at a high price, the manufacturer is able to make a second profit.

Third, by further analysis, we find that inventory holding cost plays an important role in deciding contingent sourcing quantity if the customers in the market do

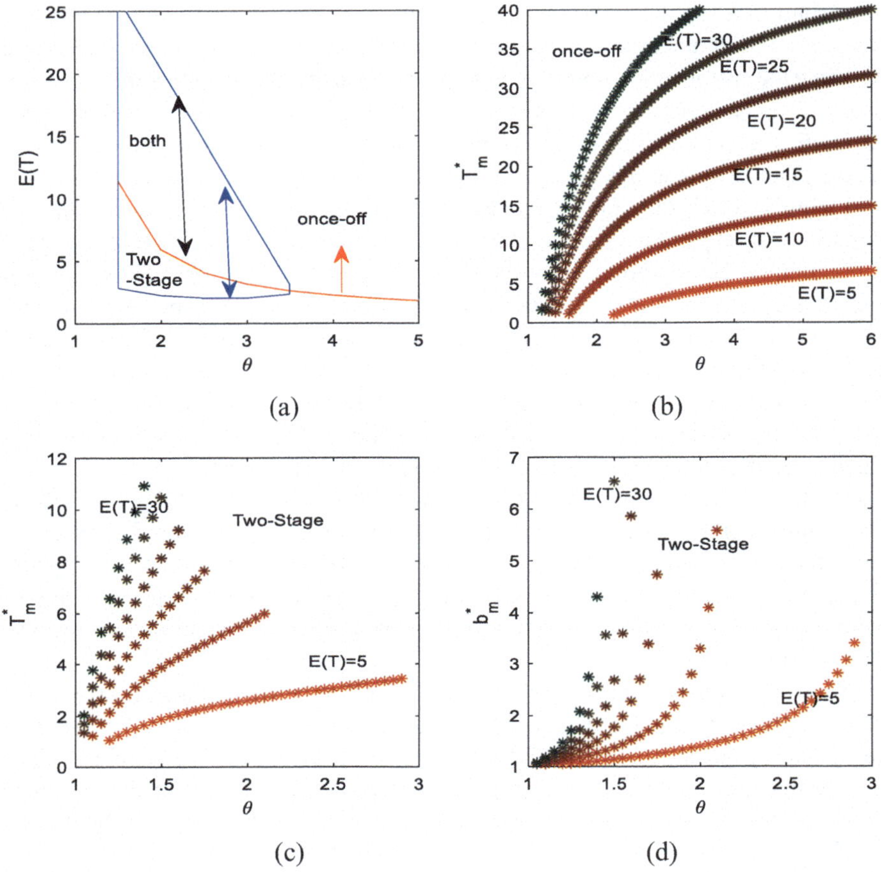

Fig. 4 Customers' once-off and two-stage stockpiling behaviour

not stockpile extra products during disruption. Conversely, if customers stock up, the manufacturer's own inventory costs become less important in the decision, regardless of which model is used.

6 Conclusions

In this paper, we consider a maker-to-order system where a manufacturer halts production due to an uncontrollable supply disruption. There are competing products in the market that customers will turn into during stockouts. Taking into account stockpiling behaviour, we investigate the contingent sourcing strategies for both the customers and the manufacturer. By doing so, we explore the ripple

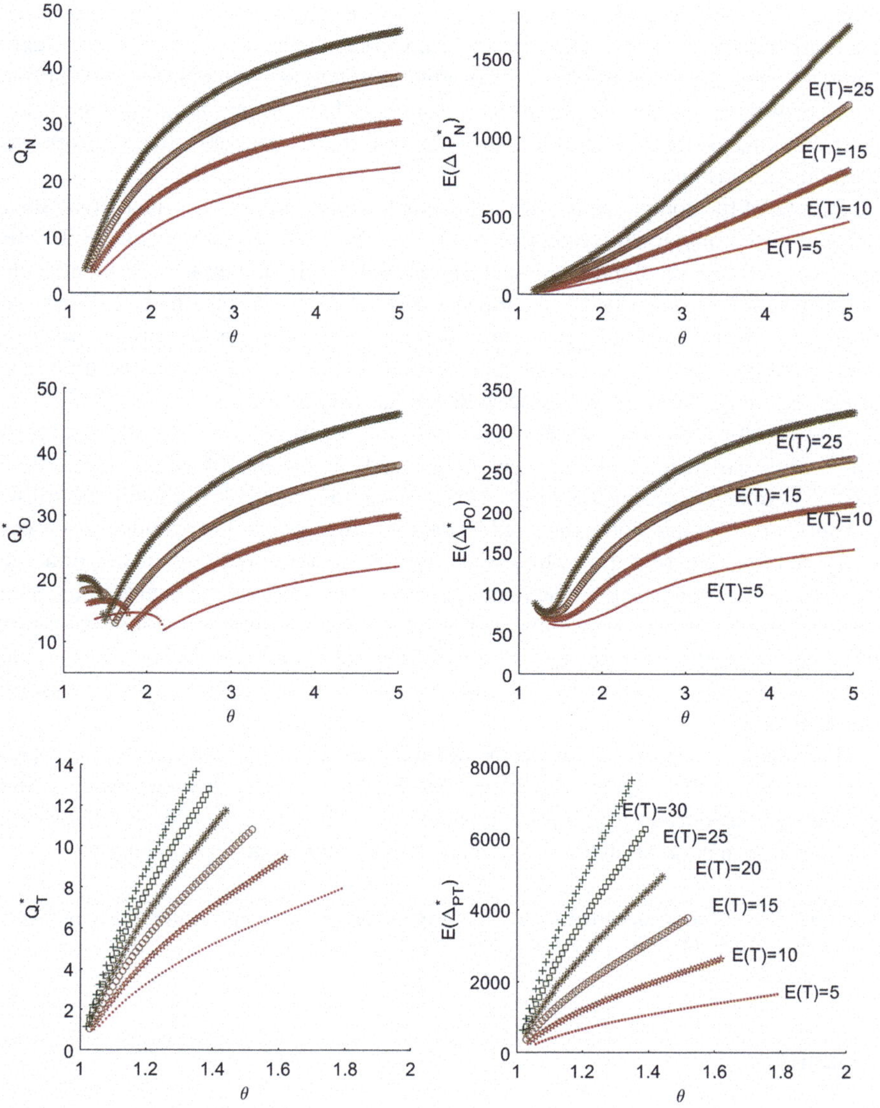

Fig. 5 Manufacturer's contingent sourcing and profit under different patterns of customers' stockpiling

effects of purchasing behaviour on the production and consumption side following a disruptive event on the supply side.

Two patterns of stockpiling modes are discussed for customers, i.e., once-off stockpiling and two-stage (continuous) stockpiling. The results present important and interesting managerial insights. First, the once-off mode is more preferable for longer disruptions triggering quick price rise. Customers' preference over products reduces their willingness of stockpiling. Conversely, continuous stocking mode leads to opposite results. The reason is understandable. Customers can quickly consume their stock and return to the original product once the random interruption ends. However, to gain that flexibility, customers have to pay a high price during their stocking process. Therefore, such mode is more advisable for interruptions with high uncertainty.

In view of the complicated cases regarding the dynamics of customer stockpiling behaviour, the manufacturer decides to order from a second supply source. To this end, we establish the contingent sourcing model for the manufacturer, identifying optimal sourcing quantity to maximize expected profit during the occurrence of disruption. Numerical analyses are provided to visually examine the ripple effects of contingent purchasing behaviours, as well as the value of utilizing extra purchases, that is, utilizing stockpiling. Further managerial insights are also provided.

The results indicate several suggestions for practitioners. First, if customers follow the two-stage stocking, there is no need for the manufacturer to realize a large contingent sourcing for production resumption. Second, if customers follow the once-off stocking mode, the results suggest a large amount of secondary sources in the markets wherein the selling price will not increase quickly. Third, how the relevant factors play a role in the manufacturer's decision-making is also influenced by the behaviours of the customers. For example, if customers do not stock up, the manufacturer's own inventory cost plays an essential role in his/her sourcing decision. However, in the cases of customer stocking up, inventory cost becomes insignificant.

This study raises several directions for future research. For example, in this paper, we consider a linear dynamic for the post-disruption price. A future study could extend by considering other possible patterns. Another area worthy of exploration is to consider a general distribution regarding the random disruption length.

Funding This work is supported by the National Natural Science Foundation of China under Grants (Nos. 72001113 and 71771053).

Appendix

Proof of Proposition 1

The customers' utility U_2 in (3) can be further specified as

$$U_2^o = U_1^o - \int_{T_m}^{T} [v_0 - p(t)]\,dt - \int_{T}^{T_m} (v - p_0)\,dt = U_1^o - \int_{T}^{T_m} [(v - v_0) + \theta t]\,dt$$

$$= U_1^o - \left[(v - v_0)(T_m - T) + \frac{1}{2}\theta\left(T_m^2 - T^2\right)\right] \quad \text{(A1)}$$

Therefore,

$$E\left(U^o\right) = U_1^o - \int_0^{T_m} \frac{(v - v_0)(T_m - T) + \frac{1}{2}\theta\left({T_m}^2 - T^2\right)}{2E(T)} dT$$
$$= \frac{1}{2}(\theta - c_h){T_m}^2 - \frac{\frac{1}{2}(v - v_0){T_m}^2 + \frac{1}{3}\theta {T_m}^3}{2E(T)}.$$

Thus $\frac{dE(U^o)}{dT_m} = \frac{1}{2E(T)} T_m \left[2E(T)(\theta - c_h) - (v - v_0) - \theta T_m\right]$. Note $\theta > c_h$. $E(U^o)$ reaches the maximum value $E\left(U^o\right)|_{T_m{}^*}$, if

$$T_m{}^* = \frac{1}{\theta}\left[2E(T)(\theta - c_h) - (v - v_0)\right] \tag{A2}$$

where $E\left(U^o\right)|_{T_m{}^*} = {T_m}^{*2}\left[\frac{1}{2}(\theta - c_h) - \frac{\frac{1}{2}(v-v_0)+\frac{1}{3}\theta T_m{}^*}{2E(T)}\right] = \frac{1}{2}\frac{{T_m}^{*2}}{2E(T)}\left[2E(T)(\theta - c_h) - (v - v_0) - \frac{2}{3}\theta T_m{}^*\right] = \frac{1}{6}\theta\frac{{T_m}^{*3}}{2E(T)}$. Clearly, $E\left(U^o\right)|_{T_m{}^*} > 0$ always holds if $T_{m^*} > 1$.

Proof of Proposition 2

First, given $T_m = bT_P$ and $\int_0^{T_m} I_c(t)dt = \frac{1}{2}(b-1)bT_P{}^2$, we see $U_1^T = \frac{1}{2}(\theta - c_h)(b-1)bT_P{}^2$. Then, given $T_n = bT$ and $\int_0^{T_n} I_c(t)dt = \frac{1}{2}(b-1)bT^2$, we see $U_3^T = -\frac{1}{2}\theta(b-1)T^2 - c_h\frac{1}{2}(b-1)bT^2 = -\frac{1}{2}(\theta + c_h b)(b-1)T^2$. Further, like (A1), we also have $U_2^T = U_1^T - \int_T^{T_m}\left[(v - v_0) + \theta t\right]dt$. Therefore,

$$E\left(U^T\right) = \int_{T_P}^{2E(T)} U_1^T f_T dT - \int_{T_P}^{T_m}\int_T^{T_m}\left[(v - v_0) + \theta t\right]dt f_T dT + \int_0^{T_P} U_3^T f_T dT$$
$$= \frac{1}{2}(\theta - c_h)(b-1)bT_P{}^2\frac{2E(T) - T_P}{2E(T)}$$
$$- \frac{\frac{1}{2}(v - v_0)(b-1)^2 T_P{}^2 + \frac{1}{6}\theta\left(2b^3 - 3b^2 + 1\right)T_P{}^3}{2E(T)} - \frac{\frac{1}{6}(\theta + c_h b)(b-1)T_P{}^3}{2E(T)}$$
$$= (1 - x){T_m}^2 F \tag{A3}$$

For convenience, we denote $F = \frac{1}{2}(\theta - c_h) + \frac{\frac{1}{3}(c_h - \theta)xT_m - \frac{1}{2}(v - v_0)(1-x) - \frac{1}{3}\theta T_m}{2E(T)} = \theta\left\{\frac{1}{2}(1-k) + \frac{\frac{1}{3}(k-1)xT_m - \frac{1}{2}\frac{\Delta v}{\theta}(1-x) - \frac{1}{3}T_m}{2E(T)}\right\}$ and $\frac{1}{b} = x$. where $k = \frac{c_h}{\theta} < 1$,

$\Delta v = v - v_0$.The first-order derivatives are determined as

$$E_x\left(U^T\right) = [-F + F_x(1-x)]\, {T_m}^2;\; E_{T_m}\left(U^T\right) = (1-x)\left(2T_m F + {T_m}^2 F_{T_m}\right) \tag{A4}$$

where $F_x = \theta \frac{\frac{1}{3}(k-1)T_m + \frac{1}{2}\frac{\Delta v}{\theta}}{2E(T)} > 0$. In other words, $E(U^T) > 0$ when $E_x(U^T) = 0$. Further, based on $E_{xx}(U^T) = (-F_x - F_x){T_m}^2 = -2F_x{T_m}^2$ and $E_{T_mT_m}\left(U^T\right) = (1-x)\left(2F + 4T_m F_{T_m}\right)$, we find Hessian matrix is negative definite when $E_{T_m}\left(U^T\right) = 0$. In all, $E(U^T)$ reaches its maximum if $E_x\left(U^T\right) = E_{T_m}\left(U^T\right) = 0$, i.e., $F = F_x(1-x) = -\frac{1}{2}T_m F_{T_m}$,that is,

$$\begin{aligned}\frac{1}{2}(1-k) + \frac{\frac{1}{3}(k-1)xT_m - \frac{1}{2}\frac{\Delta v}{\theta}(1-x) - \frac{1}{3}T_m}{2E(T)} &= \frac{\frac{1}{3}(k-1)T_m + \frac{1}{2}\frac{\Delta v}{\theta}}{2E(T)}(1-x)\\ &= -\frac{1}{2}T_m\frac{\frac{1}{3}(k-1)x - \frac{1}{3}}{2E(T)}\end{aligned} \tag{A5}$$

Firstly, we can derive $x^* = \frac{3E(T)}{{T_m}^*} - \frac{3-k}{1-k}$ from (A5). Then, substituting it into (A5), ${T_m}^* = \frac{a_2 \pm \sqrt{\Delta_1}}{2a_1} = \frac{(1-k)E(T)+2\frac{\Delta v}{\theta}\frac{2-k}{1-k} \pm \sqrt{\left[(1-k)E(T)+2\frac{\Delta v}{\theta}\frac{2-k}{1-k}\right]^2 - 12\frac{\Delta v}{\theta}(2-k)E(T)}}{2(2-k)}$ is defined by the equation $f(T_m) = 0$.

$$\begin{aligned}f(T_m) &= (2-k){T_m}^2 - \left[(1-k)E(T) + 2\frac{\Delta v}{\theta}\frac{2-k}{1-k}\right]T_m + 3\frac{\Delta v}{\theta}E(T)\\ &= a_1{T_m}^2 - a_2T_m + a_3E(T)\end{aligned} \tag{A6}$$

The following constraint is required for achieving ${T_m}^*$.

$$\Delta_1 = {a_2}^2 - 4a_1a_3E(T) = \left[(1-k)E(T) + 2\frac{\Delta v}{\theta}\frac{2-k}{1-k}\right]^2 - 12\frac{\Delta v}{\theta}(2-k)E(T) \geq 0$$

$$\leftrightarrow (1-k)E(T) - \sqrt{12\frac{\Delta v}{\theta}(2-k)}\sqrt{E(T)} + 2\frac{\Delta v}{\theta}\frac{2-k}{1-k} \geq 0$$

$$\leftrightarrow \sqrt{E(T)} \geq \frac{\left(\sqrt{3}+1\right)\sqrt{\frac{\Delta v}{\theta}(2-k)}}{1-k}, \text{ or } \sqrt{E(T)} \leq \frac{\left(\sqrt{3}-1\right)\sqrt{\frac{\Delta v}{\theta}(2-k)}}{1-k} \tag{A7}$$

Lastly, according to $1 \leq b \leq T_m$, the constraint $\frac{1}{T_m{}^*} \leq x^* \leq 1$ is required, that is, $\frac{3E(T)}{\frac{2(2-k)}{1-k}} \leq T_m{}^* \leq \frac{3E(T)-1}{\frac{3-k}{1-k}}$. By further calculation, both $T_m{}^* = \frac{a_2+\sqrt{\Delta_1}}{2a_1}$ and the first condition of (A7) can be excluded by the left-hand side of the inequality.

Proof of Proposition 3

$E(\Delta P_N)$ can be specified as $E\left(\Delta P_N\right) = \left(p_0 - c_1 - c_s\right) Q_N + \frac{1}{2}\left(\theta - c_H - \frac{p_0-c_1}{2E(T)}\right)$ $Q_N{}^2 - \frac{1}{2E(T)}\frac{1}{3}\theta Q_N{}^3$. The Q_{N^*} is achieved.

References

Babai, M. Z., Ivanov, D., & Kwon, O. K. (2023). Optimal ordering quantity under stochastic time-dependent price and demand with a supply disruption: A solution based on the change of measure technique. *Omega, 116*, 102817.

Billore, S., & Anisimova, T. (2021). Panic buying research: A systematic literature review and future research agenda. *International Journal of Consumer Studies, 45*, 777–804.

Brusset, X., Ivanov, D., Jebali, A., La Torre, D., & Repetto, M. (2023). A dynamic approach to supply chain reconfiguration and ripple effect analysis in an epidemic. *International Journal of Production Economics, 263*, 108935.

Chen, J., Huang, S., Hassin, R., & Zhang, N. (2015). Two backorder compensation mechanisms in inventory systems with impatient customers. *Production and Operations Management, 24*(10), 1640–1656.

Chen, K. B., & Yang, L. (2014). Random yield and coordination mechanisms of a supply chain with emergency backup sourcing. *International Journal of Production Research, 52*, 4747–4767.

Ching, A. T., & Osborne, M. (2020). Identification and estimation of forward-looking behaviour: The case of consumer stockpiling. *Marketing Science, 39*(4), 707–726.

Dolgui, A., Gusikhin, O., Ivanov, D., Li, X., & Stecke, K. (2024). A network-of-networks adaptation for cross-industry manufacturing repurposing. *IISE Transactions, 56*(6), 666–682.

Dolgui, A., Ivanov, D., & Sokolov, B. (2020). Reconfigurable supply chain: The X-Network. *International Journal of Production Research, 58*(13), 4138–4163.

Echefaj, K., Charkaoui, A., Cherrafi, A., & Ivanov, D. (2024). Design of resilient and viable sourcing strategies in intertwined circular supply networks. *Annals of Operations Research*. https://doi.org/10.1007/s10479-024-05873-1

Fortune. Accessed March 10 2020. https://fortune.com/2020/02/21/fortune-1000-coronavirus-china-supply-chain-impact/

Golmohammadi, A., & Hassini, E. (2020). Review of supplier diversification and pricing strategies under random supply and demand. *International Journal of Production Research, 58*(11), 3455–3487.

Guo, L., & Villas-Boas, J. (2007). Consumer stockpiling and price competition in differentiated markets. *Journal of Economics & Management Strategy, 16*(4), 827–858.

Gupta, V., He, B., & Sethi, S. P. (2015). Contingent sourcing under supply disruption and competition. *International Journal of Production Research, 53*(10), 3006–3027.

Gupta, V., & Ivanov, D. (2020). Dual sourcing under supply disruption with risk-averse suppliers in the sharing economy. *International Journal of Production Research, 58*(1), 291–307.

He, Y., Li, S., Xu, H., & Shi, C. M. (2020). An in-depth analysis of contingent sourcing strategy for handling supply disruptions. *IEEE Transactions on Engineering Management, 67*(1), 201–219.

Ivanov, D. (2020). Predicting the impacts of epidemic outbreaks on global supply chains: A simulation-based analysis on the coronavirus outbreak (COVID-19/SARS-CoV-2) case. *Transportation Research Part E: Logistics and Transportation Review, 136*, 101922.

Ivanov, D. (2024). Supply chain resilience: Conceptual and formal models drawing from immune system analogy. *Omega, 127*, 103081.

Ivanov, D., & Dolgui, A. (2019). Low-certainty-need (LCN) supply chains: A new perspective in managing disruption risks and resilience. *International Journal of Production Research, 57*, 5119–5136.

Kim, J., Giroux, M., Gonzalez-Jimenez, H., et al. (2020). Nudging to reduce the perceived threat of Coronavirus and hoarding intention. *Journal of Advertising, 49*(5), 633–647.

Kouvelis, P., & Li, J. (2012). Contingency strategies in managing supply systems with uncertain lead-times. *Production and Operations Management, 21*(1), 161–176.

Li, S., He, Y., & Chen, L. (2017). Dynamic strategies for supply disruptions in production-inventory systems. *International Journal of Production Economics, 194*, 88–101.

Li, S., He, Y., Huang, H., Lin, J., & Ivanov, D. (2024). Supply chain hoarding and contingent sourcing strategies in anticipation of price hikes and product shortages. *IISE Transactions, 56*(7), 746–761.

Li, S., He, Y., & Minner, S. (2021). Dynamic compensation and contingent sourcing strategies for supply disruption. *International Journal of Production Research, 59*(5), 1511–1533.

Putri, A., Retsan, A., Andika, H., et al. (2021). Antecedents of panic buying behaviour during the COVID-19 pandemic. *Management Science Letters, 11*(6), 1821–1832.

Rozhkov, M., Ivanov, D., Blackhurst, J., & Nair, A. (2022). Adapting supply chain operations in anticipation of and during the COVID-19 pandemic. *Omega, 110*, 102635.

Saghafian, S., & Van Oyen, M. P. (2016). Compensating for dynamic supply disruptions: Backup flexibility design. *Operations Research, 64*(2), 390–405.

Tomlin, B. (2006). On the value of mitigation and contingency strategies for managing supply chain disruption risks. *Management Science, 52*(5), 639–657.

Tsao, Y. C., Raj, P. V. R. P., & Yu, V. (2019). Product substitution in different weights and brands considering customer segmentation and panic buying behaviour. *Industrial Marketing Management, 77*, 209–220.

Yoon, J., Narasimhan, R., & Kim, M. K. (2018). Retailer's sourcing strategy under consumer stockpiling in anticipation of supply disruptions. *International Journal of Production Research, 56*, 3615–3635.

Zheng, R., Shou, B., & Yang, J. (2021). Supply disruption management under consumer panic buying and social learning effects. *Omega, 101*, 102238.

Healthcare Supply Chain Agility and Resilience: A Relational View Perspective

Rameshwar Dubey

Abstract The purpose of this chapter is to establish the relationship between relational competencies (such as transparency, trust, and collaboration) and supply chain capabilities (such as agility and resilience) through empirical research. The study focuses on the healthcare sector, particularly in light of the COVID-19 crisis, which exposed the fragility of the healthcare supply chain and resulted in significant human loss. The theoretical model is grounded in the relational view, and a pretested survey instrument was developed using a seven-point Likert scale. Data was collected from retail pharmacies in France using a database obtained from a reputable healthcare sector consulting firm, resulting in 129 usable responses. Hypotheses were tested using WarpPLS 7.0 commercial software, and the analysis revealed that transparency and trust are significant drivers of collaboration. Furthermore, collaboration among partners is a significant driver of agility and resilience. These findings indicate that relational competencies have a positive and significant influence on agility and resilience during times of crisis, supporting the relational view and explaining how these competencies drive supply chain capabilities. The chapter also acknowledges its limitations and suggests future research directions.

1 Introduction

In this chapter, I expanded on the previous chapter by Dubey (2019) in light of the COVID-19 pandemic. The crisis brought about by COVID-19 has made it necessary to reevaluate the assumptions I made in my previous work. I sought to understand how agility and resilience capabilities can assist disaster relief organisations in addressing prolonged crises. This has been a new and challenging

R. Dubey (✉)
MBS School of Business, Montpellier, France
e-mail: r.dubey@mbs-education.com

D. Ivanov et al. (eds.), *Handbook of Ripple Effects in the Supply Chain*,
International Series in Operations Research & Management Science 276,
https://doi.org/10.1007/978-3-031-85508-5_15

experience, requiring a different approach to understanding supply chains to address unprecedented crises.

The COVID-19 pandemic has substantially disrupted the supply chain, leading to shortages of essential medical supplies, equipment, and medications, significantly impacting the healthcare sector (Ivanov, 2020; Ivanov & Dolgui, 2021) as a result of the scarcity of essential healthcare supplies, including personal protective equipment (PPE), oxygen cylinders, medications, face masks, hospital capacity, and healthcare personnel (Choi, 2020; Choi, 2021). Hence, to tackle such situations, the role of agility and resilience in the supply chain to tackle the emergencies resulting from the deadly COVID-19 novel virus that took the lives of millions and the worst part was that no one was prepared for this level of crisis as it was probably a new experience for the humanity in the last five decades (Ivanov, 2022b, 2022c). The COVID-19 crisis has exposed the lack of dynamic sensing capability, speed of response to the situation, and flexibility in most organisations, including hospitals, healthcare centres, pharmaceutical companies, distributors, and healthcare workers (Spieske et al., 2022; Suresh et al., 2021; Tiwari et al., 2024). These are essential characteristics of an agile supply chain. The healthcare supply chain requires both agility and resilience to effectively respond to and recover from disruptions caused by disasters or unexpected events (Munir et al., 2022). Resilience is particularly crucial for the supply chaini's ability to swiftly restore normal operations following such events (Ivanov, 2024a, 2024b; Wieland & Durach, 2021). The current research has defined agility and resilience as two dynamic capabilities crucial for organisations to effectively manage supply chain disruptions; drawing on an ambidextrous view (Altay et al., 2018) or using the practice-based view, the agility and resilience as two desired capabilities of humanitarian supply chain essential for better humanitarian supply chain performance (Dubey et al., 2022). The existing body of literature extensively covers the topics of agility and resilience within the supply chain, particularly in addressing emergencies or crises (Ivanov, 2022b). However, there is a notable gap in understanding the significance of relational competencies in bolstering these supply chain capabilities (Bealt et al., 2016; Villena et al., 2021). In this research, my objective is to investigate how interpersonal skills and relationship-building abilities enhance an organisation's ability to adapt and withstand challenges during crises. Wieland and Wallenburg (2013) highlighted the importance of relational competencies, such as effective communication, seamless cooperation, and integration, in the context of ensuring supply chain resilience. Building on Dyer and Singh's (1998) relational view, Wieland and Wallenburg (2013) argued that developing these competencies is crucial for fostering the agility and robustness needed for supply chain resilience. In addition, Dubey et al. (2019) expanded on the resource-based view (RBV) and relational view, asserting that factors such as information sharing, connectivity, supply chain visibility, trust, and cooperation significantly impact supply chain resilience. The concepts of agility and resilience are frequently referenced by scholars in operations and supply chain management. Christopher and Peck (2004) argue that supply chain reengineering, supply chain collaboration, agility, and a risk management culture are important drivers for building supply chain resilience.

Wieland and Wallenburg (2013) define resilience as a combination of agility and robustness. In their work, Brandon-Jones et al. (2014) consider resilience and robustness as distinct outcomes resulting from a combination of resources and capabilities, while Altay et al. (2018) and Dubey et al. (2022) further argue that agility and resilience are two distinct capabilities in the context of humanitarian relief efforts. Despite varying perspectives, it is well established that agility and resilience significantly overlap, but at times, they are distinct capabilities. Hence, to settle the scholarly debates, I formulated two crucial research questions:

RQ1: What relational competencies influence agility and resilience during health crises?

RQ2: How do these relational competencies impact the agility and resilience of the healthcare supply chain?

In this study, I have developed a theoretical framework based on the relational view to understand how supply chain transparency and trust influence agility and resilience, with collaboration as a mediating factor. I propose that relational competencies play a crucial role during crises and have an impact on agility and resilience. I have limited the scope of this chapter to address two research questions and have not considered different scenarios under which these competencies may have varying effects. To test the research hypotheses, I have used a data set gathered through a survey-based instrument. The data was collected from healthcare workers' experiences during the COVID-19 pandemic, so the findings are subject to debate. However, since the study focuses on relational competencies, I have preferred perceptual data to test the hypotheses. The study includes 129 usable responses gathered from retail pharmacies in France, which, according to Kock and Hadaya (2018), is sufficient to test the research hypotheses.

The chapter is organised into six sections. The second section focuses on theory development and research hypotheses. The third section presents the research design, including survey instrument design, pretesting of the survey instrument, sampling design, survey-based data collection strategy, and data collection and non-response bias test. The fourth section presents the data analysis. The fifth section contains the discussion, and the final section concludes the chapter.

2 Theoretical Framework and Hypotheses Development

I rely on the relational view (RV) to develop a theoretical framework for addressing the research questions and hypotheses. The relational view (RV) (Dyer & Singh, 1998) proposes that organisational performance is affected by relational competencies, which are developed over time through partner collaboration (Cao & Zhang, 2011; Moshtari, 2016). Effective information exchange within the healthcare supply chain relies on transparency, which plays a crucial role in improving collaboration among the diverse stakeholders involved in managing the supply chain during times of crises, such as natural disasters or pandemics (Gualandris et al., 2021;

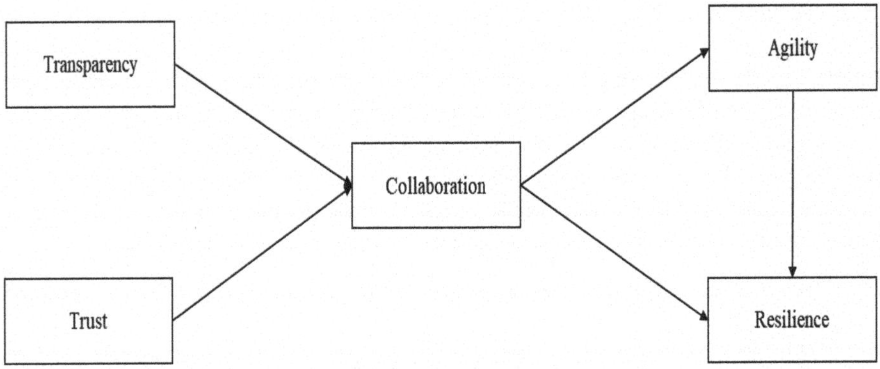

Fig. 1 Theoretical framework

Tiwari et al., 2024). This transparency allows for better coordination and decision-making, ultimately leading to more efficient and effective responses to healthcare supply chain challenges (Harland et al., 2021). In addition to transparency, trust among the actors in the healthcare supply chain is equally important for building collaboration (Duong & Chong, 2020; Scala & Lindsay, 2021). In the healthcare supply chain, transparency and trust play a crucial role in fostering collaboration among all the involved parties (Allmann & Gansterer, 2024; Brinkhoff et al., 2015). This collaboration becomes even more vital during times of crisis, as it allows for more effective coordination and response to the challenges at hand (Gabler et al., 2017; Gulati et al., 2012; Janssen & Van der Voort, 2020). The act of collaborating with others not only fosters agility, as suggested by Christopher and Peck (2004) and Dubey et al. (2021), but also aids in the development of resilience, as highlighted by Scholten and Schilder (2015). Building upon various arguments, I propose a theoretical framework (see Fig. 1) that explains how transparency, trust, and collaboration influence agility and resilience in the healthcare supply chain.

2.1 Agility

Supply chain agility has garnered considerable attention as a means of achieving a competitive edge in the market (Lee, 2002; Lee, 2004). This approach is highly valued as one of the most effective supply chain strategies, particularly when an organisation encounters high levels of uncertainty in both demand and supply (Lee, 2002). Oloruntoba and Gray (2006) assert that the ability of humanitarian aid supply chains to swiftly adapt and respond to changing circumstances is crucial for effectively reducing the suffering of victims in crisis situations. In light of the COVID-19 pandemic, there is a growing recognition of the need to reassess the distinction between emergency supply chains and healthcare supply chains (Kovács & Falagara Sigala, 2021). The crisis has underscored the critical importance of agility in healthcare supply chains during pandemics, mirroring the essential flexibility found in humanitarian relief supply chains (Nikookar et al., 2024). In the absence of comprehensive long-term planning, it is crucial for healthcare supply chains to prioritise speed and flexibility in order to effectively address and respond to the ongoing global health crisis stemming from the impact of COVID-19 (Saïah et al., 2023).

The current body of literature contains a vast amount of research that aims to provide comprehensive theoretical and operational definitions of the agile supply chain. This includes contributions from various authors such as Lee (2002, 2004), Swafford et al. (2006), Li et al. (2008, 2009), as well as Dubey et al. (2021), and Aslam et al. (2018). However, in the healthcare supply chain literature, agility is still not well developed (Müller et al., 2023). Agility is the property of a supply chain network that enables it to deal with and take advantage of uncertainties and volatility, while adaptability is used for more profound medium-term changes. Flexibility, on the other hand, is the characteristic of "*agility*" that enables supply chain networks to deal with uncertainties and volatility. Therefore, in the context of the healthcare supply chain, particularly during unprecedented times such as pandemics, it is essential for the healthcare supply chain to be agile (Munir et al., 2022). To achieve agility, the healthcare sector must develop the ability to sense and anticipate changes in both internal and external environments in order to prepare for unpredictable situations (Lee et al., 2011; El Idrissi et al., 2022). Additionally, the healthcare supply chain needs to enhance its flexibility in terms of manufacturing, sourcing, and distribution capabilities through effective collaboration (Spieske et al., 2022). Lastly, the healthcare supply chain should be able to respond promptly to changes (Lee et al., 2011).

2.2 *Resilience*

In light of recent events, the resilience of supply chains has garnered significant attention from scholars, policymakers, and practitioners (Ivanov, 2024b; Wieland & Durach, 2021). This heightened interest stems from supply chain disruptions caused by economic downturns, natural disasters, geopolitical crises, and the ongoing health crisis (Fan et al., 2022; Ivanov et al., 2017; Rinaldi et al., 2022; Sodhi & Tang, 2021). These disruptions have reverberated across various industries, including automotive, food, and healthcare, leading to shortages of essential items and inflation due to abrupt spikes in oil prices (Sodhi et al., 2023). In response to these challenges, organisations are increasingly recognising the importance of fortifying their supply chains against both short-term and long-term disruptions (Li et al., 2023). Supply chain management scholars have attempted to provide an overview of resilience and the use of the term in various contexts in the management literature (Bhamra et al., 2011; Ivanov, 2024b; Shishodia et al., 2023; Wieland & Durach, 2021).

Burnard and Bhamra (2011) explained that organisational resilience has gained significant attention following Holling's (1973) work. Holling (1973) argues that systems have two distinct features that play a critical role in their functioning: *resilience* and *stability*. Resilience is a fundamental aspect of any system, enabling it to absorb changes and remain functional despite disturbances (Ponomarov & Holcomb, 2009). This means a resilient system can withstand shocks and stresses while maintaining its functions and structures. On the other hand, stability refers to the characteristics of a system that enable it to return to its original state of equilibrium after facing an initial disturbance (Ponomarov & Holcomb, 2009).

Sheffi (2005) attempted to provide the functional definition of supply chain resilience. Supply chain resilience is defined as the property of a supply chain network that enables the chain to regain its original configuration soon after the decay of disturbing forces such as earthquakes, floods, hurricanes, tropical storms, tornadoes, tsunamis, and diseases. After a disaster, the resilience in the supply chain will determine the return to a normal phase through collaboration among the various actors in the supply chain network (Boin et al., 2010; Ivanov et al., 2013; Ivanov & Sokolov, 2013).

In light of recent global health crises, the resilience of the healthcare supply chain has become increasingly crucial (Spieske et al., 2022). It is vital to guarantee the availability of critical medical supplies, equipment, and pharmaceuticals, even during emergencies (Scala & Lindsay, 2021). This necessitates a comprehensive and interdisciplinary approach involving stakeholders across the supply chain, such as manufacturers, distributors, healthcare providers, and government agencies (Govindan et al., 2020). Effective supply chain management enables healthcare systems to respond promptly to crises, minimise disruptions, and continue delivering essential patient services. In conclusion, healthcare supply chain resilience is a multifaceted issue that demands careful attention and collaboration from all stakeholders (Zamiela et al., 2022).

2.3 Collaboration

Inter-organisational collaboration refers to a partnership in which two or more organisations share their resources, such as information, expertise, and infrastructure. They also work closely together to develop new products or designs. This collaborative effort can lead to synergies and innovative outcomes that benefit all parties involved (Cao & Zhang, 2011; Moshtari, 2016). Collaboration, as a concept, encompasses two primary facets: cooperation and coordination, as outlined by Gulati et al. (2012). Cooperation involves the process of setting collaborative goals and negotiating the partners' access to resources (Gulati et al., 2012). It also entails the establishment of shared objectives and the allocation of resources to achieve them. On the other hand, coordination involves aligning organisational tasks and actions to ensure they are in sync and working toward the achievement of cooperative outcomes. This may involve managing interdependencies, communication channels, and workflow processes to facilitate effective collaboration (Gulati et al., 2012; Moshtari, 2016).

During times of uncertainty, it is crucial for organisations to prioritise collaboration as a fundamental element in order to enhance agility and resilience within their supply chain (De Leeuw & Fransoo, 2009; Duong & Chong, 2020; Friday et al., 2021). These capabilities are vital for navigating and adapting to unpredictable circumstances (Qu & Yang, 2015). During times of healthcare crises, it is crucial for different organisations and stakeholders to collaborate effectively (Friday et al., 2021). This collaboration plays a significant role in addressing the uncertainties that result from fluctuations in both the demand for and supply of healthcare resources and services (Dolgui et al., 2024; Sathiya et al., 2023).

2.4 *Transparency*

Establishing transparency in the supply chain is crucial for creating trust and fostering effective communication and cooperation among all stakeholders, including suppliers, manufacturers, distributors, and retailers (Bui, 2024; Morgan et al., 2018). This transparency allows for better visibility into the movement of goods, helps to identify potential risks, and enables the implementation of sustainable and ethical practices throughout the supply chain (Sodhi & Tang, 2019). Supply chain transparency is defined as "*the practice of disclosing detailed information about operations and products, such as the origin and sourcing, manufacturing processes, costs and logistics*" (Montecchi et al., 2021, p. 2, c.f. Bai & Sarkis, 2020). During times of crisis, it is essential to have clear visibility and transparency in the healthcare supply chain (Finkenstadt & Handfield, 2021). This ensures that crucial medical supplies and resources can be effectively tracked, managed, and distributed where they are most needed (Harland et al., 2021; Tiwari et al., 2024). Transparency plays a crucial role in fostering collaboration during times of crisis by ensuring that all relevant information is readily available to everyone involved (Abrams et al., 2003; Ramus et al., 2017). This open and honest approach helps build trust and allows for more effective decision-making and problem-solving among all stakeholders.

2.5 *Trust*

According to Morgan and Hunt (1994), trust plays a crucial role in fostering commitment, consequently influencing cooperative behaviour. The presence of trust among partners acts as a vital adhesive, particularly during uncertain times (Das & Teng, 1998; Schilke & Cook, 2015). Building trust is a process that takes time and requires consistent effort (Luo, 2002). Trust is established as individuals consistently demonstrate reliability, honesty, and integrity over time, leading to the gradual development of trust between them (Handfield & Bechtel, 2002). Building trust within the supply chain is crucial, particularly as it involves collaborating with partners from diverse social and economic backgrounds (Han et al., 2021; Wu et al., 2014). This is especially evident in disaster relief operations, where partners may speak different languages and hold varying religious beliefs (Balcik et al., 2010; Tatham & Kovács, 2010).

During times of crisis, it is crucial to prioritise trust-building to facilitate collaborative efforts (Correia, 2024). This is especially true in the context of healthcare supply chains during pandemics, where the establishment and maintenance of trust among stakeholders significantly impacts the effectiveness of the overall system (Friday et al., 2021; Scala & Lindsay, 2021). Hence, it can be understood that building trust and maintaining transparency is crucial for enhancing collaborative behaviour. When team members trust each other and have access to transparent information, they are better equipped to work together effectively and navigate through challenging situations.

2.6 *Hypotheses Development*

In this section, I will explain the concepts shown in Fig. 1 to form research hypotheses. Based on the relational view, I have proposed how the levels of transparency (H1) and trust (H2) can influence collaborative behaviour within the supply chain. Additionally, I have suggested that collaborative behaviour plays a crucial role in enhancing the agility (H3) and resilience (H4) of the healthcare supply chain, especially during uncertain times. It is important to highlight that developing agility is vital for strengthening the overall resilience (H5) of the supply chain. In Fig. 2, I denoted H1 to represent the hypothesised relationship between transparency and collaboration, indicating that transparency in the healthcare supply chain has a positive and significant impact on improving collaboration during unprecedented crises. Similarly, I denoted H2 to represent the second research hypothesis, explaining how trust is crucial for enhancing collaboration. The third hypothesis (H3) states that collaboration among partners in times of crisis is essential for improving agility in the healthcare supply chain. The fourth hypothesis (H4) indicates that collaboration among partners is crucial for building resilience. Finally, hypothesis (H5) suggests that agility has a positive and significant effect on resilience in the healthcare supply chain.

I have formulated the hypotheses based on the principles of positivistic philosophy. The model has been developed through detailed scrutiny, including an examination of ontological, epistemological, axiological, and methodological assumptions. I won't explain these assumptions in detail here, but I suggest young researchers delve into the details while drafting their thesis. For more detailed information, I suggest that readers refer to articles for rich insights (see Bagozzi, 2011; Meredith et al., 1989).

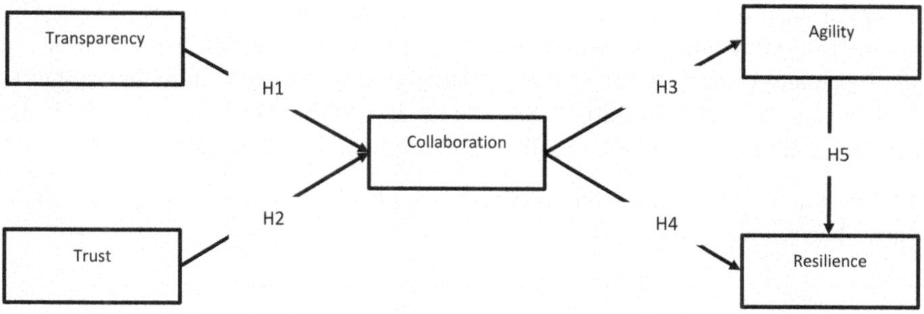

Fig. 2 Hypothesised framework

2.6.1 Transparency and Collaboration

Effective collaboration depends on transparency (Gulati et al., 2012). When partners are open and honest, they foster a clearer understanding of each other's perspectives and help build stronger commitments to shared goals (Yang et al., 2018). During times of crisis, it is crucial for partners to maintain transparency in their interactions (Grimm & Reinecke, 2024). This transparency is essential for fostering collaborative behaviour, which in turn is vital for effectively addressing and overcoming disruptions in the supply chain (Bai & Sarkis, 2020). In their studies, Harland et al. (2021) and Friday et al. (2021) have argued that transparency is critically important during the COVID-19 crisis to address supply chain disruptions in the healthcare sector. Based on their research, I hypothesise that transparency has a positive and significant effect on collaboration among the partners involved in the healthcare supply chain.

H1 Transparency has a positive and significant effect on collaboration among partners in the healthcare supply chain during crises.

2.6.2 Trust and Collaboration

Trust is essential for building commitment and enhancing cooperative behaviour (Morgan & Hunt, 1994). In times of uncertainty and during crises, it is of utmost importance for partners involved in supply chain management to have a high level of trust (Cortez & Johnston, 2020; Ireland & Webb, 2007). This trust is essential for effective collaboration and coordination among the partners (Friday et al., 2021; Tatham & Kovács, 2010). In the healthcare supply chain, it is crucial to cultivate trust among partners, but this can be particularly challenging during times of crisis (Finkenstadt & Handfield, 2021; Vann Yaroson et al., 2023; Zighan et al., 2024). This is because the intricate nature of the healthcare industry involves numerous complex obstacles that need to be overcome in order to build and maintain trust (Gilson, 2003; Jamison et al., 2019). The supply chain literature strongly supports the importance of trust and collaboration (Kwon & Suh, 2004; Sahay, 2003). The process of establishing trust and its influence on collaborative efforts in healthcare supply chains during crises is quite challenging and needs more investigation (Finkenstadt & Handfield, 2021; Spieske et al., 2022). Therefore, based on the theoretical debates, I propose the following as the second research hypothesis:

H2 Trust among partners in the healthcare supply chain positively impacts collaborative behaviour.

2.6.3 Collaboration and Agility/Resilience

Partners engage in collaborative efforts to share crucial information, expertise, and resources (Gulati et al., 2012; Moshtari, 2016). This collaborative approach is essential for developing and enhancing the organisation's sensing capabilities, which play a critical role in rapidly detecting changes in the external environment (Lee & Rha, 2016; Teece, 2007). By swiftly identifying these changes, the organisation can effectively respond to uncertainties and adapt with the desired level of flexibility (Braunscheidel & Suresh, 2009). Harland et al. (2021) and Spieske et al. (2022) noted in their studies that collaboration among partners is essential for enhancing agility in times of crisis. Therefore, based on the preceding discussions, I hypothesise that higher collaboration among the partners helps enhance agility in the healthcare supply chain during an unprecedented crisis.

H3 Collaboration among the partners has a positive and significant effect on the agility of the healthcare supply chain during the crisis.

It is crucial for organisations to collaborate with partners in order to recover and return to normal operations following disruptions (Chen et al., 2019). This collaborative effort is significantly important for rebuilding resilience in the supply chain and mitigating the impact of future disruptions (Ambulkar et al., 2015; Christopher & Peck, 2004; El Baz & Ruel, 2021). Collaboration among partners during a crisis is crucial for bolstering resilience in the healthcare supply chain (Nikookar et al., 2024). This involves sharing resources such as medical supplies, equipment, and expertise and swiftly implementing measures to restore the supply chain to its original state (Beninger & Francis, 2022). There is a substantial body of literature that supports the positive relationship between collaboration and resilience. Therefore, based on the preceding discussions, I hypothesise that collaboration among the partners in the healthcare supply chain helps enhance the supply chain's resilience.

H4 Collaboration among the partners has a positive and significant effect on supply chain resilience.

2.6.4 Agility and Resilience

In the context of supply chain management, the capabilities of agility and resilience play a crucial role in mitigating the impact of unexpected events such as natural disasters, geopolitical crises, and pandemics (Lotfi & Sodhi, 2024). According to some literature that follows the ambidexterity logic, these two capabilities are often viewed as distinct, or agility is seen as a key characteristic of resilience (Altay et al., 2018). According to Christopher and Peck (2004) and Wieland and Wallenburg (2013), agility, defined as the ability to respond quickly and effectively to unexpected disruptions, is a fundamental characteristic that contributes to the resilience of supply chains. According to Ivanov (2022a), the combination of agility, resilience, and sustainability is crucial for establishing a supply chain that can effectively adapt to changing market conditions, withstand disruptions, and operate in an environmentally responsible manner. Based on ongoing scholarly discussions, I am putting forward a hypothesis to investigate the impact of agility on resilience.

H5 Agility has a positive and significant effect on resilience in the healthcare supply chain during crises.

3 Research Design

In order to test the research hypotheses, I decided to use a survey-based approach, as suggested by Flynn et al. (1990). I made this choice because the variables used in the study are based on subjective measures, and a survey-based approach was considered the most appropriate method for collecting data to test these variables. I conducted a comprehensive literature review to find appropriate scales for measuring the reflective variables (refer to Fig. 2) and pretested them with nine experts from the healthcare sector who are responsible for managing healthcare supply chains, as well as with academics who have extensively published in reputable journals in the same field. After incorporating feedback from experts, I have made minor modifications to finalise the questionnaire. This was done to ensure that there were no ambiguous or double-barreled statements that could potentially lead to confusion or misinterpretation (Flynn et al., 1990; MacKenzie & Podsakoff, 2012).

3.1 Measures

I used a six-item reflective measure to assess transparency, which was modified following pretesting. I based the final selection of the six items on the literature review by Morgan et al. (2018). For measuring trust, I used a five-item reflective construct after modification based on pretesting. The literature I reviewed for trust measurement included work by Handfield and Bechtel (2002) and swift-trust by Dubey et al. (2019), tailored to suit the healthcare supply chain context. For collaboration, I used a four-item modified reflective construct based on Moshtari (2016). To measure supply chain agility, I utilized a five-item reflective construct from Dubey et al. (2021), and for healthcare supply chain resilience, I used a five-item reflective construct from Mandal (2017). I employed a seven-point Likert-type scale with endpoints of "strongly disagree" and "strongly agree." I can share the detailed questionnaire upon request. The objective of this chapter is for academic discourse, and the main part of the study will be published in an academic journal.

3.2 Research Setting and Sampling

The study focuses on retail pharmacies in France. I collected data with the help of a well-established consulting firm in Paris, which has extensive experience in global healthcare markets and serves over 100 clients. I obtained detailed contacts and references for over 360 retail pharmacy outlets in France. I used a questionnaire translated into French with the assistance of a translator based in Montpellier, France. I visited each retail outlet to gather data, as the pharmacies in France are usually busy and prefer to discuss before filling out the questionnaire. In the end, I gathered 129 usable responses. I selected retail pharmacies because they prepare the stock report based on the daily usage and the forecasted inventory for healthcare items. Hence, the retail pharmacies were the most suitable respondents for capturing the relationship between various organisations, including hospitals, healthcare centres, healthcare staff, and pharmaceutical companies. The measures are based on the perceptions of one key informant (Lambe et al., 2002) and were developed to examine perceptions of the dyad from the partner's perspective. The respondents were knowledgeable and had good experience in understanding collaborative efforts.

3.3 Data Collection

I received a database from a well-established consulting firm specialising in the pharmaceutical sector. I personally visited 360 retail outlets and distributed questionnaires between September 13, 2023, and February 17, 2024. I chose this method because many pharmacy retail outlets were not accustomed to filling out questionnaires and preferred discussing the objectives of data collection. This study is driven by my personal interest and is funded by Montpellier Business School. I collected 129 usable responses on May 16, 2024, after two reminders. According to Kock and Hadaya (2018), the recommended minimum sample size determination approach at a 0.7 power level suggests that the minimum sample size using the inverse square method is 122, and the gamma-exponential method minimum sample

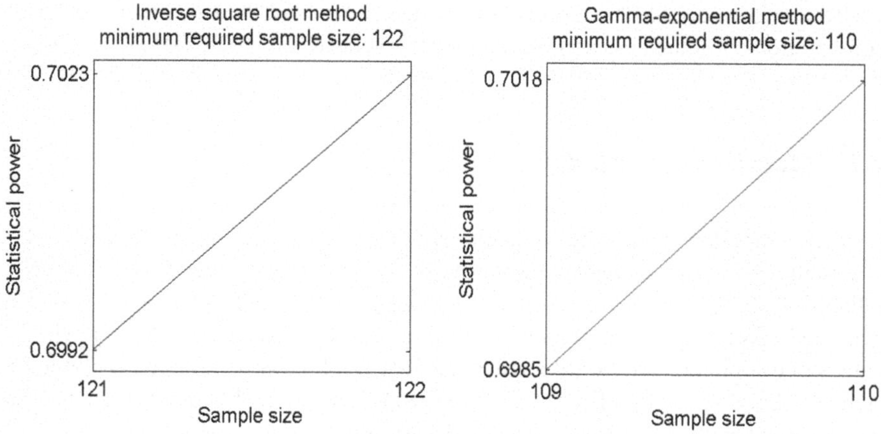

Fig. 3 Sample size (Kock & Hadaya, 2018). (source: WarpPLS 7.0 output)

size is 110. Our sample size of 129, as recommended by Kock and Hadaya (2018), was sufficient to test the research hypotheses following consistent PLSC (Dijkstra & Henseler, 2015) (Fig. 3).

Before proceeding with the statistical analyses, I conducted a non-response bias (NRB) test to ensure the integrity of the data. The data collection phase spanned nearly nine months, resulting in 129 usable responses from the 360 distributed questionnaires. Given the potential impact of non-respondents on the data, I performed wave analysis, following the methodology proposed by Armstrong and Overton (1977). To thoroughly assess non-response bias, I divided the data into halves for each item and conducted two-tailed t-tests. The analysis revealed that for each pair of items, the p-values were greater than 0.1, leading me to conclude that non-response bias is not a significant concern.

4 Data Analysis

The study used reflective constructs and WarpPLS 7.0 (Kock & Hadaya, 2018) to analyse the hypotheses. In response to criticisms of the partial least squares (PLS) method (see Dijkstra & Henseler, 2015), WarpPLS 7.0 combines factor-based PLS algorithms, incorporating the strengths of both variance-based SEM (PLS) and covariance-based SEM (Kock, 2019). Factor-based Partial Least Squares (PLS) algorithms have the same level of precision as covariance-based SEM, assuming a common factor model while having the nonparametric characteristics of traditional PLS algorithms (Kock, 2017, 2019). Additionally, factor-based PLS algorithms directly address the factor indeterminacy problem, one of the major concerns associated with traditional PLS algorithms. Traditional PLS algorithms generate composites as linear combinations of indicators, similar to factor approximations, whereas factor-based PLS algorithms provide estimates of the true factors as linear combinations of indicators and measurement errors (Kock, 2015, 2017).

4.1 Measurement Model

Each measurement item's factor loadings (λ) are all greater than 0.7, except for one item, CO4, whose absolute value is 0.36. Additionally, the scale composite reliability (SCR) for each construct is greater than 0.7, and the average variance extracted (AVEs) for each construct is more than 0.5 (Refer to Table 1). The T-values of each loading are significant at the 0.01 level (refer to Table 2 and Table 3). Hence, according to Fornell and Larcker (1981), I can argue that the constructs used in the study (see Fig. 1) possess convergent validity.

The analysis in Table 4 indicates that the square root of the average variance extracted (AVE) for each construct is greater than the correlations in the corresponding row and column. This suggests that the constructs exhibit discriminant validity.

Table 1 Loadings of the indicator variables (composite reliabilities) (AVE)

Construct	Items	Factor loadings	Variance	Error	SCR	AVE
Transparency	TRANSP1	0.87	0.75	0.25	0.94	0.72
	TRANSP2	0.87	0.75	0.25		
	TRANSP3	0.71	0.50	0.50		
	TRANSP4	0.88	0.77	0.23		
	TRANSP5	0.87	0.76	0.24		
	TRANSP6	0.88	0.77	0.23		
Trust	TRUST1	0.87	0.76	0.24	0.93	0.74
	TRUST2	0.85	0.72	0.28		
	TRUST3	0.85	0.72	0.28		
	TRUST4	0.87	0.75	0.25		
	TRUST5	0.86	0.75	0.25		
Collaboration	CO1	0.86	0.73	0.27	0.89	0.73
	CO2	0.83	0.69	0.31		
	CO3	0.88	0.77	0.23		
Agility	SCA1	0.91	0.82	0.18	0.96	0.83
	SCA2	0.93	0.86	0.14		
	SCA3	0.91	0.82	0.18		
	SCA4	0.92	0.84	0.16		
	SCA5	0.91	0.82	0.18		
Resilience	HSCR1	0.93	0.87	0.13	0.96	0.84
	HSCR2	0.94	0.88	0.12		
	HSCR3	0.91	0.83	0.17		
	HSCR4	0.93	0.86	0.14		
	HSCR5	0.88	0.78	0.22		

Table 2 T ratios for loadings

	TRANSP	TRUST	CO	SCA	HSCR
TRANSP1	9.87				
TRANSP2	11.483				
TRANSP3	8.202				
TRANSP4	11.336				
TRANSP5	12.915				
TRANSP6	13.251				
TRUST1		11.658			
TRUST2		9.928			
TRUST3		11.16			
TRUST4		11.355			
TRUST5		12.696			
CO1			10.384		
CO2			10.286		
CO3			11.676		
SCA1				13.07	
SCA2				13.216	
SCA3				11.576	
SCA4				12.235	
SCA5				12.356	
HSCR1					13.334
HSCR2					12.791
HSCR3					11.95
HSCR4					12.768
HSCR5					12.345

Additionally, in line with Fornell and Larcker's (1981) square root of the AVE test, I have also examined the cross-loadings in Table 5, as recommended by Gefen et al. (2000). The results show that each item loading in Table 5 is significantly higher on its designated construct than on the other constructs, providing further evidence of discriminant validity.

4.2 Common Method Bias

I employed a single-informant instrument based on perception, which may have introduced common method bias (CMB) stemming from factors such as consistency and social desirability (Podsakoff et al., 2003). To address this potential bias, I implemented procedural remedies following the suggestions of MacKenzie and Podsakoff (2012). These remedies aimed to ensure that respondents were well-informed and did not encounter any difficulties in providing accurate responses. Furthermore, I conducted pretesting to meticulously eliminate any potential ambiguity or double-worded statements in the instrument.

Table 3 P values for loadings (two-tailed)

	TRANSP	TRUST	CO	SCA	HSCR
TRANSP1	<0.001				
TRANSP2	<0.001				
TRANSP3	<0.001				
TRANSP4	<0.001				
TRANSP5	<0.001				
TRANSP6	<0.001				
TRUST1		<0.001			
TRUST2		<0.001			
TRUST3		<0.001			
TRUST4		<0.001			
TRUST5		<0.001			
CO1			<0.001		
CO2			<0.001		
CO3			<0.001		
SCA1				<0.001	
SCA2				<0.001	
SCA3				<0.001	
SCA4				<0.001	
SCA5				<0.001	
HSCR1					<0.001
HSCR2					<0.001
HSCR3					<0.001
HSCR4					<0.001
HSCR5					<0.001

Table 4 Correlations among constructs

	TRANSP	TRUST	CO	SCA	HSCR
TRANSP	***0.847***				
TRUST	0.815	***0.861***			
CO	0.797	0.859	***0.864***		
SCA	0.817	0.824	0.85	***0.913***	
HSCR	0.8	0.855	0.851	0.841	***0.919***

Bold values represent square root of AVEs

I have conducted a test for common method bias (CMB) using the correlational marker technique (Lindell & Whitney, 2001) in addition to the procedural remedy test. I used an unrelated variable to control for the correlations caused by the CMB. I also assessed the significance of the correlations using the equation provided by Lindell and Whitney (2001). The adjusted and unadjusted correlations showed minimal differences, and the significance of the correlations remained unchanged. Based on these findings, I conclude that the CMB cannot be completely eliminated, but its effect on the results has been minimized to some extent.

Table 5 Item loadings and cross-loadings

	TRANSP	TRUST	CO	SCA	HSCR	Type (a)	SE	*P* value
TRANSP1	**0.865**	0.038	−0.401	0.271	−0.136	Reflect	0.072	<0.001
TRANSP2	**0.865**	−0.292	0.091	−0.01	0.244	Reflect	0.072	<0.001
TRANSP3	**0.709**	−0.344	−0.153	0.177	−0.092	Reflect	0.074	<0.001
TRANSP4	**0.877**	0.235	0.213	−0.295	−0.333	Reflect	0.071	<0.001
TRANSP5	**0.873**	0.151	−0.112	0.011	0.418	Reflect	0.071	<0.001
TRANSP6	**0.88**	0.143	0.327	−0.116	−0.115	Reflect	0.071	<0.001
TRUST1	0.038	**0.874**	0.172	−0.015	−0.314	Reflect	0.071	<0.001
TRUST2	−0.055	**0.849**	−0.507	0.17	0.051	Reflect	0.072	<0.001
TRUST3	−0.015	**0.85**	−0.077	−0.136	0.35	Reflect	0.072	<0.001
TRUST4	0.132	**0.868**	−0.061	0.042	−0.137	Reflect	0.072	<0.001
TRUST5	−0.103	**0.864**	0.462	−0.06	0.061	Reflect	0.072	<0.001
CO1	−0.264	−0.005	**0.857**	0.008	0.174	Reflect	0.072	<0.001
CO2	−0.101	0.265	**0.83**	−0.17	0.182	Reflect	0.072	<0.001
CO3	0.248	0.028	**0.876**	0.353	−0.269	Reflect	0.071	<0.001
SCA1	−0.017	0.113	0.228	**0.908**	−0.075	Reflect	0.071	<0.001
SCA2	−0.044	0.163	−0.143	**0.925**	0.201	Reflect	0.071	<0.001
SCA3	0	−0.246	−0.014	**0.906**	−0.018	Reflect	0.071	<0.001
SCA4	0.043	0.011	−0.015	**0.917**	−0.136	Reflect	0.071	<0.001
SCA5	0.018	−0.045	−0.053	**0.907**	0.025	Reflect	0.071	<0.001
HSCR1	−0.207	0.253	−0.021	0.086	**0.934**	Reflect	0.07	<0.001
HSCR2	−0.086	−0.055	0.032	−0.002	**0.939**	Reflect	0.07	<0.001
HSCR3	−0.299	0.014	0.186	−0.129	**0.912**	Reflect	0.071	<0.001
HSCR4	0.114	−0.185	0.044	−0.019	**0.927**	Reflect	0.071	<0.001
HSCR5	0.5	−0.028	−0.251	0.064	**0.882**	Reflect	0.071	<0.001

Bold values represent the parsimonious structure

Podsakoff et al. (2024) highlighted the absence of a dependable statistical method to effectively tackle the common method bias (CMB) issue, despite the existence of several statistical approaches. They emphasised that employing procedural remedies represents the most effective approach to minimising the impact of CMB.

4.3 Hypothesis Testing

In Fig. 4, the estimates obtained from the PLS analysis are represented. The R^2 (R-squared) value of HSCR is 0.80, and for SCA it is 0.73. This means that the proposed theoretical model explains nearly 80% of the total variation in healthcare supply chain resilience and 73% of the total variation in healthcare supply chain agility. Additionally, the R^2 value of CO is 0.77, indicating that transparency (TRANSP) and trust (TRUST) explain nearly 77% of the total variation in collaboration. Therefore, I conclude that relational competencies such as transparency,

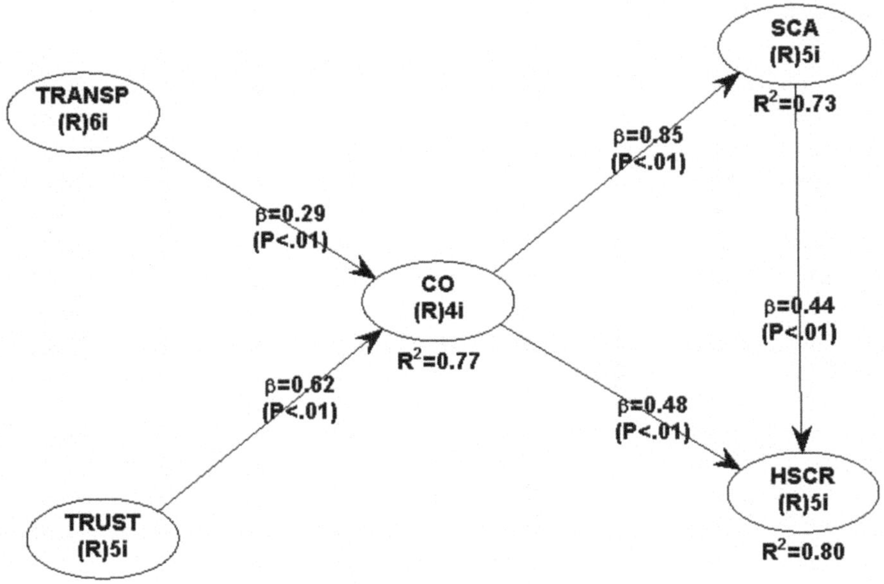

Fig. 4 Final model after PLS analysis

trust, and collaboration are strong determinants of healthcare supply chain agility and resilience during crises. Furthermore, the path coefficients (TRANSP→CO) ($\beta = 0.29, p < 0.01$), (TRUST→CO) ($\beta = 0.62, p < 0.01$), (CO→SCA) ($\beta = 0.85$, $p < 0.01$), (CO→HSCR) ($\beta = 0.48$, $p < 0.01$), and (SCA→HSCR) ($\beta = 0.44$, $p < 0.01$) indicate that hypotheses H1–H5 are well supported.

The findings from the PLS analysis confirm that the development of relational competencies plays a crucial role in fostering agility and resilience during periods of health crises. This suggests that the ability to build and maintain strong relationships is a key factor in effectively navigating and responding to challenges posed by health emergencies.

5 Discussions

The chapter's primary objective is to delve into the intricacies of agility and resilience within the healthcare supply chain. It thoroughly examines how relational competencies, such as transparency, trust, and collaboration, are pivotal in establishing and reinforcing these capabilities. The study diligently employed empirical data to validate the theoretical framework, with a specific focus on the experiences of retail pharmacies in France during the COVID-19 pandemic. To analyse the data, the study utilised perceptual measures. The statistical analysis was conducted using WarPLS 7.0, a robust tool that amalgamates the strengths of traditional PLS and covariance-based SEM. The insights derived from the analyses carry significant implications for theory, practice, and policy. Moreover, each key point is meticulously elaborated under various sub-headings to offer comprehensive insights.

5.1 *Implications for Theory*

The findings of the study provide valuable insights into the relational view (Dyer & Singh, 1998), which highlights the role of relational competencies in strengthening agility and resilience within supply chain management, particularly in the face of uncertainty and turbulence. This research builds upon previous arguments put forth by Wieland and Wallenburg (2013) and Dubey et al. (2019) to delve deeper into the mechanisms through which relational competencies contribute to the cultivation of these essential capabilities. The COVID-19 crisis caused significant disruptions, and the healthcare supply chain suffered greatly. Despite several efforts, the stakeholders, including the government, pharmaceutical companies, healthcare centres and hospitals, retail pharmacies, and patients, failed to address the disruptions during the first wave in early 2020. The root cause of the problems lies in the communication gap and the lack of transparency, which hinders visibility. These issues often occur when the stakeholders lack trust and commitment, leading to poor collaboration. Therefore, the results highlight the importance of transparency in the supply chain, the level of trust among partners, and collaboration as significant determinants of agility and resilience.

5.2 *Implications for Practice*

The research findings offer valuable insights for healthcare managers. They underscore the crucial need for establishing clear communication channels, building trust among stakeholders, and encouraging collaborative initiatives to strengthen the flexibility and resilience of the healthcare supply chain, especially during times of crisis. Transparency involves creating visibility in the supply chain, sharing information about product history, and being able to track the flow of products, information, and funds without any alterations or modifications throughout the supply chain. In many cases, lack of transparency is the missing link that leads to a lack of trust and hinders collaborative efforts. Additionally, building trust is vital, and partners should regularly meet and share information to prevent any situations of mistrust that could impact collaborative efforts. In times of crisis, it is crucial for partners to collaborate effectively. This can be achieved by aligning on shared values and actively working to minimise opportunistic behaviour, which can hinder effective collaboration.

5.3 Implications for Policy

The findings of the study offer valuable insights for policymakers and government officials, providing them with important information to effectively respond to potential future health crises. It is widely recognised that a lack of collaboration is a key factor contributing to disruptions in the healthcare supply chain. In order to foster collaborative efforts, policymakers need to prioritise the development of an ecosystem that encourages transparency and trust among all stakeholders involved in the healthcare supply chain, including healthcare providers, manufacturers, distributors, and regulatory bodies.

5.4 Limitations of Study and Future Research Directions

In the introduction section, I have already stated that the objective of this chapter is to generate academic interest among scholars who are interested in working towards agility and resilience in the healthcare supply chain from the perspective of stakeholders. The concepts of agility and resilience are quite complex, with multiple factors contributing to supply chain capabilities. Unfortunately, the existing body of literature tends to focus more on the engineering or ecological perspective, while paying less attention to the organisational perspective. Although perceptual measures have their own biases, the study's findings provide cost-effective and time-saving results. However, I recommend using longitudinal data from a reliable database to effectively test the model. Obtaining such a database is quite challenging and requires better support from various stakeholders to obtain robust results. Additionally, I have not considered the effects of contingent factors. Taking into account the influence of cultural norms, leadership styles, and the geographical context is essential for comprehending their potential effects on the ultimate results. Finally, I must note that I have not taken into consideration external variables, such as demographic factors or environmental conditions, that could potentially impact the results and conclusions of the study.

6 Conclusion

I conclude the chapter with a lot of hope and excitement that it might generate interest among the readers to appreciate the role of relational competencies in the design of supply chains. In this case, I analyse the healthcare supply chain. However, this can be replicated in the case of humanitarian supply chains or commercial supply chains in highly turbulent situations. Secondly, the importance of survey-based studies has received significant criticism. However, in the context of behavioural studies, the perceptions of the right respondents can provide much richer insight. Hence, I always believe that if the study is properly designed and the process is meticulously followed, the results are quite interesting and valuable. The chapter provides comprehensive guidance for scholars, practitioners, and policymakers, highlighting the pressing necessity to prioritize transparency, trust, and collaboration during times of crisis.

References

Abrams, L. C., Cross, R., Lesser, E., & Levin, D. Z. (2003). Nurturing interpersonal trust in knowledge-sharing networks. *Academy of Management Perspectives, 17*(4), 64–77.

Allmann, M., & Gansterer, M. (2024). Supply chain collaboration in times of crises. *International Journal of Integrated Supply Management, 17*(1), 32–72.

Altay, N., Gunasekaran, A., Dubey, R., & Childe, S. J. (2018). Agility and resilience as antecedents of supply chain performance under moderating effects of organizational culture within the humanitarian setting: A dynamic capability view. *Production Planning & Control, 29*(14), 1158–1174.

Ambulkar, S., Blackhurst, J., & Grawe, S. (2015). Firm's resilience to supply chain disruptions: Scale development and empirical examination. *Journal of operations management, 33*, 111–122.

Armstrong, J. S., & Overton, T. S. (1977). Estimating nonresponse bias in mail surveys. *Journal of Marketing Research, 14*, 396–402.

Aslam, H., Blome, C., Roscoe, S., & Azhar, T. M. (2018). Dynamic supply chain capabilities: How market sensing, supply chain agility and adaptability affect supply chain ambidexterity. *International Journal of Operations & Production Management, 38*(12), 2266–2285.

Bagozzi, R. P. (2011). Measurement and meaning in information systems and organizational research: Methodological and philosophical foundations. *MIS Quarterly, 35*(2), 261–292.

Bai, C., & Sarkis, J. (2020). A supply chain transparency and sustainability technology appraisal model for blockchain technology. *International Journal of Production Research, 58*(7), 2142–2162.

Balcik, B., Beamon, B. M., Krejci, C. C., Muramatsu, K. M., & Ramirez, M. (2010). Coordination in humanitarian relief chains: Practices, challenges and opportunities. *International Journal of Production Economics, 126*(1), 22–34.

Bealt, J., Fernández Barrera, J. C., & Mansouri, S. A. (2016). Collaborative relationships between logistics service providers and humanitarian organizations during disaster relief operations. *Journal of Humanitarian Logistics and Supply Chain Management, 6*(2), 118–144.

Beninger, S., & Francis, J. N. (2022). Resources for business resilience in a COVID-19 world: A community-centric approach. *Business Horizons, 65*(2), 227–238.

Bhamra, R., Dani, S., & Burnard, K. (2011). Resilience: The concept, a literature review and future directions. *International Journal of Production Research, 49*(18), 5375–5393.

Boin, A., Kelle, P., & Whybark, D. C. (2010). Resilient supply chain for extreme situations: Outlining a new field of study. *International Journal of Production Economics, 126*(1), 1–6.

Brandon-Jones, E., Squire, B., Autry, C. W., & Petersen, K. J. (2014). A contingent resource-based perspective of supply chain resilience and robustness. *Journal of Supply Chain Management, 50*(3), 55–73.

Braunscheidel, M. J., & Suresh, N. C. (2009). The organizational antecedents of a firm's supply chain agility for risk mitigation and response. *Journal of Operations Management, 27*(2), 119–140.

Brinkhoff, A., Özer, Ö., & Sargut, G. (2015). All you need is trust? An examination of inter-organizational supply chain projects. *Production and Operations Management, 24*(2), 181–200.

Bui, T. D. (2024). Assessing sustainable supply chain transparency practices in Taiwan semiconductor industry: A hierarchical interdependence approach. *International Journal of Production Economics, 272*, 109245.

Burnard, K., & Bhamra, R. (2011). Organisational resilience: Development of a conceptual framework for organisational responses. *International Journal of Production Research, 49*(18), 5581–5599.

Cao, M., & Zhang, Q. (2011). Supply chain collaboration: Impact on collaborative advantage and firm performance. *Journal of Operations Management, 29*(3), 163–180.

Chen, H. Y., Das, A., & Ivanov, D. (2019). Building resilience and managing post-disruption supply chain recovery: Lessons from the information and communication technology industry. *International Journal of Information Management, 49*, 330–342.

Choi, T. M. (2020). Innovative "bring-service-near-your-home" operations under Corona-virus (COVID-19/SARS-CoV-2) outbreak: Can logistics become the messiah? *Transportation Research Part E: Logistics and Transportation Review, 140*, 101961.

Choi, T. M. (2021). Risk analysis in logistics systems: A research agenda during and after the COVID-19 pandemic. *Transportation Research Part E: Logistics and Transportation Review, 145*, 102190.

Christopher, M., & Peck, H. (2004). Building the Resilient Supply Chain. *The International Journal of Logistics Management, 15*(2), 1–14.

Correia, T. (2024). Trust building in public health approaches: The importance of a "People-Centered" concept in crisis response. *Risk Management and Healthcare Policy*, 1903–1908.

Cortez, R. M., & Johnston, W. J. (2020). The Coronavirus crisis in B2B settings: Crisis uniqueness and managerial implications based on social exchange theory. *Industrial Marketing Management, 88*, 125–135.

Das, T. K., & Teng, B. S. (1998). Between trust and control: Developing confidence in partner cooperation in alliances. *Academy of Management Review, 23*(3), 491–512.

De Leeuw, S., & Fransoo, J. (2009). Drivers of close supply chain collaboration: One size fits all? *International Journal of Operations & Production Management, 29*(7), 720–739.

Dijkstra, T. K., & Henseler, J. (2015). Consistent partial least squares path modeling. *MIS Quarterly, 39*(2), 297–316.

Dolgui, A., Gusikhin, O., Ivanov, D., Li, X., & Stecke, K. (2024). A network-of-networks adaptation for cross-industry manufacturing repurposing. *IISE Transactions, 56*(6), 666–682.

Dubey, R. (2019). Resilience and agility: The crucial properties of humanitarian supply chain. *Handbook of Ripple Effects in the Supply Chain*, 287–308.

Dubey, R., Altay, N., & Blome, C. (2019). Swift trust and commitment: The missing links for humanitarian supply chain coordination? *Annals of Operations Research, 283*, 159–177.

Dubey, R., Bryde, D. J., Dwivedi, Y. K., Graham, G., & Foropon, C. (2022). Impact of artificial intelligence-driven big data analytics culture on agility and resilience in humanitarian supply chain: A practice-based view. *International Journal of Production Economics, 250*, 108618.

Dubey, R., Bryde, D. J., Foropon, C., Tiwari, M., Dwivedi, Y., & Schiffling, S. (2021). An investigation of information alignment and collaboration as complements to supply chain agility in humanitarian supply chain. *International Journal of Production Research, 59*(5), 1586–1605.

Duong, L. N. K., & Chong, J. (2020). Supply chain collaboration in the presence of disruptions: A literature review. *International Journal of Production Research, 58*(11), 3488–3507.

Dyer, J. H., & Singh, H. (1998). The relational view: Cooperative strategy and sources of interorganizational competitive advantage. *Academy of Management Review, 23*(4), 660–679.

El Baz, J., & Ruel, S. (2021). Can supply chain risk management practices mitigate the disruption impacts on supply chains' resilience and robustness? Evidence from an empirical survey in a COVID-19 outbreak era. *International Journal of Production Economics, 233*, 107972.

El Idrissi, M., El Manzani, Y., Maatalah, W. A., & Lissaneddine, Z. (2022). Organizational crisis preparedness during the COVID-19 pandemic: An investigation of dynamic capabilities and organizational agility roles. *International Journal of Organizational Analysis, 31*(1), 27–49.

Fan, D., Zhou, Y., Yeung, A. C., Lo, C. K., & Tang, C. (2022). Impact of the US–China trade war on the operating performance of US firms: The role of outsourcing and supply base complexity. *Journal of Operations Management, 68*(8), 928–962.

Finkenstadt, D. J., & Handfield, R. (2021). Blurry vision: Supply chain visibility for personal protective equipment during COVID-19. *Journal of Purchasing and Supply Management, 27*(3), 100689.

Flynn, B. B., Sakakibara, S., Schroeder, R. G., Bates, K. A., & Flynn, E. J. (1990). Empirical research methods in operations management. *Journal of Operations Management, 9*(2), 250–284.

Fornell, C., & Larcker, D. F. (1981). Structural equation models with unobservable variables and measurement error: Algebra and statistics. *Journal of Marketing Research, 18*(1), 39–50.

Friday, D., Savage, D. A., Melnyk, S. A., Harrison, N., Ryan, S., & Wechtler, H. (2021). A collaborative approach to maintaining optimal inventory and mitigating stockout risks during a pandemic: Capabilities for enabling health-care supply chain resilience. *Journal of Humanitarian Logistics and Supply Chain Management, 11*(2), 248–271.

Gabler, C. B., Richey, R. G., Jr., & Stewart, G. T. (2017). Disaster resilience through public–private short-term collaboration. *Journal of Business Logistics, 38*(2), 130–144.

Gefen, D., Straub, D., & Boudreau, M. C. (2000). Structural equation modeling and regression: Guidelines for research practice. *Communications of the Association for Information Systems, 4*(1), 7.

Gilson, L. (2003). Trust and the development of health care as a social institution. *Social Science & Medicine, 56*(7), 1453–1468.

Govindan, K., Mina, H., & Alavi, B. (2020). A decision support system for demand management in healthcare supply chains considering the epidemic outbreaks: A case study of coronavirus disease 2019 (COVID-19). *Transportation Research Part E: Logistics and Transportation Review, 138*, 101967.

Grimm, J., & Reinecke, J. (2024). Collaborating on the edge of failure: Frame alignment across multiple interaction arenas in multi-stakeholder partnerships. *Academy of Management Journal, 67*(4), 956–990.

Gualandris, J., Longoni, A., Luzzini, D., & Pagell, M. (2021). The association between supply chain structure and transparency: A large-scale empirical study. *Journal of Operations Management, 67*(7), 803–827.

Gulati, R., Wohlgezogen, F., & Zhelyazkov, P. (2012). The two facets of collaboration: Cooperation and coordination in strategic alliances. *Academy of Management Annals, 6*(1), 531–583.

Han, W., Huang, Y., Hughes, M., & Zhang, M. (2021). The trade-off between trust and distrust in supply chain collaboration. *Industrial Marketing Management, 98*, 93–104.

Handfield, R. B., & Bechtel, C. (2002). The role of trust and relationship structure in improving supply chain responsiveness. *Industrial Marketing Management, 31*(4), 367–382.

Harland, C. M., Knight, L., Patrucco, A. S., Lynch, J., Telgen, J., Peters, E., et al. (2021). Practitioners' learning about healthcare supply chain management in the COVID-19 pandemic: A public procurement perspective. *International Journal of Operations & Production Management, 41*(13), 178–189.

Holling, C. (1973). Resilience and stability of ecological systems. *Annual Review of Ecological System, 4*, 1–23.

Ireland, R. D., & Webb, J. W. (2007). A multi-theoretic perspective on trust and power in strategic supply chains. *Journal of Operations Management, 25*(2), 482–497.

Ivanov, D. (2020). Predicting the impacts of epidemic outbreaks on global supply chains: A simulation-based analysis on the coronavirus outbreak (COVID-19/SARS-CoV-2) case. *Transportation Research Part E: Logistics and Transportation Review, 136*, 101922.

Ivanov, D. (2022a). Probability, adaptability and time: Some research-practice paradoxes in supply chain resilience and viability modelling. *International Journal of Integrated Supply Management, 15*(4), 454–465.

Ivanov, D. (2022b). Viable supply chain model: Integrating agility, resilience and sustainability perspectives—lessons from and thinking beyond the COVID-19 pandemic. *Annals of operations research, 319*(1), 1411–1431.

Ivanov, D. (2022c). Lean resilience: AURA (Active Usage of Resilience Assets) framework for post-COVID-19 supply chain management. *The International Journal of Logistics Management, 33*(4), 1196–1217.

Ivanov, D. (2024a). Transformation of supply chain resilience research through the COVID-19 pandemic. *International Journal of Production Research.* https://doi.org/10.1080/00207543.2024.2334420

Ivanov, D. (2024b). Two views of supply chain resilience. *International Journal of Production Research, 62*(11), 4031–4045.

Ivanov, D., & Dolgui, A. (2021). OR-methods for coping with the ripple effect in supply chains during COVID-19 pandemic: Managerial insights and research implications. *International Journal of Production Economics, 232*, 107921.

Ivanov, D., Dolgui, A., Sokolov, B., & Ivanova, M. (2017). Literature review on disruption recovery in the supply chain. *International Journal of Production Research, 55*(20), 6158–6174.

Ivanov, D., & Sokolov, B. (2013). Control and system-theoretic identification of the supply chain dynamics domain for planning, analysis and adaptation of performance under uncertainty. *European Journal of Operational Research, 224*(2), 313–323.

Ivanov, D., Sokolov, B., & Kaschel, J. (2013). Adaptation-based supply chain resilence. In M. Essig et al. (Eds.), *Supply chain safety management* (pp. 267–287). Springer.

Jamison, A. M., Quinn, S. C., & Freimuth, V. S. (2019). "You don't trust a government vaccine": Narratives of institutional trust and influenza vaccination among African American and white adults. *Social Science & Medicine, 221*, 87–94.

Janssen, M., & Van der Voort, H. (2020). Agile and adaptive governance in crisis response: Lessons from the COVID-19 pandemic. *International Journal of Information Management, 55*, 102180.

Kock, N. (2015). A note on how to conduct a factor-based PLS-SEM analysis. *International Journal of e-Collaboration (ijec), 11*(3), 1–9.

Kock, N. (2017). Structural equation modeling with factors and composites: A comparison of four methods. *International Journal of e-Collaboration (IJeC), 13*(1), 1–9.

Kock, N. (2019). From composites to factors: B ridging the gap between PLS and covariance-based structural equation modelling. *Information Systems Journal, 29*(3), 674–706.

Kock, N., & Hadaya, P. (2018). Minimum sample size estimation in PLS-SEM: The inverse square root and gamma-exponential methods. *Information Systems Journal, 28*(1), 227–261.

Kovács, G., & Falagara Sigala, I. (2021). Lessons learned from humanitarian logistics to manage supply chain disruptions. *Journal of Supply Chain Management, 57*(1), 41–49.

Kwon, I. W. G., & Suh, T. (2004). Factors affecting the level of trust and commitment in supply chain relationships. *Journal of Supply Chain Management, 40*(1), 4–14.

Lambe, C. J., Spekman, R. E., & Hunt, S. D. (2002). Alliance competence, resources, and alliance success: Conceptualization, measurement, and initial test. *Journal of the Academy of Marketing Science, 30*(2), 141–158.

Lee, H. L. (2002). Aligning supply chain strategies with product uncertainties. *California Management Review, 44*(3), 105–119.

Lee, H. L. (2004). The triple-A supply chain. *Harvard Business Review, 82*(10), 102–112.

Lee, S. M., & Rha, J. S. (2016). Ambidextrous supply chain as a dynamic capability: Building a resilient supply chain. *Management Decision, 54*(1), 2–23.

Lee, S. M., Lee, D., & Schniederjans, M. J. (2011). Supply chain innovation and organizational performance in the healthcare industry. *International Journal of Operations & Production Management, 31*(11), 1193–1214.

Li, M. K., Sodhi, M. S., Tang, C. S., & Yu, J. J. (2023). Preparedness with a system integrating inventory, capacity, and capability for future pandemics and other disasters. *Production and Operations Management, 32*(2), 564–583.

Li, X., Chung, C., Goldsby, T. J., & Holsapple, C. W. (2008). A unified model of supply chain agility: The work-design perspective. *International Journal of Logistics Management, 19*(3), 408–435.

Li, X., Goldsby, T. J., & Holsapple, C. W. (2009). Supply chain agility: Scale development. *International Journal of Logistics Management, 20*(3), 408–424.

Lindell, M. K., & Whitney, D. J. (2001). Accounting for common method variance in cross-sectional research designs. *Journal of Applied Psychology, 86*(1), 114–121.

Lotfi, M., & Sodhi, M. S. (2024). Resilient agility under the practice-based view. *Production Planning & Control, 35*(7), 670–682.

Luo, Y. (2002). Building trust in cross-cultural collaborations: Toward a contingency perspective. *Journal of Management, 28*(5), 669–694.

MacKenzie, S. B., & Podsakoff, P. M. (2012). Common method bias in marketing: Causes, mechanisms, and procedural remedies. *Journal of Retailing, 88*(4), 542–555.

Mandal, S. (2017). The influence of organizational culture on healthcare supply chain resilience: Moderating role of technology orientation. *Journal of Business & Industrial Marketing, 32*(8), 1021–1037.

Meredith, J. R., Raturi, A., Amoako-Gyampah, K., & Kaplan, B. (1989). Alternative research paradigms in operations. *Journal of Operations Management, 8*(4), 297–326.

Montecchi, M., Plangger, K., & West, D. C. (2021). Supply chain transparency: A bibliometric review and research agenda. *International Journal of Production Economics, 238*, 108152.

Morgan, R. M., & Hunt, S. D. (1994). The commitment-trust theory of relationship marketing. *Journal of Marketing, 58*(July), 20–38.

Morgan, T. R., Richey, R. G., Jr., & Ellinger, A. E. (2018). Supplier transparency: Scale development and validation. *The International Journal of Logistics Management, 29*(3), 959–984.

Moshtari, M. (2016). Inter-organizational fit, relationship management capability, and collaborative performance within a humanitarian setting. *Production and Operations Management, 25*(9), 1542–1557.

Müller, J., Hoberg, K., & Fransoo, J. C. (2023). Realizing supply chain agility under time pressure: Ad hoc supply chains during the COVID-19 pandemic. *Journal of Operations Management, 69*(3), 426–449.

Munir, M., Jajja, M. S. S., & Chatha, K. A. (2022). Capabilities for enhancing supply chain resilience and responsiveness in the COVID-19 pandemic: Exploring the role of improvisation, anticipation, and data analytics capabilities. *International Journal of Operations & Production Management, 42*(10), 1576–1604.

Nikookar, E., Stevenson, M., & Varsei, M. (2024). Building an antifragile supply chain: A capability blueprint for resilience and post-disruption growth. *Journal of Supply Chain Management, 60*(1), 13–31.

Oloruntoba, R., & Gray, R. (2006). Humanitarian aid: An agile supply chain? *Supply Chain Management: An International Journal, 11*(2), 115–120.

Podsakoff, P. M., MacKenzie, S. B., Podsakoff, N. P., & Lee, J. Y. (2003). The mismeasure of man (agement) and its implications for leadership research. *The Leadership Quarterly, 14*(6), 615–656.

Podsakoff, P. M., Podsakoff, N. P., Williams, L. J., Huang, C., & Yang, J. (2024). Common method bias: It's bad, it's complex, it's widespread, and it's not easy to fix. *Annual Review of Organizational Psychology and Organizational Behavior, 11*(1), 17–61.

Ponomarov, S. Y., & Holcomb, M. C. (2009). Understanding the concept of supply chain resilience. *The International Journal of Logistics Management, 20*(1), 124–143.

Qu, W. G., & Yang, Z. (2015). The effect of uncertainty avoidance and social trust on supply chain collaboration. *Journal of Business Research, 68*(5), 911–918.

Ramus, T., Vaccaro, A., & Brusoni, S. (2017). Institutional complexity in turbulent times: Formalization, collaboration, and the emergence of blended logics. *Academy of Management Journal, 60*(4), 1253–1284.

Rinaldi, M., Murino, T., Gebennini, E., Morea, D., & Bottani, E. (2022). A literature review on quantitative models for supply chain risk management: Can they be applied to pandemic disruptions? *Computers & Industrial Engineering, 170*, 108329.
Sahay, B. S. (2003). Understanding trust in supply chain relationships. *Industrial Management & Data Systems, 103*(8), 553–563.
Saïah, F., Vega, D., deVries, H., & Kembro, J. (2023). Process modularity, supply chain responsiveness, and moderators: The Médecins Sans Frontières response to the Covid-19 pandemic. *Production and Operations Management, 32*(5), 1490–1511.
Sathiya, V., Nagalakshmi, K., Jeevamalar, J., Babu, R. A., Karthi, R., Acevedo-Duque, A., et al. (2023). Reshaping healthcare supply chain using chain-of-things technology and key lessons experienced from COVID-19 pandemic. *Socio-Economic Planning Sciences, 85*, 101510.
Scala, B., & Lindsay, C. F. (2021). Supply chain resilience during pandemic disruption: Evidence from healthcare. *Supply Chain Management: An International Journal, 26*(6), 672–688.
Schilke, O., & Cook, K. S. (2015). Sources of alliance partner trustworthiness: Integrating calculative and relational perspectives. *Strategic Management Journal, 36*(2), 276–297.
Scholten, K., & Schilder, S. (2015). The role of collaboration in supply chain resilience. *Supply Chain Management: An International Journal, 20*(4), 471–484.
Sheffi, Y. (2005). *The resilient enterprise: Overcoming vulnerability for competitive advantage*. MIT Press.
Shishodia, A., Sharma, R., Rajesh, R., & Munim, Z. H. (2023). Supply chain resilience: A review, conceptual framework and future research. *The International Journal of Logistics Management, 34*(4), 879–908.
Sodhi, M. S., & Tang, C. S. (2019). Research opportunities in supply chain transparency. *Production and Operations Management, 28*(12), 2946–2959.
Sodhi, M. S., & Tang, C. S. (2021). Supply chain management for extreme conditions: Research opportunities. *Journal of Supply Chain Management, 57*(1), 7–16.
Sodhi, M. S., Tang, C. S., & Willenson, E. T. (2023). Research opportunities in preparing supply chains of essential goods for future pandemics. *International Journal of Production Research, 61*(8), 2416–2431.
Spieske, A., Gebhardt, M., Kopyto, M., & Birkel, H. (2022). Improving resilience of the healthcare supply chain in a pandemic: Evidence from Europe during the COVID-19 crisis. *Journal of Purchasing and Supply Management, 28*(5), 100748.
Suresh, M., Roobaswathiny, A., & Lakshmi Priyadarsini, S. (2021). A study on the factors that influence the agility of COVID-19 hospitals. *International Journal of Healthcare Management, 14*(1), 290–299.
Swafford, P. M., Ghosh, S., & Murthy, N. (2006). The antecedents of supply chain agility of a firm: Scale development and model testing. *Journal of Operations Management, 24*(2), 70–88.
Tatham, P., & Kovács, G. (2010). The application of "swift trust" to humanitarian logistics. *International Journal of Production Economics, 126*(1), 35–45.
Teece, D. J. (2007). Explicating dynamic capabilities: The nature and microfoundations of (sustainable) enterprise performance. *Strategic Management Journal, 28*(13), 1319–1350.
Tiwari, M., Bryde, D. J., Stavropolou, F., Dubey, R., Kumari, S., & Foropon, C. (2024). Modelling supply chain visibility, digital Technologies, environmental dynamism and healthcare supply chain resilience: An organisation information processing theory perspective. *Transportation Research Part E: Logistics and Transportation Review, 188*, 103613.
Vann Yaroson, E., Breen, L., Hou, J., & Sowter, J. (2023). The role of power-based behaviours on pharmaceutical supply chain resilience. *Supply Chain Management: An International Journal, 28*(4), 738–759.
Villena, V. H., Wilhelm, M., & Xiao, C. Y. (2021). Untangling drivers for supplier environmental and social responsibility: An investigation in Philips Lighting's Chinese supply chain. *Journal of Operations Management, 67*(4), 476–510.
Wieland, A., & Durach, C. F. (2021). Two perspectives on supply chain resilience. *Journal of Business Logistics, 42*(3), 315–322.

Wieland, A., & Wallenburg, C. M. (2013). The influence of relational competencies on supply chain resilience: A relational view. *International Journal of Physical Distribution & Logistics Management, 43*(4), 300–320.

Wu, L., Chuang, C. H., & Hsu, C. H. (2014). Information sharing and collaborative behaviors in enabling supply chain performance: A social exchange perspective. *International Journal of Production Economics, 148*, 122–132.

Yang, J., Yu, G., Liu, M., Xie, H., & Liu, H. (2018). Disentangling the impact of cost transparency on cooperation efficiency in exchange partnerships. *International Journal of Production Economics, 197*, 27–34.

Zamiela, C., Hossain, N. U. I., & Jaradat, R. (2022). Enablers of resilience in the healthcare supply chain: A case study of US healthcare industry during COVID-19 pandemic. *Research in Transportation Economics, 93*, 101174.

Zighan, S., Dwaikat, N. Y., Alkalha, Z., & Abualqumboz, M. (2024). Knowledge management for supply chain resilience in pharmaceutical industry: Evidence from the Middle East region. *The International Journal of Logistics Management, 35*(4), 1142–1167.

Digital Supply Chain Twins: Managing the Ripple Effect, Resilience, and Disruption Risks by Data-Driven Optimization, Simulation, and Visibility

Dmitry Ivanov, Alexandre Dolgui, Ajay Das, and Boris Sokolov

Abstract The quality of model-based decision-making support strongly depends on the data, its completeness, fullness, validity, consistency, and timely availability. These requirements on data are of special importance in supply chain (SC) risk management for predicting disruptions and reacting to them. Digital technology, Industry 4.0, Blockchain, and real-time data analytics have the potential to achieve a new quality in decision-making support when managing severe disruptions, resilience, and the Ripple effect. A combination of simulation, optimization, and data analytics constitutes a digital twin: a new data-driven vision of managing the disruption risks in SC. A digital SC twin is a model that can represent the network state for any given moment in time and allow for complete end-to-end SC visibility to improve resilience and test contingency plans. This chapter proposes an SC risk analytics framework and explains the concept of digital SC twins. It analyses perspectives and future transformations to be expected in the transition toward cyber-physical SCs. It demonstrates a vision of how digital technologies and smart operations can help integrate resilience and lean thinking into a *resileanness* framework "Low-Certainty-Need" (LCN) SC.

D. Ivanov (✉)
Department of Business and Economics, Berlin School of Economics and Law, Berlin, Germany
e-mail: divanov@hwr-berlin.de

A. Dolgui
Department of Automation, Production and Computer Sciences, IMT Atlantique, LS2N-CNRS UMR 6004, Nantes, France
e-mail: alexandre.dolgui@imt-atlantique.fr

A. Das
Narendra Paul Loomba Department of Management, Zicklin School of Business, CUNY-Baruch, New York, NY, USA
e-mail: ajay.das@baruch.cuny.edu

B. Sokolov
St. Petersburg Federal Research Center of the Russian Academy of Sciences (SPC RAS), St. Petersburg, Russia
e-mail: sokol@iias.spb.su

D. Ivanov et al. (eds.), *Handbook of Ripple Effects in the Supply Chain*, International Series in Operations Research & Management Science 276,
https://doi.org/10.1007/978-3-031-85508-5_16

1 Introduction

Digital technologies catalyze the development of new paradigms, principles, and models in supply chain management (SCM). The Internet of Things (IoT), cyber-physical systems, and smart, connected products facilitate the development of digital supply chains (SC) and smart operations (Dolgui & Ivanov, 2022; Fazili et al., 2017; Ivanov, 2024a; Liao et al., 2017; Minner et al., 2017; Panetto et al., 2019; Qu et al., 2017; Strozzi et al., 2017; Tran-Dang et al., 2017; Yang et al., 2017). Recent surveys by Addo-Tenkorang and Helo (2016), Oesterreich and Teuteberg (2016), Gunasekaran et al. (2016, 2017, 2018), Nguyen et al. (2018), Moghaddam and Nof (2018), Choi et al. (2018), Ben-Daya et al. (2018) proposed classifications of different digital technologies and discussed their potential impacts on SCM. Such digital technologies include big data analytics, advanced manufacturing technologies with sensors, decentralized agent-driven control, advanced robotics, augmented reality, advanced tracking and tracing technologies, and additive manufacturing.

The increasing interest in digital data applications for SCM is not surprising. The quality of model-based decision-making support strongly depends on the data, its completeness, fullness, validity, consistency, and timely availability (Frazzon et al., 2021; Zheng et al., 2025; Dolgui & Ivanov, 2025). These requirements on data are of special importance in SC risk management for predicting disruptions and reacting to them (Ivanov, 2018b; Ivanov, 2020). Digital technology, Industry 4.0, Blockchain, and real-time data analytics have the potential to achieve a new quality in decision-making support when managing severe disruptions, resilience, and the Ripple effect (Frazzon et al., 2018; Ivanov et al., 2017; Ivanov & Dolgui, 2021; Ivanov et al., 2019; Li et al., 2022; Rai et al., 2021).

A combination of simulation, optimization, and data analytics constitutes a digital twin: a new data-driven vision of managing the disruption risks in SC. A digital SC twin is a model that can represent the network state for any given moment in time and allow for complete end-to-end SC visibility to improve resilience and test contingency plans (Burgos & Ivanov, 2021; Ivanov, 2018c; Ivanov, 2023; Ivanov, 2024b). This chapter proposes an SC risk analytics framework and explains the concept of digital SC twins. It analyses perspectives and future transformations to be expected in the transition toward cyber-physical SCs. It demonstrates a vision of how digital technologies and smart operations can help integrate resilience and lean thinking into a *resileanness* framework "Low-Certainty-Need" (LCN) SC (Ivanov & Dolgui, 2019).

The investigation of the interrelations between digital technology and SC risks is still at a preliminary stage of its development and requires new conceptual frameworks and taxonomies (Ivanov et al., 2019). This chapter seeks to move the discussion forward and develop a framework for a detailed analysis of SC digital technology and disruption risk effects manifested at times in *structural dynamics* (Ivanov et al., 2010) and the *ripple effect* (Dolgui et al., 2018; Dolgui et al., 2020; Elluru et al., 2019; He et al., 2019; Ivanov, 2018a, 2018b; Ivanov et al. 2014, b, 2016; Ivanov & Rozhkov, 2020; Pavlov et al., 2018; Pavlov et al., 2019; Sokolov et al., 2016). Despite initial efforts to unearth new insights about the impact of digital technologies on SC risks (Baryannis et al., 2018; Dolgui et al., 2019; Dolgui & Ivanov, 2020; Dolgui et al., 2020; Dubey et al., 2019; Ivanov et al., 2017; Ivanov et al., 2019; Papadopoulos et al., 2017; Schlüter et al., 2017; Tupa et al., 2017), the understanding of individual and interactive contributions on specific SC disruption risk management and ripple effects remains limited. This study closes this research gap by a combinatorial examination of the results gained from two isolated areas, i.e., the SC digitalization and managing the disruption risks in the SC. In particular, the focus of this chapter is directed at the data-driven decision-support systems to improve SC resilience and manage the ripple effect and disruption risks.

This chapter does not pretend to be encyclopedic but rather highlights the research that examines the relationships between SC digitalization and SC disruption risks. The objective is to identify the perspectives of digital SC twins that can be leveraged to direct future research in exploring how digital technologies affect the ripple effect and performance of the SCs and how they can be used to manage the disruption risks and to improve resilience. More specifically, this study seeks to answer the following questions:

- What relationships exist between big data analytics, Industry 4.0, additive manufacturing, Blockchain, and advanced trace and tracking systems and SC disruption risks?
- How can digitalization contribute to enhancing ripple effect mitigation and analysis?
- What digital technology-based extensions are needed in applications of quantitative analysis to ripple effect in the SC to emerge with digital supply chain twins?

2 Digital Supply Chain Technologies

Digitalization means using digitized data and digital technologies not only to improve processes, functions, and activities but also to change processes to achieve a certain benefit. The objective is to enhance revenue streams and create new business opportunities (Hagberg et al., 2016). Digitalization of operations aims to improve production and SC capability and flexibility through real-time communication and intelligent, high-resolution data systems (Reddy et al., 2016; Wamba & Queiroz, 2020). Digitalization is a continuing transformation toward a digital supply chain and progressively changes most enterprise processes.

This section reviews recent literature in four elements identified in recent surveys on digitalization applications to SCM, i.e.,

- Big data analytics
- Industry 4.0
- Additive manufacturing
- Advanced tracking and tracing technologies, Blockchain

In each of these groups, we describe the respective technology and its recent applications to SCM.

2.1 Big Data Analytics and Artificial Intelligence

Big data analytics (BDA) and artificial intelligence (AI) bring a completely new potential benefit to data-driven SC risk management. Big data has been characterized in the literature by 5Vs: volume, variety, velocity, veracity, and value (Wamba et al., 2015, 2017). Veracity and value are particularly important since data analysis shows the real value of big data.

Big data analytics (BDA) is based on knowledge extraction from vast amounts of data, facilitating data-driven decision-making (Dubey et al., 2020). The more the data from the actual production process is recorded, the more important it becomes to evaluate this data volume with the help of BDA applications. ERP systems are generally not suited to this task. One challenge is that internal and external data from smart networking products are frequently unstructured. The resulting solution is a repository that stores different data streams in their original formats. From there, the data can be reformatted and examined with descriptive, diagnostic, predictive, and prescriptive data analytics tools.

Descriptive analysis records the condition, the environment, and the functioning of the products. Diagnostic analysis analyses the reasons for reduced product performance or failure. Predictive analysis recognizes patterns that signal upcoming events (Ivanov, 2024c, 2025a, 2025b). Prescriptive analytics identifies measures to solve problems and improve outcomes (Brintrup et al., 2020; Porter & Heppelmann, 2015; Ivanov, 2024b; Jackson et al., 2024).

Analytics employs mathematical and statistical tools to collect, store, accumulate, and analyze big data volumes. The applications themselves are not new, but it is the combination with big data that brings new added value and competitive advantage. What is new is the rapid pace at which data can be captured in real time. This, in turn, extends the type and richness of data sets and offers an unprecedented opportunity for investigation. Additionally, the nature of the investigation has changed. Technological tools are continuously supplied with data and become more intelligent by using self-learning algorithms. For example, predictive analytics involves self-learning algorithms that identify and analyze relationships among variables and develop outcomes such as buyer behavior forecasts. Active human involvement is not required in this process. As a result, BDA becomes an active participant in the investigation process and can create new knowledge about unknown or buried patterns and effects. Large-scale investigations detect such patterns, turning volume data into precise insights (Sanders, 2016).

BDA has undoubtedly been the most elaborated area of digital technology application to SCM over the last decade. Johnson et al. (2016) and Simchi-Levi and Wu (2018) analyzed the application of BDA to retail. Nguyen et al. (2018) noted that optimization is the most popular approach in prescriptive analytics applications for logistics and transportation areas. Retailers strive to grow revenue, margins, and market share. Price optimization models calculate the variance of demand with price changes and combine this information with relevant cost and inventory data to recommend prices that could maximize revenue and profits. BDA applications to SCM can also be seen in procurement processes, manufacturing shop floors, promotion actions in the omnichannel model, routing optimization, real-time traffic operation monitoring, and proactive safety management (Addo-Tenkorang & Helo, 2016; Gunasekaran et al., 2016, 2017; Nguyen et al., 2018; Zhong et al., 2017). Nguyen et al. (2018) identified some additional areas where BDA can be applied to SCM in the near future. These areas include quality control in manufacturing, dynamic vehicle routing, in-transit inventory management in logistics/transportation, and order picking and inventory control systems in warehousing. Papadopoulos et al. (2017) observed that BDA can help improve SC risk management and disaster resistance. Baryannis et al. (2018) summarized recent AI applications to SC risk management and identified some future research directions in risk identification, assessment, and response. Priori et al. (2018) applied machine learning to the dynamic selection of replenishment policies according to SC environmental dynamics. Cavalcantea et al. (2019) developed a supervised machine learning approach to data-driven simulation of resilient supplier selection in digital manufacturing.

2.2 *Industry 4.0*

The intelligent networking of machines and processes with the help of digital technologies is creating autonomous, Internet-linked, self-regulating production systems, popularly termed as Industry 4.0. Industry 4.0 seeks to visualize and predict the performance of processes, plants, SCs, and product properties on the basis of information available in real time (Ivanov & Dolgui, 2021; Ivanov et al., 2019; Rai et al., 2021; Ralston & Blackhurst, 2020). For this purpose, smart sensors are applied to capture and communicate information and requirements comprehensively to any recipient in real time. Production models are implemented in the form of so-called cyber-physical production systems. Such cyber-physical production systems collect data via production-integrated sensors and measurement systems in real time, store and evaluate data, and interact actively with the physical, human, and digital world. Intra and external connectivity are provided by IoT via digital communication devices.

Industry 4.0 is a global phenomenon. There is no unique or circumscribed set of technologies or practices that define Industry 4.0. Most research considers factory concepts that share attributes of smart networking (Strozzi et al., 2017). The vision of Industry 4.0 is that the product to be manufactured carries all the relevant information about its production requirements. In addition, integrated production installations become self-organized through the collaboration of production machines, transport equipment, tools, and logistical components that can communicate with each other and exchange data via embedded systems. Digital technologies enable flexible decision-making by providing real-time data in all areas of the SC (Bonfour, 2016, p. 20). Digitalization and Industry 4.0 offers information and coordination-based competitive advantage, generates new employment opportunities, and increases visibility and control in supply chains. However, it requires long-term commitment and guarantees of data security (Porter & Heppelmann, 2015).

Industrial robots are a part of Industry 4.0, and they are found mainly in series production and warehousing applications. Robots perform high-precision tasks independently and also support employees in their work by handing over tools, for example. Their use accelerates, facilitates, and simplifies production activities. Unlike conventional industrial robots, which require time-consuming training, flexible robots learn quickly from people. They communicate with one another via the cloud and support optimal production planning. In practice, however, questions still remain on several issues such as the ownership of cloud data among SC partners or machine intercommunication protocols (Andelfinger & Hänisch, 2017).

2.3 Additive Manufacturing

Additive manufacturing technology is a design-driven manufacturing process in which components are produced from material layers on the basis of 3D data sets and a virtual blueprint. "3D printing" is often used as a synonym. The use of different materials and the elimination of previously required special tools are advantages. Furthermore, the rapid design and manufacturing process allows considerable time savings compared to conventional product development cycles (Zhang & Jung, 2018, pp. 3–5; Lario et al., 2024). Great freedom of design, low material waste, and the feasibility of economically manufactured, individualized products make additive manufacturing attractive for many industries (Li et al., 2024). The method is currently used primarily in rapid prototyping but increasingly so in series production too (Li et al., 2017). Khajavi et al. (2014), Holmström and Gutowski (2017), Feldmann and Pumpe (2017), and Li et al. (2017) described the applications of additive manufacturing to operations and SCM. Those applications range from spare part logistics to redesigning global SC production and sourcing strategy. The core of additive manufacturing applications to SCM is the usage of 3D printers at different stages in the SC to increase manufacturing flexibility, achieve shorter lead times, increase product individualization, and reduce inventory. However, mass production volumes are not commercially possible yet.

2.4 Blockchain and Advanced Tracking and Tracing Technologies

Capturing and sharing information in real time is critical to detecting faults and their extent, as well as in planning SC recovery (Sheffi, 2015). Tracking and tracing (T&T) systems aim at timely identification of deviations or danger of deviations in SCs, analysis of such deviations, alerts about disruptions that have occurred or may occur, and elaborating control actions to recover SC operability.

T&T systems combine with radio-frequency identification (RFID) and mobile devices to provide current information about process execution (Bearzotti et al., 2012). T&T systems and feedback control can be supported by RFID technology (Dolgui & Proth, 2010) and SC event management systems (Ivanov et al., 2013), effectively communicating disruptions to the SC tiers and helping revise initial schedules (Dolgui & Proth, 2010; Zelbst et al., 2012). A critical issue is detecting disruptions and their scope in real time. Embedding SC visualization and identification technology is crucial for this in practice.

In addition, emerging Blockchain applications in SCs promise enhanced scale and scope of T&T systems together with the creation of information pipeline systems and SC finance applications (Hofmann et al., 2018). The central idea is to increase visibility and efficiency based on dispersed, tamper-proof, and verifiable record-keeping in the SC.

For example, IBM and Walmart are researching how to increase food SC safety control using Blockchain technology (IBM, 2017). Recently, the applications of Blockchain technology have begun to revolutionize different aspects of SC and operations management for the development of real-time SC capabilities (Ivanov et al., 2019; Kshetri, 2018; Saberi et al., 2018; Tiwari et al., 2023). The central idea is to increase visibility and efficiency based on record-keeping in the SC. Blockchain applications to SCs become more and more important to enhance the scale and scope of digital processes along with the creation of information pipeline systems and SC finance applications (Hofmann et al., 2018). A Blockchain is a decentralized database that exists as copies in a network of computers (Crosby et al., 2016). It is a chain of blocks, because the data and information stored is captured in blocks.

Regulatory processes (e.g., customs) can be expedited using Blockchain by improving confidence in documentation. This, in turn, can result in reductions in wastage, risk, and insurance premiums. The list of all transactions is stored as copies throughout all further evolvements on numerous computers (a network of even hundreds of computers).

These and further recent examples of Blockchain technology applications to SCs (Ivanov, Dolgui, & Sokolov, 2019; Saberi et al., 2018; Sauer et al., 2022; Wu et al., 2024) support the new proposition that competition is not between the SCs, but rather between the information services and analytics algorithms behind the SCs. As such, SCs will no longer be understood as a rigid physical system with a fixed and static allocation of some processes to some firms. Instead, different physical firms will offer services in supply, manufacturing, logistics, and sales which will

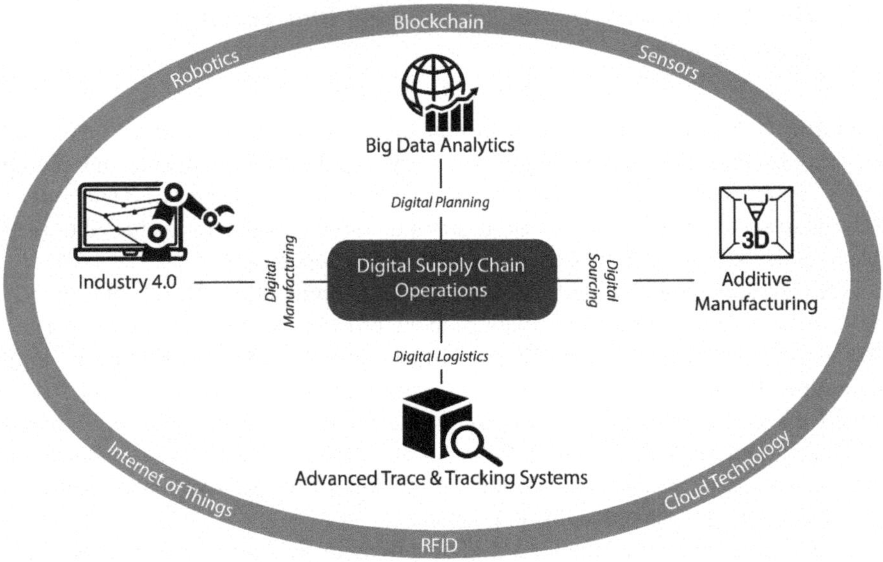

Fig. 1 Digitalization framework of supply chain risk management (Ivanov et al., 2019)

result in the dynamic allocation of processes and dynamic SC structures, forming a cyber-physical SC.

In practice, new cloud-based analytics platforms such as SupplyOn Industry 4.0 Sensor Clouds make it possible to control the SC in real time, and plan and adjust processes using up-to-date information. By simply clicking on a container type, the graphs indicate whether there has been a violation of the defined temperature or humidity limits along the time axis. The data analysis in this chart allows a quick identification of all orders where the lead time was exceeded, allowing for a quick identification of questionable transports.

In summary, the following SC digitalization framework can be presented (Fig. 1).

BDA, additive manufacturing, Industry 4.0, and advanced tracking and tracing technologies can be considered digital enablers of the four major SC processes in the SCOR model, i.e., plan, source, make, and deliver, respectively. A digital version of the SCOR model would therefore consist of digital planning, digital manufacturing, digital sourcing, and digital logistics. IoT, cloud technology, robots, and sensors form the technical guts of a digital SC. This classification will be further used in the paper for analysis of digitalization impacts on severe SC risks and the ripple effect. For each of the areas, Fig. 1 suggests possible applications of digital technology with regard to SC disruption risks. For example, additive manufacturing can reduce supply risk by creating the opportunity to replace missing materials with 3D-printed components. BDA can be used at the planning stage to identify supplier risk exposure. T&T systems can help at the reactive stage to monitor and identify disruptions. At the same time, it needs to be noted that digital technologies may have multiple applications that are not restricted to a particular SCOR process.

3 Impact of Digital Technologies on the Ripple Effect

3.1 Linking the Digital Supply Chain and Disruption Risks

Following the study by Ivanov et al. (2019), Table 1 summarizes the major drivers of digital technology applications to SCM, the respective enablers, opportunities and challenges for SCM, as well as the impact on disruption risk management and the ripple effect.

Specifically, digitalization's impact on the ripple effect, that is, the magnitude and reach (upstream and downstream) of a disruption in a part of the SC, is elaborated in Table 2.

It can be observed in Tables 1 and 2 that digitalization technologies generally have a positive impact on the ripple effect but may create a few challenges for ripple effect mitigation and control. BDA, Industry 4.0, and additive manufacturing have mixed influences on the ripple effect, while advanced T&T systems have a positive impact.

Structuring analysis in terms of the supply chain operations reference (SCOR) model, *sourcing and production* activities involving additive manufacturing and Industry 4.0 imply higher exposure to external risks and ripple effect. This could be due to an increase in complexity and probable reduction in time and demand risks due to higher flexibility and shorter lead times. Higher supply risks can be encountered if a disruption happens in the upstream SC since there is no intermediate inventory between the stages. *Delivery* process risks in the SC are alleviated by big data analytics due to better SC visibility and forecast accuracy, reduction in information disruption risks, and better quality of contingency plan activation. For integrated SC *planning*, reductions in supply and time risks can be achieved by using advanced T&T systems that enable real-time coordination and timely activation of contingency policies.

At the proactive stage, SCs are typically protected from disruptions by employing risk mitigation inventory, capacity reservations, and backup sources. This is expensive, especially if no disruption happens. Blockchain could help reduce these inefficiencies if we are able to create a record of activities and data needed for recovery in terms of synchronized contingency plans. Additive manufacturing can reduce the need for risk mitigation inventory and capacity reservations as well as for backup contingent suppliers. The decentralized control principles in Industry 4.0 systems make it possible to diversify the risks and reduce the need for structural SC redundancy, using manufacturing flexibility.

At the reactive stage, if a disruption happens, the contingency plans from the proactive stage can be deployed faster and implemented effectively if SC visibility is increased. BDA and advanced T&T systems in general, and Blockchain technology in particular, can help us to trace the roots of disruptions, to observe disruption propagation (i.e., the ripple effect), to select short-term stabilization actions based on a clear understanding of what capacities and inventories are available (emergency planning), to develop a mid-term recovery policy and to analyze the long-term

Table 1 Major digital technology applications to SCM and the impact on disruption risk management and ripple effect

Digitalization application to SCM	Enablers	Opportunities and benefits	Challenges in SCM	Impact on disruption risk management and ripple effect
Predictive analytics	Big data	Increase in promotion action quality Better demand forecasts Increase in supply chain visibility Better customer experience promotions	Data transparency and safety Coordination complexity increase in cross-channel logistics	Reduction in demand risks Reduction in information disruption risks and better quality of contingency plan activation Higher time risks because coordination complexity increases
Industry 4.0	IoT, smart products, robotics, augmented and virtual reality	Customized production system at the cost of mass production Individualized products and higher market flexibility Risk diversification Higher responsiveness Shorter lead times and better capacity utilization	Radical changes in SC and manufacturing process organization Reduction in the number of SC layers New locations close to the markets Re-qualification of employees, redesign of facility layouts Data security Increase in design and control complexity	Higher information risks Higher exposure to external risks, including unauthorized access Reduction in time and demand risks
3D printing	Additive manufacturing	Flexibility increase, Product variety, Shorter lead time Efficiency increase in MRO inventory control	Reduction in demand risks Higher exposure to external risks Higher supply risks if disruption happens in the upstream SC since no intermediate inventory in between the stages	
Advanced T&T technologies	RFID, sensors, Blockchain	Real-time identification Real-time material flow tracing Increase in data quality	Increase investments in ICT Data security	Reduction in information disruption risks and better quality of contingency plan activation Reductions in supply and time risks due to real-time coordination if activating contingency policies

Table 2 Contribution of digital technologies to ripple effect control in the SC

Reasons for ripple effect in the SC	Countermeasures	Digital technologies impact on ripple effects
Single sourcing	Multiple/Dual sourcing/Backup suppliers	Additive manufacturing tends to reduce the number of SC layers and suppliers—mitigates ripple effect Advanced T&T systems allow better SC coordination in real time—mitigates ripple effect Industry 4.0 increases sourcing coordination complexity—may delay detection and response to ripple effects BDA increases the quality of procurement processes—mitigates ripple effects
Low inventory	Risk mitigation inventory	Additive manufacturing tends to reduce the inventory in the SC—enhances ripple effects Advanced T&T systems allow inventory control in real time—mitigates ripple effects
Inflexible capacity	Postponement	Industry 4.0 and additive manufacturing increase demand and production flexibility—mitigates ripple effects
SC complexity	Global SC contingency plans	Advanced T&T systems allow better SC coordination in real time and faster contingency plan activation—mitigates ripple effects Industry 4.0 increases the SC coordination complexity—enhances ripple effects BDA contributes to an increase in supply chain visibility—mitigates ripple effects
Multistage SCs	Supplier segmentation according to disruption risks	Additive manufacturing tends to reduce the number of SC layers and suppliers—mitigates ripple effects Industry 4.0 increases the SC complexity—connectivity enhances ripple effect Advanced T&T systems allow better SC coordination in real time and faster contingency plan activation—mitigates ripple effects

performance impact of the ripple effect. Additive manufacturing has the potential to reduce disruption propagation in the SC, since the number of SC layers and the resulting complexity would be reduced.

4 Supply Chain Resileanness: Low-Certainty-Need (LCN) Framework

4.1 Conceptual Framework

The LCN SC framework suggests approaching SC disruption risk and the ripple effect field from another perspective. Rather than opposing efficiency and resilience, we suggest considering their mutual intersections to enhance each other based on synergetic effects in terms of SC *resileanness*.

Major costs of disruption management are seen in disruption prediction, protective redundancy, and reactive capabilities as a result of a higher need for certainty and the resulting higher redundancy and recovery efforts. As such, we suggest studying these areas from the perspective of efficiency and resilience complementarity (Fig. 2).

According to Fig. 2, structural complexity, process inflexibility, and non-flexible usage of resources, and insufficient parametric redundancy increase uncertainty and disruption risk propagation in the SC. The ultimate objective of the LCN SC design is to develop the ability to operate according to planned performance regardless of environmental changes. As such, the LCN SC design possesses two critical capabilities, i.e.,

- Low need for uncertainty consideration in planning decisions
- Low need for recovery coordination efforts

Structural variety, process flexibility, and parametrical redundancy ensure disruption resistance and recovery resource allocation and allow for SC operation in a broad range of environmental states. This means that planning activities in the LCN SCs do not heavily rely on uncertainty prediction and proactive protection investments. Similarly, recovery coordination efforts are reduced to a minimum. Note that the LCN SC design does not necessarily imply higher costs but rather seeks for an efficient combination of lean and resilient elements.

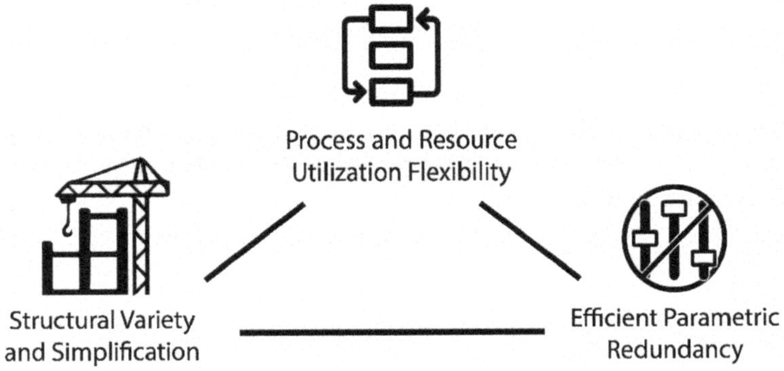

Fig. 2 Low-certainty-need supply chain framework

Let us discuss the principles of implementing the LCN SC framework in practice using digital technology.

4.2 *Process and Resource Utilization Flexibility*

Process and resource utilization *flexibility* means in a wider sense an establishment of universal, very flexible workstations such as those postulated in Industry 4.0 systems. Similarly, the usage of universal materials can be considered with regard to recovery flexibility in the SC. Additive manufacturing technology can also positively influence product and process flexibility, resulting in a combination of efficiency and resilience. Additive manufacturing can reduce the need for backup contingency suppliers. The decentralized control principles in Industry 4.0 systems make it possible to diversify the risks with the help of increased manufacturing flexibility. New research directions can be seen with regard to the impact of digitalization on SC design resilience (Ivanov et al., 2019). For example, Big Data analytics and advanced Trace &Tracking systems in general, and Blockchain technology in particular, can help to trace the roots of disruptions, observe disruption propagation (i.e., the ripple effect), select short-term stabilization actions based on a clear understanding of what capacities and inventories are available (emergency planning), develop a mid-term recovery policy, and analyze the long-term performance impact of the ripple effect. Additive manufacturing has the potential to reduce disruption propagation in the SC since the number of SC layers and the resulting complexity would be reduced.

4.3 *Non-expensive Parametric Redundancy*

Non-expensive parametric redundancy targets the efficient reservations of capacity, inventory, and lead time. More specifically, those reservations need to be considered not as a non-used redundancy, but rather for use in normal operation modes as well. Network redundancy optimization can be viewed as a new research topic in this area. Another aspect of parametric redundancy is its efficient allocation. A new research direction extending the existing value-stream mapping techniques toward SC resilience can be considered. Efficient redundancy can be implemented by using additive manufacturing that helps to reduce the need for risk mitigation inventory and capacity reservations. Finally, new material classification schemes need to be developed subject to material criticality and risk exposure in terms of the efficiency and resilience of SC design.

5 Digital Supply Chain Twin: Data-Driven Optimization and Simulation to Manage the Disruption Risks

5.1 Supply Chains as Cyber-Physical Systems

Today and looking at the near future, the SC will be as good as the digital technology behind it. *The recent examples of digital technology applications to SCs allow for the new proposition that the competition is not between SCs but rather between SC services and the analytics algorithms behind the SCs.* The services may be ordered in packages or as individual modules (Fig. 3).

Examples of SC and operations analytics applications include logistics and SC control with real-time data, inventory control, and management using sensing data, dynamic resource allocation in Industry 4.0 customized assembly systems, improving forecasting models using Big data, machine learning techniques for process control, SC visibility, and risk control, optimizing systems based on predictive information (e.g., predictive maintenance), combining optimization and machine learning algorithms, and simulation-based modeling and optimization for stochastic systems.

Success in SC competition will become more and more dependent on analytics algorithms in combination with optimization and simulation modeling. Initially intended for process automation, business analytics techniques now disrupt markets and business models and have a significant impact on SCM development. *As such, new disruptive SC business models will arise where SCs will be understood not as rigid physical systems with a fixed and static allocation of some processes to some firms. Instead, different physical firms will offer services of supply, manufacturing, logistics, and sales which will result in a dynamic allocation of processes and dynamic SC structures.* Recent literature documented the possibility of modeling such integrated service-material flow SCs (Ivanov et al., 2014; Sokolov et al., 2023).

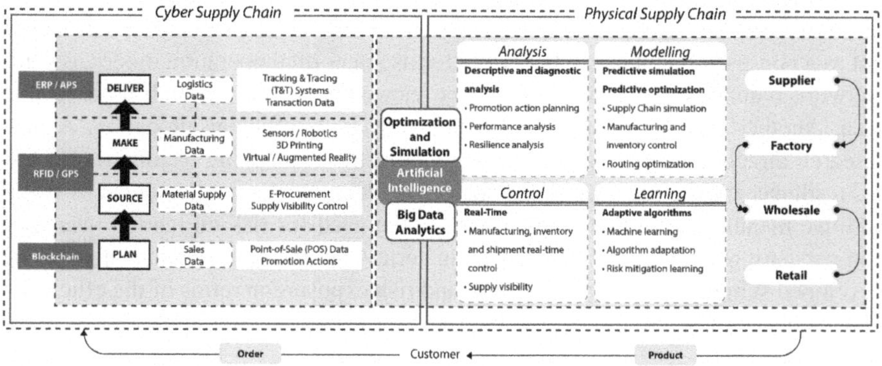

Fig. 3 Service and material flow coordination in the cyber-physical supply chain

5.2 Supply Chain Digital Twins

Dunke et al. (2018) underline that digitalization and Industry 4.0 may significantly influence the optimization techniques in the SC domain as well as disruption propagation impacts on SC performance. With the help of optimization and simulation approaches, current research generates new knowledge about the influence of disruption propagation on SC output performance, considering disruption location, duration and propagation, and recovery policies. New digital technologies create new challenges for the application of quantitative analysis techniques to SC ripple effect analysis and open new ways and problem statements for these applications.

In the past decades, simulation and optimization have played significant roles in solving complex problems. Successful examples include production planning and scheduling, SC design, and routing optimization, to name a few. However, many problems remain challenging because of their complexity and large scale, and/or uncertainty and stochastic nature. In addition, the major application of optimization and simulation methods in the last decades was seen in decision support, meaning that decision-makers were to manually provide the model input and interpret the model output. On the other hand, the rapid rise of business analytics provides exciting opportunities for Operations Research and the reexamination of these hard optimization problems, as well as newly emerging problems (Fig. 4).

Sourcing, manufacturing, logistics, and sales data are distributed among very different systems, such as ERP, RFID, sensors, and Blockchain. Big data analytics integrates this data into information used by AI algorithms in the cyber SC and managers in the physical SC. As such, a new generation of simulation and optimization models is arising. The pervasive adoption of analytics and its integration with Operations Research shows that simulation and optimization are key, not only in the modeling of physical SC systems but also in the modeling of cyber SC systems and learning from them.

An example of a decision-support system that combines simulation, optimization, and data analytics is shown in Fig. 5.

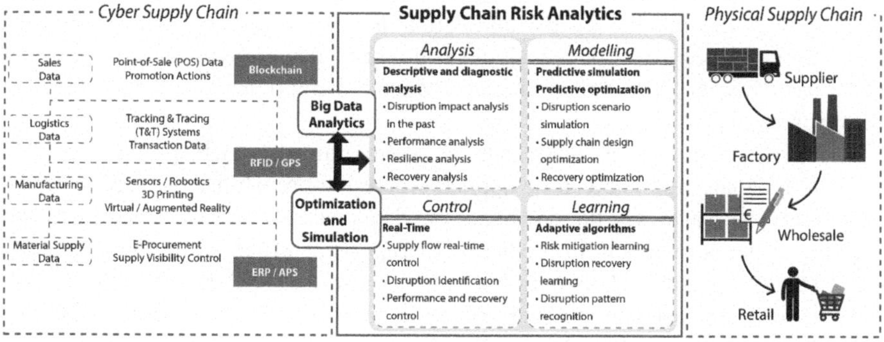

Fig. 4 Digital supply chain risk analytics framework

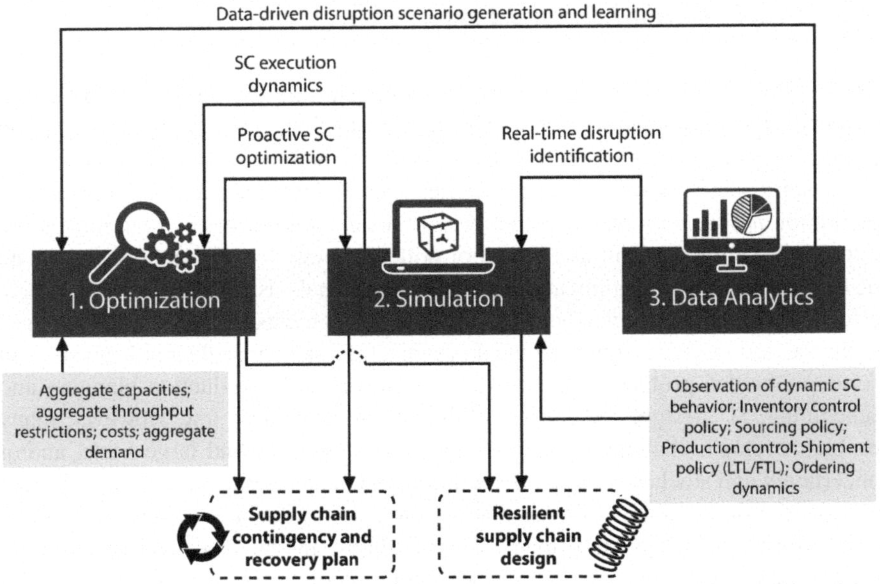

Fig. 5 Concept of a decision-support system for supply chain risk analytics (Ivanov et al., 2019)

The decision-support system for SC risk analytics aims at proactive, resilient SC design in anticipation of disruptions and structural–parametrical adaptation in the case of disruptions. *The decision-support system is based on a concept that combines simulation, optimization, and data analytics.* The simulation–optimization part of the system is intended to provide proactive, resilient SC optimization and simulation of SC dynamic behavior in the event of possible disruptions or disruption scenarios. In addition, this supports reactive, predictive simulation of disruption impacts on SC performance and recovery policies, which are subsequently optimized in a prescriptive manner using an analytical model. The data analytics part of the system is applied to disruption identification in real time using process feedback data, e.g., from sensors and RFID. In addition, this aims at automated data input of disruption data into the reactive simulation model for recovery policy simulation and optimization. Finally, data analytics is used as a data-driven learning system at the proactive stage, helping to generate adequate disruption scenarios for resilient SC design and planning.

At the proactive level, mathematical programming models produce notable insights for managers and can be applied where the probability of disruption can be roughly estimated. On the one hand, big data analytics and advanced trace and tracking systems may help in predicting disruptions and providing more accurate data to build sophisticated disruption scenarios for resilient SC design analysis. Digital technologies open new problems for resilient SC design. For example, additive manufacturing changes SC designs whereby new resilient sourcing problems may arise. This area can further be enhanced using collaborative purchasing platforms.

At the reactive level and with regards to mitigation strategies and identifying disruption impact on finance and operational performance, digital technologies can be extensively used to obtain real-time information on the scope and scale of disruptions, their propagation in the SC, and to simulate possible recovery strategies. In addition, at the reactive level, adaptation is necessary for achieving desired output performance by ensuring the possibility of changing SC plans and inventory policies. Adaptation processes in ripple effect control can be supported by feedback and adaptive control methods using decentralized agent techniques with the help of digital technologies (Levalle & Nof, 2017). Visualizing these processes through virtual reality-supported simulation has not yet been done extensively to model the ripple effect in the supply chain. For this, simulation models, along with new digital technologies, can improve tools that are already used in developing SC agility and visibility in terms of disruption velocity.

A combination of simulation and optimization can extend the scope of both. Combining the methods enables:

- Network optimization to minimize total SC cost
- Dynamic analysis of ordering, production, inventory, and sourcing control policies using simulation

Simulation is a newer tool and is especially powerful when combined with optimization. More SC managers are now adopting the practice of using these techniques together.

What can a typical SC simulation-optimization model include, and what factors can it account for when working on risk analysis?

5.2.1 Network Design and Geographical Information

Network design, with regard to the geographical location of sites, is the core of most SC simulation models. GIS maps are used in simulation models to locate the sites and calculate distances, routes, and travel times along real roads. In addition to geospatial calculations, they provide visualization and transparency in a model.

5.2.2 Operational Parameters

Inventory control policies, back-order rules, production batching, and scheduling algorithms, as well as shipment rules and policies, need to be defined in the model and balanced against each other for both normal and disrupted operation modes. Modern SC simulation tools enable visual modeling of these policies and do not require programming skills.

5.2.3 Disruptions and Recovery

The duration of random or scheduled disruption events can be modeled with the probability distribution. As to recovery, analysts can set individual recovery policies for different sites and define the rules of policy activation depending on when the event occurs, the expected duration, and the severity of the disruption.

5.2.4 Performance Impact

The direct impact of the ripple effect is reflected in changes to KPIs. Revenue, sales, service level, fill rate, and costs are typically calculated. Unlike analytical models that usually focus on a particular metric (e.g., costs/profit), simulation enables the simultaneous measurement of all metrics in the same model. Their values can be checked at any chosen moment of the time period modeled. This way, disruption duration can be modeled, performance impact measured, and mitigation policies evaluated for efficiency.

A simulation model that considers all of these factors can be the basis for building a successful *digital twin* of a physical SC that can be used for complex analysis of SC risks, the development of contingency plans, and more efficient operational management.

A *digital SC twin* can support decision-making about the physical SC on the basis of data. At each point in time, the digital twin mirrors the physical SC: the actual transportation, inventory, demand, and capacity data and can be used for planning and real-time control decisions. The combination of simulation, optimization, and data analytics constitutes a full stack of technologies which can be used to create an SC digital twin—a model that always represents the state of the network in real time (Fig. 6).

As stated, a digital twin reflects the current state of an SC, with the actual transportation, inventory, demand, and capacity data. For example, if there is a strike at an international logistics hub, this disruption can be spotted by a risk data monitoring tool and transmitted to the simulation model as a disruptive event. Then, simulation in the digital twin can help forecast possible disruption propagation and quantify its impact. In addition, simulation enables efficient testing of recovery policies and the adaptation of contingency plans—for example, alternative network topologies and backup routes can be reconsidered on the fly. These screenshots are taken from any Logistix™ software and show the map-based model animation and the model-building editor.

The output data from a digital twin simulation can be transferred to an ERP system or a business intelligence (BI) tool to analyze the performance impact of the disruptions. Additionally, a simulation model can activate BI algorithms. For example, if the service level in a simulation model decreases to a certain level, the digital twin might activate a BI algorithm to search for the cause of the problem. Interacting with other SCM tools, a digital twin provides a control tower for end-to-end SC visibility.

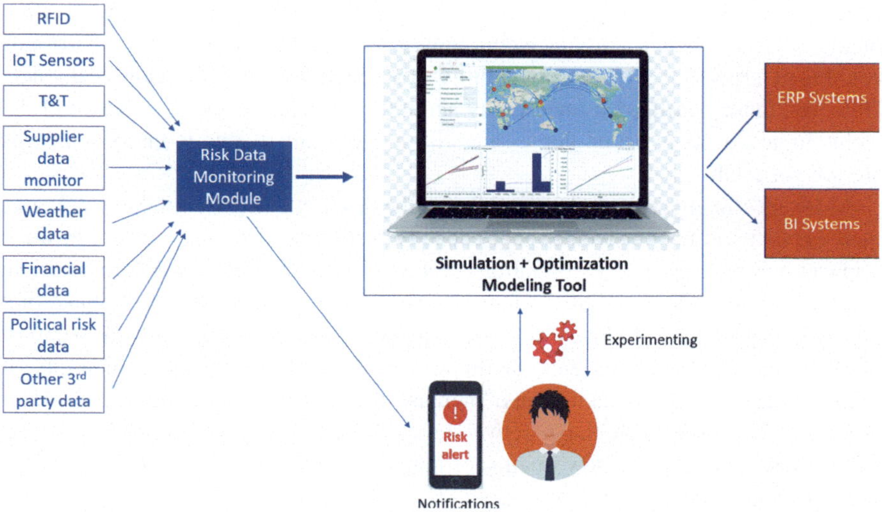

Fig. 6 Supply chain digital twin (Ivanov, 2018c)

6 Conclusions

The impact of digitalization and Industry 4.0 on the ripple effect in the SC has been studied in this chapter. Despite some partial efforts to uncover new insights into the impact of digital technologies on SC risks, the understanding of the individual contribution and the interplay of different digital technologies on specific SC disruption risk management and ripple effect is still vague. This study contributes to the body of knowledge in the field by combining the results gained from two isolated areas, i.e., the impact of digitalization on SCM, and managing the ripple effect in the SC.

Digitalization is expected to increasingly penetrate the industry in the coming years, greatly changing operating and business systems, as well as the economy. Such potential offers new approaches to SC risk management that bring both opportunities and challenges. The fusion of the digital world with industrial processes is the so-called digital transformation. In addition to internal and cross-company processes in production and logistics, this also applies to the products and services offered to customers that need to be refined through the use of digital technologies. This chapter explained how digital technologies can be used to manage SC disruption risks and the ripple effect.

The trend toward the application of digital technologies goes beyond the manufacturing company. The supplier network, the customer network, and the logistics service providers must also install and develop digital technologies to make the entire SC in nonstop delivery flexible. For this reason, the focus must be on risk management for every SC actor in the event of more frequent incidents such as natural catastrophes or supplier disruptions. The sources and handling process of risks need to be understood to facilitate the successful application of digital technologies. Digital technologies can potentially offer SCs enormous benefits in terms of transparency, visibility, cost reduction, efficiency, and resilience. However, there is still great uncertainty about the application and acceptance of the technologies, as many technologies are still in development, and industry standards are not yet established.

More specifically, this study found that at the *proactive* stage, digital technologies increase demand responsiveness and capacity flexibility. This may have a positive impact on reductions in risk mitigation inventory in ripple effect control. In addition, shorter lead times due to additive manufacturing enhance the impact of digitalization on inventory control. Industry 4.0 and additive manufacturing, with the support of BDA and T&T technologies, facilitate a new quality of proactive planning of risk management infrastructure and increase the ability to reconfigure resources at the recovery stage. At the *reactive* stage, Blockchain, T&T technologies, and BDA allow a principally new quality of data coordination and SC visibility when simulating and activating recovery policies.

In terms of the SCOR model, *sourcing and production* activities can be adversely affected by additive manufacturing and Industry 4.0, which carry higher exposure to external risks and ripple effect. A plausible explanation is the increase in complexity and the reduction in time and demand risks that occur, driven in turn by greater flexibility and shorter lead times. Higher supply risks can be encountered if a disruption happens in the upstream SC, since there is no intermediate inventory between the stages. The risks in the *delivery* processes are influenced by big data analytics with regard to a reduction in demand risks due to better SC visibility and forecast accuracy, reduction in information disruption risks and better quality of contingency plan activation. Reductions in supply and time risks in integrated SC *planning* can be achieved by using Blockchain and advanced T&T systems that provide real-time coordination while activating contingency policies. Designing a resilient SC can be influenced by higher information risks, higher exposure to external risks and a reduction in time and demand risks on the basis of Industry 4.0 technology and additive manufacturing.

A number of directions for simulation and optimization applications to SCM have been identified for digital technology application. BDA and advanced T&T systems may help in predicting disruptions and providing more accurate data to build sophisticated disruption scenarios for resilient SC design analysis. Digital technologies can be used extensively to obtain real-time information on the scope and scale of disruptions and their propagation in the SC and to simulate possible recovery strategies. In addition, at the reactive level, adaptation is necessary for achieving the desired output performance by ensuring the possibility of changing SC plans and inventory policies. Adaptation processes in ripple effect control can be supported by feedback and adaptive control methods using decentralized agent techniques with the help of digital technologies. Visualizing these processes through virtual reality-supported simulation has not yet been done extensively to model the ripple effect in the SC.

Future decision-support systems will extensively utilize digital technologies and the digital SC twin, i.e., a computerized model of an SC updated with actual data in real time.

Notwithstanding the rapid developments in SCs and their digital twins, a number of questions arise:

- Is the SC as resilient as the digital technology behind it?
- If yes, what will provide the most competitive advantage in the future: physical SCs or their digital twins?
- Will SC resilience be managed by humans, artificial intelligence, or a hybrid of both?
- What will be the role of future SC risk managers?

There is much research and practical potential with regard to the questions stated above. These can hopefully motivate new insightful developments in research on the ripple effect and disruption risk.

References

Addo-Tenkorang, R., & Helo, P. T. (2016). Big data applications in operations/supply-chain management: A literature review. *Computers & Industrial Engineering, 101*, 528–543.

Andelfinger, V., & Hänisch, T. (2017). *Industrie 4.0: Wie cyber-physische Systeme die Arbeitswelt verändern*. Springer Gabler.

Baryannis, G., Validi, S., Dani, S., & Antoniou, G. (2018). Supply chain risk management and artificial intelligence: state of the art and future research directions. *International Journal of Production Research, 57*(7), 2179–2202.

Bearzotti, L. A., Salomone, E., & Chiotti, O. J. (2012). An autonomous multi-agent approach to supply chain event management. *International Journal of Production Economics, 135*(1), 468–478.

Ben-Daya, M., Hassini, E., & Bahroun, Z. (2018). Internet of things and supply chain management: A literature review. *International Journal of Production Research, 57*(15–16), 4719–4742.

Bonfour, A. (2016). *Digital future, digital transformation: From lean production to acceluction*. Springer.

Brintrup, A., Pak, J., Ratiney, D., Pearce, T., Wichmann, P., Woodall, P., & McFarlane, D. (2020). Supply chain data analytics for predicting supplier disruptions: A case study in complex asset manufacturing. *International Journal of Production Research, 58*(11), 3330–3341. https://doi.org/10.1080/00207543.2019.1685705

Burgos, D., & Ivanov, D. (2021). Food retail supply chain resilience and the COVID-19 pandemic: A digital twin-based impact analysis and improvement directions. *Transportation Research – Part E: Logistics and Transportation Review, 152*, 102412.

Cavalcantea, I. M., Frazzon, E. M., Forcellinia, F. A., & Ivanov, D. (2019). A supervised machine learning approach to data-driven simulation of resilient supplier selection in digital manufacturing. *International Journal of Information Management, 49*, 86–97.

Choi, T. M., Wallace, S. W., & Wang, Y. (2018). Big data analytics in operations management. *Production and Operations Management, 27*(10), 1868–1883.

Crosby, M., Pattanayak, P., Verma, S., & Kalyanaraman, V. (2016). Blockchain technology: Beyond bitcoin. *Applied Innovation, 2*, 6–10.

Dolgui, A., & Ivanov, D. (2020). Exploring supply chain structural dynamics: New disruptive technologies and disruption risks. *International Journal of Production Economics, 229*, 107886.

Dolgui, A., & Ivanov, D. (2022). 5G in digital supply chain and operations management: Fostering flexibility, end-to-end connectivity and real-time visibility through internet-of-everything. *International Journal of Production Research, 60*(2), 442–451.

Dolgui, A., & Ivanov, D. (2025). Internet of behaviors: Conceptual model, practical and theoretical implications for supply chain and operations management. *International Journal of Production Research, 63*(1), 1–8.

Dolgui, A., Ivanov, D., Potryasaev, S., Sokolov, B., Ivanova, M., & Werner, F. (2020). Blockchain-oriented dynamic modelling of smart contract design and execution control in the supply chain. *International Journal of Production Research, 58*(7), 2184–2199.

Dolgui, A., Ivanov, D., & Rozhkov, M. (2020). Does the ripple effect influence the bullwhip effect? An integrated analysis of structural and operational dynamics in the supply chain. *International Journal of Production Research, 58*(5), 1285–1301.

Dolgui, A., Ivanov, D., Sethi, S., & Sokolov, B. (2019). Scheduling in production, supply chain and Industry 4.0 systems by optimal control: Fundamentals, state-of-the-art, and applications. *International Journal of Production Research, 57*(2), 411–432.

Dolgui, A., Ivanov, D., & Sokolov, B. (2018). Ripple effect in the supply chain: An analysis and recent literature. *International Journal of Production Research, 56*(1–2), 414–430.

Dolgui, A., & Proth, J. M. (2010). *Supply chain engineering: Useful methods and techniques*. Springer.

Dubey, R., Gunasekaran, A., Childe, S. J., Bryde, D. J., Giannakis, M., Foropon, C., Roubaud, D., & Hazen, B. T. (2020). Big data analytics and artificial intelligence pathway to operational performance under the effects of entrepreneurial orientation and environmental dynamism: A study of manufacturing organisations. *International Journal of Production Economics, 226*, 107599.

Dubey, R., Gunasekaran, A., Childe, S. J., Wamba, S. F., Roubaud, D., & Foropon, C. (2019). Empirical investigation of data analytics capability and organizational flexibility as complements to supply chain resilience. *International Journal of Production Research, 59*(1), 110–128.

Dunke, F., Heckmann, I., Nickel, S., & Saldanha-da-Gama, F. (2018). Time traps in supply chains: I optimal still good enough? *European Journal of Operational Research, 264*, 813–829.

Elluru, S., Gupta, H., Kaur, H., & Singh, S. P. (2019). Proactive and reactive models for disaster resilient supply chain. *Annals of Operations Research, 283*, 199–224.

Fazili, M., Venkatadri, U., Cyrus, P., & Tajbakhsh, M. (2017). Physical internet, conventional and hybrid logistic systems: A routing optimisation-based comparison using the Eastern Canada road network case study. *International Journal of Production Research, 55*(9), 2703–2730.

Feldmann, K., & Pumpe, A. (2017). A holistic decision framework for 3D printing investments in global supply chains. *Transportation Research Procedia, 25*, 677–694.

Frazzon, E. M., Freitag, M., & Ivanov, D. (2021). Intelligent methods and systems for decision-making support: Toward digital supply chain twins. *International Journal of Information Management, 57*, 102281.

Frazzon, E. M., Kück, M., & Freitag, M. (2018). Data-driven production control for complex and dynamic manufacturing systems. *CIRP Annals–Manufacturing Technology, 67*(1), 515–518.

Gunasekaran, A., Papadopoulos, T., Dubey, R., Wamba, S. F., Childe, S. J., Hazen, B., et al. (2017). Big data and predictive analytics for supply chain and organizational performance. *Journal of Business Research, 70*, 308–317.

Gunasekaran, A., Tiwari, M. K., Dubey, R., & Wamba, S. F. (2016). Big data and predictive analytics applications in supply chain management. *Computers & Industrial Engineering, 101*, 525–527.

Gunasekaran, A., Yusuf, Y. Y., Adeleye, E. O., & Papadopoulos, T. (2018). Agile manufacturing practices: The role of big data and business analytics with multiple case studies. *International Journal of Production Research, 56*(1–2), 382–397.

Hagberg, J., Sundstrom, M., & Egels-Zandén, N. (2016). The digitalization of retailing: An exploratory framework. *International Journal of Retail & Distribution Management, 44*(7), 694–712.

He, J., Alavifard, F., Ivanov, D., & Jahani, H. (2019). A real-option approach to mitigate disruption risk in the supply chain. *Omega, 88*, 133–149.

Hofmann, E., Strewe, U. M., & Bosia, N. (2018). *Supply chain finance and blockchain technology*. Springer.

Holmström, J., & Gutowski, T. (2017). Additive manufacturing in operations and supply chain management: No sustainability benefit or virtuous knock-on opportunities? *Journal of Industrial Ecology, 21*(1), 21–24.

IBM. (2017). Retrieved November 20, 2017, from https://www-03.ibm.com/press/us/en/pressrelease/50816.wss

Ivanov, D. (2018a). Revealing interfaces of supply chain resilience and sustainability: A simulation study. *International Journal of Production Research, 56*(10), 3507–3523.

Ivanov, D. (2018b). *Structural dynamics and resilience in supply chain risk management*. Springer.

Ivanov, D. (2018c). Managing risks in supply chains with digital twins and simulation. Retrieved from https://www.anylogistix.com/resources/white-papers/managing-risks-in-supply-chains-with-digital-twins/.

Ivanov, D. (2020). Predicting the impacts of epidemic outbreaks on global supply chains: A simulation-based analysis on the coronavirus outbreak (COVID-19/SARS-CoV-2) case. *Transportation Research Part E: Logistics and Transportation Review, 136*, 101922.

Ivanov, D. (2023). Intelligent digital twin (iDT) for supply chain stress-testing, resilience, and viability. *International Journal of Production Economics, 263*, 108938.

Ivanov, D. (2024a). Digital supply chain management and technology to enhance resilience by building and using end-to-end visibility during the COVID-19 pandemic. *IEEE Transactions on Engineering Management, 71*, 10485–10495.

Ivanov, D. (2024b). Conceptualisation of a 7-element digital twin framework in supply chain and operations management. *International Journal of Production Research, 62*(6), 2220–2232.

Ivanov, D. (2024c). *Introduction to supply chain analytics*. Springer Nature. ISBN 978-3-031-51240-7.

Ivanov, D. (2025a). When is the supply chain resilient? Customer and operational perspectives. *International Journal of Production Research*, 1–16. https://doi.org/10.1080/00207543.2025.2454331

Ivanov, D. (2025b). Comparative analysis of product and network supply chain resilience. *International Transactions in Operational Research*. https://doi.org/10.1111/itor.13612

Ivanov, D., & Dolgui, A. (2019). Low-certainty-need (LCN) supply chains: A new perspective in managing disruption risks and resilience. *International Journal of Production Research., 57*(15-16), 5119–5136.

Ivanov, D., & Dolgui, A. (2021). A digital supply chain twin for managing the disruptions risks and resilience in the era of Industry 4.0. *Production Planning and Control, 32*(9), 775–788.

Ivanov, D., Dolgui, A., & Sokolov, B. (2013). Multi-disciplinary analysis of interfaces "Supply Chain Event Management – RFID – Control Theory". *International Journal of Integrated Supply Management, 8*, 52–66.

Ivanov, D., Dolgui, A., & Sokolov, B. (2019). The impact of digital technology and industry 4.0 on the ripple effect and supply chain risk analytics. *International Journal of Production Research, 57*(3), 829–846.

Ivanov, D., Dolgui, A., Sokolov, B., & Ivanova, M. (2017). Literature review on disruption recovery in the supply chain. *International Journal of Production Research, 55*(20), 6158–6174.

Ivanov, D., & Rozhkov, M. (2020). Coordination of production and ordering policies under capacity disruption and product write-off risk: An analytical study with real-data based simulations of a fast moving consumer goods company. *Annals of Operations Research, 291*(1-2), 387–407.

Ivanov, D., Sokolov, B., & Dilou Raguinia, E. A. (2014). Integrated dynamic scheduling of material flows and distributed information services in collaborative cyber-physical supply networks. *International Journal of Systems Science: Operations & Logistics, 1*(1), 18–26.

Ivanov, D., Sokolov, B., & Dolgui, A. (2014). The Ripple effect in supply chains: Trade-off 'efficiency-flexibility-resilience' in disruption management. *International Journal of Production Research, 52*(7), 2154–2172.

Ivanov, D., Sokolov, B., Dolgui, A., Werner, F., & Ivanova, M. (2016). A dynamic model and an algorithm for short-term supply chain scheduling in the smart factory industry 4.0. *International Journal of Production Research, 54*(2), 386–402.

Ivanov, D., Sokolov, B., & Kaeschel, J. (2010). A multi-structural framework for adaptive supply chain planning and operations control with structure dynamics considerations. *European Journal of Operational Research, 200*(2), 409–420.

Ivanov, D., Sokolov, B., & Pavlov, A. (2014). Optimal distribution (re)planning in a centralized multi-stage network under conditions of ripple effect and structure dynamics. *European Journal of Operational Research, 237*(2), 758–770.

Ivanov, D., Tsipoulanidis, A., & Schönberger, J. (2019). *Global supply chain and operations management: A decision-oriented introduction into the creation of value* (2nd ed.). Springer Nature.

Jackson, I., Ivanov, D., Dolgui, A., Namdar, J. (2024). Generative artificial intelligence in supply chain and operations management: A capability-based framework for analysis and implementation. *International Journal of Production Research, 62*(17), 6120–6145.

Johnson, K., Lee, A. B. H., & Simchi-Levi, D. (2016). Analytics for an online retailer: Demand forecasting and price optimization. *Manufacturing and Service Operations Management, 18*(1), 69–85.

Khajavi, S. H., Partanen, J., & Holmström, J. (2014). Additive manufacturing in the spare parts supply chain. *Computers in Industry, 65*(1), 50–63.

Kshetri, N. (2018). Blockchain's roles in meeting key supply chain management objectives. *International Journal of Information Management, 39*, 80–89.

Lario, J., Mateos, J., Psarommatis, F., & Ortiz, Á. (2024). A cost model for the investment feasibility of quality inspection technologies in the zero defect manufacturing era. *International Journal of Production Research, 1–16*. https://doi.org/10.1080/00207543.2024.2383780

Levalle, R. R., & Nof, S. Y. (2017). Resilience in supply networks: Definition, dimensions, and levels. *Annual Reviews in Control, 43*, 224–236.

Li, G., Xue, J., Li, N., & Ivanov, D. (2022). Blockchain-supported business model design, supply chain resilience, and firm performance. *Transportation Research – Part E Logistics and Transportation Review, 163*, 102773.

Li, J., Jia, G., Cheng, Y., & Hu, Y. (2017). Additive manufacturing technology in spare parts supply chain: A comparative study. *International Journal of Production Research, 55*(5), 1498–1515.

Li, W., Sun, H., Tong, M., Mustafee, N., & Koh, L. (2024). Customizing customization in a 3D printing-enabled hybrid manufacturing supply chain. *International Journal of Production Economics, 268*, 109103. https://doi.org/10.1016/j.ijpe.2023.109103

Liao, Y., Deschamps, Y., de Freitas, E., Loures, R., & Ramos, L. F. P. (2017). Past, present and future of industry 4.0–a systematic literature review and research agenda proposal. *International Journal of Production Research, 55*(12), 3609–3629.

Minner, S., Battini, D., & Çelebi, D. (2017). Innovations in production economics. *International Journal of Production Economics, 194*, 1–2.

Moghaddam, M., & Nof, S. Y. (2018). Collaborative service-component integration in cloud manufacturing. *International Journal of Production Research, 56*(1–2), 676–691.

Nguyen, T., Zhou, L., Spiegler, V., Ieromonachou, P., & Lin, Y. (2018). Big data analytics in supply chain management: A state-of-the-art literature review. *Computers & Operations Research, 98*, 254–264.

Oesterreich, T. D., & Teuteberg, F. (2016). Understanding the implications of digitisation and automation in the context of industry 4.0: A triangulation approach and elements of a research agenda for the construction industry. *Computers in Industry, 83*, 121–139.

Panetto, H., Iung, B., Ivanov, D., Weichhart, G., & Wang, X. (2019). Challenges for the cyber-physical manufacturing enterprises of the future. *Annual Reviews in Control, 47*, 200–213.

Papadopoulos, T., Gunasekaran, A., Dubey, R., Altay, N., Childe, S. J., & Wamba, S. F. (2017). The role of big data in explaining disaster resilience in supply chains for sustainability. *Journal of Cleaner Production, 142*(2), 1108–1118.

Pavlov, A., Ivanov, D., Dolgui, A., & Sokolov, B. (2018). Hybrid fuzzy-probabilistic approach to supply chain resilience assessment. *IEEE Transactions on Engineering Management, 65*(2), 303–315.

Pavlov, A., Ivanov, D., Pavlov, D., & Slinko, A. (2019). Optimization of network redundancy and contingency planning in sustainable and resilient supply chain resource management under conditions of structural dynamics. *Annals of Operations Research*. https://doi.org/10.1007/s10479-019-03182-6

Porter, M. E., & Heppelmann, J. E. (2015). How smart, connected products are transforming companies. *Harvard Business Review*.

Priori, P., Ponte, B., Rosillo, R., & de la Fuente, D. (2018). Applying machine learning to the dynamic selection of replenishment policies in fast-changing supply chain environments. *International Journal of Production Research, 57*(11), 3663–3677.

Qu, T., Thürer, M., Wang, J., Wang, Z., Fu, H., Li, C., et al. (2017). System dynamics analysis for an internet-of-things-enabled production logistics system. *International Journal of Production Research, 55*(9), 2622–2649.

Rai, R., Tiwari, M. K., Ivanov, D., & Dolgui, A. (2021). Machine learning in manufacturing and Industry 4.0 applications. *International Journal of Production Research, 59*(16), 4773–4778.

Ralston, P., & Blackhurst, J. (2020). Industry 4.0 and resilience in the supply chain: A driver of capability enhancement or capability loss? *International Journal of Production Research, 58*(16), 5006–5019. https://doi.org/10.1080/00207543.2020.1736724

Reddy, G. R., Singh, H., & Hariharan, S. (2016). Supply chain wide transformation of traditional industry to industry 4.0. *Journal of Engineering and Applied Sciences, 11*(18), 11089–11097.

Saberi, S., Kouhizadeh, M., Sarkis, J., & Shen, L. (2018). Blockchain technology and supply chain management. *International Journal of Production Research, 57*(7), 2117–2135.

Sanders, N. R. (2016). How to use big data to drive your supply chain. *California Management Review, 58*(3), 26–48.

Sauer, P. C., Orzes, G., & Culot, G. (2022). Blockchain in supply chain management: A multiple case study analysis on setups, contingent factors, and evolutionary patterns. *Production Planning & Control, 35*(9), 901–916. https://doi.org/10.1080/09537287.2022.2153078

Schlüter, F., Hetterscheid, E., & Henke, M. (2017). A simulation-based evaluation approach for digitalization scenarios in smart supply chain risk management. *Journal of Industrial Engineering and Management Science, 1*, 179–206.

Sheffi, Y. (2015). Preparing for disruptions through early detection. *MIT Sloan Management Review, 57*, 31.

Simchi-Levi, D., & Wu, M. X. (2018). Powering retailers digitization through analytics and automation. *International Journal of Production Research, 56*(1–2), 809–816.

Sokolov, B., Ivanov, D., Dolgui, A., & Pavlov, A. (2016). Structural quantification of the ripple effect in the supply chain. *International Journal of Production Research, 54*(1), 152–169.

Sokolov, B., Spesivtsev, A., Sukhoparov, A., & Zakharov, V. (2023). Meaningful and formal problem statement of the technologies synthesis and programs of grass feed production proactive management. *Smart Innovation, Systems and Technologies, 331*. https://doi.org/10.1007/978-981-19-7780-0_29

Strozzi, F., Colicchia, C., Creazza, A., & Noè, C. (2017). Literature review on the 'smart factory' concept using bibliometric tools. *International Journal of Production Research, 55*(22), 6572–6591.

Tiwari, S., Sharma, P., Choi, T. M., & Lim, A. (2023). Blockchain and third-party logistics for global supply chain operations: Stakeholders' perspectives and decision roadmap. *Transportation Research Part E: Logistics and Transportation Review, 170*, 103012.

Tran-Dang, H., Krommenacker, N., & Charpentier, P. (2017). Containers monitoring through the physical internet: A spatial 3D model based on wireless sensor networks. *International Journal of Production Research, 55*(9), 2650–2663.
Tupa, J., Simota, J., & Steiner, F. (2017). Aspects of risk management implementation for Industry 4.0. *Procedia Manufacturing, 11*, 1223–1230.
Wamba, S. F., Akter, S., Edwards, A., Chopin, G., & Gnanzou, D. (2015). How 'big data' can make big impact: Findings from a systematic review and a longitudinal case study. *International Journal of Production Economics, Elsevier, 165*, 234–246.
Wamba, S. F., Ngai, E. W. T., Riggins, F., & Akter, S. (2017). Transforming operations and production management using big data and business analytics: Future research directions. *International Journal of Operations & Production Management, 37*(1), 2–9.
Wamba, S. F., & Queiroz, M. M. (2020). Industry 4.0 and the supply chain digitalisation: A blockchain diffusion perspective. *Production Planning & Control*, 1–18.
Wu, X. Y., Pu, X., & Fan, Z. P. (2024). Blockchain technology adoption strategies of a supply chain considering possible product defects. *International Journal of Production Research, 1–27.* https://doi.org/10.1080/00207543.2024.2368086
Yang, Y., Pan, S., & Ballot, E. (2017). Innovative vendor-managed inventory strategy exploiting interconnected logistics services in the physical internet. *International Journal of Production Research, 55*(9), 2685–2702.
Zelbst, P. J., Green, K. W., Sower, V. E., & Reyes, P. M. (2012). Impact of RFID on manufacturing effectiveness and efficiency. *International Journal of Operations & Production Management, 32*(3), 329–350.
Zhang, J., & Jung, Y. (2018). *Additive manufacturing*. Elsevier Science & Technology.
Zheng, G., Ivanov, D., & Brintrup, A. (2025). An adaptive federated learning system for information sharing in supply chains. *International Journal of Production Research.* https://doi.org/10.1080/00207543.2024.2432469
Zhong, R. Y., Xu, C., Chen, C., & Huang, G. Q. (2017). Big data analytics for physical internet-based intelligent manufacturing shop floors. *International Journal of Production Research, 55*(9), 2610–2621.

The manufacturer's authorised representative in the EU is Springer Nature Customer Service Centre GmbH, Europaplatz 3, 69115 Heidelberg, Germany. If you have any concerns regarding our products, please contact ProductSafety@springernature.com

Printed and bound by CPI Group (UK) Ltd, Croydon, CR0 4YY
30/09/2026
02225598-0012